Introduction to Biosemiotics

Introduction to Biosemiotics

The New Biological Synthesis

Edited by

Marcello Barbieri
University of Ferrara, Italy

Springer

A C.I.P. Catalogue record for this book is available from the Library of Congress.

ISBN-10 1-4020-4813-0 (HB)
ISBN-13 978-1-4020-4813-5 (HB)
ISBN-10 1-4020-4814-9 (e-book)
ISBN-13 978-1-4020-4814-2 (e-book)

Published by Springer,
P.O. Box 17, 3300 AA Dordrecht, The Netherlands.

www.springer.com

Printed on acid-free paper

To Thomas Sebeok (1920–2001)
and to his vision of a synthesis between Biology and Semiotics

TABLE OF CONTENTS

EDITORIAL

MARCELLO BARBIERI

THE CHALLENGE OF BIOSEMIOTICS

Semiotics, literally, is the study of signs and initially it was thought to be concerned only with the products of culture. Mental phenomena, however, exist also in animals, and cultural semiotics can be regarded as a special case of *biological semiotics*, or *biosemiotics*, a science that started by studying semiotic phenomena in animals and then was gradually extended to other living creatures. Eventually, the discovery of the genetic code suggested that the cell itself has a semiotic structure, and the goal of biosemiotics became the idea that all living creatures are semiotic systems. But what is a semiotic system? According to cultural semiotics, there are two answers to this question. One is the model proposed by Ferdinand de Saussure, who defined a semiotic system as a duality of "signifier and signified" or "sign and meaning". The other is the model of Charles Peirce, who pointed out that interpretation is an essential component of semiosis and defined a semiotic system as a triad of "sign, object and interpretant".

In 1974 Marcel Florkin argued that "signifier and signified" are equivalent to "genotype and phenotype" and proposed for biosemiotics the dualistic model of Saussure. He underlined however that *"in linguistics the sign is arbitrary with reference to the relation between its two faces. In molecular biosemiotics, on the other hand, signifier and signified are in a necessary relation imposed by the natural relations of material realities".* According to Florkin, in other words, in molecular biosemiotics there is no arbitrary relationship between *signifier* and *signified*, and he explicitly declared that *"A bioseme carries no* Bedeutung *or* Sinn *(no meaning). Its signifier is an aspect of molecular configuration and its signified is an aspect of biological activity".*

In the 1960s and 70s Thomas Sebeok started a lifelong campaign in favor of the idea that language has biological roots, but rather than following Saussure he adopted the triadic scheme of Peirce first in zoosemiotics, in 1963, and then in the more general field of biosemiotics (Sebeok, 1963, 1972, 1986). According to Sebeok, any semiosis is necessarily a triadic relationship, and interpretation is its *sine qua non* condition. He insisted that *"there can be no semiosis without interpretability"* and that interpretation is *"a necessary and sufficient condition for something to be a semiosis"* (Sebeok, 2001). The Peirce-Sebeok model of semiosis has become extremely popular and it has been adopted by most biosemioticians, in particular by the Copenhagen-Tartu school (Claus Emmeche, Jesper Hoffmeyer and Kalevi Kull) and (in a *hermeneutic* version) by the Prague school (Anton Markoš).

A third model of semiosis, however, does exist and was suggested in the 1980s by the theory that the cell is a triad of genotype, phenotype and *ribotype*, where the ribotype is the ribonucleoprotein system of the cell and represents its "codemaker", i.e. the seat of the genetic code (Barbieri 1981, 1985, 2003). This amounts to saying that the cell contains a "codemaker" but not an "interpreter", because the rules of the genetic code do not depend on interpretation. They are virtually the same in all living systems and in all circumstances, and that has been true for almost the entire history of life on Earth. In this framework, the simplest semiotic system is a triad made of "sign, meaning and code" and the origin of semiosis (the *semiotic threshold*) does not coincide with the origin of interpretation (the *hermeneutic threshold*).

Another approach to biosemiotics has been proposed by Howard Pattee who investigated the physical conditions that are necessary for codes and symbolic controls. Pattee introduced the concept of *epistemic threshold*, the boundary region where local matter has not only its intrinsic physical properties governed by universal laws, but where it is also "about" something else. Epistemic matter, in other words, "stands for" something, and the "standing for" relation is usually considered an *emergent* process that leads necessarily to a triadic Peircean relationship of "matter, interpreter and referent" (Pattee, 1969, 1972, 2001).

Today, in short, we have four different models of biological semiosis and at least four different schools of biosemiotics. Despite all that, biosemiotics remains an isolated discipline and many perceive it as a small field that lies at the outskirts of science, somewhere between biology and linguistics. This is because there is a very strange paradox at the heart of modern science. On the one hand it is acknowledged that the genetic code is the bedrock of life, and on the other hand it is underlined that it is not a real code. The argument is that the genetic code would be real only if it was associated with the production of *meaning*, but modern science does not deal with meaning and is bound therefore to deny any reality to the code of life.

According to the dominant paradigm, the genetic code is fundamentally a *metaphor*. It is a linguistic construction that we use in order to avoid long periphrases when we talk about living systems, but no more than that. It is like those computer programs that allow us to write our instructions in English, thus saving us the trouble to write them in binary digits. Ultimately, however, there are only binary digits in the machine language of the computer, and in the same way, it is argued, there are only physical quantities at the most fundamental level of Nature.

This conclusion, known as *physicalism*, or *the physicalist thesis*, has been proposed in various ways by a number of scientists and philosophers (Chargaff, 1963; Sarkar, 1996; 2000; Mahner and Bunge, 1997; Griffiths and Knight, 1998; Griffith, 2001, Boniolo, 2003). It is probably one of the most deeply dividing issues of modern science. Many biologists are convinced that the genetic code is a real and fundamental component of life, but physicalists insist that it is real only in a very superficial sense and that there is nothing fundamental about it because *it must be* reducible in principle, to physical quantities.

This, in fact, is the only answer that allows people to say that there are no signs and meanings at the basis of life, and that semiotic processes are not fundamental events. But what a price to pay! It is perfectly right to mention the genetic code practically in every single problem of biology, provided one keeps in mind that *it is not meant to be serious*. On the face of this, many biologists prefer to avoid the issue altogether, which is hardly surprising. But some do not. The issue is not so much the physicalist thesis as the nature of life itself. The experimental reality is that proteins are manufactured by molecular machines based on the rules of the genetic code and there is little point in saying that the code *must be* a metaphor simply because modern science does not know how to cope with meaning. That is the challenge of biosemiotics: the codes are a fundamental reality of life and we simply have to learn how to introduce signs and meanings in science.

This book is addressed to students, researchers and academics who are not familiar with biosemiotics and want to know more about it. It is a highly qualified introduction to this new field because it is written by many of its major contributors. At the same time, it contains the most recent developments in the basic issues of biosemiotics and provides therefore a fairly accurate portrait of the present state of the art.

The book is divided into three parts. The first is dedicated to a brief historical account and the last to a few research applications, whereas the central, and longest, part of the book is devoted to theoretical issues.This is because the real obstacle to biological progress, today, is not lack of data but a pervasive theoretical paradigm that continues to deny the semiotic nature of life, or to pay only lip-service to it, thus depriving the biological codes of all their revolutionary potential.

Biosemiotics is truly a new biological "synthesis" because it brings together biology and linguistics, and effectively brings down the old divide between the "Two Cultures". Its main challenge, as we have seen, is to introduce *meaning* in biology, on the grounds that organic codes and processes of interpretation are fundamental components of the living world. Biosemiotics has become in this way the leading edge of the research in the fundamentals of life, and is a young exciting field on the move. This book wants to bring it out of the small niche in which it has been developed so far and make it available to all those who are prepared to accept the challenge raised by the discovery of the genetic code and of biological meaning.

ACKNOWLEDGEMENTS

I wish to thank Jacco Flipsen, at Springer-Netherlands, for proposing the project of this book in the first place, and Catherine Cotton for organizing the reviewing of its chapters and for taking care of countless editorial problems. I am also grateful to Catherine for bringing the ideas of this book to the attention of science journalists and for taking part in the 6th Gathering in Biosemiotics, thus proving to be not only a highly professional editor but also an enthusiastic supporter of the emerging field of biosemiotics.The actual production of the book has been directed by Ria Kanters at Springer Netherlands with wonderful care and I wish I could adequately express my gratitude for the splendid job that she has done.

REFERENCES

Barbieri M. (1981). The Ribotype Theory on the Origin of Life. *Journal of Theoretical Biology*, 91, 545–601.

Barbieri M. (1985). *The Semantic Theory of Evolution*. Harwood Academic Publishers, London and New York.

Barbieri M. (2003). *The Organic Codes. An Introduction to Semantic Biology*. Cambridge University Press, Cambridge, UK.

Boniolo G. (2003). Biology without Information. *History and Philosophy of the Life Sciences*, 25, 255–273.

Chargaff, E. (1963). *Essays on Nucleic Acids*. Elsevier, Amsterdam.

Emmeche, C. and Hoffmeyer, J. (1991). From language to nature: the semiotic metaphor in biology. *Semiotica* 84(1/2), 1–42.

Florkin, Marcel (1974). Concepts of molecular biosemiotics and molecular evolution. In *Comprehensive Biochemistry,* vol. 29 part A (Comparative Biochemistry, Molecular Evolution), Marcel Florkin and Elmer H. Stotz (eds.), Elsevier, Amsterdam, 1–124.

Griffith, P.E. (2001). Genetic Information: A Metaphor in Search of a Theory. *Philosophy of Science*, 68, 394–412.

Griffith, P.E. and Knight, R.D. (1998). What is the developmental challenge? *Philosophy of Science*, 65, 276–288.

Hoffmeyer, J. and Emmeche, C. (1991). Code duality and the semiotics of nature. In M. Anderson and F. Merrell, eds., *On Semiotic Modeling*. Mouton de Gruyter, New York.

Kull, K. (2001). Jakob von Uexküll: An Introduction. *Semiotica*, 134 (1/4), 1–59.

Mahner, M. and Bunge, M. (1997). *Foundations of Biophilosophy*. Springer Verlag, Berlin.

Markoš, A. (2002). *Readers of the Book of Life: Conceptualizing Developmental Evolutionary Biology*. Oxford University Press, Oxford.

Pattee, H. H. (1969). The physical basis of coding and reliability in biological evolution. In C. H. Waddington (ed.) *Toward a Theoretical Biology* Vol. 1, Edinburgh Univ. Press, 1969, pp. 67–93.

Pattee, H. H. (1972). Laws and constraints, symbols and languages. In C. H. Waddington (ed.) *Towards a Theoretical Biology* Vol. 4, Edinburgh Univ. Press, 1972, pp. 248–258.

Pattee, H. H. (2001). The physics of symbols: bridging the epistemic cut, *BioSystems*, 60, 5–21.

Peirce, Charles S. (1931–1958) *Collected papers of Charles Sanders Peirce*. Harvard University Press, Cambridge Massachusetts.

Sarkar, S. (1996). Biological Information. A Skeptical Look at some Central Dogmas of Molecular Biology. In S. Sarkar (ed.) *The Philosophy and History of Biology*. Kluwer Academic Publishers, Dordrecht, 187–231.

Sarkar, S. (2000). Information in Genetics and Developmental Biology. *Philosophy of Science*, 67, 208–213.

Saussure, Ferdinand de (1916). *Cours de linguistique générale*. Payot, Paris.

Sebeok, T.A. (1963). Communication among social bees; porpoises and sonar; man and dolphin. *Language*, 39, 448–466.

Sebeok, T. A. (1972). *Perspectives in Zoosemiotics*. Mouton, The Hague.

Sebeok, T. A. (1986). The doctrine of signs. *Journal of Social and Biological Structures* 9, 345–352.

Sebeok, T. A. (2001). Biosemiotics: Its roots, proliferation, and prospects. *Semiotica*, 134, 68.

CHAPTER 1

THE EVOLUTIONARY HISTORY OF BIOSEMIOTICS

DONALD FAVAREAU[1]

University Scholars Programme, National University of Singapore, favareau@gmail.com

The key question lying at the root of all this is: How could natural history become cultural history? Or, to put it another way... How did something become "someone"? – Jesper Hoffmeyer, *Signs of Meaning in the Universe* (1996:viii).

A PERSONAL PRELUDE: MY STROLL THROUGH THE WORLDS OF SCIENCES AND SIGNS

Having spent the last six years in regular correspondence with the world's small but steadily growing population of "biosemioticians," I feel warranted in saying of this diverse group of molecular biologists, neuroscientists, zoologists, anthropologists, psychologists and philosophers, that while each one more or less found their way into this common project alone – proceeding from vastly different starting points and through drastically varying routes – it might yet not be too broad a claim to say that a growing discontent with what was being offered as (or in lieu of) "explanation" regarding the nature of empirically observed, real-world sign processes in their respective fields of origin appears to be the single most common impetus setting the majority of these researchers on their respective paths to what has now converged to become the growing interdisciplinary project of *biosemiotics.*

Indeed, my own entry into this field came as the result of my growing discontent with the inability of cognitive neuroscience to confront issues of experiential "meaning" at the same level that it was so successful in, and manifestly committed to studying the mechanics of those very same electro-chemical transmission events by which such meanings were being asserted (but not explained) to, be produced. For the 1990s were declared (by fiat of an actual act of Congress) to be "The Decade of the Brain" in the United States – and, reservations about the seriousness of such self-aggrandizing hyperbole aside, this period did indeed see a great explosion of ideas and energy emanating out of such newly minted hybrid research projects

M. Barbieri (ed.), Introduction to Biosemiotics, 1–67.

as neurophilosophy, evolutionary psychology, dynamic systems theory, cognitive neuroscience, and Artificial Intelligence/Artificial Life.

What intrigued me about this research then (and now) was the fact that at the heart of these disparate research projects lay the primordial scientific question: "What is the relation between mental experience, biological organization, and the law-like processes of inanimate matter?" However, and for reasons that should become clear as this "history" progresses, that ancient and comprehensively articulated question progressively became re-formulated (and ultimately replaced) by the much narrower and more presumption-driven question: "How does the human brain produce the mind?" And this is a very different question – making some very different assumptions – from the prior formulation, as we soon shall see.

However, even to this perhaps less optimally formulated latter question, many interesting analyses were made, hypotheses proposed and theories advanced – though none proved fully satisfactory, even on the theoretical level, and as the inquiry began taking on its institutionally funded form, fewer and fewer of the major participants in the debate took the opportunity to reflect publicly on whether the question of "how do brains produce minds" was not itself framed in such a way that there could never be provided for it a satisfactory answer.

For with the object of study itself being invariably conceptualized either in a modified Cartesian sense (i.e. – "mind" as a immaterial system property either emergently produced by, or actually reducible to, the activity of an material brain) – or as an "inherently unknowable" phenomenon (McGinn 1999) or as to outright category mistake of "folk psychology" (Churchland 1984) – it's hard to see how any progress *could* be made on this issue, given the artificially barren parameters within which the search was set.

Towards the end of the century, the application of dynamic systems theory to neuroscience was promising to open up a third alternative to the "dualist-and-immaterialist versus reductionist-and-determinist" impasse, and several visionary brain researchers (e.g. Edelman 1992, Damasio 1994, Freeman 2000, Llinás 2001, Fuster 2003) were suggesting interesting syntheses that, although not phrased this way by their proponents, attempted to preserve the interdependent reality of both sign relations and material relations in their explications of the phenomenon of "mind."

These searches still, however, proceeded from the yet too partial understanding that somehow "the mind produces sign relations" – when, in point of fact, it was not until neuroscientist and bio-anthropologist Terrence Deacon (1997) suggested a new way of looking at the problem of language origins through the lens of Charles S. Peirce's architectonic of "sign relations" *per se* that it became clear at least to some people (myself included) that a potentially more viable approach to the conventional mind-brain question might be to *not* begin that study by using the uniquely human manifestation of mental experience as the archetypal example of the system needing explanation, as if it – alone among the products of the natural world – somehow arose *ex nihilo* and persists today *sui generis* – but to inquire first, instead, into the far more fundamental relationship of all purposive organisms to subjective experience (a term which itself, it turns out, denotes a far broader set of natural relations and phenomena than are indicated when one begins *a priori* with the stipulation that "subjective experience" is something that arises wholly out of, and remains forever locked away within, brains).

Coming upon Deacon's provocative synthesis of Peirce re-set the fundamental terms of inquiry for me, and soon led me to discover the work of Danish biosemiotician Jesper Hoffmeyer – and it is from this point that I date my own decision to become part of the yet nascent interdiscipline of biosemiotics. And to this day, I can still recall the precise moment of my casting of this die. I had only gotten as far as page 40 in Hoffmeyer's (1996) *Signs of Meaning in the Universe* when I came upon a passage wherein he discusses the concept of self-reference in a system. This he compares to the perpetual creation of "a map which is so detailed that the cartographer and the map that he is making are swept up into it." This elegant little description so perfectly captured the paradox that most contemporary neuroscientific theory both entailed and yet was simultaneously denying and/or attempting to run away from, that at the end of my reading of that passage – one of many such delightful asides ornamenting the profound and seriously consequential ideas argued for in this book – I e-mailed Jesper Hoffmeyer in Copenhagen, and found myself in Denmark a few weeks later debating the relations between intersubjectivity and mirror neurons at the First Annual International Gathering in Biosemiotics.

Researchers from 18 different countries were present at that initial Gatherings, with backgrounds ranging from physics and molecular biology to animal ethology, robotics, evolutionary psychology and philosophy of semiotics and of mind. And since entering the interdisciplinary project that this group was in the process of creating, I have since learned much about the understandings attained by the various disciplines from which each of my colleagues has been informed – as well as about the longer tradition of "theoretical biology" that remains relatively (and detrimentally) untaught as part of a scientific education in the United States.

For even today, it is more the norm than the exception for university life science majors to be instructed right at the outset of their studies that "science only studies observable phenomena. It functions in the realm of matter and energy [and therefore] it is a serious mistake to think that the methods of science can be applied in areas of investigation involving other aspects of human experience, e.g., matters of the mind" (Miller and Harley's *Zoology*, 1994:11) and that "most neuroscientists and philosophers now take for granted that all biological phenomena, including consciousness, are *properties of matter*" as writes Nobel Prize winner Eric Kandel at the conclusion of his authoritative *Principals of Neural Science* (2000:1318, italics mine). "Is the 'problem of consciousness' *real*, however?" he then asks, noting far too accurately that "some philosophers and *many neuroscientists* believe that *consciousness is an illusion*" (ibid).

And so the question accordingly arises: How did modern science – the communal knowledge-generating system *par excellence* – arrive at this sterile impasse whereat the investigation of individual knowledge-generating systems *as* knowledge-generating systems *per se* came to be seen, at best, as a vexingly paradoxical riddle and, at worst, as falling outside the scope of legitimate scientific inquiry entirely?

It is in the hope of providing at least partial illumination of the historical processes by which this particular explanatory Gordian knot was tied, that the first half of this history of biosemiotics begins.

A BRIEF HISTORY OF THE TYING OF A GORDIAN KNOT

The resistance to studying "subjective experience" *qua* "subjective experience" (and not just studying the interactions of its material substrate) has a long and principled history in science – and it precisely this history that we need to understand first, if we are ever to understand how something as oddly named as "biosemiotics" is not only not an anti-science nor a pseudo-science, but is genuinely a proto-science aimed at scientifically distinguishing and explaining the use of sign relations both between and within organisms.

Accordingly, it might seem at first that an examination into the uniquely influential works of René Descartes (1596–1650) would be a logical place to start this discussion – Descartes' work being emblematic of the "bifurcation" between modernity and pre-modernity in both the sciences and in philosophy, as well, of course, as the bifurcation between the mental and material "realms" that we continue to travel, in better and worse ways, today. And, indeed, it will be necessary to discuss Descartes' role in shaping the trajectory of modern science if the history of biosemiotics is to make sense within its larger narrative.

Like all of us, however, Descartes too played his role informed by a set of prior narratives that are themselves contingent products of history as well. So if we are to understand the relationship of biosemiotics to the modern science from which it proceeds and to some extent challenges, we must also understand the relation of modern science to the practices and understandings about the natural world from which *it* proceeded and, for the most part, not merely challenged but actively proposed to supplant.

Thus, the first difference between the two projects of "biosemiotics vis-à-vis modern science" and of "modern science vis-à-vis everything that preceded it" can be clearly stated. For as we shall see shortly, the goal of biosemiotics is to *extend* and to *broaden* modern science, while adhering strictly to its foundational epistemological and methodological commitments – it does not seek in any genuine sense of the term to "oppose" much less "supplant" the scientific enterprise, but, rather, to continue it, re-tooled for the very challenges that the enterprise itself entails, if not demands.

The same cannot be said of the relation of modern science to its parent traditions of ancient Greek thought and medieval scholasticism however. For the founders of modern science, as again we shall see shortly, were adamant in their declarations that all such older traditions were to be renounced wholesale and were to be replaced with something "entirely new". And so, if we are to make sense of the narrative that is the history of biosemiotics only within the larger narrative that is the history of science, we now find that yet an even earlier history and an even larger embedding narrative has to be recounted here. Thus, if we are to understand the history of science, it will be critical to understand the intellectual traditions from which modern science, and the whole modern age of the 16th and 17th centuries, was self-consciously rebelling. But to understand what joins those two older traditions, one first and foremost has to understand Aristotle.

And what better place to begin a history of biosemiotics than with the West's first genuine biologist? For by tracing the winding evolutionary path that begins in the ancients' observational thinking about life processes, and continues through the heavily mediated symbolic thinking of the medievals about sign processes, we may at last begin to get a clearer view of the conceptual entanglements between signs and nature that Descartes sought to resolve not with a yet more entangled synthesis, but with an Alexandrain cleaving that would leave the two halves disconnected and the thread that once was their unity forever cut.

And as it is the job of biosemiotics to begin weaving this thread together again, we must first discover just how and why it got tangled up in the way that it did in the first place. We begin our history proper, then, before it ever occurred to anyone to tie such a knot out of the naturally occurring continuum bearing humans, nature, animals and culture.

PHASE ONE: SEMIOTICS WITHOUT SCIENCE

It is said that the ancient Greeks had no real vocabulary for, nor philosophical interest in, discussing the distinction between "natural" and "cultural" signs. Rather, reports historian John Deely (2001), even up until the Roman period, σημείο (L: *semeion*) for the Greeks remained primarily a medical term – roughly akin to the modern concept of *symptom* – that referred only to the outward manifestations of an internal condition or overall state of affairs. And it is from this word *semeion,* of course, that the word "sign" – "something that suggests the presence or existence of some other fact, condition, or quality"[2]– proceeds.

We can notice at the outset, however, that a close association between something present and observable (pallor, a rash, a swelling of the ankles) with either something else that is present and unobservable (an organ malfunction or a tapeworm within) or with something else that is non-present and unobservable (overindulgence of food and drink, an accident that happened previously) imbue this notion of *semeion* with at least the two following characteristics: (1) The phenomenon of *relation* is inextricable from the concept; the *semeion* is always a symptom "*of*" something other than itself or just itself (and is so, presumably, whether or not any physician or patient is looking at it, as it necessarily pre-exists both of their awarenesses of it). (2) Something "unobservable" is being *educed* in the process of observing.

AUGUSTINE: THE "NATURAL" AND "GIVEN" SIGNS

These two broad characterizations alone will be sufficient for several centuries worth of "realist" versus "idealist" debate arising, self-consciously or not, from the question of whether (2) or (1) should be given primacy in one's understanding of what it is to be a "sign." And this is a debate whose lack of historical resolution deeply informs our own difficulties with the concept of a "sign" today. Paralleling our own deepening understanding in biosemiotics, perhaps, Augustine of Hippo (354–430) in his treatise of 387 known as *De Dialectica*, first assigned ontological priority to the second of these two characteristics, by codifying the extra-medical notion of the sign (the Latinized *signum*) as constituting "anything *perceived*, which in so doing, causes something other than itself to come into *awareness* (*Signum est quod se ipsum sensui et praeter se aliquid animo ostendit*)."

Returning to the topic twelve years later, however, in *De Doctrina Christiana* (398), Augustine refined his definition by emphasising the first of the sign relation's two aspects. There, he writes that "*a sign is something* which, offering itself to the senses, conveys something other to the intellect (*Signum...est res praeter speciem quam ingerit sensibus, aliud aliquid ex se faciens in cogitationem venire*)." The shift in emphasis from the semiotic capacity of the agent to the semiotic capacity of the sign vehicle *per se* is a subtle one, but as Deely (2001:215) notes, the implication of Augustine's general formulation is that there exists a mode of actuality in the real world that contains and is thus a higher-order category of both "mind-dependant relations" and "mind-independent relations" – i.e., the category of relations *qua* relations. And this actually constitutes the first recorded consideration of "sign relations" *per se*. Yet this was not the implication that was taken up and developed subsequently, but turned out to be exactly that most potentially fruitful part of Augustine's formulation that lay dormant for the majority of the Middle Ages, and, indeed, for most of modernity as well, as we shall see.

Rather, and unfortunately, it was not Augustine's posit of the sign relation's unification – but of its distinctions – that would turn out to be perplexingly consequential for the history of sign study. For it is in Augustine, too, that we first find the beginning of the philosophic tradition that distinguishes between "natural signs" (*signa naturalia*) and what might be called "cultural signs" – but that Augustine himself calls "given signs" (*signa data*). *Signa naturalia*, for Augustine, are those signs that, "apart from any intention or desire of using them as signs, do yet lead to the knowledge of something else" – one might think of the relations of physical contiguity, such as the relation of smoke to fire, or of the footprint in the ground to the animal that made it, or of a fossil. "Given signs" (*signa data*), on the other hand, are "those [signs] which living beings mutually exchange in order to show, as well as they can, the feelings of their minds, or their perceptions, or their thoughts" – such as, presumably, talk and gestures and the marks on this page and Augustine's *Confessions*.[3]

Subsequent inquirers into the notion of sign relations will come to realize, however, that Augustine's distinctions raise as many questions as they propose to answer. And among these many questions are: *For whom* do such natural signs "lead to knowledge of something else"... *other than* those with the "intention or desire for using them" as such? And must the given signs that "living beings mutually exchange in order to show... the feelings of their minds" be *deliberately* and *expressly* "exchanged" – or may they be subconsciously *performed* and *registered*? Do animals use *signa naturalia* or *signa data*? And in what relation towards each other do these two categories of "natural" and "given" sign relations ontologically stand? Perhaps most importantly of all: Is it "perception" and "awareness" on the part of some agent that *gives* a sign its representational efficacy – or does the agent merely "apprehend" a relation in the world that is already there, regardless of its apprehension?

Not because he did not recognize these sorts of questions, but because they were extraneous to his purpose of examining how sacrament and scripture function as the revealed signs of God, did Augustine more or less leave the discussion of signs *qua* signs at this point (Deely, 2001:22). Still, as Meir-Oeser (2003) writes, "despite all the internal ruptures and inconsistencies, Augustine's doctrine of signs is based on a definition that, for the first time, intends to embrace both the natural indexical sign and the conventional linguistic sign as [but two sub-]species of an all-embracing generic notion of sign, thus marking a turning point in the history of semiotics."[4]

Certainly, from a *history of biosemiotics* standpoint, Augustine's early formulation of a sign as primarily being constituted by a relation between one aspect of the natural world and another (one which just so happens to be constituted as a "perceiver") is so manifestly commonsensical and unencumbered with specially-created dichotomies, that had the contingencies of history been otherwise, and had sign study proceeded from Augustine's definitions, rather than from a radically disemboweled version of Aristotle, as we shall soon see it do – we may not have found ourselves here today still trying to establish as a general understanding the idea that the world of sign relations *per se* did not start with the advent of *homo sapiens* – and that a sign relation is not something that was created *ex nihilo* by the minds of human beings – but rather, that the minds of human beings are themselves the product of a *de novo* use of absolutely natural and biological sign relations.

ARISTOTLE: *ON LIFE* AND *ON INTERPRETATION*

The contingencies actualized by history have *not* been otherwise, however, and thus the understandings about sign relations that came to be most generally accepted next are ones that were to have dire consequences for subsequent centuries' attempts at incorporating the resulting notion of "sign" relations with the modern project of science. For that version of the scientific project that we have inherited today, of course, proceeds in a fairly straight line from the experimentalist instrument of Francis Bacon's *Novum Organon* – which is itself an historically situated reaction to what had been the primary "instrument of logic" and investigation about the natural world for the scholastics of the medieval world – i.e., Aristotle's six books on logic known collectively as the *Organon*. But in calling for a revolution in the approach of scientific investigation from the deductive to the inductive, Bacon and his contemporaries yet inherited an impoverished notion of "sign relations" that would devolve into a literally irreconcilable mind-body dualism at the hands of René Descartes a mere twenty-one years later.

This assumption of an essential dualism between material relations and sign relations continues to inform the practices and premises of modern science up until the present day. And because of this, it is incumbent upon us to spend the necessary amount of time here retracing the historical trajectory that precluded for centuries even the possibility of a science devoted to investigating the myriad ways in which material relations could come to function as sign relations in the lives of living beings.

For in the seven centuries that followed Augustine, the churchmen studying his doctrine of signs did so only in the sacred context in which it was intended. For examinations into the natural world, they turned, of course, to The Philosopher, Aristotle. But the Aristotle of the early Middle Ages was only a partial Aristotle, at best, consisting only of the six books translated into Latin by Boethius (480–524) in the sixth century C.E. These six books on logic, thought to have been collected by Andronicus around 40 B.C. so as to present the reader with a structured system of logic, would come to be *the* standard text of non-Biblical learning in the thousand years between the fall of Rome and the beginnings of the modern era – so much so that they became collectively known as just the *Organon* – the "instrument" of knowledge and well-ordered thought.

Critically, however, these six books were only one small part of Aristotle's overall understanding about the logic of human reasoning *and* the logic of the natural world. The rest of Aristotle's works – and the ones through which one can get an understanding of how the logic of human relations *comes out of and fits in with* the logic of the natural world (a 'biosemiotic' understanding, as it were) – these were lost to the West for over a thousand years. And from these impoverished initial conditions, a magnificent edifice that was yet only half-informed was constructed over the course of the next ten centuries.

For the centrality of the Aristotelian *Organon* as the primary "instrument of logic" throughout the whole of the Middle Ages – without the corresponding Aristotelian texts on nature and biology – meant that the focus of the next dozen centuries, at least as far as the investigation into "sign relations" is concerned, would proceed from Aristotle's meditations of the sign *exclusively* as it is manifested in human experience. Indeed, *De Interpretatione* – that book of the *Organon* that deals most specifically with semantics, hermeneutics and propositional logic – focuses entirely on the relations of "words" and "sentences" and begins thus:

> Spoken words are the symbols of mental experience and written words are the symbols of spoken words. Just as all men have not the same writing, so all men have not the same speech sounds, but the mental experiences, which these directly symbolize, are the same for all, as also are those things of which our experiences are the images.

This latter notion – that "those things of which" *our experiences are the images* are tied in some deep way to "what all men have" in their very constitution *as* men (or, more properly, as human beings and as animals) – Aristotle declines to expand upon in *De Interpretatione*, mentioning suggestively that it "has been discussed in my treatise about the soul, [and] belongs to an investigation distinct from that which lies before us here" (330 BC /1941:38). Having access to the thought of Aristotle only through Boethius's translation of the six books of the *Organon*, the first six centuries of monastic scholars, however, had no access to this referenced "treatise about the soul" and were thus literally prevented from seeing how the arguments of *De Interpretatione* could be understood as a but a particular subset of those in *De Anima* (and in *De Sensu et Sensibilibus*).

De Anima, of course, is about life, and the translation of "anima" as "soul" can be a misleading one to modern English speakers who are not philosophers. Because anything resembling the body-separable, spirit-like "soul" of the Platonic, Christian and (later) Cartesian traditions is antithetical to what Aristotle is referring to by the term ψῦχή (Latinized as *anima*) in this work. And, in some ways, the understandings of our current science are closer to Aristotle's ideas about *anima* than has been the case at any time since his rediscovery in the West in the 11th century – so much so that a modern gloss on Aristotle's famous dictum that *"the soul is the first actuality of a natural body that is potentially alive"* might today read: *"life is the emergent system property of the interactions of a self-catalyzing system that can adapt to its environment to persevere"* and the basics of his hylomorphism to state that *the biological "form" of such life is the product of its (for us: evolutionary and ontogenetic) embedding in the world, and itself consists of those particular sets of systemic relations that serve to organize a material substrate into a particular kind of organism.*

Thus, to the extent that even this (highly oversimplified) gloss is representative of the interdependent recursivity of Aristotle's biology, we can see that: (1) animal form is shaped in regard to organisms' interaction with the world, and vice-versa (anticipating Darwin, although, of course Aristotle was assuming the fixity of these systemic organism-world arrangements, and not their evolution); (2) the organisms' actions upon the world (which subsequently change that world) are both enabled by and constrained by the organism's systemic biological constitution, including its perceptual capacities (anticipating von Uexküll); and (3) it follows that as the result of (1) and (2) there is a both a "realism" to sign relations and a deep necessity for the joining together of the extra-biological relations of external reality to the embedded biological relations within organisms such that "what *occurs* in the case of the perceiving [system] is conceivably analogous to what holds true in that of the things themselves" (*De Sensu vii*).[5]

In perception, as well as in imagination, in other words, "it is not the stone which is present in the soul but its form" (*De Anima viii*). Understood within Aristotle's overarching conceptual system of hylomorphism, and again translated for modern ears (especially those conversant with dynamic systems theory), this means that *there exists a structural coupling between the relations constituting organisms and the relations constituting the external world that ensures a veridical alignment between the two* that holds across the *scala naturae*. And again, we can see how the development of evolutionary theory two millennia later (as well as the study of animal perceptual worlds *qua* perceptual worlds that we will be discussing later) can further inform this conjunction between bio- and semiotic- reality, making the prospects of a either a nominalist or a Cartesian divorce between knowers and the world they know the bewitchment of a symbolic overcoding system that itself no longer recognizes its own grounding in the relations of the material world (cf. Deacon 1997, Hoffmeyer 1996).

Thus, the breaking apart of the subordinate study of human words and propositions in *De Interpretatione* from the superordinate study of animal organization and interaction in the world that Aristotle develops in *De Anima* – a more or less accidental bifurcation owing to the contingencies of history – became the starting point of a developmental pathway whose alternative trajectory would remain *terra incognita* long after the end of the Middle Ages and right up to the last half of the twentieth century. For the result of the ever-widening bifurcation in the scholastic period between the investigations of bio-logic and the investigation of semeio-logic resulted in the assumption that it is what the scholastics called the "mental word" (*verbum interius*) – or what we might designate more precisely today as "linguistically mediated experience" – that was to be the natural starting point and, eventually, the exclusive focus of "sign" study.

Yet this would prove to be a guiding assumption that is at the same too broad and too narrow, in that understanding the essence of a "sign" *per se* to be an object that is mediated through the mental experience of human beings, conflates what is merely one example of the superordinate category of "sign relations" into the definition of the entire category itself. Doing so thus accomplishes a logical conflation and an explanatory reduction at one time – a misstep that would have profound consequences for the next dozen centuries of philosophic inquiry, and by extension, for the subsequent foundation of modern scientific thought.

For only centuries later would be reclaimed the evolutionarily coherent notion that the appearance of humans with their unique kind of "mental experience" is itself the product of a legacy of sign relations arising out of animals' interactions with each other and with the external world. And that in order for even these most primitive multicellular animals to come into being, processes of organization whereby living cells could co-ordinate their interactions with each other (and, again, with the external environment that they had to somehow come to negotiate in order to survive), proto-semiotic "substitution relations" – biologically instantiated processes whereby detection of the presence of *x* becomes a reliable indicator of *y* – had to evolve.

SIGN DIVERGENCE AND CONVERGENCE IN THE LATE MIDDLE AGES

This is not to say, however, that the Middle Ages was entirely bereft of thinkers dedicating their considerable intellects to an examination of the role of sign relations in life. Indeed, both Roger Bacon (1214–1293) and his contemporary Robert

Kilwardby (1215–1279) independently called for, and made explicit attempts at establishing, a "science of signs" (*scientia de signis*) "in terms of a universal notion abstracted from the [phenomena of] particular signs" (Meier-Oeser 2003). Both projects floundered, however, given the prevailing interests and valences of their time, and were unable to resist the gravitational pull towards misunderstanding human symbol use as the archetypal relation that one studies when one studies "sign relations." For then, as now, the attempt to understand more general and fundamental sign processes through the application of criteria that only apply to more specific and derivative sign processes, resulted in an unrecognized "Orwellian rewriting of the evolutionary past in terms of the present"[6] that, not surprisingly, failed to satisfactorily account for the possibility of *any* sign relations emerging out of the world of nature at all.

Yet thus did almost all investigations into the nature of "sign relations" throughout the Middle Ages take as the object of their inquiry not a triad of relations bringing together the extra-mental world of agents, actions and objects in the first instance, but a triad of relations joining "mental speech" (*oratio mentalis*) and its relations to the rest of the intellect (*intellectus*) on the one hand, and to the grammar of the spoken word (*vox verbi*) on the other. Propositions, human mentation, psychological states, linguistic relations and their resulting (and often unacknowledgedly linguaform) conceptual understandings – these were the first principles and paradigmatic assumptions from which the "sign sciences" of the Middle Ages – and, indeed, of most of the modern age – set forth but could not proceed.

For even with the recovery of the lost texts of Aristotle from the Arab world in the 13th century, the much needed re-reading of *De Interpretatione* in light of *De Anima* never occurred (and, indeed, has not truly occurred yet). Not surprisingly, then, did William of Ockham (1287–1347) exacerbate the incipient dualism between extra-mental relations and sign relations by asserting that the universal properties of things were merely the universalizing mental *signum* (signs) of human minds. In such ground did the seeds of an increasingly mentation-centric *nominalism* flourish, and the self-reinforcing "humanification of the sign" progress.[7]

Indeed, it was only towards the absolute twilight of scholasticism and the dawn of the modern period that a minority of thinkers, primarily those associated with the Iberian University of Coimbra, would attempt a reconceptualization of the sign as a relation that may supercede any given human way of being – and this conclusion was only reached through their hermeneutic reconsideration of Augustine's original assertion that "a sign, in every case, imports 'something relative to something else' (*aliquid stans pro alio*)" (Deely 2001:426).

It was proceeding from this investigation that the most prescient of these Iberians, John Poinsot (1589–1644) in his *Tractatus de Signis,* refuted both Platonic realist and Ockhamist nominalist understandings of sign relations with his conclusion that: "the most formal rationale of a sign *consists in being something substituted* for a significate, whether as an object external, *or* as representable within" (Poinsot 1632/1985:163, italics mine).

In ways which we will have both time and need to expand upon more fully later in this history, Poinsot's understanding of the sign as being something that is in its very essence a triadic *relation* of *x* as *y* to *z* in its first instance[8] – and only derivatively any actually instantiated realization of such a relation (e.g., of a mental sign to a human knower, or a odorant molecule to an opossum, or in the exchange of Ca+ as a second messenger in the incessant interaction between living cells) – resuscitates the naturalistic Aristotelian understanding of a world of creatures whose internal organization give rise to their external interactions and vice-versa. In such a world, mind-dependant relations and mind-independent relations are tightly woven.

Thus, philosopher historian John Deely claims that Poinsot's muddle-clearing "identification of *signs* with pure relations as such [constitutes] medieval semiotics' highest point of development… as the question of whether signs can be identified with any definite class of things able to exist [independently], whether as physical or as psychological realities, is definitely answered in the negative" (2001:434). Rather, writes Deely, "in every case, the sign as such, *consisting in the relation* between sign-vehicle and object-signified, is something suprasubjective" to the yet necessary participation of them both, in any system capable of acting upon the "things" of the material world so as to be able to actively transform them into the "objects" of triadic relation (2001:434).

This means, claims Deely, that "those 'things' or 'perceived objects' that we [mistakenly] *call* "signs" – things such as traffic lights, barber poles, words, [thoughts], and so on, are not, technically speaking, *signs* but the *vehicles* of signification" (ibid) – an understanding which, if adopted widely, would constitute a radical corrective to the futile attempts to discover what it is about neurons (or about nucleotides, or second messenger molecules, or spoken sounds or the ink marks on this page) *per se* that "signifies" or is a "sign" of anything.

Instead, the discoverable relevant relations of system *x as well as* those of entity, state or event *y during the course of interaction* whereby *y* is acted upon as a sign of *z* for *x* becomes the focus of investigation – and while this may sound like a task only feasible within the massively complex calculations of advanced dynamics systems theory, one should bear in mind that this was exactly the kind of principled scientific, naturalistic "sufficient explanation" that Aristotle was calling for when he wrote that the relations proper to *biologically* organized systems are "enmattered formulable essences" partaking of an interdependent, but absolutely non-mysterian and scientifically examinable "double character" that any full explanation of such system has to include in its account:

> Hence a physicist would define an affection of soul differently from a dialectician… the latter assigns the material conditions, the former the form or formulable essence… Thus the essence of a house is assigned in such a formula as 'a shelter against destruction by wind, rain, and heat'; while the physicist would describe it as 'stones, bricks, and timbers'; but there is a third possible description which would say that it was *that form in that material with that purpose or end.* Which, then, among these is entitled to be regarded as the genuine physicist? The one who confines himself to the material, or the one who restricts himself to the formulable essence alone? Is it not rather the one who combines both in a single formula? (*De Anima: i*).

And from Poinsot's formulation to our current understanding about the generative, recursive dynamics of autopoetic systems, it is only one small step to realizing that one of the implications Aristotle's assertion about the "double character" of "enmattered formulable essences" is that *sign relations* are those genuinely existing, materially manifested relations that *join* system-internal and system-external relations into a web of utilizable experience for *all* organisms – and, indeed, this is one of the founding premises of what today calls itself "biosemiotics". Yet having progressed the understanding of the fundamental nature of sign relations to this point, one would hardly think that the time was ripe to abandon the progress made thus far altogether and to assert an even more radical separation of mind-dependent relations from everything else.

Such a discontinuous and divisive posit would itself constitute a schism between the classic-scholastic tradition of thinking and, well, everything else. Yet such a schism is, indeed, precisely what René Descartes had in mind when he announced his project to renounce all prior knowledge, and build the edifice of understanding completely anew, in 1641, in his nightgown, by the fire.

PHASE TWO: SCIENCE WITHOUT SEMIOTICS

"What is a man? Shall I say a reasonable animal? Certainly not; for then I should have to inquire what an animal is, and what is reasonable; and thus from a single question I should insensibly fall into an infinitude of others more difficult; and I should not wish to waste the little time and leisure remaining to me in trying to unravel subtleties like these."[9]

So wrote René Descartes in 1641, expressing his resistance to the prospect of becoming a biosemiotician, right at the outset of modernity – a modernity that this particular resistance not only helped to shape, but to actually bring into being.

The subsequent history of this resistance would fill many volumes. However, no understanding of the current state of biosemiotics or of the conditions which made its emergence necessary, if not inevitable, would be genuinely intelligible without a brief re-telling of an oft-told tale regarding yet another decisive turn in the road that has led us to our present pass – children of a hostile, and yet impossible, divorce between not only mind and body, nature and culture but, now too, unexpectedly, between scientific explanation and ordinary human understanding – a tale that the philosopher Bruno Latour has christened "The Strange Invention of an "Outside" World" (1999:3).

And, indeed, so absolutely ordinary does it feel to us as the inheritors of Descartes' legacy to set the terms of our understanding in the form of an "experiential debate" between that which is "in the world" independent of any minds – and that which is "in our minds" independent of what is in the world – that it often goes unnoticed that smuggled into the very terms of this debate, the latter stands in relation to the former as a kind of impotent Platonic shadow or blind mendicant – and the mind becomes the glass through which we see the world darkly, rather than face-to-face.

This understanding, like all others, no matter how infrequently considered, has a history of its own, for Descartes by no means came upon his radical ideas *ex nihilo*, regardless of how he would have us understand him doing so (or, indeed, as he himself may have understood himself as doing so) in the *Meditations* of 1641.

DESCARTES: BIFURCATING THE NATURAL WORLD INTO BODY AND SPIRIT

For by 1641, both the scholastic tradition and the hegemony of Aristotelian explanation of natural phenomena had all but passed into eclipse in Europe. Modern mathematical notation – one of the primary instruments with which both Newton and Descartes would revolutionize our ideas of what it is to "do science" – made its belated arrival on the continent only in the preceding century (where its initial denunciation by Church authorities as a "pagan notion" of the Arabs and the Hindus, and thus to be resisted, stemmed exactly as little of the rising tide of secularity as did their subsequent denunciations of the works of Copernicus, Galileo and Kepler, and for much the same reason: in an exponentially individualistic and mercantile society, the calculus of utilitarian efficiency trumps the zero-sum game of static absolutism).

Yet while the gradual defenestration of Aristotelian physics had already begun in earnest with the works of Buridan (1300–1358) and Oresme (1323–1382) two centuries earlier, equally critical to the spirit of Descartes' project (and to the successful way it resonated through the ensuing three centuries) was the turn away from received authority and toward the autonomy of the individual that was the *zeitgeist* of the later Middle Ages. Humanism, the Renaissance, a burgeoning urbanite and merchant population, the Reformation, anticlericalism, the rise of the universities and the antagonism between change and conservatism that marks any such period of rapid development all formed the backdrop against which Descartes would "autonomously" resolve to "abandon the study of the letter, and to seek *no knowledge other than that which could be found in myself* or else in the great book of the world" (*Disc 1:9*).[10]

This was a move that was to prove critical for the subsequent history of Western thought, for what Descartes reports he finds when he looks inside himself is not an *innenwelt* of referential relations reaching out into the world and structured through participation in a ubiquitous human culture of symbolic reference stretching back at least 12,000 years to the establishment of human settlement (to pick an inarguably late but, because of that, uncontestable date in the evolution of symbolic culture). Rather, and bizarrely, he finds instead an immaterial solipsist who suspects he's being lied to.

"I suppose, then," Descartes writes, "that all the things that I see are false; I persuade myself that nothing has ever existed of all that my fallacious memory represents to me. I consider that I possess no senses; I imagine that body, figure, extension, movement and place are but the fictions of my mind . . . and of my former opinions I shall withdraw all that might even in a small degree be invalidated by the reasons which I have just brought forward, in order that there may be nothing at all left beyond what is absolutely certain and indubitable" (1641 [1973:150]).

Descartes' project, of course, is a quest for "absolute" (read: non-relative) certainty – and the discovery of at least one contextless and necessarily true axiom or assertion that will serve as the foundation for a sturdy system of reliable and correct knowledge to be constructed. Having already devised one such sturdy knowledge-bearing system – that of analytic geometry and its Cartesian co-ordinate system – in 1637, Descartes now embarks on a radical version of the subtractive method in order to successfully discover a single Archimedean point of truth.

Thus convinced of the need to reject the *entirety* of received opinion from the past – as well as to renounce belief in the primacy of embodied sense experience as being the most fundamental route to "knowing" – Descartes decides to consider as "false until proven otherwise" the entirety of both tradition *and* sensation and to seek absolute certainty in the only place then left available to him – i.e., in "the thoughts *which of themselves* spring up in my mind, and which were not inspired by anything *beyond my own nature alone" (ibid)*.

This decision to assume that methodological solipsism could serve as the foundation for the construction of a veridical, empirical science was, indeed, a "bifurcation" from the understandings of an inherently embodied cognition that had been assumed from antiquity and developed continually, if variously, by the scholastics right up until the time of Descartes himself (e.g., in the works of the Iberian school and, especially, John Poinsot).

Moreover, Descartes' attempt to "build anew from the foundation [and in so doing] establish a firm and permanent structure in the sciences" (1641 [1973:144]) by first razing to the ground the edifice of inherited error and by then sterilizing himself against the deception of bodily interface with the world by denying the efficacy of embodied relations was ultimately only considered a completely *constructive* success by Descartes – who then goes on to build his edifice for the securing of absolute certainty anew upon his *cogito*, and its corollary proof of the prerequisite existence of a benevolent and non-deceiving God.

Yet, "having abjured history as a means to truth," writes philosopher of science Alisdair MacIntyre, "Descartes recounts to us his own history as the medium through which the search for truth is to be carried on" (1974:59). And as it is this account that set the course of the next three centuries of thinking about "knowing" in the West, it is worth considering MacIntyre's analysis of Descartes' history-changing enterprise in full:

> "Descartes starts from the assumption that he knows nothing whatsoever until he can discover a presuppositionless first principle on which all else can be founded. [In so doing] he invents an unhistorical self-endorsed self-consciousness and tries to describe his epistemological crisis in terms of it. Small wonder that he misdescribes it. ... [for first among the many features of the universe and about his own historically embodied being] he does not recognize that he is *not* putting in doubt is his own capacity to use the French and Latin languages ....[as well as] what he has inherited in and with these languages: namely, a way of ordering both thought and the world expressed in a set of meanings. These meanings have a history... [but] because the presence of his languages was invisible to Descartes [he does not realize that] how much of what he took to be the spontaneous reflections of his own mind is in fact a repetition of sentences and phrases from his school textbooks – even the *Cogito* is to be found in Saint Augustine" (1974:60).

Inspired by the reformationist and revolutionary *zeitgeist* of his time, however, Descartes was not the only one of his contemporaries agitating for a clean break with the medieval past. That feeling had been growing, rather, at least since Petrarch retroactively designated the thousand years between the collapse of the Roman Empire and his own 14th century Italy to have been "the Dark Ages" of human thought. The multiple European "Renaissances," the Protestant Reformation, the rise of mercantilism and the rapid advancement of printing, lens and machine technologies, all played their parts in laminating this retrospective construction of a "backwards" time from which humanity was finally emerging – an idealization of the individual "over and above" history and nature without which the self-conscious seeding of a "scientific revolution" in the first part of the 17th century could hardly have fallen upon fertile ground. But if we see the coalescing of this scientific revolution, as most historians rightly do, as one of the major branching-off nodes in the cladistic history of Western thought – and, more importantly, as the branch on which we yet now reside – it will do well for us to examine what Descartes and his radical contemporaries may have left behind at this consequential forking of the roads ... as it just may be something we are going to have to go back and retrieve today if we are to carry on that very vision of scientific progress that Descartes and his contemporaries have bequeathed to us.

For in "asking how an isolated mind could be *absolutely* as opposed to relatively sure of anything in the outside world," notes historian and anthropologist of science Bruno Latour, Descartes "framed his question in a way that made it impossible to give the only reasonable answer... [i.e.,] that we are *relatively* sure of the many things with which we are daily engaged... [But] by Descartes' time, this sturdy relativism, based on the number of *relations* established *with* the world, was already in the past, a once-passable path now lost in a thicket of brambles" (1999:4).

Medievalist John Deely echoes Latour's point, expanding upon it even more precisely when he observes that "if we put [late-medieval thinker John] Poinsot's claim that the doctrine of signs transcends *in its starting point* the division of being into *ens reale* and *ens rationis* into contemporary terms, [then] what is being asserted is that semiotic [whereby the worlds of mind-dependent relations and mind-independent relations are bridged for the cognitive agent through the mediating relation of sign use] transcends the opposition of *realism* to *idealism*" that has come to define the "mind-body" and the "knowledge/fact" debates initiated by René Descartes and persisting to this very day (2001:483)

With Descartes, rather, "the priority of *signs to objects* becomes lost to view, and *objects of experience* become not a partial revelation of surrounding nature and culture, but a screen separating the mind from things" (Deely 2001:520, italics mine). But Descartes, of course, was not alone in seeing the need for a "radical surgery" that would separate *res cogitans* off from the rest of *res extensa* and come to see it as inhabiting its own little private world – an immaterial world that would quickly be recognized as a scientifically unexaminable world, no less, and yet the only world, supposedly, in which something as equally ghostly as "sign relations" could appropriately be thought to dwell.

"NOTHING LOST": MODERNITY PROCEEDS APACE

Certainly, William of Ockham (1285–1349) may have helped forge the blade for Descartes' radical surgery with his own wholesale denial of the existence of mind-independent universal relations and the reduction of our apprehension as such to "only thought-objects in the mind (*objectivum in anima*)" (1323 [1991]). This is a considerable ontological demotion of Aquinas' (1225–1274) far more subtle (and biosemiotic) understanding of the apprehension of such relations – like all sign relations – as partaking of "a dual being: one in singular things, another in the soul, and both [contribute their respective] accidents to it" (1252 [1965]). Here, again, it can be seen that at the heart of Ockham's cutting away is a dissection that offers no complementary implement for then suturing mind and world back together again.[11]

The more immediate precedent for Descartes' dualism, however, was undoubtedly Francis Bacon's *Novum Organon* – the "new instrument" that, in 1620, announced the inherent futility of reliance on "a mind that is already, *through the daily intercourse and conversation of life,* occupied with unsound doctrines and beset on all sides by vain imaginations" (1620 [1863], italics mine). Instead, and again very much in the spirit of his age, Bacon would proclaim twenty years before Descartes that: "Our only remaining hope and salvation is to begin the whole labour of the mind again…[and] that the entire work of the understanding be commenced afresh" (ibid).

Like Descartes, Bacon saw "error" as a ubiquitous product of the men both of his time and of all time before him – and, like Descartes, rather than understanding fallibility to be an intrinsic aspect of the effective functioning of symbolic reasoning – sought a "mechanism" designed to subtract it out from the human repertoire out entirely.[12] "The mental operation which follows the act of sense I for the most part reject," declared Bacon, anticipating Descartes' dream argument (though not his ball of wax). "There thus remains but one course for the recovery of a sound and healthy condition – namely, that the entire work of the understanding be commenced afresh" – again, prefiguring Descartes here, but now advancing the completely contradictory prescription that: "the mind itself be from the very outset not left to take its own course, but guided at every step; and the business be done as if by machinery" (ibid).

Bacon's mind-correcting machinery would come from outside: in the communally objective project of empirical experimentalism and induction. Descartes' mind-correcting machinery would come from within: in the irrefutable and eternal truths of mathematics and logical deduction. Abduction – the mind-producing process of acting upon what is presently given in an exploratory fashion, equipped only with the underdetermined understandings that have proved most effective thus far – was out of the picture for the interim (at least "officially" and in the symbolically self-reporting human world; the animals, we may assume, were going about their business as they always do: abductively, but not self-reflectively so).

And though neither Bacon's error-reducing inductive method, nor Descartes' error-reducing deductive method, succeeded in being adopted by their contemporaries *in toto*, the enacted *synthesis* of their mathematical-experimentalist methodologies – when coupled to the engine of generatively recursive collectivism initiated by the Royal Society in 1660 and still self-developing healthily to this day – would prove to be the single most effective technology for the securing of veridical knowledge ever developed by the mind of man.

Descartes' radical bifurcation, then, was not a failure – rather, in some sense it succeeded far too well. Which is to say that at least half of the severance was successful and went on to succeed beyond any reasonably foreseeable expectation. For after Descartes, the study of "bodies" would proceed entirely independently of the study of "mind" – their realms, after all, were separate in their essences – and thus the truth claims made by science need not be accountable to the truth claims made by the humanities, and vice-versa. And why should the science of Descartes' time have seen this liberation as in any way undesirable? As the more foundational of the two enterprises – in that the object of its study are those organizational principles of the world that exist extra-mentally and can only derivatively be "known" by human beings – why assume the additional burden of having to explain how it is and in what way a human being can come to "know" anything to begin with?

Bacon's experimentalism was vindicated by Robert Boyle's (1627–1692) foundation of "public science" and the establishment of the Royal Society made it clear: the laboratory would be the theatre of evidence, and what could not be shown there was outside the realm of science proper. To this domain of the visible and the material, the pure truths of mathematics would be admitted by Isaac Newton (1643–1727), thereby rightly vindicating Descartes. Thus armed with the error-correcting mechanisms of induction and deduction – and with the exponential power of a group of interacting agents pursuing individual ends within the *telos* of a formalized system – the study of "bodies" and their material relations would allow human beings to actually leave the planet and return to it in less than another 300 years.

The other half of Descartes' bifurcation, unfortunately, did not fare as well. Amputated from the natural world of material and logical relations from which it came, "the mind" and all of *its* internal relations – sensation, perception, subjective experience, knowledge and, in the singular case of human beings, language and symbolic thought – was increasingly ruled unfit as an object of genuine scientific inquiry, and was as such left to hobble down an increasing impoverished back-lane of abstraction, speculation, and pure, virtually ungrounded symbol use. For one of the more unfortunate effects of Descartes and his contemporaries' uniquely influential attempts to cure subjective error was that the "subject" began disappearing from scientific inquiry altogether.

But what needs to be foregrounded here is that it has never been the absolutely natural property of living organization called "the mind" (or, as neuroscientist Rudolfo Llinás (2001) is quick to clarify, "the property of being minded") *per se* that is to blame for this sad state of affairs. This condition is found everywhere throughout the animal world, once one realizes that the biological system property of "mind" is no more *synonymous* with "human (symbolic, linguaform) mind" than the term "body" is synonymous with "human (biped, mammalian) body – and that those creatures lacking language and the ability for abstractive thought are no "less" minded in the functional and biological sense than those lacking opposable thumbs (or, for that matter, gills or wings) are any "less embodied." Here, as everywhere in the natural world, huge differences in capability, capacity, and the structures which have evolved to meet the real-world challenges of life vary extraordinarily across species. But respiration remains respiration; digestion, digestion; locomotion, locomotion; and reproduction, reproduction regardless of whether we are talking about live birth and sexual copulation, egg-laying practices, pollination strategies or spore formation. There, and rightly so, the whole range of relevant and incommensurable differences is openly acknowledged, *in the full acceptance and understanding* that these species-specific adaptations are all serving precisely the one same biologically analogous end.

The single most compelling reason that the biological function of "knowing" is not likewise included in the list of universal attributes of living organisms is *not* because it isn't happening (and happening as variously and as species-specifically as does every other biological universal), but because our very *idea* of what *constitutes* "knowing" has been warped by Descartes' conflation of "mindedness" *per se* with "human mindedness" and "knowing" *per se* with "symbolic cognition" (again, see Deacon 1997 for a very clear discussion distinguishing between these two *very* different life processes whereby organisms "know" the world).

Persistently, in the back of our minds (which might explain something right there!), we equate "mind" and "knowing" *only* with our particular form of adaptation to this universal biological need.[13] And this, of course, presents us with a two-fold problem: First, if all of the fine-tuned purposive, responsive, evasive, interactive and anticipatory behavior that we observe taking place *ubiquitously* throughout the animal world cannot be calling "knowing," then what shall we call it when a previously motionless copepod reacts to the sudden presence of a quickly approaching predator by discharging a bioluminescent "depth charge" that is time-delayed so as to burst into illumination far from its site of origin in the copepod itself, instantly alerting the predator and sending it off on a false line of pursuit while the copepod swims safely away? Are we to say that the self-reflexive ability to symbolize its own experience and articulate that set of symbols to another *constitutes the criteria* for "knowing" *per se*? If so, then the bee can never "know" what flower to land on, the deer can never "know" which other animals in its surround to mate with and which to flee from, the penguin can never "know" which chick is her offspring, and – in fact – all other living beings except the human essentially the input/output automatons that Descartes claimed they are.

The second problem that this raises, of course, is this: *If* all animals other than human beings are now and have always been mind-less, how did the human being "evolve" its own mind *ex nihilo*? The problem is a classic *reductio ad absurdum*, once "supernatural" explanations are deemed illegitimate (and remember, it was and *is* supernatural explanation that allows Descartes to assume his bifurcation in the first place: God imparts to man a bit of His own Divine essence – "mind" – *and* sees to the organization of all the animal's lives "for them" by building into the mechanics of their mindless input/output actions His own Divine plan. It all seems a "bit much" to accept so uncritically at this late date, but not deliberately going back to examine one's inherited and critically unexamined starting assumptions often results in such odd effects... as Descartes himself well knew!).

Finally, the unexamined conflation of "mind" with "human mind" leaves the entire question of the species-specific peculiarity *of* that kind of human "mindedness" untouched. *If* we are dealing with yet another product of biological evolution, what *is* it that allows the human mind to engage in abstractive, symbolic reasoning, self-reflective intellection, "language games" of all kinds and the ability to imaginatively manipulate reality "off-line" as it were? What is the *nature* of this kind of cognition and sign use – and in what ways is it similar to and different from its functional counterparts in the lives of the termite and the baboon? Should we look for its source in the physical structure of the brain, as we look for the source of generating the ultrasound of echolocation in the larynx of the bat? Or should we, as Andy Clark (1997) suggests, look also in the distributed cognitive prostheses of the surrounding environment where we "off-load" our symbolic representations for cognitive exploitation in the way that the bluefin tuna exploits the very water vortices it produces in order to propel it along at speeds its own body could never accomplish on its own?

Few of these questions had even been *asked* prior to the last ten years – and far too few of them are being asked today, precisely because of the persistence of the still far too institutionally enshrined Cartesian conflation of "mind" with its specifically species-particular form of linguistic representation and symbolic reference – and, in some cases, its even less intellectually defensible notion as a disembodied and somehow self-realizing autonomous "entity." This persistent Cartesian misconception has been perhaps the single greatest "block [upon] the road to inquiry" (as Charles Peirce would say), steering natural scientists away from the problem for centuries, and causing the subsequent "investigations" into its nature by philosophers after Descartes' time to become the embarrassingly fruitless project that it has been ever since.

And this is the reason why we have spent so much time discussing this particular fork in the road. For with the explanatory surgery of Descartes' "mind-body bifurcation" now strongly in place by the end of the 17th century, the unparalleled success of the "body" sciences – including the "body" aspects of the biological sciences – were all but officially absolved from worrying about questions of subjective knowing in general, and thus felt no real pressing need to "waste what little time and leisure remaining... in trying to unravel subtleties like these" (Descartes 1641:[1973:80]). Equally unhappily, those thinkers who did pursue the issue, an increasing lack of need to consult, or eventually to even be conversant in, science. In short, Descartes' divorce between "material reality" (*res extensa*) and "knowing reality" (*res cogitans*) had worked too well, and the subsequent "history" of natural science – a science that must include beings that *both* know *and* are material – was explanatorily the worse for it.

"History," however, is a notion that comes to us from the Latin word for "narrative" (*historia*) which itself derives from the ancient Greek word for "witness" (Ἵστωρ). Thus, unlike the linear record of geological change, history – even scientific history – has actors and, to paraphrase Chekhov, "if there's a gun on stage in act one, chances are that it is going to go off in act three" (1904). And this is precisely what happened next.

FROM DYADIC TO TRIADIC RELATIONS: "INFORMATION" INVADES THE SCENE

Running off the momentum of the newly institutionalized *Novum Organon* of the Royal Society, the 17th through 19th centuries saw an explosion of biological knowledge made possible by Leeuwenhook's deployment of the microscope, the cellular structures of plants and animals, the exchange of nutrients and gases, the developmental stages of life from inception to death, and the synthesis of organic compounds from inorganic materials all were relatively amenable to the then-available physical and chemical understandings. It is only with Wilhelm Johannsen's (1857–1927) introduction of the "gene" concept in 1909 that "information" *per se* becomes something that is going to have to be accounted for by science.

But "information," under the Cartesian schema, could only be one of two things: either a relation proper only to the mind – in which case it was scientifically unexaminable *perforce* – or a pure product of material interactions, operating under mathematico-logical conditions – in which case it was not truly "information *of*" something, but merely whatever it happened itself to materially be (e.g., a catalyst, an agonist, etc). von Baer's (1792–1876) discovery of the epigenetic development of the fertilized ovum into structures expressing hereditary traits, however, made both these definitions equally unsatisfactory.

Thus, in coining the word "gene" to denote "the functional unit of heredity" – *whatever* it might turn out to be – Johannsen, much like today's biosemioticians, merely *thematized* – and by so doing explicitly *problematized* – what was implicitly being "discussed but not discussed" with the acceptance of von Baer's non-preformationist germ layer theory of embryonic development in 1827. For if preformationism is wrong, and an organism's cellular structure is not pregiven but developmental – as von Baer's experiments in comparative embryology showed it to be – then some "information exchange" is taking place within the developing embryo in order for undifferentiated cells to become differentiated tissue and thereafter, the resulting structures of arms, brains, livers and limbs.

Johannsen, again, had no more insight into what precisely this "unit of heredity" might turn out to be, nor how it functioned as it did, than did Darwin, Galton, Mendel, Flemming or Weismann – all of whom also posited germ theories of inheritance that, at their core, remained wholly unexplicated with regard to exactly *how* the interaction of this "germ" with the rest of the cellular material could result in the development of absolutely novel structure when at all points there is only...the germ and the material.[14] That it does so is clear – but just what the process *is* that explains its ability to do so was a question that the science of Johannsen's time had not even a coherent vocabulary for conceptualizing.

In delineating the distinction between hereditary genotype and metabolic phenotype, however, and in assigning a property to this gene that was in essence *informational* (in that it served to function for the creation of something other than itself for the system that it was embedded in as, materially, itself), Johannsen opened up the "problem of information" in a science that, since Descartes, had had nothing but success in dealing with things that acted merely as what they were materially – and not things that acted both as what they were materially and what they were not, but could be used to functionally "stand for."

And accordingly, while great strides have subsequently been made in our understanding of the purely material relations underlying "the genetic code" – conceptually unclear still is the absolutely scientifically legitimate question of in what just sense *information* – defined not just as the inanimate sequence of nucleotides themselves but as the *functional relation* of those nucleotides *to a system* for which they serve as "sequences of code"– can be thought to be a property of *things*. For Francis Crick, articulating his "central dogma" of genetic inheritance, "information" was synonymous with "the sequence of amino acid residues" *per se* (1985:1) while for Claude Shannon and Warren Weaver, "information" was the diminution of uncertainty in a system absolutely without regard to cognitive or semantic considerations (1949:8). Both Crick's notion of "information flow" from gene to protein and Shannon and Weaver's mathematical theory of "communication across a channel" thus explicitly deny that the "information" that they are talking about "means" anything in the sense that we associate with the word "meaning."

But here again, we see the intransigent Cartesian conflation still subtly, and perniciously, at work – undermining even the possibility that material relations and symbolic relations might stand in any other relation to one another than that of matter and anti-matter. Because the rationality driving Crick's, and Shannon and Weaver's denials is based upon the assumption all "true meaning" is *symbolic* meaning – the kind of relation that human beings are exploiting when they talk using language or think in terms of abstract representations. And in this sense, of course, neither nucleotides nor electronic pulses along a length of wire in themselves have any "*symbolic* meaning." Why, then, use this strange word "information" to describe them?

The move made here – as with the concepts "signal" in molecular biology, "message" in neuroscience and "communication" in animal ethology – is to understand that "*the code for talking about* the *genetic* code is that the term "genetic *code*" is only a metaphor, and should not be understood as denoting what it would denote in everyday usage" (cf. Griffith 2001; also Barbieri 2003a and this volume). But in all three instances, the question reasserts itself – a metaphor for what? "Processes we do not yet understand," certainly. But what *kind* of processes? Ultimately, informational, representative, and meaningful ones – so once again, we are back to having to confront the existence, in the *biological* world, or genuinely *semiotic* processes.

And the refusal to cross this self-imposed Rubicon inherited from the Cartesian legacy – a refusal born out of fear, generally, that if one does engage with issues of "meaning" one has automatically "crossed over" and *out* of the realm of doing real science – prevents theorists like Watson and Crick and Shannon and Weaver from seeing in what way their intuitions to use the words "information" and "communication" can point both them and us to a deeper understanding of those terms, one which is neither eliminative and reductionististic, nor mystical and unfalsifiable, but utterly naturalistic through and through – *if* we remain open to the understanding (that our dedication to science demands of us) that *all* things in the natural world evolved out of that natural world and nowhere else.

If, in other words, there are biological creatures that are alive today that use symbols, and exchange meanings, and have culture, and can deal in counterfactuals and can think abstractly – as undoubtedly there are – *and* there are other living beings living under the same physical conditions who have evolved from virtually the exact same genetic processes, and who have developed myriad other capabilities, but just not those particular ones last listed – *and* one denies at the outset that all of this is the result of a Divine miracle – *then* thoughts, meanings, symbols, culture and everything else that we associate today with the human mind are *grounded* in the structures, events, principles and relations that constitute the natural world. Understanding this, the research questions then become: what particular *relations* in the naturally occurring world does human symbolic understanding exploit differently, say, than primate indexical understanding does, or that the iconic relations chemotaxis affords for the amoeba? Are the earlier processes still at work in the later ones? How much and what kind of environmental restructuring is necessary for the full functioning of each? And *is* there a primitive organizational sense whereby the digital "differences" in electronic pulses down a length of wire (or, in the biological case, an axon), and the sequential differences in base pairs affixed to the phosphate backbone of a DNA molecule really *do* in-form the immediate next moment of consequential change in a living system? How does all this work? And how does all this work *together?* These are the questions that biosemiotics will wind up asking, seeking *not* a reductionist anthropomorphism of "all things in nature as human" but just the opposite: a principled evolutionary and biological understanding of how all things in human (and in animal) life are natural – including "knowing", including "meaning", including "thought" and, because of these last three, including "signs."

Interestingly enough, however, it is not the biosemioticians who stand ready to reject the notion that the biological set of relations constituting "sign processes" are, in fact, massively complex, organically organized material interactions – most biosemioticians would rejoice at such a discovery – conversely, it is far too often the committed physicalist who so closely (and so incorrectly) equates the entire category of "semiotic processes" with the one limiting case of symbolic, human mental processes, that to talk about the former *is* to talk about the latter – in which case, of course, they are completely right to reject the initial premises. Descartes' bifurcation, in other words, is continuing to keep the sciences of material interaction and the sciences of semiotic interaction apart.

But if biosemiotics has any one single most constructive message to give the mainstream scientific community, surely it is precisely this: a semiotic process is not a ghostly, mental, human thought process. Rather, it is, in the first instance, nothing more nor less mysterious than that natural interface by which an organism actively negotiates the present demands of its internal biological organization with the present demands of the organization of its external surround. And the fact that this is done incessantly – by all organisms, and by us – should not blind us to the significant fact that such moment-to-moment activity is always and perpetually an *enacted accomplishment* – and thus one that is going to have to be explained, if we are ever to understand the bio-logical side of living organisms' material interactions.

Yet so scandalous and counter-intuitive was this notion of genuine sign relations in nature – so drenched with and indistinguishable from, as it were, their singular symbolic manifestation as "mental thoughts and human words" – Descartes' *divine birthright of human intelligence* – that when Darwin's contemporary George Romanes (1848–1894) presented anecdotal evidence in support of even the possibility of animal intelligence, Edward Thorndike (1874–1949) announced that the goal of his own work would be dedicated to disprove "the despised theory that animals reason" (1898:39). How human intelligence could ever have "evolved" out of a world of absolutely non-semiotic animal relations then becomes something of a paradox – and, in fact, J. B. Watson (1878–1958) and B. F. Skinner (1904–1990) drew out the logical entailments of this view to eventually argue that human mental states, likewise, were "an illusion" – a position implicitly endorsed by the approach of many manifestly competent neuroscientists, and explicitly argued for in the "eliminative materialism" of Paul and Patricia Churchland (1984; 2002)...still victims of Descartes' destructive dualism, even after all these years.

Not surprisingly, then, do we begin to see at the dawn of the twentieth century, cracks and fissures arising in the scientific edifice out of internal tensions generated by the need to keep "subjectivity" out of science not only in its methodology, but also *as a focus of investigation* – *despite* the absolutely undeniable facts that: (1) the natural world is full of subjective agents, (2) the natural world itself must have produced these subjective agents once one rules out the possibility of supernaturalism as a legitimate scientific explanation, and (3) it is the subjective experiences of these agents that *leads* them to act upon the natural world in ways that materially *change* that world (and in so doing change the substrate that world then becomes for the evolution of subsequent subjective agents). Yet all generatively and recursively of this undeniable natural phenomena only becomes denied *as* "natural" phenomena with the adoption of the quite unnatural bifurcation insisted upon by Descartes that puts the entirety of human "mind" – along with every kind of "knowing" operation one might conceivably be tempted to assign to the purposive behavior of non-human animals – into the ghostly realm of the absolutely immaterial... and, again, despite the overwhelming evidence to the contrary of the existence of a plentitude of knowing, material, purposively acting, biological beings.

Moreover, not only was Descartes' legacy of ontological bifurcation causing cracks and fissures to appear in the explanations being offered for any researcher in the biological sciences who looked too closely at the obviously enacted subjective experience of living organism and the informational capacity of the genetic code, but it was also exerting a complementary tectonic pressure on the long line of philosophers, humanists, and researchers in the social sciences, who found themselves on the other side of the Cartesian divide, trying ever unsuccessfully to meet a challenge that, by its very premises, could never be met. Eventually, a few of the most frustrated – which may be really be to say the most committed – members of these two groups started pushing back against their respective fields' Cartesian boundaries and began scouting around in distant coastlines in an effort to more effectively redraw the inherited, but prohibitively unrealistic, map.

It is to just that group of interdisciplinarians that we now turn, for their work will provide our entrée into the current state of the field, constituting, as it does, the most recent evolutionary turn in the natural history of *biosemiotics*.

PHASE THREE: SCIENCE WITH SEMIOTICS

Because the current cohort of scholars constitutes the "first generation" of self-identified biosemioticians, the history of this cohort as a whole would have to consist of the history of each member, as he or she – faced with the internal contradictions or explanatory evasions of their home discipline – made their own unique pilgrimage to a place where biology and semiotics merged as one. Although doubtlessly fascinating, it would be impossible here to recount all these individual journeys from Istanbul and Los Angeles, Helsinki and Bologna, Toronto and St. Petersburg, Sao Paolo and Prague, and describe the many and various disciplinary sites of origin spanning across biochemistry and philosophy departments, dynamic systems research labs and anthropological field sites, and the lifelong private research investigations of individual scholars, many of whose final destinations are, as of this writing, unknown.[15]

What we must do here instead, in order to bring coherence to this account, is to focus on just those few figures most responsible for bringing this diverse group of scholars together. These would be the outspoken interdisciplinary organizers whose explicitly stated program of coalescing semiotics and biology increasingly attracted similarly inclined scholars into their orbit, and whose journals, conferences and book projects would come to constitute the gravitational center around which the interdiscipline of *biosemiotics* would gradually coalesce. And of this handful of "outspoken interdisciplinary organizers" perhaps none was more outspoken, more interdisciplinary, and more organizationally active and astute than the late Thomas A. Sebeok (1920–2001), without whom the current interdiscipline of biosemiotics would not have taken shape in its present form.

JOINING SIGN SCIENCE WITH LIFE SCIENCE: THOMAS A. SEBEOK

While a growing number of isolated scholars working in widely-separated disciplines were all toiling away at various *independent* lines of inquiry into the problems of information processing, intercellular communication, behavioral psychology, neurobiology and animal ecology – and long before the birth of such self-consciously "interdisciplinary fields" as "artificial intelligence" "dynamic systems research" or "cognitive neuroscience" – an academic polymath who once described himself as something akin to an "*Apis mellifera,* who darts solitary from flower to flower, sipping nectar, gathering pollen [and] serendipitously fertilizing whatever he touches" (Sebeok 1995) was to pioneer the practices that the modern-day university refers to as "interdisciplinarity" in the course of founding the project that today bears the title of *biosemiotics.*[16]

This self-described *apis,* Thomas A. Sebeok (1920–2001), left his native Hungary at age sixteen to study at Cambridge University, before immigrating to the United States where he received his doctorate in linguistics from Princeton in 1945, while simultaneously doing research under both Roman Jakobson at Columbia University and Charles Morris at the University of Chicago (Bernard 2001). A specialist in Finno-Ugric languages, Sebeok's appointment as the head of the Department of Uralic and Altaic Studies at Indiana University led to decades worth of extensive fieldwork not only investigating the internal organization of linguistic systems *per se,* but also in investigating the higher-order manifestations of such systems, in the form of anthropology, folklore studies and comparative literature (*ibid*).

Sebeok's growing interest in the organization of semiotic systems in general, combined with his aforementioned polymath intellect, led him to carry out some of the first computer analyses of verbal texts; to investigate the use of nonverbal signs in human communication; and to establish, with Charles Osgood, the pioneering interdisciplinary field of *psycholinguistics* in 1954. Six years later, during a fellowship at Stanford University's Center for Advanced Studies in the Behavioral Sciences, Sebeok indulged his lifelong passion for biology, studying both nonverbal human sign behavior as well as the communication practices of animals, both in the wild and as domesticated by human trainers (Tochon 2000). From this intense period of study came his programmatic call for the founding of the study of *zoösemiotics* – "a discipline within which the science of signs intersects with ethology, devoted to the scientific study of signaling behaviour in and across animal species" (1963).[17]

Sebeok's commitment to what he considered to be the two fundamental academic virtues of "*publishing and teaching as much as possible*; and, equally importantly, *doing one's best to facilitate the success of one's colleagues*" (1995:125, as cited in Kull 2003) led to his refashioning of Indiana University's Research Center for Anthropology, Folklore, and Linguistics – of which he had been appointed chair – into the Research Center for Language and Semiotic Studies in 1956, and to the co-founding of the International Association for Semiotic Studies in Vienna in 1969. In his activities for both these institutions, Sebeok's reputation as a tireless book editor, indefatigable conference convener, inveterate journal contributor, and all around facilitator of academic interaction across continents and disciplines became (and remains) legendary among his peers.

Thus it was Thomas Sebeok who would be responsible, more than anyone else, for bringing practitioners from the life sciences and the social sciences into dialogue with each other for the course of the next almost forty years, resulting in the collaborative interdisciplinary project that we today know as *biosemiotics.* The *Approaches to Semiotics* book series that he founded in 1964 eventually ran to 112 volumes over the course of its almost thirty year run; he was editor-in-chief of the journal *Semiotica* from its inception in 1969 until his death in 2001; and the list of international conferences Sebeok played a role in initiating with the express goal of bringing scientists and semiotician together would constitute a document – and, indeed, it is one that has yet to be compiled.

SEBEOK'S SYNTHESIS OF SCHOLARSHIP EAST AND WEST

Catalysts, by definition, enable reactions to occur much faster because of changes that they induce in their *reactants*. And so, too, it was with Sebeok who, in the course of building an interdisciplinary network (or symbiotic niche, as he might call it), must in addition be credited as the key figure most responsible for integrating both the current works and the rich theoretical traditions of otherwise forgotten academics toiling in the Soviet East into Western academia's collective consciousness. This he did often through his own smuggling of desperately proffered manuscripts across mutually antagonistic Cold War borders in the 1960s and 1970s (as memorably recounted in Sebeok 1998 and 2001) – actions which themselves serve as a wonderfully apt metaphor for his own "bee-like" approach to the unnaturally antagonistic disciplinary partitioning between the science and the humanities that he devoted his entire career to cross-pollinating.

These trips darting in and out from behind what was then called the Iron Curtain turned out to have particular significance for the development of Sebeok's *zoosemiotics* program into what he would later call *biosemiotics* (Sebeok 1998). For while acknowledging his debt of understanding to both Swiss "zoo biologist" Heini Hediger (1908–1992) for his pioneering work on the communicative practices of animals (and between animals and humans in the practices of animal training) – as well as to the Italian oncologist Giorgio Prodi (1929–1988) for his equally bold investigations into the semiotics of immunology and call for a comprehensive program of "natural semiotics" investigating the genetic, metabolic, neural and immunological sign-exchange processes of the human body (a program that Sebeok would later characterize as "endosemiotics" (Sebeok 1976) – it was Sebeok's 1970 trip to the then "forbidden city" of Tartu in the Estonian Soviet Socialist Republic to meet the Russian semiologist Juri Lotman (1922–1993) – a trip that he would later describe as "a singular Mecca-like field for us pilgrims laboring in the domain of semiotics" (Sebeok 1998) – that would forge the link between Sebeok's lone bee-like investigations and an entire rich tradition of semiotic thought that was virtually unknown of in the West.[18]

It was a difficult time for such East-West mutual collaboration, however, and Soviet émigré Vyacheslav Vsevolodovitch Ivanov (1929-) recalls that many scholars' works that were forbidden to be published in Moscow at this time had to be surreptitiously channeled to Lotman to be published in Tartu (1991:36). In turn, it was Sebeok who was entrusted by Juri Lotman with his seminal biosemiotic manuscript *O Semiosfere* for translation and publication in the West (Sebeok 1998).[19] Lotman's delineation of the realm of sign relationships permeating human life is, of course, a cognate of the word *biosphere* – the organizational space wherein living beings occur and interact – and was designed to foreground the autopoetic nature of *sets* of sign relations (such as "language" and "culture") as "modeling systems" for embodied action in the world of things by agents. And in this sense – the sense that Jesper Hoffmeyer will later use the same term, though unaware of its prior use by Lotman – it is a deeply biosemiotic notion.

Sebeok, however, found Lotman's early explication of the concept – which largely restricts its scope of inquiry to the human and symbolic realm of interactions that Vernadsky called the *noosphere*[20] – to describe a *necessary* concept for understanding our species-specific use of, and immersion in, a world of materially consequential sign relationships – but not an explanatorily *sufficient* one for doing so.

For it is one thing merely to assert, as he does himself some time later, that the human being is "a joint product of both natural and cultural forces" (Sebeok 1986:xi). But in and of such an assertion in itself, it remains unclear if what is being talked about are two mutually exclusive and possibly antagonistic forces, or some kind of symbiotic relationship, or merely two largely artificially designated extremes along what is, in fact, a continuum. Still left fully unexplicated then, felt Sebeok, was an explanation of how the set of sign relations constituting the human symbolic semiosphere emerged from – and in some sense remained dependant on – our simultaneous existence as biological beings. For that explication, Sebeok would have to look elsewhere.

Thus Sebeok continued his own decades-long study into animal communication both via the majority of research literature then extant and through his hands-on work with zoobiologist Heini Hediger (Sebeok 2001a). And as he did so, he became increasingly convinced that the sign relations taking place in animal communication and those in human language – while belonging to a common genus of interaction in the natural world – were yet divergent species that also needed to be understood on their own terms. Starting in 1977 and continuing well into the 1990s, Sebeok published in-depth critiques of the various underestimations, overestimations, anthropomorphisms and machino-morphisms being then attributed to animal cognition (e.g. Sebeok 1977, 1980, 1988, 1990).

In these writings, Sebeok is particularly adamant in insisting that what "ape language trainers" such as Duane and Sue Savage-Rumbaugh (1977, 1986), Allen and Beatrix Gardner (1979, 1989) and David and Ann Premack (1977, 1984) were attempting – or at least what they were succeeding at – should not be confused with the idea that these apes had acquired the ability to use "language" in the true sense of the word. Rather, felt Sebeok, such researchers were merely shaping the animals' behavior along purely iconic and indexical (stimulus-response) levels – without themselves having a discriminating enough understanding of sign relations to understand the underlying difference between the two phenomena. Thus, by pronouncing the resulting Skinnerian chain of purely associative reflexes to be the equivalent of "language," Sebeok concluded, these researchers were "looking in the destination for what should have been sought in the source" (as he succinctly titled his 1980 paper reviewing this work).

Sebeok was convinced that approaching animal communication as a truncated form of human language, just as Terrence Deacon would argue later in an elegant book-length consideration of language origins, "inverts evolutionary cause-and-effect" (1997:53). For to Deacon – and to Sebeok – the proper question is not: "Do animals have language the way that we do, and if not, why not?" but rather: "As animals ourselves, how is it that we have language?" For what makes human "language" unique, both Sebeok and Deacon agree, is not the mere ability to map sounds or gestures onto physically co-present things as referents in the first instance – but the far rarer ability to be able to flexibly, systematically and effectively manipulate representations of non-present, impossible and counterfactual conditions in the knowledge that we *are* "manipulating representations" (and not the things themselves) in doing so.

Yet what modern semiotician ever thinks of signs as other than exclusively human cultural products? And what modern scientist ever thinks of biological organization as itself perfused with signs?

The explication that Sebeok was to find was one that he himself had to help to create. And so to understand the synthesis that Sebeok was to propose as the "starting point" for his proposed interdiscipline joining the life sciences with the sign sciences – his *biosemiotics,* as he was soon to call it – one must first understand how Sebeok's long-standing study of the semiotic logic of relations explicated by the American philosopher and scientist Charles S. Peirce (1839–1914) served as the substrate upon which Sebeok's later rediscovery of the research into the perceptual worlds of animals undertaken by the then all-but forgotten Estonian biologist Jakob von Uexküll (1864–1944) would act as a powerfully synthesizing reagent.

SEBEOK'S SYNTHESIS OF CHARLES S. PEIRCE

A laboratory trained chemist, astronomer, mathematician and logician, Peirce advanced a logic of sign relations – or "semeiotic" (as Peirce had called it) – that Sebeok was well conversant in, having studied under at least three self-proclaimed epigones of Peirce – i.e., C.K. Ogden, (1889–1957), Roman Jakobson (1896–1982) and Charles Morris (1901–1979).[21] And because Peirce's "architectonic of triadic logic" deeply informs so much of Sebeok's work, it would be impossible to continue this particular "line of emission" in the history of biosemiotics without providing here at least a summary overview (or brief flash, as it were) of this, one of the two main sources of incandescence illuminating Sebeok's biosemiotic vision.[22]

A scientist by training and by temperament, "sign" relations for Peirce are a species of a larger genus of relations whereby potentiality becomes actualized, and the actualized interacts with other likewise realized actualities so as to result in pattern. This, of course, sounds extremely abstract on first glance – but in point of fact, nothing could be more natural (literally) to those, bought up in the scientific faith. On the inanimate level, for example, the very "beginning" of our contemporary cosmos was a single point of undifferentiated energy (if, indeed, "energy" is not already too sophisticated a term) whose "development" into our current universe is nothing other than the *history* of its successive recursive change as, at each point, literal physical possibilities are made available only as the result of immediately preceding action, and as one of those possibilities is actualized, a new and slightly changed set of possibilities (and constraints) come into being. Thus, we see (retrospectively), and here only schematically the uncoupling of the unified force, which results in the generation of quarks that then makes possible to generation of hadrons, the results of whose interaction in the rapidly cooling universe gives rise to the existence of neutrons, that can then later join together with protons to form the universe of atoms that…*ad infinitum* (or *finitum*, as the case may eventually be).

The point is that there is nothing mysteriously "metaphysical" about Peirce's notions of what he calls *firstness, secondness,* and *thirdness.* Rather, these relations refer, in a radically fundamental sense, to the scientifically examinable (and scientifically necessary) relations of *possibility, existence* and *law.* That the more recent conceptualizations of chaos and complexity theory have given us a better vocabulary with which to talk about such utterly natural phenomena (e.g., iteration, interaction, emergence, downward causation and – in the case of living organisms – autopoesis) attests to the richness of Peirce's "logic of relations."

A major part of Peirce's logic is his "semeiotic" – or logic of sign relations. Here, the last-most-current or "given" state of affairs in the world to the perceiving agent is present to that agent *in its firstness* as an unlabeled "raw feel" (what others have termed its *qualia*). Of all the things that unlabeled sensation "could be," the agent – given the set of existing possibilities and constraints made possible at that moment by its own biological organization and set of prior associations – "experiences" that set of feels *as x* (hunger, the color red, a flower, etc), and this *secondness* of experience builds a web of brute sensations into a web of meaningful perceptions.

And, finally, once not just the sensations and the perceptions but the *relations* within that web (i.e., of sensations *to* perceptions, and of perceptions *to* other perceptions) become representable *as signs* in their own right (e.g., as in musical notation, mathematical notation, linguistic notation, etc), the malleable conventionality of *thirdness* becomes available to living organisms for the re-contextualization of both *firstness* and *secondness* (sensation and perception) into what we generally refer to as symbolic *understanding.* And it was precisely the mystery of how and why it is that human beings have become such "savants" in the use of thirdness, while the majority of other species have not, that drove Sebeok to search beyond the elegant theoretical logic of Peirce and into the cacophonous real world of animals and their sign behavior.

SEBEOK'S SYNTHESIS OF JAKOB VON UEXKÜLL

Sebeok himself recounts how he had come across what is largely considered to be an execrable translation of an early version of von Uexküll's *Theoretische Biologie* decades earlier, while still a student at Cambridge, but found it both "bafflingly murky" and "beyond doubt over his head" – as well not at all germane to his then-current project as a sixteen year old Hungarian student attempting to learn English (2001:64; 1998:34).

Thirty years later, Sebeok would read von Uexküll's fully finished version in the original German – and this re-reading, in the words of contemporary biosemiotician Marcello Barbieri, "was a kind of fulguration on the road to Damascus" for Sebeok (Barbieri 2002:285). There is some truth in this characterization, as we shall see. For in his rediscovery of von Uexküll, Sebeok felt that he had not only found the long missing piece of the puzzle that he had been looking for – but he was also convinced that he had found what so many other laborers in so many other fields should have been looking for all this time as well – i.e., an absolutely naturalistic way of understanding the link between the human world of signs and the animal world of signs. So it is to Jakob von Uexküll and his study of the perceptual worlds of animals that we now turn.

Cited by both Konrad Lorenz (1903–1989) and Nico Tinbergen (1901–1972) as the founder of the modern discipline of ethology, Estonian-born German biologist Jakob von Uexküll (1864–1944) devoted his entire life to the study of animals, training first as a zoologist and afterwards going on to the pioneering work in muscular neurophysiology that would result in *Uexküll's law of neuromuscular regulation,* often cited as the "first formulation of the principle of negative feedback [and thus reafferent control] in living organisms" (Lagerspetz 2001:646). A dedicated physiologist *and* biologist, Uexküll drew a distinction between the two projects that is worth quoting in the words of his contemporary archivist in full:

"Already in his first monograph Uexküll (1905) assigned different roles to physiology and biology. Physiology should organize the knowledge about organic systems by looking for causalities. Having preserved the advantage of the experimental method, physiology should help to [inform] biology. In distinction to physiology, biology has to use the scientific method to go beyond the investigation of causalities by exploring the laws that [account for] the purposefulness of living matter. Therefore biology should study organisms not as objects, but as active subjects, thus focusing on the organism's purposeful abilities that provide for the active integration into a complex environment. Biology therefore had to deal with holistic units and to maintain a broader scope than physiology in order to grasp the interactive unity of the organism and the world sensed by it. For describing this unity Uexküll introduced the term *Umwelt* (1909)" (Rüting 2004).

A forerunner and conceptual pioneer of the study of feedback and reafferent control in the workings of what will later come to be called complex, adaptive self-organizing systems, Uexküll was not privileged, of course, to the rich vocabulary of "autopoetic" explanation that his own groundbreaking work would engender almost a full century later. Yet Ludwig von Bertalanffy's (1901–1972) "general systems theory" – as well as its increasingly sophisticated descendents (i.e., cybernetics, catastrophe theory, chaos theory and complexity theory) – all issue from von Uexküll's notion of the *Funktionskreis* or "functional cycle" of perception and action that effectively "couples" the ever-changing system that is the organism to the ever-changing system that is the world.

Thus, in the discussion about "causalities" above, Uexküll has no recourse to the vocabulary of "emergent system properties" "recursive downward causation" "dynamic instability" or "autopoesis" that would allow him to delineate for his readers the distinction being drawn between the purely material and efficient relations of brute physiological stimulus-response regularities and the higher-order "systemic" relations between world and organism (as well as organism parts to whole) that are the bottom-up product of – as well as the top-down shaper of – those physiological interactions (both in real-time and in evolutionary time) to begin with.

That said, both Uexküll's pioneering work on marine biology, as well as his prescient conceptualization of feedback systems, leave him only dimly remembered, if at all, in the two fields he most directly spawned (animal ethology and dynamic systems research).[23] And this is undoubtedly due to von Uexküll's Baerian resistance to the Darwinism of his time and to the somewhat "telic" understanding of what he calls "Nature" (1934/1957; 1940/1982). Like von Baer, Uexküll felt that Darwin "treated the concept of causality incorrectly and did not consider the internal [component in the active self-organization and creation] of individual organisms" (cited in Kull 1999c). Given that all these men – Darwin, no less than von Baer and von Uexküll – were working long before the development of modern genetic knowledge, it is perhaps no surprise that each of them sees more clearly the "reverse but complementary" side of the picture that the other neglects.

It can be seen, however, that von Uexküll was working very much within a nineteenth century Romantic intellectual culture that was still vibrant in Estonia, while the science of Darwin's England was increasing utilitarian, mechanistic and Malthusian. Thus, the original quote in the paragraph reads that "Darwin did not consider the internal *strive for perfection* of individual organisms" in lieu of the less teleologically "loaded" description of what is essentially same concept of proximate, systemic interaction towards homeostatic optima observable in all living organisms that I have substituted for it in the brackets above.

However, and as Hoffmeyer notes, just as Darwinism needed Mendelian genetics for its full coherence, "it is only through *integration with* the theory of evolution that [von Uexküll's] *umwelt* theory can truly bear fruit" (1996:58). For just as transmutation of species needs a shuffling mechanism to allow for the variations which are then acted upon in natural selection, so too does von Uexküll's Kantian notion – that, as regards the subjective experience of living creatures, "it is utterly in vain to go seeking in the world for causes that are independent of the subject; we always come up against objects which owe their construction to the subject" (1926:xv) – needs to be supplemented with a theory of how such subjects themselves have come to be so peculiarly constructed. For von Uexküll, however, this was not seen a problem, but as the very mark of the limits of Kantian reason. "There lies concealed, eternally beyond the reach of knowledge, the subject – Nature" concludes von Uexküll and, *sic passim*, "nature's sovereign plan" (1934/1957:80).

An epigone of Peirce himself, however, Sebeok believed that in science, as in *umwelt*, such ubiquitous law or "plan" is precisely what calls out to be explained. And thus Sebeok began to undertake the long interdisciplinary project of attempting to introduce into the framework of mainstream science and evolutionary theory, an operationalizable synthesis of the Peircean logic of sign relations with the Uexküllian naturalistic research project of *Umweltforschung*. With now a clear vision that the abyss between sign study and biology had found its bridge, Sebeok began the project that we today call *biosemiotics* – a project whose goal was nothing less than a scientific understanding of how the subjective experience of organisms – as enabled differently by each species' particular biological constitution – comes to play a genuinely causal role in the ongoing co-organization of nature.

A PROJECT OF MASS CROSS-POLLINATION: SEBEOK'S SYNTHESIS OF RESEARCHERS

As can be inferred from the above, the rediscovery of von Uexküll's work had a profound effect on Sebeok's subsequent work and thought. Two decades later, he would recollect that his re-reading of Uexküll's *Theoretische Biologie* after thirty intervening years studying human and animal communication practices "unfolded a wholly unprecedented, innovative theory of signs, the scope of which was nothing less than semiosis in life processes in their entirety" (1998:7).

Apt, then – though requiring a little further clarification – is Barbieri's (2002) claim that upon the "fulguration" that von Uexküll's *umwelt* theory was to him, Sebeok "decided to end his individual search and start an active campaign of proselytization" (285). For unlike the Biblical Saul, Sebeok knew full well that his search was not truly "over" – and that in von Uexküll's *umwelt* theory, he had found but *one* critically important tool for understanding with which he could proceed in his ongoing investigations. In this sense, Sebeok was more like Isaac Newton – or perhaps even Francis Bacon, with his own newfound *Novum Organon,* as we have discussed above – than like the spiritually completed Saul.

But Barbieri's charge of "proselytization" is on the whole a fair one – though happily enough, as it turns out, for many of the scholars laboring in the fields of biosemiotics today, many of whom find von Uexküll's articulation of *umwelt* to be a manifestly helpful terminological tool – and many of whom were, quite directly, brought to this realization in the person of Thomas Sebeok himself.

For "Sebeok would often point out that the list of those who did semiotics without knowing it would fill the pages of an infinitude of books" writes Sebeok's frequent collaborator Marcel Danesi as part of his mentor's obituary: "If we recall correctly, he referred to this state of affairs as the "Monsieur Jourdain syndrome." Monsieur Jourdain was, of course, the character in Molière's *Bourgeois Gentilhomme* who, when told that he spoke good prose, answered by saying that he didn't know he spoke in prose. Analogously, Sebeok would point out to some scholar in a field such as psychology, anthropology, or medicine that he or she was, like Monsieur Jourdain, doing something of which he or she was not aware – semiotics. The number of "converts" he made for semiotics in this way are innumerable" (Danesi 2002).

Thus, in 1977, Sebeok delivered his now-famous speech on "Neglected Figures in the History of Semiotic Inquiry: Jakob von Uexküll" at the *Third Wiener Symposium on Semiotics* in Vienna. Thure von Uexküll (1908–2004), Jakob's son, was in the audience and not long after, Sebeok – along with Thure, a physician whose pioneering work on the semiotics of psychosomatic disorder and treatment virtually single-handedly raised the field of psychosomatic medicine to prominence in Germany, where it is now part of the mandatory curriculum for all medical students (*BMJ* 2004), and Giorgio Prodi (1929–1988), an oncologist studying the "endosemiotics" of immunological self-organization and cell signaling – "conducted a week-long open-ended seminar, so to speak, on the practical and conceivable ins and outs of biosemiotics" (Sebeok 1998:8).

These "intensive triadic brainstorming" sessions, as Sebeok characterized them, "led directly to the series of pivotal seminars held annually in the late 1980s and early 1990s" at the Glotterbad Clinic for Rehabilitative Medicine near Freiburg (*ibid*). Drawing an ever widening circle of biologists, physicians, philosophers and semioticians into his orbit, Sebeok in effect fashioned what he would later come to term an interdisciplinary "cybernetic loop" or "self-excited circuit" (Sebeok 1998:9).

A steady stream of international conferences, monographs, journal articles, special issues, and book collections followed (see Kull 1999 and 2005 for two excellent overviews) – most of them either initiated by, or with significant contributions from, Thomas Sebeok himself. Indeed, in his memorial remembrance of Sebeok, seminal biosemiotician Jesper Hoffmeyer remarks that "without Sebeok's enormous influence and prestige to pave the way, the growth of biosemiotics might well have been seriously hampered through the usual territorial defense mechanisms released more or less automatically in academia whenever somebody attempts crossing [its] Cartesian divides" (Hoffmeyer 2002:385).

The resulting "yet even more modern synthesis" of Peircean semiotics with Uexküllian *umwelt* theory in the overarching framework of dynamic systems theory that underpins much of modern biosemiotics and that is the direct result of Thomas Sebeok's "proselytization" in the years following his rediscovery of von Uexküll is not a synthesis that I have either the space for, nor have been commissioned to, explicate in any minimally sufficient detail here (but see Baer 1987, Brier 2003, Danesi 1998 and 2000, Deely 1995, Pertrilli and Ponzio 2001, and Sebeok and Umiker-Sebeok 1992 for thoroughgoing discussions thereof).

Suffice it to say, though, that even those colleagues-in-biosemiotics who today reject the Peircean perspective, *or* the primacy of perceptual *umwelt*, *or* the entire undertaking of the project of biosemiotics itself (and there are some loyal skeptics who most helpfully hold this view, as we shall see) – even these scholars are no longer scattered researchers working in sterile isolation from one another and in utter ignorance of each other's work, but are instead today "colleagues" in a field called "Biosemiotics" *because* of the tireless proselytization (and cross-pollination) efforts of Thomas A. Sebeok throughout the 1970s, 1980s, 1990s and those years of the 21st century ending only with his death.

In effect, summarizes Barbieri, "the making of biosemiotics [in the form of the field that we see it as today] has been heretofore a 40-year-long affair which can be divided into two phases: the first (1961–1977) was a period of uncoordinated attempts, often of utterly isolated initiatives, while the second (1977–2001) was a period in which individual ideas could fall on a more receptive ground and contribute, under the discreet supervision of Thomas Sebeok, to the collective growth of the field" (2002:286).

And, indeed, his obituary states that of all of his accomplishments, "he was most proud of having brought into being a group of theoretical biologists and semioticians to pursue this field of investigation" (*SLIS* 2002). It is the ongoing coalescence of this group that we will turn our attention to now – for the history of this ongoing coalescence *is* the extant "history" of Biosemiotics *per se* (though what will come of this coalescence and where that history will lead remain, of course, matters of pure *potentia* at this time).

SEBEOK'S LEGACY AND THE CONTINUATION OF THE BIOSEMIOTIC PROJECT

One of the many tributes paid to Sebeok in obituary was made by his long-time colleague Marcel Danesi, who – summing up a lifetime's work in fields as diverse as anthropology, linguistics, computer science and zoology, reiterated the claim that what Sebeok himself was most proud of was his having "transformed semiotics back into a "life science" – having taken it back, in effect, to its roots in medical biology [and specifically, the uninterrupted tradition of symptomology found in all cultures]. In other words, he uprooted semiotics from the philosophical, linguistic, and hermeneutic terrain in which it has been cultivated for centuries and replanted it into the larger biological domain from where it sprang originally" (2002).

Interestingly enough – and perhaps a tribute to Sebeok's underlying vision all along – it is not "semioticians" *per se* that one finds attending the conferences and penning the journal articles in the field called biosemiotics today, but molecular biologists, embryologists, philosophers of science, zoologists, roboticists, neurobiologists, psychologists and dynamic systems theorists instead. Most of these scholars have found their way into the field through their own unique and surreptitious pathways, and many hold a variety of views regarding the relationship of signs to biology that in no way derive from the works of Peirce or von Uexküll, much less than those of Sebeok himself.

For it turned out that the nerve that was ready to be hit by the promise of a scientifically informed biosemiotics was not at all one that was calling out for excitation in the academic world of semiotics (with a few conspicuous exceptions of course, which will be discussed below). Rather, the priming was taking place variously, but steadily, over the last 50 years of scientific advance and inquiry in the West.

For while Sebeok was busy building networks in Scandinavia and Eastern Europe, back in the West individual researchers in the fields of neurobiology, clinical psychology, molecular biology, artificial intelligence, and philosophy of mind (to name a few) were busily engrossed in their own attempts at either resolving or undoing the disastrous Cartesian dichotomy separating bodies and minds. In neurobiology, for example, one saw the works of Gerald Edelman (1992) Antonio Damasio (1994) Walter Freeman (2000) and Joaquin Fuster (2003)[24] among many others; in AI, the "distributed cognition" theories of Andy Clark (1997), Rodney Brooks (1999), Marvin Minsky (1988), and Douglas Hofstadter (1979); in biology proper, there were the critiques of Walter M. Elsasser (1998), Richard Lewontin (1992), Robert Rosen (1991), and Howard Pattee (1982, 1988); and in dynamic systems theory, the works of Edward Lorenz (1963) René Thom (1989), Ilya Prigogine (1984), Susan Oyama (1985) and Stuart Kauffman (1995, 2000) – again, just to mention some few of the most obvious.

But these researchers (and many more, some of whom will eventually make their way into the interdiscipline of biosemiotics and whom we will be discussing presently) were, as said, largely pursuing their own independent research agendas, working and exchanging ideas amongst their own disciplinary colleagues, and were not actively involved in constructing a network of researchers from widely divergent academic backgrounds in the sense that Thomas Sebeok was.[25]

Some small interdisciplinary networking groups *were* independently breaking out here and there at this time, however. Kull recalls three regular series seminars on theoretical biology that arose independently in the Soviet east during the 1970s – one in St. Petersburg led by Sergei Chebanov, one in Moscow led by Aleksei Sharov, and one in Tartu, Estonia led by himself that "all later made a shift towards biosemiotics" (2005:21). In the West, geneticist Conrad Hal Waddington (1905–1975) held a series of conferences entitled *Towards a Theoretical Biology* each year from 1966–1969 that attracted such participants as Lewis Wolpert, Brian Goodwin, R.C. Lewontin, David Bohm, W.L. Elsasser, René Thom, Howard Pattee, Ernst Mayr and John Maynard Smith. Yet while all of these participants undoubtedly both contributed to, as well as came away from, these conferences with an enriched notion of the phenomenon of "self-organization" in complex systems, these conferences did not result in the creation of any one coherently ongoing "group" or specifically focused collective agenda, such as can be found in the current project of biosemiotics.

Instead, the major line of development that would result in the constitution of the field of biosemiotics as it exists today were a series of informal but increasingly productive seminars emerging from the University of Copenhagen beginning in the 1980s and culminating in the ongoing international Gatherings in Biosemiotics conferences which have been held annually since 2001. And by almost every account, the figure at the center of this activity was then and remains now the man whose name is most closely associated with the field of biosemiotics, the Danish molecular biologist and public intellectual Jesper Hoffmeyer (1942–).

JOINING LIFE SCIENCE WITH SIGN SCIENCE: JESPER HOFFMEYER

Trained and hired as a biochemist by the University of Copenhagen in 1968, Hoffmeyer had been active in Danish public life since his days as a student activist in the mid-1960s. Son of a social reformist physician who had co-edited an antifascist periodical called *Kulturkampen (The Struggle for Culture)* in the 1930s, Hoffmeyer's own deep interest in the intersection of nature and culture led to his founding of a journal entitled *Naturkampen (The Struggle for Nature)* in the 1970s.

A prolific science writer and journalist as well as a working university professor and molecular biologist, "by the 1980s, Jesper Hoffmeyer had become one of the most visible intellectuals in the debate on technology and society in Denmark" write his biographers (Emmeche et al 2002:38). Deeply inspired by the work of cybernetician and anthropologist Gregory Bateson (1904–1980), Hoffmeyer had been struggling to articulate a non-reductionist understanding of the relationship of organisms to their genomes at a time when the rapid advancement of gene sequencing technology was promising a yet more reductionist understanding of the same, and Richard Dawkins was capturing the popular imagination (as well as that of some scientists) with his notion of "the selfish gene" (1976).

Recalling this period, Hoffmeyer writes that in 1984 it had occurred to him that "the historical consequence of making dead nature [i.e., physics] the model of nature at large was that all the talking—and all mindfulness—went on exclusively in the cultural sphere. As a result we now suffer the divided existence of the two great cultures, the humanities and the scientific-technological culture" (2002:99).

Finding it intuitively unnatural to attempt an explanation of the hereditary efficacy of DNA in isolation from the DNA-organism system in which it always appears, Hoffmeyer claims that he wanted to invoke in his scientific colleagues of that time "a new kind of curiosity, a curiosity directing its attention towards, what we might call 'the wonder of the code' and which does not put that wonder aside by the enclosure of the codes into one or the other state space [of deterministic physics] or life-world [of pure subjective experience]. For it is the nature of the 'code' to point outside of its own mode of existence—from the continuous to the discontinuous message, from the physical and therefore law bound message [of the nucleotide sequence] to the more free message [of the organism whose actions in the world will or will not result in that nucleotide sequence's eventual evolution and survival], and back again in an unending chain" (2002:99).

"For it is the nature of the 'code' to point outside of its own mode of existence." Almost certainly unaware then of the maxim of St. Augustine, much

less of the obscure late-scholasticism of John Poinsot, Hoffmeyer's common-sense appreciation of the profoundly important distinction between material organization and that same material organization in its use as a sign for something other than itself led him, like Sebeok before him, to an investigation into the semiotic logic of relations between organisms and their environment (1984), between organisms and each other (1988), within organisms (1992) and in the triadic logic of the nineteenth century scientist-philosopher Charles S. Peirce.

By 1985, Hoffmeyer was committed to the idea of developing "a semiotics of nature, or biosemiotics as he chose to call this effort, [that could intelligibly explain how] all the phenomena of inherent meaning and signification in living nature – from the lowest level of sign processes in unicellular organisms to the cognitive and social behavior of animals – can emerge from a universe that was not [so] organized and meaningful from the very beginning" (Emmeche et al 2002:41).

And in this, again like Sebeok – whose path he would not yet cross for several more years – Hoffmeyer's personal passion for, and dedication to, this project – as well as, more importantly, the kind of work on the subject that he began producing – drew an ever-growing coterie of like-minded individuals into his orbit. In 1984, his initial formulation of a theory of analog-digital "code-duality" in biology was published, and soon thereafter he began his intensive series of collaborations with biologist Claus Emmeche, who would later go on to head the Center for the Philosophy of Nature and Science Studies at the University of Copenhagen, and to become a major figure in biosemiotics in his own right – in addition to authoring a body of related work on dynamic systems theory (1992, 2000a), artificial intelligence (1991, 1994), and the history and philosophy of science (1999, 2002).

By 1986 both Hoffmeyer and Emmeche were attending a Copenhagen study circle with the physicist Peder Voetmann Christiansen wherein the semiotics of Peirce were much discussed. Philosopher and literary analyst Frederik Stjernfelt joined this group (known then as the "Helmuth Hansen Study Circle" after the Danish philosopher) soon thereafter, eventually inviting French mathematician and theoretical biologist René Thom – whose work also drew heavily upon Peircean semiotics and Uexküllian *umwelt* theory – to deliver a lecture on his development of catastrophe theory (Stjernfelt 2002:58).

Microbiologist Mogens Kilstrup would later find his way into Hoffmeyer's circle, as would the biologist and cyberbetician Søren Brier (1995, 1998, 2001), who would several years later establish the interdisciplinary journal *Cybernetics and Human Knowing* in which many of the Helmuth Hansen group would publish seminal articles.[26] During this time, too, Hoffmeyer continued to publish his ideas on code-duality and self-description, now drawing also upon the works of biophysicist and systems theorist Howard Pattee (1969, 1972, 1982).

In 1989, Hoffmeyer published a seminal article on "the semiosis of life" in Danish, and this was followed by his founding of the proto-biosemiotic journal *OMverden* (roughly: "*Umwelt*") in 1990. "The journal was an intellectual success," writes his biographers, "but a [financial] failure for the publishing company, so its life was brief" (Emmeche et al. 2002:41). The journal did find its way into the hands of both Thure von Uexküll and Thomas A. Sebeok, however, and when Hoffmeyer went to attend a conference on psycho-neuro-immunology in Tutzing later that same year, he met both of these men for the first time – having spotted Sebeok walking around the conference with a copy of *OMverden* protruding from his jacket pocket (Hoffmeyer 2002:384).

The joining together of "Sebeok's people" with "Hoffmeyer's people" was a signal event in the development of the contemporary field of biosemiotics. From Sebeok's "semioticians exploring biology" side came such accomplished scholars as John Deely (1986), Myrdene Anderson (1990), Floyd Merrell (1996), and Martin Krampen (1981)[27]– while from Hoffmeyer's "biologists exploring semiotics" side came himself, Claus Emmeche, Søren Brier, Mogens Kilstrup, Frederik Stjernfelt and Peder Voetmann Christensen. It was in the aftermath of this meeting that Sebeok was to declare the investigations of the life sciences and the sign sciences must be co-extensive if either was to proceed (1990), and from this point on, the term *biosemiotics* is used to refer to this project by all parties involved.

Less than one year later, Hoffmeyer and Emmeche's seminal two-part paper on code-duality appeared in Anderson and Merrell's anthology *On Semiotic Modeling* (Hoffmeyer and Emmeche 1991) and in Sebeok's international journal *Semiotica* (Emmeche and Hoffmeyer 1991), winning the publisher's top annual award, and bringing the work of the Helmuth Hansen group to an international audience. In 1992, the volume *Biosemiotics: The Semiotic Web* was published, to which no less than twenty-seven authors contributed. This exposure served to establish an ever-growing interface with other biologists and semioticians whose research was converging along these lines. The internationally-minded Danish Society for the Semiotics of Nature was also officially established at this time, with the express purpose of bringing together researchers from around the world who were interested in pursuing this new line of inquiry.

It was also in 1992 that theoretical biologist Kalevi Kull, a convener of some of the earliest conferences on semiotic approaches in theoretical biology taking place in the Soviet Union in the 1970s and curator of the Jakob von Uexküll Centre at the University of Tartu in Estonia, would meet Jesper Hoffmeyer at Thure von Uexküll's Glottertal conference near Freiburg – and from then on become the *de facto* historian both of biosemiotics in the Sebeok-Hoffmeyer tradition and of the tradition of Eastern European theoretical biology in general. Kull would also begin presenting an annual lecture course in biosemiotics at the University of Tartu in 1993 that continues to this day, and has been instrumental in arranging the annual International Gatherings in Biosemiotics, in addition to his own considerable contributions in advancing the field (e.g. Kull 1998, 2000, 2001).

In 1993, Jesper Hoffmeyer published his definitive work on biosemiotics *En Snegl Pa Vejen: Betydningens naturhistorie (A Snail on the Trail: The Natural History of Signification)*, was later translated into English the book that as *Signs of Meaning in the Universe* (1996). It is this exceedingly readable book, perhaps more than any other, that provides most newcomers their entry – and, in many cases, their impetus – into the field, and that most clearly lays out the project of biosemiotics as an attempt to situate culture in nature without reducing either to the blind forces of purely mechanical efficient causation.

Written in the attempt to popularize the ideas of biosemiotics to the widest possible audience, the following passage conveys much of the flavor of the work. After discussing the evolution of single-celled life, multicelluars, and the increasing variety of animals' sensory capacities, Hoffmeyer turns to the evolution of human cultural cognition and writes:

> Among all the roles in the ecological theatre there was one pertaining to creatures with lengthy life histories and an especially well-developed talent for capitalizing on their experiences. Often theses creatures, the apes, had developed brains capable of accommodating an extremely complex image of their surroundings, a very sophisticated *umwelt*. [And eventually] there came a day when this creature realized that it was itself an *umwelt* builder; that its role was, in act, a role; that other creatures performed other roles and had different kinds of *umwelt*; that the world was one thing and *umwelt* another; and that, when one died, this *umwelt* would actually disappear while the world as such would carry on. ... [Yet over time, this creature was able to] create a bond of a quite unprecedented nature: a double bond founded on the need to share the *umwelt* with one another, i.e., making private experiences public property, turning the subjective into the objective. To cut a long story short, this creature ... invented the spoken word. (1996:34–35)

With its provocative ideas cloaked in the simplest of languages, the English language publication of Hoffmeyer's *Signs of Meaning* was enthusiastically reviewed (1998) and remains as of this writing probably the single most widely read and frequently cited text on biosemiotics. Its impact on scholars internationally continues as each year new biosemioticians come into the fold as a result of their "stumbling upon" this work (a tale frequently recounted at the annual International Gatherings in Biosemiotics).

And, indeed, directly as a result of the reception to the work's international availability in 1996, Hoffmeyer found himself "communicating with a cross-disciplinary audience of scientists, philosophers and scholars from various specialties [and was] invited to conferences in the fields of systems theory, self-organizing complex systems, cognitive science, general semiotics, media and communication theory and, of course, an increasing number of workshops and symposia devoted specifically to biosemiotics and its relations to other fields of semiotics and biology" (Emmeche et al. 2002:42).

A slew of journal articles and conference presentations on biosemiotics by the members of the Helmuth Hansen group and their growing coterie of international colleagues followed (see particularly the special issues of *Semiotica* of 1998 (Vol 120 3/4) and 1999 (Vol 127 1/4), as well as the *Annals of the New York Academy of Sciences* 2000 (Vol 901) and, for a more extensive list of publications covering this period, Kull 2005:20). Eventually these second-generation heirs of Sebeok's Glottertal conferences were able to bring together a growing group of younger researchers for whom the idea of dynamism in autopoetic systems was no longer a "radical proposal" – but was, instead, the starting point from which to proceed to try to build a coherent interdiscipline. And by the middle of the year 2000, the first annual International Gatherings in Biosemiotics was being planned.

A DIVERSE ECOSYSTEM OF RESEARCHERS: THE GATHERINGS IN BIOSEMIOTICS

Thomas Sebeok was most content, it seems, when he was bearing many torches – and after his death at age eighty-one in 2001, each of these had to be picked up and passed on to a successor.

Already by this time, however, the center of gravity for the biosemiotics project had been establishing itself at the University of Copenhagen under the auspices of Jesper Hoffmeyer and Claus Emmeche who, along with theoretical biologist Kalevi Kull and cybernetician Søren Brier, established the Biosemiotics Group at the University of Copenhagen in the early 1990s. And it was this group that, in 2001, finally succeeded in inaugurating an annual international conference devoted exclusively to biosemiotics.

Quite unsure at the time about who, if anyone besides themselves, would show up, the first International Gatherings in Biosemiotics turned out to be an unprecedented success. Held on May 24-27, 2001 at the Institute for Molecular Biology at the University of Copenhagen (in the very room, it was noted, that Wilhelm Johannsen first introduced the word "gene" into science in 1909) the first of these annual conferences was attended by over 30 presenters from 18 countries and produced papers in neurobiology, zoology, artificial intelligence, linguistics, molecular biology, cybernetics, meta-systems transition theory, and the history and philosophy of science.[28]

The international Gatherings have been held five times since then, and while not every researcher working in the field of biosemiotics attends these annual meetings, many – if not most – of the principal contributors to the field do. There, the second-generation heirs of Sebeok's Glottertal conferences bring together a growing group of formerly independent researchers and their younger colleagues for whom the idea of dynamism in autopoetic systems is no longer a "radical proposal" – but is, instead, the starting point from which to proceed to try to build a coherent interdiscipline. In addition, with the inaugural publication of the peer-reviewed *Journal of Biosemiotics* and the establishment of the long-planned International Society for Biosemiotic Study in 2005, this "third phase" in the growth and development of biosemiotics promises dramatic changes to the field – most of the more interesting ones, of course, being unforeseeable.

Even from this early standpoint, however, we can discern certain patterns and currents that are sure to play a role. The following selective list of just the most regular of the international conference's participants gives a flavor of the interdisciplinary convergences – and divergences – of approach in the quest to articulate a truly comprehensive science of life and sign processes.[29]

CONTRIBUTIONS FROM OUTSIDE THE COPENHAGEN-TARTU NEXUS

One of the approaches that does not come strictly out of the Copenhagen-Tartu lineage is represented at these conferences by Prague cell physiologists Anton Markoš and Fatima Cvrčková (2002, 2002a, 2002b) who advance an understanding of living systems that is fundamentally *hermeneutic*. Representatives of a growing interdisciplinary movement towards theoretical biology and interdisciplinary study in the Czech Republic,[30] Markoš and Cvrčková view the current work being done within the contemporary biological paradigm (including their own work) to be an effective – but by necessity only partial – illumination of processes that exceed the potential of formalized representation to exhaustively map them.

Taking an approach towards living organisms that owes as much to the "historically effected hermeneutics" of Hans-Georg Gadamer (1900–2002) as it does to the self-regulatory symbiotic systems theories of Lovelock (1996) and Margulis (1987), Markoš writes that: "[Since the moment of its inception,] life has never ceased to exist and has again and again been confronted by actual conditions, by memory, by forgetting, and by re-interpretations of the remembered" (2002:163). As Markoš reminds us in his masterful exegesis of scientific study *Readers of the Book of Life*, the living organization of an organism changes itself and its relations to its surround on a moment-to-moment (as well as on an evolutionary) basis in a way that no machine logic or mathematical formalization could ever predictively account for. Indeed, and it is this very embodiment of a possibility-collapsing "non-logic" that allows a living system to effectively explore and to creatively exploit novel state spaces, giving it "the characteristics of a field, a culture, a statement, and of course, [only] sometimes also of a machine" (2002:163).

With Gadamer, Markoš asserts that "the nature of knowledge is hermeneutical and is rooted in experience, history and in structures" that are themselves ever-changing as each new moment is changed as a result of the actions taken in the one prior. Attempting to reduce this rich world of living-acting-perceiving-and-signifying onto the "necessarily incomplete, reduced, flattened" descriptions of the objectivist scientific model (Cvrčková 2002:184) would be akin to attempting to realize Hoffmeyer's self-referential notion of creating "a map which is so detailed that the map maker and the map that he is making are swept up into it" – something that not even the world-modeling organism itself can ever fully objectify, much less make static (1996:40). Working biologists "just like any others," Cvrčková and Markoš's work yet reminds us never to lose sight of Korzybski's admonition that "the map is not the territory" – lest we find ourselves taking seriously such map-sensible but experientially-nonsensical claims as "the genetic code is just a metaphor" and "consciousness is an illusion" (...an "illusion," one should always ask of such a pronouncement, *of what*?).

Yağmur Denizhan and Vefa Karatay (1999, 2002), a dynamic systems engineer and a molecular biologist, respectively, from Boğaziçi University in Istanbul, build upon the work of theoretical physicist and computer scientist Valentin Turchin's

(1931) meta-systems transition theory in order to model the dynamics of self-increasing complexity in embedded systems, and the subsequent emergence of bottom-up system properties that then come to function recursively as top-down biases and constraints.

Physicists Edwina Taborsky (1998) and Peder Voetmann Christensen (2000), almost alone among biosemioticians, have sought to explicate Peirce's own understanding of his semeiotic as a being a subset of a logic of relations that can be used to understand how *any* set of relations hold together. Peirce's highly complex architectonic regarding (roughly) *possibility, being, and law* may yet prove to be a rich mine for physicists, as well as for biosemioticians, and Taborsky and Christensen are among the first to be blazing this trail.

And while physicists Christensen and Taborsky are approaching the organization and interactions of energy and matter from a triadically interactive perspective, biophysicist Howard Pattee has devoted the last 37 years of his life to the study of "precisely those dynamical aspects of physics (time, energy) that are necessary to implement codified instructions" – or, in other words: What are the physics necessary (if not sufficient) for semiosis? (Umerez 2001).

One of the original attendees at Waddington's "Towards a Theoretical Biology" conferences of 1969–1972, Pattee was forecasting as early as 1965, to those few who would listen, that "we may expect that the origin of life problem will shift away from the evolution of the building blocks and the elementary operations of joining them together, to the more difficult problem of the *evolution of control* in complex organizations. This problem is more difficult because the idea of 'control' is not defined in the same sense as we can define biochemicals [*per se*]...A live cell and a dead collection of the identical biochemicals in the same structural organization differ essentially in the amount of *intermolecular control* that exists in each unit (1965:405–406).

Like so many whose work we've had the occasion to overview here, Pattee's precisely articulated questions would in time help generate the conceptual frameworks and vocabularies needed for addressing them. Thus, the general principles behind such bottom-up and top-down "intermolecular control" would later be codified as "*autopoiesis*" by Maturana and Varela (1973, 1974) and as "*dissipative structure*" by Ilya Prigogine (1969), while for Pattee, the concepts of the *epistemic cut* and *semantic closure* are necessary to a complete understanding of how and in what scientifically examinable way, matter can come to "stand for" something other than itself in and to a system – the ultimate research question of biosemiotics (see Pattee 2005 and this volume).

DEVELOPMENTS AND CHALLENGES 2001–2005

Indeed, it would require a book-length monograph of its own to detail the interdisciplinary research interests and data presented at the annual International Gatherings in Biosemiotics, all of which, in one way or the other, are devoted to this central question of the non-mystical role of "representation" and its "meaning" in the organization and interactions of living organisms. In lieu of that, I will merely direct the reader to pursue on his or her own the representative list of cited publications corresponding to some of the more regular attendees to the annual Gatherings as referenced below.

In the areas of animal studies, ethology and zoology, Dominque Lestel (2002), Timo Maran (2003), Mette Böll (2002), Dario Martinelli (2005) and Aleksei Turovski (2000) are all pursuing biosemiotic lines of investigation in their work. Examination into the relations of intercellular signaling processes are molecular biologists Luis Emilio Bruni (1997, 2001), Mia Trolle Borup (2005), Mogens Kilstrup (1997) and Abir Igamberdiev (1999), as well as immunologist Marcella Faria (2005), embryologists Johannes Huber and Ingolf Schmid-Tannwald (2005), and pharmacologist Sungchul Ji (2002).

Researchers into dynamic systems theory who are incorporating biosemiotics into their models include Hernán Burbano (2005), Stephen Pain (2002), Toshiyuki Nakajima (2005), Assen Dimitrov (2004), Wolfgang Hofkirchner (2002), João Queiroz (2005), Charbel Niño El-Hani (2005), László Hajnal (2003), and Karel Kleisner (2004).

"Neurosemiotic" approaches to brain research and consciousness studies have been proposed by Andreas Roepstorff (2004), Anton Fürlinger (1998), Sidarta Ribeiro (2003), Alessandro Villa (2005) and the author (Favareau 2001, 2002); while a biosemiotically informed approach to Artificial Intelligence and cognitive robotics has been undertaken by Tom Ziemke (2003) and Noel and Amanda Sharkey (1999, 2002).

Maricela Yip (2005), Pierre Madl (2005), and Almo Farina (2004) all apply a biosemiotic approach to their research into sustainable ecosystems, Yair Neuman (2003) applies it to theoritical immunology while anthropologists Myrdene Anderson (1999), Thierry Bardini (2001), Cornelius Steckner (2004), Andreas Weber (2002), and Mark Reybrouck (2005) focus on the cultural semiotics of human-to-human interaction.

Enriching and informing all of this discussion is the work of semioticians and linguists Tuomo Jämsä (2005), Sergey Chebanov (1994) and Adam Skibinski (2004), Han-Liang Chang (2005) and Juipi Chien (2003), philosopher Giinter Witzamy (2000), biosemiotic theorists Andres Luure (2002), Aleksei Sharov (2002) and Kaie Kotov (2002), Gregory biographer Peter Harries-Jones (1995), and archivist for the Jakob von Uexküll Institute for *Umweltforschung*, Torsten Rüting (2004).

As the result of this intense collaboration and international exchange of ideas, the biosemiotic project of examining the sign processes in life processes is becoming more interdisciplinary and more international every year. In 2005, the International Society for Biosemiotic Study that Thomas Sebeok had proposed over a decade earlier was officially founded; and in the same year, the first issue of the international *Journal of Biosemiotics* appeared.

And as the surest sign of growth, principled divisions within the biosemiotic project are already beginning to appear. The reach of biosemiotics is growing and bringing into its orbit those from farther fields. No longer can it be assumed that a self-identified "biosemiotician" necessarily believes that the semiotic categories of Peirce – or even the *Umweltforschung* of von Uexküll – are the optimal starting points on which to build a scientific articulation of sign processes in biology.

Rather, in the five years since Sebeok's death, the annual international Gatherings in Biosemiotics have been blessed with a steady stream of external challengers and internal self-critique. Tommi Vehkavaara (2002, and forthcoming) and Stefan Artmann (2005, and this volume) have been most vocal, and most productive, in challenging the assumptions of the consensus articulation in informed and informative ways. Such informed criticism is of inestimable value to a growing field whose members spend the majority of their year responding to uniformed criticism ("No, it's not sociobiology; no, it isn't spiritualist or vitalist; no, we don't think that an amoeba has thoughts; or that you can attract a spouse using subliminal Neuro-Linguistic-Programming techniques...") resulting from a lack of familiarity with the field.

For as productive as these Gatherings have been for the exchange of ideas and the development towards a common goal, equally important is the fact the international biosemiotic conferences and journal articles have also resulted in a series of penetrating critiques. Coming from within the circle of those who have spent considerable time with the published materials (as opposed to those critics from the outside who, upon hearing the name "biosemiotics," simply conflate the project with "sociobiology," "anthropomorphism" or some variant of New Age pseudo-philosophy and then proceed – as they should, were the equation to be correct – to dismiss it out of hand as psuedo-science), these internally informed even as they critiques highlight both the existing shortcomings as well as the possibly inherent problematics in the current articulation of the biosemiotic project *per se*, even as they point to alternative possible ways to develop a semiotically-informed biology without reliance on the ideas of von Uexküll or Peirce.

Philosopher of science Stefan Artmann, for example, sees biosemiotics as an example of a consilience-seeking "structural science" which he defines (with Küppers 2000) as: any "transdisciplinary formalization programme that tries to discover abstract analogies between research problems of different empirical sciences in order to contribute to their solution" (2005:234). Along with the majority of biosemioticians, Artmann believes that the more such work is successful, the faster biosemiotics will become just an uncontroversial part of everyday biology. "This is the ironic fate of every productive structural science," writes Artmann, "It begins as educated analogizing, constructs step by step an interdisciplinary bridge between disciplines, transforms their way of thinking, supports the progress of scientific knowledge with the help of its transdisciplinary formal reasoning – and eventually becomes superfluous" (2005:238).

Such an evolution, I feel justified in asserting, is exactly what most proponents of biosemiotics are hoping for – the "best case scenario" resulting from all their efforts to articulate the natural history, and the natural constitution, of the use of sign relations in the biological world. Unlike the practitioners of what he suggestively calls the "Copenhagen interpretation" of biosemiotics (e.g., Hoffmeyer,

Emmeche, Kull, et al.), however, Artmann (2005) proposes that a "model-theoretic" approach incorporating mathematical representations of sign relational possibilities (somewhat akin to the formalisms of Artificial Intelligence/Artificial Life research) will be critical if the field is to move forward – yet Artmann finds a strong resistance among the Peirceans towards "reducing" sign relations in this way.[31]

Philosopher Tommi Vehkavaara similarly objects that "Charles Peirce's and Jacob von Uexküll's concepts of sign assume an unnecessarily complex semiotic agent" (2003:547) and that in order for these concepts to be naturalizable for use in an effective biology, they must be shown as arising out of "more primitive forms of representation" (2002:293). For Vehkavaara, "the minimal concept of representation and the source of normativity that is needed in its interpretation can be based on the 'utility-concept' of function" in a self-maintaining system that is able to switch "appropriately between two or more means of maintaining itself" while in continuous interaction with its environment (2003:547). Vehkavaara thus urges the adoption of concepts from Mark Bickhard's (1999, 2003) "interactivist" models of autonomous agency as prerequisites to the emergence of the kind of triadic sign relations discussed in higher animals by von Uexküll and, *mutatis mutandis,* by Peirce.

Without a doubt, though, the most radical challenge to the Peircean approach to understanding the sign relations of living systems comes from embryologist and *Systema Naturae* (and now also *Journal of Biosemiotics*) editor Marcello Barbieri, who posits an alternative biosemiotic paradigm that is not organicist and qualitative in its origins, but mechanist and quantitative through and through.

MARCELLO BARBIERI: NOT INTERPRETATION, BUT ORGANIC CODES

A molecular biologist and experimental geneticist for over thirty years, Barbieri first proposed his "ribotype theory" of the origin of life in 1981. Working in the tradition of Manfred Eigen (1977), Freeman Dyson (1985) and Graham Cairns-Smith (1982), Barbieri realized from his work in embryology that just as the epigenesis of embryonic development requires an "endogenous increase in complexity" that "reconstructs" the phenotype from the "incomplete projection" of information that is the genotype (2003:213-215), so, too must have this embodied logic or "convention" have had to evolve for doing so "at the time when the *esopoesis* of precellular molecular aggregation was evolving into the *endopoesis* of polymerizing ribosoids (and, eventually, into the true *autopoesis* of 'cells')" (2003:142).

For Barbieri, this naturally evolved "convention" – though interactive always in a triadic relationship of genotype, phenotype and ribotype – is not to be explained (or non-explained, as he would argue) as being so fundamentally coextensive with life that it – like growth, metabolism, and self-initiated movement – it is merely assumed to be a "first principle" of living organization from which the rest of the investigation of biology is to proceed – a position that he feels the Copenhagen school is guilty of perpetuating.[32]

Rather, posits Barbieri, the earliest macromolecular precursors to tRNA not only predated, but actually brought into existence cellular genotypes and phenotypes, through their own physical constitution's ability to establish a reliable correspondence between freestanding nucleic and amino acid aggregates. "Any organic code is a set of rules [or conventions] that establish a correspondence between two independent worlds, and this necessarily requires molecular structures that act like *adaptors*, i.e., that perform two independent recognition processes," writes Barbieri, "This gives us an objective criterion for the search for organic codes, and their existence in nature becomes therefore, first and foremost, an experimental problem" (2005:119).

"The cell is the unity of life," claims Barbieri, "and biosemiotics can become a science only if we prove that the cell is a semiotic system."[33] And since at least 1981, this is exactly what Barbieri has been proposing. "Historically we are still very much in a period of DNA supremacy," he wrote back then, "and it will take perhaps a new generation of biologists to realize that genes alone could not have started life on earth any more than proteins alone could. The reason for this is that we are imbued with the concept that a cell is essentially a throwaway survival machine built by the genes, and a genuinely new attitude toward the origin of life will become popular only when this view is replaced by a different one" (1981:571).

Highlighting the introduction of yet another limiting and still far-too-consequential dichotomy into the narrative of Western science, Barbieri argued in his 1981 article that Wilhelm Johannsen did for molecular biology exactly what Descartes did for traditional biology, divorcing genotype from phenotype just as Descartes divorced mind from body – and in so doing introduced an impossible dualism incompatible with the biological reality of interacting levels of organization.

For, argues Barbieri, "the very definition of phenotype leads us to conclude that the genotype-phenotype duality cannot be a complete theoretical description of an organism. It is a didactic concept which was introduced by Johannsen in 1909 to differentiate between hereditary and phenomenological characteristics, and it was only an unfortunate accident that the duality has been elevated to the status of a theoretical category" (1981:577).

Indeed, "the real distinction between genotype and phenotype is based on the distinction between the one-dimensional world of information and the three-dimensional world of physical structures. The critical point is that there is no *direct* communication between these two dimensions of reality. A gene cannot build a protein any more than a protein can instruct a gene. The central dogma states that information does flow from genes to proteins, but only because it has been 'taken for granted' that a third party exists which can actually implement the transition. What is not usually emphasized is that such an intermediary cannot be either another group of genes or another group of proteins" (*ibid*).

In pointing to the need for a triadic explanation of not just genes and proteins, but genes (one-dimensional information sequences), proteins (three-dimensional physical structures) and *whatever it is that joins them explanatorily*, and that *uses* genes to *make* proteins, Barbieri was not just calling for a new way of thinking about how living cells operate today – but also of how living cells came to be in the first place. Thus was a *semantic theory of evolution* necessary, along with a *semantic theory of the cell* – and from 1981 to 1985, Barbieri worked virtually in isolation to articulate them both.

The gene-carrying cell that we know today, he posited, may have begun as a colony of ribonucleoproteins engaged in producing other colonies of ribonucleoproteins. Proposed before the Cech and Altman's Nobel prize-winning discovery of ribozymes in 1989, Barbieri had already forseen the possibility of – and, perhaps more importantly, the need for – something that would play the rolc of a *polymerizing ribosoid* in 1981. This "ribotype" as he dubbed it, itself had the character of a primitive RNA molecule, yet also had the capacity to catalyze a peptide bond between amino acids. It thus served to *bring together* the previously distinct worlds of RNA molecules and amino acids, introducing into the world the genotype, phenotype and ribotype relation that today *constitutes* the self-replicating cell. Overlooked as a derivative "intermediary" in its modern instantiation as "transfer RNA," such primitive ribotypes were, in fact, the seat of the genetic code and the first "codemakers" to appear in the history of life. Thus, claims Barbieri, "there was no real discontinuity between precellular and cellular evolution. Only the acquisition of sophisticated replication mechanisms brought about by the evolution of quasi-replication mechanisms which had been developed by the ancestral ribosoids to produce other ribosoids" (1981:573–574)

In what would be considered to be a revolutionary re-thinking of both the origin of cellular life and of its ongoing internal relations even today, in 1985 Barbieri expanded upon these ideas in a work entitled *The Semantic Theory of Evolution* that was enthusiastically received both by mathematician René Thom (1923–2002) and by philosopher of science Sir Karl Popper (1902–1994). Eighteen worth of theoretical refinement later, Barbieri would present the mature form of his theories in his 2003 masterwork *The Organic Codes: An Introduction to Semantic Biology*. There he would lay out the empirical evidence that has been gathered in the interim for the existence of a whole array of organic codes that he postulated in the earlier work, including RNA splicing codes (97–100), intercellular signal transduction and integration codes (101–108), cellular migration and adhesion codes (112–114), and cytoskeletal arrangement codes (172–173).

In these codes, as in the genetic code, there is no physical or chemical necessity between, say, the release of a certain neurotransmitter and the cascade of events that follow *save* the presence of the set of conventional internal relationships that have been selected evolutionarily and are embodied in the form of the complex of mediating molecules joining the so-called "first" and "second" messengers. This set of physically realized, biological relationships *is* the extra-genetic code whereby biological specificity is ensured. Thus, argues Barbieri, we have to add the processes of *natural conventions* in addition to the processes of *natural selection* to our study and understanding of the organization and evolution of the natural world (2003:153).

In its triadicity and interactivity, Barbieri's semantic theory of the cell and its evolution seem to fall well within the biosemiotic perspective we have been discussing above. Yet Barbieri has a challenge for the Peircean-von Uexküllian tradition of Sebeok and Hoffmeyer, in that primordially, for Barbieri, "meaning" is "completely accounted for by objective and reproducible entities" (this volume). In fact, for Barbieri, "*any* time that we discover that the link between two organic worlds [read: between two dissimilar sets of internally convergent or autopoetic relations] requires not only catalysts but also *adaptors*, we are very likely to be in the presence of an organic code, and therefore of organic meaning" (2002:293).

This focus on the endogenous organization of organisms as the primordial site of meaning-making – and the corollary conclusion that such meaning-making is, in its first instance, mechanical and derivative, rather than experiential and primitive – leads Barbieri to posit a semiotic/hermeneutic threshold in the evolution of living beings:

"The first semiotic structure that appeared in the history of life was the [ribonucleoprotein] apparatus of protein synthesis, and the genetic code [joining nucleotides to amino acids] was the first code, but not the only one. The evolution of semiosis was essentially due to the appearance of other organic codes, especially in eukaryotic cells, and it was these new codes that increased the complexity of the eukaryotes and eventually allowed them to produce semiotic systems capable of interpretation, i.e. *hermeneutic* systems. The model of Peirce and Sebeok, therefore, is still valid but only for hermeneutic systems. The origin of semiosis (the *semiotic threshold*) and the origin of interpretation (the *hermeneutic threshold*) were separated by an extremely long period of evolution, because interpretation is dependent on context, memory and learning, and probably evolved only in multicellular systems. The history of semiosis, in short, was a process that started with context-free codes and produced codes that were more and more context-dependent. Today, our cultural codes are so heavily dependent on context that we can hardly imagine semiosis without interpretation, and yet *these are distinct processes* and we need to keep them apart if we want to understand their origin and their evolution in the history of life" (2006: forthcoming).

The subjective experience of animals interpreting their surrounds as highlighted by von Uexküll, and even the triadic logic of relations developed by Peirce, claims Barbieri, can only function as "descriptive sciences, not explanatory ones…[for in this framework] semiosis requires three basic elements – object, interpreter and sign – which are *preconditional* and therefore *primitive* entities. [As] consubstantial agents of semiosis …they are the starting point [whereby a sign relation comes into being] and therefore cannot be reduced any further" (2002: 291–292).

Thus, although the Peircean/Uexküllian tradition shows us *that* sign relations are critical to the organization and interaction of the biological world, claims Barbieri, they do not show us *how* the underlying physical mechanisms work. For that, he suggests that biosemiotics needs to turn away from qualitative organicism in its approach and instead adopt "good rational, old-fashioned machine-like models" in the investigation of the roles of codes, signs, and meaning in living systems (202:294).

Such machine models, Barbieri stresses, do not have to be eliminative-reductionist ("for a machine is a machine not when it is reduced to pieces, but precisely when it is put together into a functioning whole"), not does they have to be physically constructed (e.g., a Turing machine), nor necessarily a set of mathematical equations. "Natural selection," writes Barbieri, "is a mechanistic model which is entirely expressed in words. The important point is that the model has the *logic of a machine*" (2002:289).[34]

In so arguing against the organicist orientation of the Copenhagen school, Barbieri aligns himself with the mechanistic tradition of "Descartes, Newton, Lamarck, Darwin ... and Jacques Monod" over and against the representative group of biosemiotic precursors cited by Stjernfelt: "Saint-Hilaire, von Baer, D'arcy Thompson, Spemann ... Brian Goodwin, René Thom and Stuart Kauffman" (Barbieri 2002:284; Stjernfelt 2002:79).

It remains an open and ongoing question as to whether Barbieri will be successful in his efforts to refashion the primary biosemiotic articulation from one of "signs" to "codes" – or if, indeed, contrary to Barbieri's own current position, a coherent synthesis between his articulation and the presently predominant Peircean-Uexküllian articulation can be achieved. Untold more possibilities exist, of course, for as Hoffmeyer reminds all newcomers to biosemiotics in the introduction of his seminal work, "To be decent scientists, we must take one another's realities seriously enough to try to eliminate the contradictions" (1996:ix). Biosemiotics, he continues, "suggests one way of doing this" – and then he adds with characteristic humanist-scientist understanding, "There may, of course, be other ways" (*ibid*).

A PARTING PROLOGUE: THE FUTURE HISTORY OF BIOSEMIOTICS

A heuristic formula for the development of any kind of scientific inquiry might consist in successive initial phases of: observation, intuition, articulation and experimentation – which, if felicitous, then begin to cycle into one other generatively and recursively. If this formulation can serve us as even a rough guide to the progression of scientific inquiry, then biosemiotics today is surely well past phases one and two, and is working diligently within phase three with a look to the arrival of phase four, at which time it will no longer be a "revolutionary science" in the Kuhnian sense, but quite simply, part of the background assumptions and paradigm of the everyday "normal science" of biology.[35]

Whether or not this day will come, only the history written *after* this history can tell. Certainly, the study of sign processes within life processes cannot be forestalled forever, as the more we learn about the former, the more we find ourselves confronted with the latter. Eventually, the "blind faith" that these sign processes can be studied *only* in their material aspects and not *also* in their aspects as signs *qua* signs for the systems that are using them as such will be forced to give way under the weight of empirical evidence that is even now pouring in daily from the research being done in every area of the life sciences.

Yet many working scientists do not feel comfortable toiling at a "science" that is still in its articulation phase. For the claim that "articulation" must come *before* "experimentation" so as to arrive at "understanding" may seem strange to those scientists who are working in long-established fields where the defining and fundamental articulations have already been settled – and, indeed, fields that may already be well into their third and fourth re-articulations, as in physics. Yet MacIntyre (1974) has argued well that the history of all sciences have followed this chronology of observation, intuition, and articulation before experimentation – for, indeed, how would one know what one was experimenting "on" or "for" if one did not already have in place at least a provisional articulation of what one has intuited based on observation? And success in science has long followed the path, from the pre-Socratics to Copernicus, Newton to Darwin, Einstein and Bohr to Watson and Crick.

"You won't look for something if you don't believe it's even there," Marcello Barbieri reminds us frequently, and in his (2003) *The Organic Codes*, he relates how:

> In the 1950s, it became clear that protein synthesis required a transfer of information from nucleic acids to proteins, and people realized that such a process must necessarily use a code. The existence of the genetic code, in other words, was predicted *before* doing the experiments that actually discovered it, and the results of those experiments were correctly interpreted as proof of the code's existence. [Contrarily,] in the case of signal transduction, the experiments were planned from the very beginning as a means of studying the biochemical steps of the phenomenon, and not as a search for codes, and the biological reactions of that field were regarded *a priori* as normal catalyzed processes, not as codified processes. *No code had been predicted, therefore no code was discovered*.... [and this is how molecular signal transduction] has been studied ever since" (2003:233).

Moreover, the fact that researchers were "looking for" a genetic code at all has its roots in the process of observation, intuition and articulation that led Wilhelm Johannsen to propose the existence of a "gene" in the first instance. For there again, an *observation* – about familial sameness – led to an *intuition* – about material transmission – that had to be articulated – as the "theoretical unit of heredity" (*whatever* that might turn out to be ... and some candidates were: cell, protein, blood, vapor and many others) – to be called, for *articulatory* purposes – a "gene" – *before* researchers started conducting experiments to find out if this so-called "unit of heredity" actually existed and, if so, what it physically *was* and *how it worked.*

The twisted ladder of the double-helix DNA molecule, could Johannsen or any of his contemporaries had somehow seen it back then, would never have suggested itself as anything other than just a spirally molecule – which, of course, on one level, it is. But its *function* is something more, and that is not something that can be ascertained just by looking at its material form alone. Rather, only by looking at its material form in a context of explanation – an articulation, or provisional theory – can one begin to do the experiments that will lead to the warranted conclusion that this molecule *functions* as the "unit of heredity" in this particular set of material interactions that *is* "reproduction between organisms."

Precisely analogous to this is the current state of neurobiological research with which we opened this discussion on page one. There – as in genetics, as in pharmacology, as in animal behavior study – if one is not *looking for* the biological construction of a "sign relation" within the set of material interactions that is brain activity, one can "see" all the chemical-electrical activity there is to be seen – but one will never know how to explain it *as* any particular *instance* of "sign activity" until one has a provisional theory – or *articulation* – of in just what a "biological sign category" consists. The microscope can only *present* – it cannot "make sense of" or explanatorily "reveal." That takes a theory – which is an articulation, based on intuition and observation – which is then subject to rigorous experiment.

For a neuron will remain a neuron no matter what, and its chemical and electrical properties – which we already understand quite well today – are not going to change. But whether or not we ever even look to see if this particular neuron's activation is functioning as part of an indexical circuit, an iconic one, or a symbolic one – to this *kind* of question, we will never get an answer, so long as "sign processes" remain misunderstood as equivalent to "human cultural constructs" and not the fundamental biological relations that biosemiotics insists that they are.

Yet one can only get an answer to those questions that it is "legitimate" to ask – and thus the job of biosemiotics right now is to articulate its questions about sign processes in biology to the point that they become taken up by the larger scientific community as being legitimate questions to ask. For many of these questions are often yet intuited as being "not quite legitimate" questions to ask, even now – and even with the continual insistence of virtually everyone involved in the biosemiotic project, that what is being asked for is *not* a retreat into mysticism, supernaturalism, immaterialism, or reification of some scientifically unexaminable thing or element called "the sign" *per se* – but, rather, the same type of rigorous, repeatable, falsifiable examinations into a set of naturally-occurring relations in the world that living beings both need (internally) and use (externally) in order to survive.

One can examine these phenomena in their aspects as sign phenomena (i.e., in their aspects as substitution relations for some non-immediately present other) and still be doing actual science – this is the biosemiotic "message" in a nutshell. But the long legacy of Cartesian reductionism that has allowed modern science to examine the inanimate aspects of the world (Descartes' *res extensa*) so successfully, has kept it closed off from the equally natural product of nature that is "knowing relations" or "cognition" (Descartes' *res cogitans*).

Thus, despite all the problems that Cartesian body-mind dualism keeps increasingly forcing upon life scientists, the majority of experiments being done today – in neuroscience, molecular biology, immunology, pharmacology, etc. – are all informed by a theory that precludes, under the very terms of its bifurcated ontology, even the possibility of coherently – much less scientifically – understanding the phenomena under investigation: phenomena like messaging, signaling, representation, communication, understanding, and sign. Biosemiotics has thought these matters through from both their biological and their semiotic sides and as come to the conclusion that the problem is not in the phenomena, but in the unnecessary restrictiveness of the informing theory.

Biosemioticians would argue the absolutely legitimate fear of contaminating science with spiritualism, vitalism, anthropomorphism and anti-scientism of every stripe has had the unintended consequence of forcing life-science into the unnatural and reactionary position of materialist reductionism – and that this has diminished it and closed off its explanatory possibilities towards system phenomena that cannot be so reduced – not because such phenomena are spiritual or immaterial, but simply because of their nature as agent-object-action relations of a biological organism. For a system that is alive must maintain itself in a constant state of self-reconstruction – this means that it must simultaneously and incessantly negotiate the ordering of both of its own internal set of intra-system relations as well as its macro-system level interactions with an externality that is constituted by a whole other set of internal relations of its own. To do this, with a third set of "mediating" relations at the interface between the two becomes necessary. Merely to *survive* this incessant triadic existential demand (much less to *evolve* within it) necessarily introduces into the phenomena under examination the proximate and system-centric relations of *function, use, purpose,* and *goal* – as well as the superordinate relation needed to achieve all of these relations, the relation of substitution or "standing for" – i.e., the biological relation of *sign.*

But again: because biosemiotics is *not* challenging in any way the absolute need for, and manifest success of, examining the material aspects of these phenomena *qua* those material (and not "material and also relational") aspects, doing lab experiments now will not "advance" the biosemiotic understanding any further. Biosemioticians will see a neuron firing and say that is a "sign" whose vehicle is this chemical-electrical event – while mainstream neuroscientists will see the same neuron firing and say that parsimony demands we say no more than just: this is a chemical-electrical event. But to the organism that neuron is firing in, which of these two understandings is the more inclusive and veridical? And is it not this organism – this system of interactions – that we are ultimately trying to understand in all its fullness?

Left *only* with what can be seen "iconically," we are back to seeing DNA before there is a coherent theory of genetic inheritance in place. The results of lab experiments will always be the same for both of us in our capacities as "object-ive" observers, and thus the burden of proof, quite rightfully, is now on the biosemioticians to articulate why the biosemiotic insistence that the same phenomenon must also be explicated from the "subject-ive" standpoint of the system under examination is not only possible and warranted, and worthy of the development of new scientific conceptual tools – but is also the understanding that may prove to be more predictive, more knowledge-generating, and more explanatorily sufficient than the current biological models that are now in use.

Like Aristotle's ideal naturalist who was able to successfully capture *both* the material nature of a phenomenon as well as its "meaning" in the lives of the organisms involved with it, *without losing the essential aspects of either,* biosemiotics strives for an explanatory subjective knowledge/object knowledge synthesis in order to explain *nature's* genuine subject /object syntheses. But whether or not anyone currently working in the field of biosemiotics can actually accomplish this, of course, remains to be seen. Thus far, the majority of our effort has been expended trying to convince our colleagues in the sciences and the humanities that such a synthesis is even necessary. And as premier biosemiotician Claus Emmeche reminds us, while the biosemiotic understanding of sign relations as genuine relations of the natural world may seem to its adherents as a "robust, sophisticated, coherent, well founded, fruitful and comprehensive scheme of thought...in the long run, it cannot escape being judged by its fruits – and we do not yet know the historical result of that judgment" (2000b:224).

And thus we end this brief overview of the ongoing history of biosemiotics as we started it – *in media res.* For while Thomas Sebeok (2001) referred to the 1970s as the "prehistory" of biosemiotics, and Marcello Barbieri (2002), writing of the 1990s, opined that biosemiotics was as yet still coming into its "adolescence" – it is difficult not to feel as we end this as-yet preliminary "history" that both the reader and I have arrived here at the present moment just as the *real* history of biosemiotics is about to get underway.

That said, all that is now left for me to do as a historian of this project and a member of this community is welcome all our readers to this thriving young interdiscipline and, on behalf of my colleagues in biosemiotics everywhere, invite you to actively contribute to its ongoing history.

NOTES

[1] I am extremely indebted to Dr. Stefan Frazier of San Jose State University for his incalculable assistance and support in reading the early drafts of this manuscript. I also wish to thank Dr. Barbara Ryan of the National University of Singapore for her assistance in the copyediting of this tent.

[2] An absolutely ordinary – but quite profound, it turns out – definition from the American Heritage Dictionary (Houghton Mifflin), 2006.

[3] Augustine: *De doctrina christiana* II, 2 (1963: 34) in: *Sancti Augustini Opera*, ed. W. M. Green, CSEL 80, Vienna. Cited in Meir-Oeser (2003).

[4] Meir-Oeser Stephan, "Medieval Semiotics" *The Stanford Encyclopedia of Philosophy (Winter 2003 Edition)*, Edward N. Zalta (ed.). Available at: *http://plato.stanford.edu/archives/win2003/entries/semiotics-medieval.*

[5] A more comprehensive comparison between Aristotle's ideas and those of biosemiotics, however, is the project of another day. Interested readers are heartily encouraged to begin this investigation on their own, however, and note that the project to resuscitate a wholly non-spiritual, non-mystical, scientific notion of local *system teleology* based on Aristotle's subtle and widely-misunderstood notion of *formal causation* is one that Stanley Salthe (1993, 2006) has been pursuing for some time. (See also Jesper Hoffmeyer's notion of *semiotic causation,* this volume, as well as John Deely's penetrating discussion of Aristotelian "relation" in 2001:226–231)

[6] This wonderfully insightful phrase is from Terrence Deacon's equally insightful *The Symbolic Species* (1997:53), a highly recommended entry point into biosemiotics – for although Deacon does not identify himself as a biosemiotician *per se*, many biosemioticians draw inspiration from his work.

[7] It is germane to note here that Deely observes that it is precisely those aspects of Ockham's writings called the *via nominalia* that were "presciently called the *via moderna*" by his successors at Oxford as the High Middle Ages were coming to a close (2001:395)

[8] More precisely: *x* registered simultaneously both as itself (i.e., *x* and not nothing; *x* and not *y*) and as a significate pointing to something other than itself (non-*x*, or *x* not only as *x*, but as *y*) – even if that y is "other instances of *x*", as in the iconic organization of categorical perception.

[9] René Descartes, Meditations on First Philosophy, *Meditation Two: On the Nature of the Mind,* 1641 [1973:80].

[10] *Discourse on the Method of Rightly Conducting One's Reason and Seeking the Truth in the Sciences*, 1637 [1973:24]

[11] Note, too, that it would yet be several centuries after Descartes' attempt to describe the non-minded world of animals as "mere clockwork mechanisms" (1649/1991: 365-6, 374) – and almost 100 years

after Lloyd Morgan would deploy his Occamite Canon – that biologist Francis Crick would note that: "While Occam's razor is a useful tool in the physical sciences, it can be a very dangerous implement in biology" given that evolution does not organize living beings "parsimoniously" in any straightforward kind of sense. "It is thus very rash to use simplicity and elegance as a guide in biological research" warns Crick (1988).

[12] The tradition of seeing the human being as the perpetually duped and deceived animal – *homo decipi*, as it were – would turn out to be one of the most enduring, if unfortunate, tropes of all modernity, snaking its way out of Plato's cave through the "revolutionary" pronouncements of Marx and Freud and to the "revelationary" pronouncements of neuronal and genetic eliminative materialism on the one hand, and the pseudo-postmodernism of "radical deconstructionism" on the other. As I have argued elsewhere (Favareau 2001a), nothing could be more diametrically opposed to the understandings advanced by biosemiotics than this self-regarding yet internally-contradictory stance that I hereby dub "the Fallacy fallacy."

[13] This joke commonly attributed to comedian Steven Wright captures the dilemma well: "Last night I was all alone in my room and I started thinking, "You know, the human brain is probably the most magnificent structure ever created in nature." ... but then I thought: "Wait a minute. Who's *telling* me this?"

[14] Again, we are in an analogous position when we try to understand how "signs" of any kind – the ink marks on this page, the waggle dance of bees, a voltage change generated in a cortical neuron – comes to signify something other than itself, when there is only, physically, itself. And the answer of course, here and on the genetic level, is that we must look at "information-bearing" things not in their material isolation – where they are, in fact, nothing but themselves – but also in the function that they serve in the system that makes use of them *as* signs, in order to see how they can be both "nothing but themselves" and "standing for something other than themselves" in the operation of that system. Exploring this logic of relations within the scientific paradigm is, of course, the *raison d'être* of biosemiotics.

[15] Moreover and by necessity, not every attempt at a science of biological sign-use undertaken even in the last half century can be included in this short history. Such a survey would, of course, be impossible given the space available and would, by necessity, involve long discussions on the history and major figures of comparative psychology, cognitive science, molecular biology, Artificial Intelligence, pharmacology, cognitive neuroscience and much much more. And it is only because of such space limitations that even the individual accomplishments of such generally accepted "proto-biosemioticians" as Elia Sercarz (1988), Sorin Sonea (1988), Günter Bentele (1984), Yuri Stepanov (1971), F.S. Rothschild (1962), and Marcel Florkin (1974) are not discussed at length in this text. This is not to say, however, that the works of these researchers is insignificant to the larger project whose narrative is recounted here. Florkin, Stepanov and Rothschild – a molecular biologist, a text semiotican, and a psychologist, respectively – each independently coined the term "biosemiotic" to describe where they wanted their investigations to be heading. But because no inter-disciplinary movement resulted from these individual efforts, I have made the purely editorial decision to refrain from any in-depth discussion of them here. No slight on my part is intended by these purely editorial decisions, and those wishing to consult the original works are directed to the bibliography, as well as to the more inclusive "pre-histories" of Sebeok (1998, 2001) and Kull (1999, 1999a, 1999b, 2005).

[16] As is evident from the footnote above, Thomas A. Sebeok was not the first to coin the compound noun joining "bio" with "semiotics" (again, see Kull 1999 for a detailed history of the use of the term) – however, it is the specific project that Sebeok initiated and christened as such that is the subject of this history and this book.

[17] Deely notes that it was Margaret Mead who, at the end of a contentious conference about animal communication that Sebeok had organized in 1962, proposed the specific form of the word "semiotics" to denote "patterned communication in all modalities, [whether] linguistic or not" (Deely 2004) – an understanding perfectly congruent with Sebeok's growing conviction that human language "was not much more than that realm of *nature* where the logosphere – Bakhtin's dialogic universe – impinges in infant lives and then comes to predominate in normal adult lives" (Sebeok, 2001).

[18] Mihaly Csikszentmihalyi's (1934) distinction between a "field" and a "domain" remained one central to Sebeok's life and thought. In short: A *domain* refers to an intellectual culture of shared meanings, definitions, assumptions, rules and evidentiary procedures (such as "science," or more finely, "medical science"), while a *field* comprises "all the individuals who act as gatekeepers to the domain...[and who decide] whether a new idea...should be included in the domain" (Csikszentmihalyi 1997:27–8). And in 1970, Juri Lotman's Tartu-Moscow Semiotic School was by far the closest thing resembling an established *field* of disciplinary gatekeepers for the nascent world of international semiotic study. (Cf. Sebeok 1998, Kristeva 1994 and Kull 1999b).

[19] The history of this manuscript's subsequent loss at the hand of a translator is recounted in Sebeok 1998. Suffice it to note for our purposes that it would not be until twenty years after the event, in 2005, that the English language translation of Lotman's manuscript would appear in the journal that Lotman himself founded in 1964, *Trudy po znakovym sistemam* – now known in English as *Sign Systems Studies*, Volume 33.1

[20] Lotman himself resisted this equivalence (1989:43), insisting that the ability of cognitive agents to shape the material surround of their environment (Vernadsky's *noosphere*) differed from the purely "abstract" cognitive interactions of the *semiosphere*. The distinction that Lotman fails to draw here – as is so often the case in such discussions about "mind and world" that yet accept the assumptions of Cartesian dualism on some fundamentally under-examined level – is the failure to differentiate between the *symbolic* level of embodied, biologically based sign processing, and its equally biological iconic and indexical substrates, with which it is on an experiential continuum. Such delineations are critical to the project of a scientifically sound biosemiotics that can yet account for the realities of abstraction and counterfactual reasoning, and we will have much more to say about these delineations presently. For an edifying discussion of the Lotman/Vernadsky controversy, see Chang 2002 and Kull 1999b.

[21] Later, Sebeok himself would be instrumental in tracking down the author of an obscure unpublished doctoral dissertation on Peirce and commissioning him to revise the all-but-forgotten manuscript thirty years later for publication. This work (Brent 1993) has since become the definitive biography of Peirce.

[22] For more in-depth overviews, see Colapietro 1989 1996, Deely 1990 2001, Deledalle 2000, Parmentier 1994, Savan 1976, and the e-resource for all things Peircean, *Arisbe* at: *http://members.door.net/arisbe/arisbe.htm*

[23] We pass over here, in the interest of space, Uexküll's influence on the then-developing field of neuroscience, and especially his influence upon one of its principal founders, Charles Scott Sherrington (1857–1952), who credits von Uexküll frequently and whose work on the neurobiology of reflex, posture and muscle movement was a direct outgrowth of von Uexküll's earlier experiments (Lagerspetz 2001:646). Suffice it to say that the notion of the "neural net" is already prefigured in Uexküll (1928:106) – and while many contemporary neuroscientists and roboticists take these notions as their starting points, few have worked their way back to von Uexküll for the purposes of either further enlightenment, nor for the acknowledgement of a debt (but see Fuster 2003 and Ziemke and Sharkey 2001 for exceptions).

[24] "Bio-semiotic" premises are implicitly discoverable – though never fully articulated as such – in all of these neurobiologists' works to some extent, though none save Fuster show any acquaintance with the work of von Uexküll or Peirce that informs much of contemporary biosemiotics.

[25] Though perhaps it would be fair to say that Stuart Kauffman eventually did also pursue such a deliberately interdisciplinary project, via his long-standing participation in the Santa Fe Institute.

[26] A journal dedicated to the study of "second-order cybernetics, autopoesis and cyber-semiotics" – roughly, the role of feedback and generative recursion in the organization of observing systems, self-maintaining systems, and sign-using systems – Brier's journal is deeply influenced by the work of biologists Humberto Maturana and Francisco J. Varela (1987), cyberneticians Heinz von Foerster (1982) and Ernst von Glasersfeld (1987), as well as the pioneering interdisciplinarity of cybernetician/anthropologist/psychologist Gregory Bateson (1973).

[27] These four, along with Sebeok, Thure von Uexküll and Joseph Ransdell, issued a polemical call for a "new paradigm" of semiotically informed science (and vice-versa) at just about the same time that Hoffmeyer was independently coming to the same conclusion in 1984. (See Anderson et al. 1984).

[28] Many of these papers have since been published in *Sign Systems Studies*, Vol 30.1 (2002).
[29] This history would not be complete without mentioning those related researchers who, while not regular attendees at the Gatherings, continue to produce work that has particular relevance for most biosemioticians. Among these scholars must surely be included Stanley Salthe (1993), Kochiro Matsuno (1999), Luis Rocha (2001), Peter Cariani (2001), Robert Ulanowicz (1986), Mark Bickhard (1999), John Collier (1999), Merlin Donald (1991), David Depew (1996), Bruce Weber (2000) and perhaps most of all Terrence Deacon (2003), whose 1997 *The Symbolic Species* is perhaps the clearest and most compelling application of Peircean semiotic to evolutionary biology yet produced. And while Deacon does not identify himself as a biosemiotician *per se*, seminal biosemiotician Claus Emmeche spoke for many when he remarked at the recent Gregory Bateson Centennial Symposium in Copenhagen that "Many biosemioticians consider themselves not only Peirceans, but Deaconians as well."
[30] This movement also includes biologist and philosopher of science Zdeněk Neubauer, systems theorist Ervin Laszlo, cognitive scientist Ivan Havel and geologist Václav Cílek. An excellent English language introduction to their ideas can be found in Havel and Markoš (2002), which collects the proceedings of a conference that also features contributions from Giuseppe Sermonti, Pier Luigi Luisi, and Mae-Wan Ho.
[31] It should also be noted here that many of the "non-Peirceans" from outside of the Copenhagen school – such as Prague physiologists Anton Markoš and Fatima Cvrčková – also eschew the idea that formalized equations between "digital signs and bodily (or analog) entities [could] be reduced to an unequivocal correspondence" (Cvrčková and Markoš 2005:87). Rather, for the majority of more complex organisms (and certainly for mammals), the action of interpretation upon a sign is "its own shortest description" (a la the incompressible algorithms discussed by Kauffman 2000).
[32] In all fairness, not all members of the so-called Copenhagen tradition subscribe to this line of thinking – Taborsky (2001) and Christiansen (2002), for example, certainly do not – nor, indeed, did Peirce himself. Artmann (*in preparation*) and Barbieri (2001, this volume) have argued convincingly, however, that the assumption that true *sign* processes start with life (and, for all practical purposes, vice-versa) is retrievable in the works of Hoffmeyer, Emmeche, Kull, et al., and I do believe that this assertion is a reasonable one.
[33] Personal correspondence with the author April 21, 2006.
[34] It is precisely this assertion that, I think, is most strenuously argued against in Anton Markoš' *Readers of the Book of Life*, as discussed above (see also Markoš 2002a:136, 2002:221; 2005:87). Hoffmeyer (1996:38,95) and Emmeche (2001:659) have similarly voiced their opposition to this idea.
[35] Bruno Latour (1987) distinguishes these two phases in the construction of knowledge as, first, "science in the making" – which is characterized by uncertainty, debate, personality, happenstance and abduction – followed by "ready made science" – which is characterized by relatively uncontentious induction using formulae, models, vocabulary, theories, methodologies and technologies that have been vetted in the earlier phase. The layperson's notion of "science" is generally the latter; the scientist's experience, the former – but as Latour argues against Kuhn, the relation between the two enterprises is not revolutionary struggle, but evolutionary dialectic.

WORKS CITED

Anderson, Myrdene. (1990). "Biology and semiotics." In *Semiotics in the Individual Sciences.* W.A. Koch (ed.), 254–281. Bochum: Universitätsverlag Dr. N. Brockmeyer.

Anderson, Myrdene; Deely, John; Krampen, Martin; Ransdell, Joseph; Sebeok, Thomas A. and Thure von Uexküll (1984). "A semiotic perspective on the sciences: Steps toward a new paradigm", *Semiotica* 52–1/2, 7–47.

Aristotle. *De Anima.* (330 BC /1941) Translator: J. A. Smith. In McKeon, Richard. *The Basic Works of Aristotle,* New York, Basic Books, pp. 533–603.

Aristotle. *De Interpretatione.* (330 BC /1941) Translator: E. M. Edghill. In McKeon, Richard (1941). *The Basic Works of Aristotle,* New York, Basic Books, pp. 38–61.

Aristotle. *De Sensu et Sensibilius.* (c350 /2004). Translator: J.I. Beare. Available online as e-book at: http://etext.library.adelaide.edu.au/a/aristotle/sense/

Artmann, Stefan. (2005). "Biosemiotics as a structural science: Between the forms of life and the life of forms." *Journal of Biosemiotics 1*, pp. 229–265.

Artmann, Stefan. (*forthcoming*). "Computing codes versus interpreting life: Two alternative ways of synthesizing biological knowledge through semiotics." *Journal of Biosemiotics* (to appear).

Augustine of Hippo: (387 /1975) *De dialectica*, ed. Jan Pinborg, translation with introd. and notes by B. Darrel Jackson, Dordrecht: Reidel.

Augustine of Hippo (389 /1963): *De Doctrina Christiana*. In *Sancti Augustini Opera*, ed. W. M. Green, CSEL 80, Vienna. Cited in Meir-Oeser (2003).

Aquinas, Thomas (1252 /1965). "De Ente et Essentia (On Being and Essence)." In Bobik, Joseph (Trans.) *Aquinas on Being and Essence: A Translation and Interpretaion*. Notre Dame: Notre Dame University Press.

Bacon, Francis. (1620 /1863). *Novum Organum*. In James Spedding, Robert Leslie Ellis, and Douglas Denon Heath in Francis Bacon's *Works (Vol. VIII)* Boston: Taggard and Thompson. Available Online at: http://www.constitution.org/bacon/nov_org.htm

Baer, Eugen. (1987). "Thomas A. Sebeok's Doctrine of Signs." In Krampen, Martin (Ed). *Classics of Semiotics*. New York, Plenum Press, pp.181–210.

Barbieri, Marcello (1981). "The ribotype theory of the origin of life." *Journal of Theoretical Biology* **91**: 545–601.

Barbieri, Marcello. (1985). *The Semantic Theory of Evolution*. New York: Harwood Academic.

Barbieri, Marcello. (2002). "Has biosemiotics come of age?" *Semiotica, 139*(1/4), 283–295.

Barbieri, Marcello. (2003). *The Organic Codes: An Introduction To Semantic Biology*. Cambridge, UK.; New York: Cambridge University Press.

Barbieri Marcello. (2003a). "Biology with information and meaning." *History and Philosophy of the Life Sciences*, 25, 243–254.

Barbieri Marcello. (2005). "Life is artifact-making." *Journal of Biosemiotics 1*, 107–134.

Bardini, Thierry. (2001) *Bootstrapping: Douglas Engelbart, Coevolution, and the Origins of Personal Computing*. Stanford: Stanford University Press.

Bateson, Gregory. (1973). Steps To An Ecology Of Mind; Collected Essays In Anthropology, Psychiatry, Evolution, And Epistemology. San Francisco,: Chandler Pub. Co.

Bentele, Günter. (1984). Zeichen und Entwicklung: Vorüberlegungen zu einer genetischen Semiotik. Tübingen: G.Narr.

Bernard, Jeff (2001) *Obituary: Thomas A. Sebeok November 9, 1920 – December 21, 2001.* Available at the website for the International Association for Semiotic Studies page: http://www.uni-ak.ac.at/culture/withalm/semiotics/AIS/sem-people/sebeok/TASebeok-obit.html

Bickhard, Mark. H. (1999). "Representation In Natural and Artificial Agents." E. Taborsky (Ed.) *Semiosis. Evolution. Energy: Towards a Reconceptualization of the Sign.* (15–25). Aachen: Shaker Verlag.

Bickhard, Mark. H. (2003). "The Biological Emergence of Representation." In T. Brown, L. Smith (Ed.) *Emergence and Reduction: Proceedings of the 29th Annual Symposium of the Jean Piaget Society.* (105–131). Erlbaum.

Böll, Mette (2002). *The evolution of empathy in social systems.* Paper Presented at the International Gatherings in Biosemiotics Conference 2, the University of Tartu, Estonia. June 14–17.

Brent, Joseph. (1993).*Charles Sanders Peirce: A Life.* Bloomington: Indiana University Press.

Brier, Søren ed. (1995). "Cyber-semiotics: On autopoesis, code-duality and sign games in biosemiotics. *Cybernetics and Human Knowing 3* (1), pp. 3–14.

Brier, Søren ed. (1998). "Cyber-semiotics: A transdisciplinary framework for information studies." *Biosystems* 46, pp. 185–191.

Brier, Søren ed. (2001). "Cybersemiotics and Umweltlehre." *Semiotica 134* (1/4), pp. 779–814.

Brier, Søren ed. (2003), "Thomas Sebeok and the Biosemiotic Legacy.", Special Memorial Issue of *Cybernetics and Human Knowing* 10.1.

British Medical Journal. (2004) "Obituary: Thure von Uexküll." (October 30) 329:1047. Available at: *http://bmj.bmjjournals.com/cgi/content/full/329/7473/1047*

Brooks, Rodney. (1999). *Cambrian Intelligence: The Early History of the New AI.* Boston: MIT Press.

Bruni, Luis Emilio. (1997). "Cultura, determinismo económico y la confusión del mapa con el territorio." *Venezuela Analitica Revista Electrónica Bilingüe 13.* Available at: http://www.analitica.com/archivo/vam1997.03/lit2.htm

Bruni, Luis Emilio. (2001). "Biosemiotics and ecological monitoring." *Sign System Studies 29*(1), 293–312.

Burbano, Hernán A. (2005). *Determinism, Indeterminism and Semiotic Election.* Paper Presented at the International Gatherings in Biosemiotics Conference 5, Urbino University, Italy. July 20–24.

Campbell, Donald T. (1974). "Evolutionary Epistemology." In P. A. Schilpp (Ed.) *The Philosophy of Karl R. Popper*, LaSalle, IL: Open Court. pp. 412–463.

Cariani, Peter (2001). "Symbols and dynamics in the brain." *Biosystems.* Vol 60, pp. 59–83.

Chalmers, David (1995). "Facing Up to the Problem of Consciousness" *Journal of Consciousness Studies* 2(3):200–19.

Chang, Han-liang. (2002). *Is Language a Primary Modeling System?–On Juri Lotman's Semiosphere.* Paper presented at the International Conference on *Cultural Semiotics: Cultural Mechanisms, Boundaries, Identities* University of Tartu, Estonia, February 26 – March 2, 2002.

Chang, Han-liang. (2005) "Biosemiotics: Nature in Culture or Culture in Nature?" *Chung Wai Literary Monthly* 34 (7).

Chebanov, Sergei V. (1994). "Man as participant to natural creation: Enlogue and ideas of hermeneutics in biology." *Rivista di Biologia 87*(1), 39–55.

Chekhov, Anton. (1904). *Teatr I iskusstvo.* No. 28, 11 July, p. 521.

Chien, Juipi Angelina (2003): *Diagramming as a Convergence of C. S. Peirce, Jakob von Uexküll, and E. Gombrich.* Paper Presented at the International Gatherings in Biosemiotics Conference 3, University of Copenhagen, Denmark. July 11–14.

Christiansen Peder Voetmann (2000) Macro and Micro-Levels in Physics In *Downward Causation: Minds, Bodies and Matter.* Peter Bøgh Andersen, Claus Emmeche, Niels Ole Finnemann and Peder Voetmann Christiansen (Eds). Århus: Aarhus University Press. pp. 51–62.

Churchland, Paul. (1984) "Eliminative Materialism" in *Matter and Consciousness.* Cambridge, Mass: MIT Press. pp. 43–49.

Churchland, Patricia. S. (1986). *Neurophilosophy: Toward A Unified Science Of The Mind-Brain.* Cambridge, Mass.: MIT Press.

Churchland, Patricia. S. (2002). *Brain-Wise: Studies In Neurophilosophy.* Cambridge, Mass.: MIT Press.

Clark, Andy. (1997). *Being There: Putting Brain, Body, And World Together Again.* Cambridge, Mass.: MIT Press.

Colapietro, Vincent. (1989). *Peirce's Approach to the Self: A Semiotic Perspective on Human Subjectivity State.* Univ of New York Press.

Colapietro, Vincent and Thomas Olshewsky. (1996). (Eds.) *Peirce's Doctrine of Signs.* Berlin: Mouton de Gruyter.

Collier, John (1999) "The Dynamical Basis of Information and the Origins of Semiosis.", in Edwina Taborsky (ed) *Semiosis. Evolution. Energy Towards a Reconceptualization of the Sign.* Aachen: Shaker Verlag:111–136

Crick, Francis. (1958). "On Protein Synthesis." *Symposium for Experimental Biology XII,* pp. 139–163.

Crick, Francis. (1988). *What Mad Pursuit: A Personal View of Scientific Discovery*, New York, New York: Basic Books.

Csikszentmihalyi, Mihaly. (1977). *Creativity: Flow and the Psychology of Discovery and Invention.* New York: HarperCollins.

Cvrčková, Fatima (2002). "The Darwinian purpose enters the post-genomic era: A case study." In Havel, Ivan and Markoš, Anton (Eds.) *Is There a Purpose in Nature? How to Navigate Between the Scylla of Mechanism and the Charybdis of Teleology.* Czech Republic: Vesmir. pp. 175–184.

Cvrčková, Fatima and Markoš, Anton. (2005). "Beyond bioinformatics: Can similarity be measured in the digital world?" *Journal of Biosemiotics 1*, pp. 87–105.

Damasio, Antionio. (1994). *Descartes' Error: Emotion, Reason, And The Human Brain*. New York: Putnam.

Danesi, Marcel. (1998). (Ed.) *The Body in the Sign: Thomas A. Sebeok and Semiotics*. Ottawa: Legas.

Danesi, Marcel. (2002). "Thomas A. Sebeok (1920–2001) Remembered." *SIGN Journal*, Vol 3.1, Available at: http://vicu.utoronto.ca/courses/semiotics/SIGN_Danesi_TAS.htm

Danesi, Marcel. (2000). "The biosemiotic paradigm of Thomas A. Sebeok." In Tarasti, E. *Commentationes in Honorem Thomas A. Sebeok Octogenarii*. Imatra:ISI. pp. 5–29.

Deacon, Terrence. W. (1997). *The Symbolic Species: The Co-Evolution Of Language And The Brain*. New York: W.W. Norton.

Deacon, Terrence. W. (2003). "The heirarchic logic of emergence: Untangling the interdependence of evolution and self-organization." In B. H. W. a. D. J. Depew (Ed.), *Evolution and Learning: The Baldwin Effect Reconsidered* (pp. 273–308). Cambridge, Mass.: MIT Press.

Deely, John. (1990). *Basics of Semiotics*. Bloomington: Indiana University Press.

Deely, John. (1995). *Thomas A. Sebeok: Bibliography 1942–1995*. Bloomington: Eurolingua.

Deely, John. (2001). *Four Ages Of Understanding: The First Postmodern Survey Of Philosophy From Ancient Times To The Turn Of The Twenty-First Century*. Toronto: University of Toronto Press.

Deely, John. (2004) "Thomas Albert Sebeok, "Biologist Manqué.", *International Association for Semiotic Studies 2004 World Congress, Lyon*. (Sebeok Memorial Essay) available at: http://carbon.cudenver.edu/~mryder/itc/sebeok.html

Deely, John N., Williams, B. and Kruse, F.E. (eds.) (1986). *Frontiers in Semiotics*. Bloomington: Indiana University Press.

Deledalle, Gérard. (2000) *Charles S. Peirce's Philosophy of Signs*. Bloomington: Indiana University Press.

Denizhan, Yağmur and Karatay, Vefa. (2002). Evolution of the "window." *Sign System Studies 30*(1), pp.259–270.

Depew, David. (1996). *Darwinism Evolving: Systems Dynamics and the Genealogy of Natural Selection*. Boston: MIT Press.

Descartes, René. (1637 /1973). "Discourse on the Method of Rightly Conducting the Reason." In *The Philosophical Works of Descartes*, translated by E. Haldane and G. Ross, Cambridge: Cambridge University Press, pp. 79–130.

Descartes, René. (1641 /1973). "Meditations on First Philosophy." In *The Philosophical Works of Descartes*, translated by E. Haldane and G. Ross, Cambridge: Cambridge University Press. pp. 131–200.

Dimitrov, Assen I. (2004). *Intelligence: Not Seeking Algorithm, But Meaning*. Faber: Bulgarian Academy of Sciences.

Donald, Merlin. (1991). Origin of the Modern Mind: *Three Stages in the Evolution of Culture and Cognition*. Cambridge: Harvard University Press.

Edelman, Gerald. (1992). *Bright Air, Brilliant Fire: On The Matter Of The Mind*. New York, N.Y.: BasicBooks.

Edelman, Gerald & Tononi, Guilio. (2000). *A Universe Of Consciousness: How Matter Becomes Imagination*. New York, NY: Basic Books.

Elsasser, Walter M. (1998). *Reflections on a Theory of Organisms*. Baltimore: The Johns Hopkins University Press.

Emmeche, Claus. (1991). "A Semiotical Reflection on Biology, Living Signs and Artificial Life." *Biology & Philosophy, 6*(3), 325–340.

Emmeche, Claus. (1994). *The Garden In The Machine: The Emerging Science Of Artificial Life*. Princeton, N.J.: Princeton University Press.

Emmeche, Claus. (1998). "Defining life as a semiotic phenomenon." *Cybernetics & Human Knowing, 5*(1), 3–17.

Emmeche, C. (1999). The biosemiotics of emergent properties in a pluralist ontology. In E. Taborsky (Ed.), *Semiosis, Evolution, Energy: Towards a Reconceptualization of the Sign*. Aachen: Shaker Verlag, pp. 89–108.

Emmeche, Claus. (2000a). "Closure, Semiosis, Autopoiesis, and Autocatalysis – Closure, Function, Emergence, Semiosis, and Life: The Same Idea? Reflections on the Concrete and the Abstract in

Theoretical Biology." *Annals of the New York Academy of Sciences, 901*, 11.

Emmeche, Claus. (2000b). "Transdisciplinarity, theory-zapping and the growth of knowledge." *Semiotica, 131*(3/4), 217–228.

Emmeche, Claus. (2001). "Does a robot have an Umwelt? Reflections on the qualitative biosemiotics of Jakob von Uexküll." *Semiotica* 134 (1/4): 653–693.

Emmeche, Claus. (2002). "Taking the semiotic turn, or how significant philosophy of biology should be done." *Sats, The Nordic Journal of Philosophy, 3*(1), 155–162.

Emmeche, Claus and Hoffmeyer, Jesper (1991). "From language to nature: the semiotic metaphor in biology." *Semiotica* 84(1/2), 1–42.

Emmeche, Claus; Kull, Kalevi and Stjernfelt, Frederik (2002). *Reading Hoffmeyer, Rethinking Biology.* Tartu: Tartu University Press.

Faria, Marcella (2005). *RNAs as code makers: A biosemiotic view of RNAi and cell immunity.* Paper Presented at the International Gatherings in Biosemiotics Conference 5, Urbino University, Italy. July 20–24.

Farina, Almo (2004): *Eco-field versus habitat: Shifting a paradigm in developing a cognitive ecology.* Paper Presented at the International Gatherings in Biosemiotics Conference 4, Charles University, Prague. June 1–5.

Favareau, Donald (2001). Beyond self and other: The neurosemiotic emergence of intersubjectivity. *Sign Systems Studies,* 30 (1), pp. 57–101.

Favareau, Donald (2001a). "Phenomenal Epiphenomena: Deconstructing the Fallacy Fallacy." Invited seminar presentation, *First International Gatherings in Biosemiotics*, University of Copenhagen, Denmark. May 24–27.

Favareau, Donald. (2002). Constructing representema: On the neurosemiotics of self and vision. *Semiotics, Evolution, Energy and Development Journal* 2 (4), 1–39.

Florkin, Marcel (1974). "Concepts of molecular biosemiotics and of molecular evolution." In *Comprehensive Biochemistry* 29A, 1–124.

Freeman, Walter. J. (2000). *Neurodynamics: An exploration in Mesoscopic Brain Dynamics.* London; New York: Springer.

Fürlinger, Anton. (1998). "Locomotion and Cognition: From fish to hominids." *Proceedings of the 26th Goettingen Neurobiology Conf. 1998* Vol 1

Fuster, Joaquin. M. (2003). *Cortex And Mind: Unifying Cognition.* Oxford; New York: Oxford University Press.

Griffith, Paul.E. (2001). "Genetic Information: A Metaphor in Search of a Theory." *Philosophy of Science*, 68, 394–412.

Hajnal, László (2003): *A new model for biology?* Paper Presented at the International Gatherings in Biosemiotics Conference 3, the University of Copenhagen, Denmark. July 11–14.

Harries-Jones, Peter. (1995). *A Recursive Vision: Ecological Understanding and Gregory Bateson.* Toronto: University of Toronto Press.

Havel, Ivan and Markoš, Anton (2002). *Is There a Purpose in Nature? How to Navigate between the Scylla of Mechanism and the Charybdis of Teleology* Czech Republic: Vesmir.

Hoffmeyer, Jesper. (1984). *Naturen I Hovedet. Om Biologisk Videnskab.* København: Rosinante.

Hoffmeyer, Jesper. (1988). Bioinformation techniques and the view of nature. In Thill, G., and Kemp, P. (eds). *The Triumph of Biotechnologies: The Domestication of the human Animal.* Namur: Presse Universitaires de Namur. pp. 107–115.

Hoffmeyer, Jesper. (1992). "Some semiotic aspects of the psycho-physical relation: The endo-exosemiotic boundary." In T. S. a. J. U. Sebeok (Ed.), *Biosemiotics: The Semiotic Web* (pp. 101–124). Berlin: Mouton de Gruyter.

Hoffmeyer, Jesper. (1996). *Signs Of Meaning In The Universe.* Bloomington: Indiana University Press.

Hoffmeyer, Jesper. (1998). "Semiosis and Biohistory: A Reply." In *Semiotics in the Biosphere: Reviews and Rejoinder.* Special issue of *Semiotica* Vol. 120. no. 3/4, pp. 455–482, 1998.

Hoffmeyer, Jesper. (2000a). "The Biology of Signification." *Perspectives in Biology and Medicine, 43*(2), 252–268.

Hoffmeyer, Jesper. (2000b). "Life and reference." *Bio systems, 60*(1), 8.

Hoffmeyer, Jesper. (2001). "Seeing virtuality in nature." *Semiotica, 134*(1–4), 18.
Hoffmeyer, Jesper. (2002). "Code Duality Revisted." *Semiotics, Evolution, Energy and Devlopment Journal* Vol 2 (1), pp. 98–117
Hoffmeyer, Jesper and Emmeche, Claus (1991). "Code-duality and the semiotics of nature." *On Semiotic Modeling*, Myrdene Anderson and Floyd Merrell (eds.), 117–166. Berlin, Mouton de Gruyter.
Hofkirchner, Wolfgang (2002) "The Status of Biosemiotics." *Semiosis, Energy, Evolution And Development Journal* 3, pp. 4–15
Hofstadter, Douglas. (1979). *Gödel, Escher, Bach: An Eternal Golden Braid.* Middlesex: Penguin.
Huber, Johannes and Schmid-Tannwald, Ingolf. (2005). *Epigenetic mechanisms following mammalian fertilization reveal basic principles of constructivist epistemology.* Paper Presented at the International Gatherings in Biosemiotics Conference 5, Urbino University, Italy. July 20–24.
Igamberdiev, Abir. (1999). "Foundations of metabolic organization: Coherence as a basis of computational properties in metabolic networks." *BioSystems* 50(1): 1–6.
Jämsä, Tuomo (2005). *Language and Nature.* Magisterial Lecture Presented at the International Gatherings in Biosemiotics Conference 5, Urbino University, Italy. July 20–24.
Ji, Sungchul (2002). "Microsemiotics of DNA." *Semiotica* 138 (1/4):15-42.
Kandel, Eric; Schwartz, James and Jessell, Thomas. (2002). Principles of Neural Science. New York: McGraw-Hill.
Karatay, Vefa and Denizhan, Yağmur. (1999). "The non-decreasing character of complexity: A biological approach." *Proceedings of the Third International Conference on Computing Anticipatory Systems.* Volume 5. Liège: Belgium.
Kauffman, Stuart. (1995). *At Home In The Universe: The Search For Laws Of Self-Organization And Complexity.* New York: Oxford University Press.
Kauffman, Stuart. (2000). *Investigations.* Oxford; New York: Oxford University Press.
Kilstrup, Mogens. (1997). "Biokemi og semiotik: Molekylær biosemiotik." In: *Anvendt Semiotik*, pp. 97–123.
Kleisner, Karel (2004): *GENES – MEMES – SEMES. Towards the new concept of mimicry.* Paper Presented at the International Gatherings in Biosemiotics Conference 4, Charles University, Prague. June 1–5.
Kotov, Kaie. (2002)."Semiosphere: A chemistry of being." *Sign Systems Studies,* 30 (1), pp. 41–56.
Krampen, Martin (1981). "Phytosemiotics." *Semiotica*, 36.3/4, 187–209; reprinted Deely, Williams and Kruse. 96–103.
Kristeva, Julia. (1994). "On Juri Lotman." *Publications of the Modern Language Association 109* (3), pp. 375–376.
Kull, Kalevi. (1998). "Organism as a self-reading text: anticipation and semiosis." *International Journal of Computing Anticipatory Systems* 1, 93–104.
Kull, K. 1998. "On semiosis, Umwelt, and semiosphere." *Semiotica* 120(3/4), 299–310.
Kull, Kalevi (1999). "Biosemiotics in the twentieth century: a view from biology." *Semiotica* 127(1/4), 385–414.
Kull, Kalevi (1999a). "On the history of joining bio with semio: F.S.Rothschild and the biosemiotic rules." *Sign Systems Studies* 27, 128–138.
Kull, Kalevi (1999b). "Towards biosemiotics with Yuri Lotman." *Semiotica* 127(1/4): 115–131.
Kull, Kalevi (1999c). "A Teleology of the Estonian Research Tradition" talk given for the Baltic and Finnish Studies Association at Indiana University, November 9, 1999. Summary here: http://www.iub.edu/~bafsa/articles.html
Kull, Kalevi. (2000a). "An introduction to phytosemiotics: Semiotic botany and vegetative sign systems." *Sign Systems Studies, 28*, 326–350.
Kull, Kalevi. (2000b). "Organisms can be proud to have been their own designers." *Cybernetics & Human Knowing, 7*(1), 45–55.
Kull, Kalevi. (2000c). "Part V Closure, Information, and Thermodynamics – Active Motion, Communicative Aggregations, and the Spatial Closure of Umwelt." *Annals of the New York Academy of Sciences, 901*, 8.
Kull, Kalevi. (2001). "Jakob von Uexküll: An introduction." *Semiotica* 134(1/4): 1–59.

Kull, Kalevi. (2001). "Living forms are communicative structures, based on the organic codes." *Cybernetics and Human Knowing* 8(3): 91–94.
Kull, Kalevi (2003). "Thomas A. Sebeok and biology: Building biosemiotics." *Cybernetics and Human Knowing* 10(1): 47–60.
Kull, Kalevi. (2005). "A brief history of biosemiotics." *Journal of Biosemiotics* Vol 1 pp. 1–34.
Küppers, Bernd-Olaf. (2000). Information and the Origin of Life. Cambridge: MIT Press.
Lagerspetz, Kari. (2001): Jacob von Uexkull and the origins of cybernetics. *Semiotica 134* (1–4): 643–651.
Latour, Bruno. (1987). *Science in Action: How to Follow Scientists and Engineers through Society.*. Harvard University Press. Cambridge. Mass.
Latour, Bruno. (1999). *Pandora's Hope: Essays on the Reality of Science Studies.* Harvard University Press. Cambridge. Mass.
Lestel, Dominique. (2002). "Human/animal communications, language and evolution." *Sign Systems Studies, 30*(1), 201–212.
Lewontin, Richard. (1992). "The Dream of the Human Genome." *New York Review of Books,* May 28, pp. 31–40.
Llinás, Rodolfo. R. (2001). *I Of The Vortex: From Neurons To Self* (1st ed.). Cambridge, Mass.: MIT Press.
Lotman, Juri. (1984 /2005). "On the Semiosphere." *Sign Systems Studies, 33*(1), 215–239.
Lorenz, Edward. (1963). "Deterministic nonperiodic flow." *Journal of Atmospheric Sciences.* Vol.20 : 130–141.
Luure, Andreas. (2002). "Understanding life: Trans-semiotic analogies." *Sign Systems Studies,* 30 (1), pp. 315–326.
MacIntyre, Alisdair C. (1974 /1980). "Epistemological crises, dramatic narrative, and the philosophy of science." In G. Gutting (Ed.), *Paradigms and Revolutions: Appraisals and Applications of Thomas Kuhn's Philosophy of Science.* (pp. 54–74). Notre Dame: University of Notre Dame.
Maran, Timo (2003). "Mimesis as a phenomenon of semiotic communication." *Sign Systems Studies* 31.1, pp. 191–215.
Margulis, Lynn, & Sagan, D. (1987). *Microcosmos: Four Billion Years Of Evolution From Our Microbial Ancestors.* Boston: Allen &Unwin.
Markoš, Anton. (2002a). "Evolution, purpose, teleology." In Havel, Ivan and Markoš, Anton (Eds.) *Is There a Purpose in Nature? How to Navigate Between the Scylla of Mechanism and the Charybdis of Teleology.* Czech Republic: Vesmir. pp. 113–128.
Markoš, Anton. (2002b). "Purpose and biology." In Havel, Ivan and Markoš, Anton (Eds.) *Is There a Purpose in Nature? How to Navigate Between the Scylla of Mechanism and the Charybdis of Teleology.* Czech Republic: Vesmir. pp. 42–64.
Markoš, Anton. (2002c). *Readers Of The Book Of Life: Contextualizing Developmental Evolutionary Biology.* Oxford; New York: Oxford University Press.
Martinelli, Dario (2005). "A whale of a sonata – Zoomusicology and the question of musical structures." *Semiosis, Energy, Evolution And Development Journal* 2005 (1), p. 2–29.
Matsuno, Kochiro & Swenson, R. (1999). "Thermodynamics in the Present Progressive Mode and its Role in the Context of the Origin of Life" *BioSystems* 51, 53–61.
Maturana, Humberto & Varela, Francisco ([1st edition 1973] 1980). "Autopoiesis and Cognition: the Realization of the Living." Robert S. Cohen and Marx W. Wartofsky (Eds.), *Boston Studies in the Philosophy of Science* 42. Dordecht: D. Reidel Publishing Co.
McCord Adams, Marilyn. (1987). *William Ockham.* Notre Dame: Notre Dame Press.
McGinn, Colin. (1999). *The Mysterious Flame: Conscious Minds in a Material World.* New York: Basic Books.
Meier-Oeser, Stephan (2003), "Medieval Semiotics.", *The Stanford Encyclopedia of Philosophy (Winter 2003 Edition),* Edward N. Zalta (ed.). Available at: http://plato.stanford.edu/archives /win2003/entries/semiotics-medieval.
Merrell, Floyd (1996). *Signs Grow: Semiosis and Life Processes.* Toronto: University of Toronto Press.

Miller, Stephen A. and John P. Harley. (1994) *Zoology.* 2nd Ed. Dubuque, Iowa: Wm. C. Brown Publishers.

Minsky, Marvin. (1988). *Society of Mind.* New York: Simon and Schuster.

Nakajima, Toshiyuki (2005). *Managing uncertainty of events by semiosis in living systems.* Paper Presented at the International Gatherings in Biosemiotics Conference 5, Urbino University, Italy. July 20–24.

Ockham, William. (1323 /1991). *Quodlibetal Questions,* Alfred J. Freddoso and Francis.

E. Kelley, trans., 2 vols., *Yale Library of Medieval Philosophy*, New Haven: Yale University Press, 1991.

Neuman, Yair (2003) *Processes and Boundaries of the Mind: Extending the Limit Line.* New York: Kulwer Academic.

Oyama, Susan (1985). The Ontogeny of Information: Developmental Systems and Evolution. Cambridge: Cambridge University Press.

Pain, Stephen (2002) - *Introduction to biorhetorics: applied rhetoric in the life sciences.* Paper Presented at the International Gatherings in Biosemiotics Conference 2, the University of Tartu, Estonia. June 14–17.

Parmentier, R. J. (1994). *Signs In Society: Studies In Semiotic Anthropology.* Bloomington: Indiana University Press.

Pattee, Howard. (1965). Experimental Approaches to the Origin of Life Problem. In F.F. Nord (ed.) *Advances in Enzymology, Vol. 27.* New York: Wiley. pp. 381–415.

Pattee, Howard (1969). "The physical basis of coding and reliability in biological evolution." in *Towards a Theoretical Biology* 1, C. H. Waddington, ed. Edinburgh Univ. Press, pp. 67–93.

Pattee, Howard (1972). "Laws and constraints, symbols and languages." in *Towards a Theoretical Biology* 4, C. H. Waddington, ed., Edinburgh Univ. Press, pp. 248–258.

Pattee, Howard. (1977). "Dynamic and Linguistic Modes of Complex Systems." *International Journal for General Systems.* 3: 259–266.

Pattee, Howard. (1982). "The need for complementarity in models of cognitive behavior." In W. Weimer and D. Palermo (Eds.), *Cognition and the symbolic process*, Vol 2 (pp. 21–34). Hillsdale, NJ: Erlbaum.

Pattee, Howard. (1982a). "Cell psychology: An evolutionary approach to the symbol-matter problem." *Cognition and Brain Theory* 5(4), 325–341.

Pattee, Howard. (1988). "Simulations, realizations, and theories of life." In C. Langton (Ed.), *Artificial Life* (pp. 63–77). *Santa Fe Institute Studies in the Sciences of Complexity.* Reading, MA: Addison-Wesley.

Pattee, Howard (2001). "The Physics of Symbols: Bridging the Epistemic Cut." in *BioSystems* 60, 5–21.

Pattee, Howard (2005). "The physics and metaphysics of biosemiotics." *Journal of Biosemiotics 1*, 281–301.

Peirce, Charles. S. (1931–1935). *Collected papers of Charles Sanders Peirce* (Vol. 1–6). Cambridge, MA: Harvard University Press.

Peirce, Charles. S. (1958). *Collected papers of Charles Sanders Peirce* (Vol. 7–8). Cambridge, MA: Harvard University Press.

Petrilli, Susan, and Ponzio, Augusto (2001). *Thomas Sebeok and the Signs of Life.* Great Britain: Icon Books.

Ponzio, Augusto. (1990). *Man as a Sign: Essays on the Philosophy of Language.* Berlin: Mouton de Gruyter.

Poinsot, John (1632 /1985). *Tractatus de Signis.* With English translation by John Deely. Berkeley: University of California Press.

Prigogine, Ilya. (1969), *Structure, Dissipation and Life. Theoretical Physics and Biology*, Versailles, 1967. North-Holland Publ. Company, Amsterdam.

Prigogine, Ilya, & Stengers, I. (1984). *Order Out Of Chaos: Man's New Dialogue With Nature.* Toronto; New York, N.Y.: Bantam Books.

Queiroz, João, Emmeche, Claus and El-Hani, Charbel Niño (2005). "Information and Semiosis in Living Systems: A Semiotic Approach." *Semiosis, Energy, Evolution And Development Journal* (1), p. 60–90.

Reybrouck, Mark (2005). *The musical code between nature and nurture: Biosemiotic and ecological claims.* Paper Presented at the International Gatherings in Biosemiotics Conference 5, Urbino University, Italy. July 20–24.

Ribeiro, Sidarta, Queiroz, João, and Ivan de Araújo (2003): *The emergence of referential symbolic process in non-human primates communication: A zoosemiotic analysis based on the Peircean extended theory of sign.* Paper Presented at the International Gatherings in Biosemiotics Conference 3, the University of Copenhagen, Denmark. July 11–14.

Robert E. Ulanowicz (1986) *Growth and Development: Ecosystems Phenomenology.* San Jose, CA: Excel Press.

Rocha, Luis M. (2001). "Evolution with Material Symbol Systems." *Biosystems.* Vol. 60, pp. 95–121. LAUR 00–1604.

Roepstorff, Andreas. (2004) "Cellular Neurosemiotics: Outline of an interpretive framework." In Joachim Schult (ed.) *Studien zur Theorie der Biologie*, Vol. 6: *Biosemiotik - Praktische Anwendung und Konsequenzen für die Einzeldisziplinen.* VWB - Verlag für Wissenschaft und Bildung. Berlin.

Rosen, Robert (1991). *Life Itself: A Comprehensive Inquiry into the Nature, Origin, and Fabrication of Life.* New York: Columbia University Press.

Rothschild, Friedrich S. (1962). "Laws of symbolic mediation in the dynamics of self and personality." *Annals of New York Academy of Sciences* 96: 774–784.

Rüting, Torsten. (2004) *Jakob von Uexküll: Theoretical Biology, Biocybernetics and Biosemiotics.* Available at: www.math.uni-hamburg.de/home/rueting/UexECMTB.doc.

Rüting, Torsten (2003): *Uexküll's "Institut für Umweltforschung" - Biosemiotics in action?* Paper Presented at the International Gatherings in Biosemiotics Conference 3, the University of Copenhagen, Denmark. July 11–14.

Salthe, Stanley N. (1993). *Development and Evolution: Complexity and Change in Biology.* Cambridge (Mass.): The MIT Press.

Salthe, Stanley N. (2006). "Purpose in Nature." Available online at: http://www.nbi.dk/~natphil/salthe/Purpose_In_Nature.pdf.

Savan, David (1976). *Peirce's Semiotic.* Toronto: Victoria College.

Sebeok, Thomas A. (1963). "Communication among social bees; porpoises and sonar; man and dolphin." *Language* 39: 448–466.

Sebeok, Thomas A. (1969). "Semiotics and ethology." In: *Approaches to Animal Communication*, Thomas A. Sebeok and Alexandra Ramsay (eds.), 200–231. The Hague: Mouton.

Sebeok, Thomas A (1976). *Contributions to the Doctrine of Signs.* Lisse, Netherlands: The Peter de Ridder Press.

Sebeok, Thomas A. (1977). "Ecumenicalism in semiotics." In *A Perfusion of Signs*, Thomas A. Sebeok (ed.), 180–206. Bloomington: Indiana University Press.

Sebeok, Thomas A., ed., (1977). *How Animals Communicate* (Bloomington, IN: Indiana University Press).

Sebeok, Thomas A. (1977). "Neglected figures in the history of semiotic inquiry: Jakob von Uexkull." In *The Sign and Its Masters* (pp. 187–207). Lanham, MD: University Press of America.

Sebeok, T. A. (1980). *Speaking of Apes: A Critical Anthology of Two-Way Communication with Man.* New York: Plenum Press.

Sebeok, Thomas. A. (1986). *Encyclopedic Dictionary Of Semiotics.* Berlin; New York: Mouton de Gruyter.

Sebeok, Thomas A. (1986). *I Think I Am A Verb: More Contributions To The Doctrine Of Signs.* New York: Plenum Press.

Sebeok, T. A. (1988). 'Animal' in biological and semiotic perspective. In T. Ingold (Ed.), *What is an Animal?* (pp. 63–76). London: Unwin Hyman.

Sebeok, Thomas A. (1990) *Essays in Zoosemiotics.* Toronto: Toronto Semiotic Circle.

Sebeok, Thomas A. (1990a). "Sign science and life science." In J. Deely (Ed.), *Semiotics 1990* (pp. 243 – 252). Lanham, MD: University Press of America.

Sebeok, Thomas A. (1995). Into the rose-garden. In J. Deely (Ed.) *Thomas A. Sebeok: Bibliography 1942–1995* (pp.116–125). Bloomington: Eurolingua.

Sebeok, Thomas A. (1998). "The Estonian connection." *Sign Systems Studies* 26, 20–41.

Sebeok, Thomas A. (1999). "Biosemiotica I." ed. with "Biosemiotica II." guest-ed. Jesper Hoffmeyer and Claus Emmeche, Special Issue *Semiotica* 127-1/4.

Sebeok, Thomas A. (2001). "Biosemiotics: Its roots, proliferation, and prospects." *Semiotica, 134*(1), 18.

Sebeok, Thomas A. (2001a) *The Swiss Pioneer: In Nonverbal Communication Studies, Heini Hediger (1908–1992)*. Ottawa: Legas Press.

Sebeok, Thomas A. and Umiker-Sebeok, Jean, (1980). *Speaking of Apes: A Critical Anthology of Two-Way Communication with Man*. New York: Plenum.

Sebeok, Thomas A. and Rosenthal, Robert, eds., (1981). *The Clever Hans Phenomenon: Communication with Horses, Whales, Apes, and People*. New York: The New York Academy of Sciences.

Sebeok, Thomas A and Umiker-Sebeok, Jan. (eds.) (1992). *Biosemiotics: The Semiotic Web 1991*. Berlin: Mouton de Gruyter.

Sercarz, Eli E., Celada, Franco, Mitchison, N. Avrion and Tada, Tomio (eds.) (1988). *The Semiotics of Cellular Communication in the Immune System*. Berlin: Springer.

Shannon, Claude and Weaver, Warren. (1949): *The Mathematical Theory of Communication*. Urbana: University of Illinois Press.

Sharkey, Amanda J.C. (1999) *Combining Artificial Neural Nets: Ensemble and Modular Multi-Net Systems*. Springer-Verlag.

Sharkey, Noel. E. (2002), "Biologically Inspired Robotics." In M.Arbib (Ed), *Handbook of Brain Theory and Neural Networks*, Cambridge: MIT press.

Sharov, Aleksei. (1992). "Biosemiotics: A functional-evolutionary approach to the analysis of the sense of information." In *Biosemiotics: The Semiotic Web 1991*, T.A.Sebeok and J.Umiker-Sebeok (eds.), 345–373. Berlin: Mouton de Gruyter.

Sharov, Aleksei. (2002). "Pragmatics and biosemiotics." *Sign System Studies 30* (1), pp. 245–258.

Skibinski, Adam. (2004). "Metalogy: A Commentary on Mind, Recursion and Topological Inference." *Semiosis, Energy, Evolution And Development Journal* 2004 (1), p. 70–90:

SLIS Press News (2002) *Thomas A. Sebeok, Senior Fellow at SLIS, Passes On*. School of Library and Information Science News, Indiana University. Available at: http://www.slis.indiana.edu/news/story.php?story_id=364.

Sonea, Sorin (1992). "Half of the living world was unable to communicate for about one billion years." In: *Biosemiotics: The Semiotic Web 1991*, T.A. Sebeok and J. Umiker-Sebeok (eds.), 375–392. Berlin: Mouton de Gruyter.

Spade, Paul Vincent, (1994). *Five Texts on the Mediaeval Problem of Universals: Porphyry, Boethius, Abelard, Duns Scotus, Ockham*. Indianapolis, Ind.: Hackett. Includes a complete translation of Ockham's discussion of universals from *Sent*. I.2.4–8.

Steckner, Cornelius (2004): *Environmental misfit in vision and grasp*. Paper Presented at the International Gatherings in Biosemiotics Conference 4, Charles University, Prague. June 1–5.

Stepanov, Yuri S. (1971). *Semiotika*. Moscow: Nauka.

Stjernfelt, Frederik (1992). "Categorical perception as a general prerequisite to the formation of signs." In: *Biosemiotics: The Semiotic Web 1991*, T.A. Sebeok and J. Umiker-Sebeok (eds.), 427–454. Berlin: Mouton de Gruyter.

Stjernfelt, Frederik. (2000). "Mereology and semiotics." *Sign Systems Studies, 28*, 73–98.

Stjernfelt, Frederik. (2002). "Recollections." In Emmeche, Claus; Kull, Kalevi and Stjernfelt, Frederik (eds). *Reading Hoffmeyer, Rethinking Biology*. Tartu: Tartu University Press. pp. 57–60.

Taborsky, Edwina. (1998). *Architectonics Of Semiosis* (1st ed.). New York: St. Martin's Press.

Taborsky, Edwina. (2001a). *The internal and the external semiosic properties of reality*, from http://www.library.utoronto.ca/see/SEED/Vol1-1/Taborsky-Journal1.html.

Taborsky, Edwina. (2001b). "What is a sign?" *Journal of Literary Semantics, 30*(2), 12.

Thom, Rene (1989) *Structural Stability and Morphogenesis an Outline of a General Theory of Models*. Boulder, Co.: Westview Press.

Thompson, D'Arcy Wentworth (1917). *On Growth and Form*. Cambridge: University Press.

Thorndike, Edward. L. (1898). "Animal Intelligence: An Experimental Study of the Associative Processes in Animals." *Psychological Review, Monograph Supplements*, No. 8. New York: Macmillan.

Tochon, François V. (2000). "Review of The Swiss Pioneer in Nonverbal Communication Studies: Heini Hediger (1908–1992) by Thomas A. Sebeok." *International Journal of Applied Semiotics*, Volume 3, Number 1, pp. 137–143.

Trolle Borup, Mia (*nee* Krause, Mia) (2005) *Genotype-phenotype relations. On the modeling of biological phenomena in genetics and molecular biology.* Doctoral Dissertation, Niels Bohr Institute, University of Copenhagen.

Turovski, Aleksei. (2000). "The semiotics of animal freedom." *Sign Systems Studies, 28*, 380–387.

Umerez, Jon (2001). "Howard Pattee's theoretical biology: a radical epistemological stance to approach life, evolution and complexity" *Biosystems*. Vol 60, pp. 159–177.

Varela, Francisco J.; Maturana, Humberto R.; & Uribe, R. (1974). "Autopoiesis: the organization of living systems, its characterization and a model." *Biosystems* 5 187–196.

Vehkavaara, Tommi. (2002). "Why and how to naturalize semiotic concepts for biosemiotics." *Sign Systems Studies 30* (1), pp. 293–313.

Vehkavaara, Tommi. (2003). "Natural self-interest, interactive representation, and the emergence of objects and *Umwelt*: An outline of basic semiotic concepts for biosemiotics." *Sign Systems Studies 31* (2), pp. 547–587.

Vehkavaara, Tommi. (forthcoming). "Limitations on applying Peircean semeiotic." *Journal of Biosemiotics.* (to appear).

Villa, Alessandro E.P. (2005). *The Neuro-Heuristic paradigm.* Paper Presented at the International Gatherings in Biosemiotics Conference 5, Urbino University, Italy. July 20–24.

Von Foerster, Heinz (1982) *Observing Systems.* Intersystems Publications.

von Glasersfeld, Ernst (1987) *Construction of Knowledge.* Intersystems Publications.

von Uexküll, Jakob. (1905). *Leitfaden in das Studium der Experimentellen Biologie der Wassertiere.* Weisbaden: J.F. Bergmann Verlag.

von Uexküll, Jakob. (1909). *Umwelt und Innenwelt der Tiere.* Berlin: Julius Springer Verlag.

von Uexküll, Jakob. (1926). *Theoretical Biology.* Trans. By D.L. Mackinnon. London: Kegan Paul.

von Uexküll, Jakob. (1928). *Theoretische Biologie (Second Edition.)* Berlin: Julius Springer Verlag.

von Uexküll, Jakob. (1934 /1992). "A stroll through the worlds of animals and men: A picture book of invisible worlds." In *Semiotica* 89(4), 319–391.

von Uexküll, Jakob. (1940 /1982). "The theory of meaning." *Semiotica, 42*(1), 25–87.

von Uexküll, Thure. (1986). Medicine and Semiotics. *Semiotica* 61:201–217.

Waddington, Conrad Hal, 1905–1975, ed. *Towards a theoretical biology. An IUBS Symposium.* Chicago: Aldine Pub. Co. 1968–1972.

Weber, Andreas. (2001). "Cognition as expression: The autopoetic foundations of an aesthetic theory of nature." *Sign Systems Studies, 29*(1), 153–168.

Weber, Bruce & Deacon, Terrence. (2000). "Thermodynamic cycles, developmental systems, and emergence." *Cybernetics & Human Knowing, 7*(1), 21–43.

Witzany, Günther (2000) *Life: The Communicative Structure. A New Philosophy Of Biology.* Norderstedt: Libri Books.

Yip, Maricela and Madl, Pierre (2005). *Semiosis aspects of ecosystems of the invasive Caulerpa taxifolia.* Paper Presented at the International Gatherings in Biosemiotics Conference 5, Urbino University, Italy. July 20–24.

Ziemke, Tom (2003). "Robosemiotics and embodied enactive cognition." *Semiosis, Energy, Evolution And Development Journal* (3), p. 112–124:

Ziemke, Tom and Sharkey, Noel (2001). "A stroll through the worlds of robots and men: Applying Jakob von Uexkull's theory of meaning to adaptive robots and artificial life." *Semiotica* 134 (1/4): 701–746.

CHAPTER 2

SEMIOSIS IN EVOLUTION

TUOMO JÄMSÄ

Universities of Joensuu and Oulu, Finland, t.jamsa@dnainternet.net

Abstract: The essay puts forward an adaptation of the Peircean model of semiosis. The interpretant is a higher monitor in the model watching over the semiosic communication between the object and the representamen, integrating the intrinsic asymmetry (cf. Lotman) between the two and giving the interpretations. To justify the broad range of semiosis, it is important to ground the hypothesis of the cosmological evolution and to show how life emerges from the physicochemical basis. Semiosis in life processes is dealt with in more detail. The assumption of the semiotic closure and epistemic cut by Pattee is accepted, in principle, but mainly reduced to the intrinsic dissymmetry between the object that represents potentials in the model and the representamen that stands for the actual changes and things in it. The dichotomy of energy and matter roughly demonstrates the dialectic between object and representamen at the physical level

Keywords: semiosis, semiotic model of nature, semiotic closure, epistemic/ontic cuts

BIOSEMIOTICS FROM A PERSONAL POINT OF VIEW

Biosemiotics has no clear date and place of birth. In my view, the year 1940 might be regarded as the first great landmark in the field. Jakob von Uexküll's *Bedeutungslehre* ('The theory of meaning,' 1982) was published then. The German title of the book, without any definite or indefinite article, suggests that the author has possibly thought of his contribution as a general theory of meaning. von Uexküll was a biologist and his scientific career is reflected in the book: the basic scheme of communication stems from his concept of 'Funktionskreis' (function cycle) that he developed further from 1909 to his small masterpiece mentioned above through his death in 1944. His two editions of *Theoretische Biologie* from 1920s stand for his ambition to see meaning as the key concept of life. *Bedeutungslehre* summarizes his theory of meaning in a bit over 60 pages.

M. Barbieri (ed.), Introduction to Biosemiotics, 69–100.

It fulfills the most important function of biosemiotics: it accounts for the events of life as an interpretation of sign and meaning. The posterity has placed a high premium on von Uexkülls's work. At the turn of the millennia a special issue of *Semiotica* with over 800 pages and about 40 articles was published in honor of him (Kull 2001).

Jakob von Uexküll did not know semiotics; he simply made up his own biological theory of meaning apparently without having even the faintest idea of the ties between his theory and semiotics. He had a great precursor, Charles Sanders Peirce. Peirce's life work was to construct an all-embracing theory of semiotics on the basis of logic and philosophy. A contemporary of his was the father of structuralism, Ferdinand de Saussure, who expanded the analysis of language into all kinds of signs. On my scale, Peirce is definitely one of the few giants in philosophy. He was so far ahead of his time that it would take half a century before his originality was properly realized. As a theorist of language, de Saussure was in a class of his own. Compared to Peirce, he was yet narrower.

The first to explicitly apply semiotics into animal behavior was Thomas A. Sebeok. He had learnt to know semiotics and Peirce already in the 1930s, from his professor in linguistics, Charles W. Morris, and then from Roman Jakobson. Jakobson is often characterized as the linguist of the century, alongside with de Saussure, of course. After his move to the United States, he delivered his first series of lectures in his new home country soon after the outset of World War II. The lectures were mainly focused on semiotics and Peirce. One great aim in Jakobson's life in North America was, he has told himself, to introduce both to the American academics. He made his rounds at various universities giving a great part of his lectures on Peirce. It is partly down to him that little by little, Peirce was known still better.

In his two-volume book from 1971, Günther Tembrock used the term 'Biokommunikation' for animal behavior. Sebeok had published his *Animal Communication* in 1968 but dubbed the semiotic research of it 'zoosemiotics.' In the mid-80s, I worked as a visiting associate professor for Finnish language and culture in Bloomington Campus of Indiana University. Sebeok was originally a Finno-Ugrist and had started the teaching of Finnish in Bloomington in the early 1940s. Later, he had founded a Research Center for Semiotics and headed it ever since. He still belonged to the faculties of our Department of Ural-Altaic Studies, an exotic department with Finnish, Estonian, Hungarian, Turkish, Mongolian and Tibetan staffed by experts on each field. One of the professors for Tibetan was Dalai Lama's brother. Practical life in the Peabody House, the home of the department, was one kind of global semiotics. Sebeok was starring in his research center and was famous for his brilliance as a teacher. He knew in person lots of high rank intelligentsia, Nobel Laureates included, and his lectures were inspiring, full of memories from the encounters and conversations with the 'high society,' anecdotes that gave a deeper experiential understanding of the thread of the lecture than sheer theoretical generalizations could ever have done. At that time, he seemed to be fully-booked for lectures abroad and now and then, the assistants deputized him.

Thomas A. Sebeok has been to semiotics a great gathering figure. He kept at the center of his worldwide semiotic web organizing connections and above all editing new publications of semiotics. As an editor, the number and quality of his publications is no doubt one kind of a world record. Zoosemiotics and later biosemiotics were his favorites. The success that biosemiotics has had since the early 90s, the particular shape standing out in semiotics, is partly due to Sebeok's share. He supported biosemiotics with all his heart. The last book of his, compiled together with Marcello Danesi (2000), deals with modeling primarily from a biosemiotic point of view. The book is one of the great mileposts in general semiotic theory.

Biosemiotics in the modern sense of the word originates in the 1980s. The first herald of it is Marcello Barbieri's article from 1981. Barbieri has been von Uexküll's kindred spirit. Like von Uexküll, he has developed a rich and unique semantic theory of biology alone without obviously knowing anything about semiotics or biosemiotics.

I would like to mention two key figures in the current biosemiotic movement: Jesper Hoffmayer and Kalevi Kull. They have organized meetings and coordinated biosemiotics in general and are much credited for the fact that biosemiotics has formed in the margin of biology and semiotics a discipline of its own. Semiotics typically works on boundaries. On one side, the location of biosemiotics in the margin has meant a joint ground of semiotics and biology. Thanks to semiotics, largely an empirical science has escalated into a new paradigm.

My first encounter with biosemiotics took place in the IASS conference at Berkeley in June 1994. By then I had not even heard about applying semiotics into the interpretation of the events in nature. Jesper Hoffmayer chaired the Berkeley session in his elegant style. The best were the comments and questions after the papers. Biosemiotics was to me, a semanticist, fascinating and completely new. Among the biologists, I felt an associate member but amateurism has given me much, too. I have taken part in the meetings and gatherings of the biosemioticians now and then in Imatra, Tartu, Copenhagen and Urbino.

CONCEPTUAL DEMARCATION

Above is a graph of the Peircean semiotics. All suchlike figures are adaptations because Peirce himself never introduced his model graphically. Besides, the origin of the figure is controversial because we cannot be confirmed the figure is authentic. Whatever the case, the graph serves us as an introduction to semiosis, especially in its usual sense.

As the two forks above the vertical in the graph indicate, the subject's interpretation concerns the process between meaning and sign. The definition of meaning, such as it appears in the diagram, follows the conception of Wittgenstein in his *Tractatus* (2001[1922]: 3.203]), "A name means an object. The object is its meaning.", but differs from the views of what the later Wittgenstein (1964: 69) wrote down, "The use of [a] word in practice is its meaning." The meaning in use is the cardinal semantic doctrine Wittgenstein proclaims in his most important

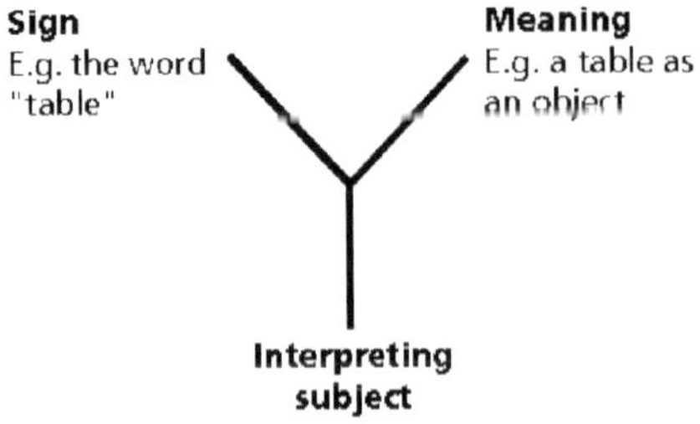

Figure 1. The Peircean model of semiosis according to Jesper Hoffmeyer, put forward by Arne Stjernholm Madsen (2006 [1999])

posthumous work *Philosophical Investigations* (1999 [1953]). As for the figure 1, equalizing meaning and object (cf. *Tractatus*) raises questions, from a semiotic point of view. Instead, the connection of sign and meaning is characteristic of semiotics.

The term 'interpreting subject' in the graph gives to understand that the biosemiotic premises hold true for the communication between an organism and its habitat. Jakob von Uexküll has introduced the name 'Umwelt' to represent the world an organism lives in. The life can be described as a continuous dialogue in the function cycle between an interpreting subject and its umwelt. Interpretation is composed of 'writing and reading,' of encoding meanings into signs and decoding signs into meanings. From the point of view of a reader not familiar with the Peircean doctrine of signs, it is necessary to point out that signs are not only morphemes (meaningful items) of language or paralinguistic traits of human or animal communication but also patterns or things in the umwelt called icons (images or the like) and indices (all kinds of entities inside and outside an organism). Semiotics terms the dialogue 'semiosis.' The Hoffmeyer graph (such as Stjernholm Madsen presents it) goes well for true semiosis between the subject and its umwelt. This sense of animal semiosis has been on view since Sebeok coined the term 'zoosemiotics' in the 1960s. The term 'zoosemiotics' is not behind the times, however. Sebeok and Danesi (2000) frequently use it in the context of organisms but not, of course, in that of molecules.

In the history of semiotics, physical and chemical phenomena are usually left outside while biological ones are included (see for instance Eco 1976: 6–). Sebeok (1986: 15) highlights the threshold between the animate and inanimate as a boundary between information and the true semiosis. The threshold of information and semiosis Sebeok speaks about is, however, questionable. How to divorce information from semiosis and meaning? Logicians like Yehoshua Bar-Hillel (1964: 221–310) and Jaakko Hintikka (1968) have applied the mathematical rules of information theory into language. In current information theory, codes can be understood as systems of information and a code may be defined as "a semiotic resource—a meaning potential—that enables certain kinds of meanings to be made" (Thibault 1998).

There seems to be no decisive dividing line between the presemiotic and semiotic worlds. All processes of the universe are events that realize in a certain context each. Therefore, they must be logically definite in the space-time they take place. The contents and rules of meanings have a general basis, composed of physical, chemical, logical, mathematical, etc. rules. We can call the foundation of meanings 'information' and the mathematical theory of meaning 'information theory.' It does not, though, change the fact: all events in the universe are necessarily contextual and represent semiosis.

The name 'Interpreting subject' in the above figure is relevant only in the relationship between an organism and its umwelt. At the utmost, it represents the view that genuine semiosis is exclusively based on a dialogue between subject and object, the physicochemical events are radically different and the interplay between physical and chemical entities is not semiosis but merely information. By all accounts, Sebeok distinguished semiosis from information in the spirit of his zoosemiotics. In case semiosis, by definition, concerns only living beings in the interactive roles with their umwelts, then it would be trouble-free to exclude everything outside the dialogue from semiosis and call it information.

Current biosemiotics deals with biophysics and biochemistry, with molecular processes of life that are ultimately based on subatomic (quantum mechanical) and chemical substances and the biological factors maintaining life. We cannot draw a clear divide between the 'dead nonsemiosis' and the semiosis of life. Life represents no doubt a new emergent level but it has not come into being from scratch and there is a certain link between the physicochemical and biological processes.

The prevailing biosemiotic paradigm seems to imply that everything in the context of life, down to molecular biology, biochemistry and biophysics, would meet the criterion of semiosis. When it comes to merely physicochemical processes apart from life most semioticians obviously leave them out of sign discourses. As for this issue, one may come up against uncertainty at least if not even determined resistance.

The boundary between the animate and the inanimate is blurred. According to one estimate, there are more than 400 different definitions of life (Adams 2004: 232–236). Organisms, plants and fungi stand for organic processes and life but they represent also the inorganic world out of which life has emerged and into which, in season, it falls back. The close interplay of the live and dead straddles all organisms. The vital bodily functions have a chemical and physical foundation. The organic and inorganic are inseparably intertwined. The interconnectedness of the two will be seen from a more coherent point of view in the next section.

LIFE IN THE FRAMEWORK OF COSMOLOGY

The next analysis is based on a hypothesis, a thought experiment. Traditionally, the only distinct boundary between semiosis and nonsemiosis (or information, as Sebeok characterized it) is regarded to mark the territory where organisms, plants and fungi are engaged in their dialogues with the umwelts. The name 'interpreting subject' of the above figure can be – as already mentioned – interlinked exclusively with the lords of their habitats.

The biochemical processes biosemiotics examines, as well, are different. They don't have any interpreter watching the process concerned and seem not to make decisions to reach a certain result; they are self-sustaining and mostly automatic and as such don't decisively differ from chemical and physical events apart from life. To withstand the test imposed by the general scholarly demands, the premises of biosemiotics aimed at the study of biochemical processes principally or in a certain respect alike to the subject–object dialogue should be clarified.

While following the current chemical and molecular trends in biosemiotics this essay tries to expand the semiotic principles of explanation into the evolution of our universe as a whole. The evolution of the universe cannot be distinguished from the single kinds of evolution called cosmological, physical, chemical and biological evolutions. All kinds of evolution are inseparably bound together. The different names reflect the difference of professional angles on the research subjects. The autonomy of each science cannot be impugned: there are the universe, the physical laws, the elements, the compounds and life. From a semiotic point of view, sciences form a whole, anyway.

The cosmological evolution follows the physical laws and there would be no universe without matter and energy. Life is possible only in a space-time continuum. The three dimensions of space and the dimension of time can be realized merely in the world of matter and energy. So does life, too. Amino acids are the necessary building blocks of life. In the lab, amino acids come in mirror-image pairs. Both forms – the left- and the right-handed ones – behave identically in most chemical reactions but biological processes use only the left-handed variety. This may refer to water in which amino acids and life emerged. Water still favors the left-handedness, probably due to the magnetic properties of it (Shinitzky and his research group 2006). The title of the research and the plausible background of it elegantly show how physical, chemical, biological and implicitly even cosmological properties are deeply interwoven.

I will try to evidence that the present kind of semiotic procedure applied into the processes of life relevantly coheres with the analysis of cosmological, physical and chemical events. Therefore, it is well grounded to show how the biological evolution – though on a new emergent step of development – results from the intrinsic implications out of the background of the earlier cosmological evolution with its physical and chemical underpinnings. There seems to be a predisposition to physical, chemical and even biological evolutions since the moment of singularity with the Big Bang: the unfolding of the universe from the size of a subatomic flicker to billions of light years across within the first trillionth of a second after its cataclysmic birth (WMAP 2006). The predisposition to life can be predicted from the inherent features of the universe at the beginning of its story.

The idea of the fine-tuned universe is based on the assumption that there are physical constants relating to each other in the fashion required for the universe's hospitality for life. Any small change in the twenty or so physical constants would make the universe radically different and life impossible. For example, if the strong nuclear force had been 2% stronger, it would have made the stellar development completely unthinkable and, of course, prevented the universe from setting up life. The anthropic cosmological principle (Barrow & Tipler 1986) points out how the constants seem to be tailored for intelligent life forms to exist. If any of the fundamental physical constants were sufficiently different, then no life would have emerged. The life such as we know it is based on the organic chemistry of carbon but it concerns only the life such as we know it. According to an alternative biochemistry, the functions of life might be similarly based on inorganic chemistry and on elements like silicon, nitrogen or phosphorus.

The emergence of life is bound with the laws of physics and chemistry. As astrophysicist Fred Adams (2002: chapt. 6 & 7) argues, life is a natural outcome of complexity supported by our universe. Sometimes in future is the next step of complexity, far beyond our grasp, probable to come. The Milky Way seems to harbor a myriad of planets with the same physicochemical conditions of habitability as Earth. Adams draws the conclusion that the number of viable habitats is truly enormous.

The key to the evidence of the interconnectedness between the physicochemical properties of substances and the emergence of life probably lies in chemosynthesis. In chemosynthesis, inorganic compounds are synthesized into organic amino acids, lipids, sugars etc. called chemosynthetic autotrophs or chemotrophs. They use hydrogen as a source of electrons for reducing carbon dioxide to food and giving off methane ("marsh gas," CH_4) as a byproduct. Hydrothermal vents, cracks in the deep ocean floor where chemosynthesis sustains the life of such organisms like tube worms, yellow mussels, clams and pink sea urchins are known since 1977.

There are geothermally heated basins of water beneath the sea floor, too, and plenty of chemosynthetic life forms in them. Prokaryotic biomass of bacteria and Archea exceeds 10^5 microbial cells/cm^3 in deep marine sediments even at depths close to 1,000 m below the seafloor. Extrapolation of these numbers to a global scale indicates that these deeply buried cells may represent one-tenth to one-third of living biomass on Earth. The researchers state how "despite the vast contribution of living biomass, relationships between the microbial community structure and distribution and the geophysical and geological conditions in subseafloor environments remain largely unknown." These habitats thrive thanks to geothermic heat. The proportions of the biomass on seafloor and beneath give a strong evidence for where life originates from (Inagaki & al. 2006).

U.S. Geological Survey scientist Frank Chapelle and his research group (2002) studied an Archaea community deep in the subsurface source of a hot spring in Idaho – in the first documented case of such a microbial community. No other known life form could exist under that kind of harsh conditions. Archaea live on a diet of hydrogen and carbon dioxide in anaerobic circumstances. Methanogens degrade contaminants at chemical and oil spills and produce methane. Scientists think Archaea-type microorganisms could exist in worlds without organic carbon (e.g., Mars and one of its moons, Europa). Archaea are prevalent in Earth's subsurface where oxygen does not exist. They might occur also on Enceladeus, one of the tiniest moons of Saturn. In March 2006, the orbiting Cassini spacecraft spotted water geysers on the icy moon. It has been added to the list of places within the solar system most likely to harbor extraterrestrial life. Besides water, life needs also a source of heat. Such a source is so far unknown on Enceladeus.

Research suggests that life has emerged deep inside Earth (Pueschel. 2006), the oceans above and in interstellar space. Hyperthermophilic methanogens, members of Archea, can exist in a biosphere up to 110°C six kilometers deep. Hyperthermophilic methanogens belong to the domain of Archea and Archea with their simple genetic systems could have originated at a high temperature deep inside the Earth. Hyperthermophiles like Crenarcheota can grow at 121°C (the same temperature in the autoclaves used to sterilize, for instance, surgical instruments). Adams (2004: 229) refers to species of Archea that survive even at the temperature of 170°C. Archea might have formed one of the earliest long-lived habitats. The symbiosis of methane gas and Archea hints at the possibility of subsurface life among other planets of the solar system. Many of them (like Mars and Europa) seem to have suitable subsurface conditions although totally inhospitable surfaces.

The prokaryotes and Archea were found in the context of the first 'black smokers,' the undersea vents. Initially, Archea were defined as bacteria. The sequence of their ribosomal RNA showed, however, that they bore a close relationship to the eukaryotes, the cell line even higher animals like humans belong to. There is certain evidence for the assumption that chemosynthesis, the link between the inorganic and organic, the inanimate and animate, might have come into being in the life of Archaea.

In March 2006, New Scientist published a cover story entitled "Alien Rain. Every drop contains life but not as we know it." The story (Muir 2006: 34–37) focuses on the red rains in the Indian state of Kerala, in 2001, and backgrounds them with references to earlier rains alike in the English port of Great Yarmouth and in Mexico. Godfrey Luis, a physicist at New Delhi, links them with microbes hitching a ride to Earth on a comet. In a preprint of their article, Luis and a colleague of his, A. Santhosh Kumar (2006), state that the major elements in the rain particles are like cells with carbon and oxygen. But they are not ordinary cells because they don't have any DNA. The analysis is still unfinished but it may support our claim of the universality of life in space. The hypothesis of the origin of life in space is confirmed lately. Chemical compounds appropriate to life have been detected in the analysis of interstellar dust clouds (Ehrenfreund & Menten 2003). The constellation of the Swan, for instance, includes chains of carbon, carbon monoxide and dioxide and water.

Life seems to be much more common in the universe than usually anticipated. Recently, the mindset has gained new ground. Freeman Dyson (1979, see also Adams & Laughlin 2002: 29) has put forth a scaling hypothesis for abstract life forms. In principle, some kinds of life forms thrive at any temperature. The term 'scaling' refers here to the assumption that the rate of energy use and the type of consciousness is in proportion to the prevailing temperature. Dyson's hypothesis is in agreement with one of his basic principles, "Life resides in organization, not in substance" (Dyson 1985 quoted in Harold 2003: 12).

EVOLUTIONARY SEMIOSIS

My semiotic application departs, as already stated, from the assumption that the idea of evolution concerns not only life but also the evolution of the universe in general. As given to understand above, the cosmological evolution is in key and includes – despite the fact that the physicochemical circumstance cannot, in fact, be separated from it – the physical, chemical and biological kinds of evolution. The semiotic model appears in figure 2 and is faithful to the Peircean thought. The interpretation may be, though, different because the semiosis in focus takes place in the cosmological framework. The object of the semiotic application is theoretical. The following discourse tries to answer a big question: What is our world and the life of it generally like, from a semiotic point of view? It goes without saying that the reply does not relate to the character of nature as such but to the assumed kinds of semiosis in it.

The three segments in figure 2 stand for three vectors that make up a space. The vectors – called 'object,' 'representamen' and 'interpretant' – don't represent a single event that would move on from object to representamen across the interpretant but an infinite cyclic process of events. The three take part in the cycle equally and, in practice, simultaneously and the processes have no fixed direction. The role of the interpretant is more crucial than the other ones. The interpretant takes control over semiosis. The weight of the interpretant is illustrated in the figure by the fact that the object and representamen vectors are branches of the interpretant vector. The location of the interpretant simultaneously depicts the logical status the interpretant has in relation to object and representamen. The interpretant involves and implies the two.

The terms in the figure below are originals of the Peircean model. Peirce (CP 2.228) characterized them in a fashion semioticians know almost by heart: "A sign, or *representamen*, is something which stands to somebody for something in some respect or capacity. It addresses somebody, that is, creates in the mind of that person an equivalent sign, or perhaps a more developed sign. That sign which it creates I call the *interpretant* of the first sign. The sign stands for something, its *object*. It stands for that object, not in all respects, but in reference to a sort of idea, which I have sometimes called the *ground* of the representamen." Peirce

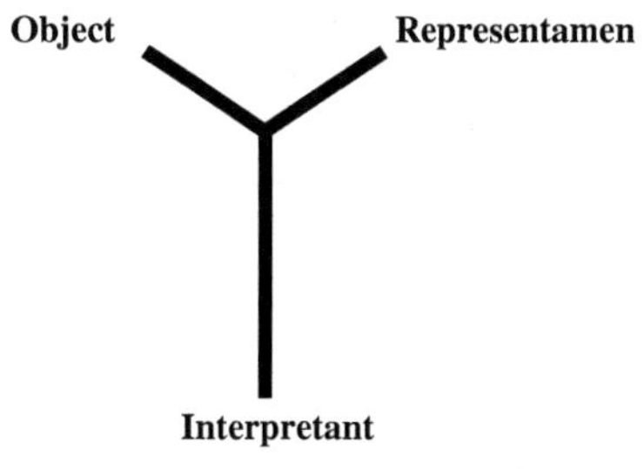

Figure 2. An adaptation of the semiotic model of Peirce

himself widened the term 'mind' by speaking about 'quasi-mind, for instance in his letters to Lady Welby. The 'Quasi-Mind' gave him a welcome opportunity to see the whole university as a seat of semiosis. In his writings from 1906 (GP 4.551) is included a most telling passage, "Thought is not necessarily connected with a brain. It appears in the work of bees, of crystals, and throughout the purely physical world; and one can no more deny that it is really there, than that the colors, the shapes, etc., of objects are really there."

In the following analysis, the traditional Peircean terms are used, though the meaning definitions of them may differ. Semiotics is a tool of interpretation that, in turn, is determined by the way we piece together the meanings underlying each term and apply these meanings into the discourse to be analyzed. In this adaptation, the object and the representamen are depicted in parallel vectors. The fork of vectors refers to a central idea in the Aristotelian philosophy, the dialectics between the potential and the actual, and is quite close also to Peirce himself who was influenced by Aristotle. The object is, by definition, something potential while the representamen (representation) is an actual sign or simulacrum of it. We could illustrate the relationship between the two by comparing it to a sentence written by an author or articulated by someone at a meeting. The meanings of the sentences concerned reflect in the minds of the author or panelist as 'objects.' They are potential ideas. The sentences written or spoken are actual representamens of the potential objects. Crucially, the process between object and representamen is a transition of something potential into something actual.

When confronting the meaning capacity a written or spoken sentence may have with the actual sentences we concretely come to realize the differences between the potential meaning and the representation of it in a sentence. Sentences are things (like roads) in the traditional philosophical sense of the word; meanings again are hidden entities of mind – "opaque, elusive, minimal meanings," as John Simon, an American theater critic and author, once has put it. Meanings remind of the world in the state of evolving while their actual representations reside in the state of being; metaphorically, we might speak about the actual representamens as if they were mummified in a thing-like silence. To illustrate the dialectics between the potential and the actual by confronting the meaning of a sentence with the real sentence is an illustration of the world in a continuous process of creation. The example matches well to the interplay between object and representamen in general and suggests how evolution in many cases moves on like groping for right words. In line with this kind of interpretation, representamens would be closer to 'objects' in the traditional fixed sense of the word. Instead, the term 'representamen' is an excellent designation for representation. The representamen is something that stands for the original wealth of the semiotic object.

In addition to being potential, semiotic objects are also intentional. By 'intentionality' of nature I mean that all natural processes are directed towards something that is in harmony of the inherent rationality, the intrinsic order in a broad sense, of the universe. The processes of nature are intentional because they fulfill the boundary conditions and in the spirit of them, adopt a certain direction in every context. As for the object of the evolution, the intentionality is realized in a very narrow sense at the beginning. The systemic implicature provides that the grades of freedom multiply in the course of evolution. From this point of view, the birth of life has been a decisive event. In sum, intentionality is an omnipresent quality that comes from the integer in the context, in which the semiotic object exists. As we have just mentioned semiotic objects are something that are coming into being, not objects in the traditional sense. The intentional composition largely determines how they are going to shape up. To go on with our example of writing or speaking a sentence, the integer of the context is the author's or panelist's will to say something. Humans are at the level able to make fine-tuned choices, such as the writing or speech act will presume, not only the on–off decisions (to express himself or not) that are points of departure in making choices of all kinds of organisms, plants included, although the spectrum of decision-making is far larger.

The term 'autopoiesis' was coined by Maturana and Varela in 1973 (89), "- - - the space defined by an autopoietic system is self-contained and cannot be described by using dimensions that define another space. When we refer to our interactions with a concrete autopoietic system, however, we project this system on the space of our manipulations and make a description of this projection [cf. Dyson's maxim of life quoted above]." Autopoiesis is usually confined to living self-maintained systems but can be applied also to the universe. Ultimately, the autopoietic integer determines how the universe will work. The universal context and the integer of it comprise all other contexts and their integers. Thus, we cannot break the cosmological rules and their inclusive physicochemical and biological rules in the context, in which we aim to write or say a statement.

The idea of semiosis, in which semiosis is realized by the representations of initial potential intentional objects of the primary logical level of Peirce (Firstness) makes the semiotic model deep and applies particularly well to evolutionary processes. The whole story of evolution can be perceived as an interplay between the potential intentional objects, most often not visible at all but hidden and underlying, and the actual representations of them. The evolutionary travel of the potential objects into representamens may take eons to realize. Apparently, there are a lot of potentials that will never be materialized.

Semiosis is comparable to some kind of play. The interpretant includes the manuscript but the manuscript is far away from the usual dramas scripted for the stage. The manuscript of the play of evolution has no scripted dramatis personae and no fixed lines but the actors seem to improvise on the story all the time. There are no intermissions but great turning points instead, catastrophes for instance when the dinosaurs went extinct in the Jurassic. Not only individuals and species die but also whole kingdoms. Earth is 4.6 billion years old and the oldest known sediments of fossils near the South African coast show that life was thriving well almost four billion years ago (Adams & Laughlin 1999: 62–63). About 800 million years ago was born the 'Ediacara fauna,' named after the best finding place of the fossils in Ediacara Hills, Australia, relics of multicellular organisms, like pillows or air mattresses. Then suddenly, 540 million years ago, the Cambrian explosion took place and almost all the phyla known today emerged.

Death and life are perceived as opposites in individual biographies. In the drama of evolution, death and catastrophe are ultimately positive forces that contribute to the great idea of life whatever it might be. The manuscript of the drama of evolution is open. If we search for the thread in the drama of evolution in retrospect we strike on one plausible great idea – the idea of a development toward increasing complexity. It seems to be a built-in tendency not only in matter and energy but also and above all in life.

The interpretant contains the instructions for semiosis. The physical constants are included in them. They are really vital for our universe and the cosmological evolution of it and speak for the evidence of rational principles in evolution. The constants could be explained on the basis of the assumption of the multiuniverse. Tegmark (2004) deals with parallel universes. In addition to them, we could assume also successive ones and consider our home universe an offspring of a parent universe. This is, of course, possible but so far, we don't have any empirical evidence for it. That's why, we have to abandon it so far. The existence of the physical constants is beyond dispute and refers to the obvious fact that they are kept in the archive of the interpretant, in its memory.

The assumption of the existence of memory in the processes of life is put forward in the code theory of Barbieri and dates back to von Neumann's (1966) idea of self-replicating automata. Barbieri (2003: 82–91) speaks about the memory matrix and the memory algorithms. In addition to memory, we could attach also learnability, a correlate of memory, to the equipment of the interpretant. As an implication of our self-centerdness, we are habituated to think that only higher animals like us can have memory. Barbieri challenges this view. The recent brain research has found mirror neurons that are explained to be bound to our self-awareness. The cells responsible for memory seem to have the same capacity, not only in relation to the idea of self but also to the experiences of events.

The semiotic model seems to clearly show how the primordial universal circumstances and regularities have given rise to the natural history as a whole. This state of affairs is crucially due to the interpretant, which is, of course, included in everything, from Big Bang to the history of life. Organisms seem to have developed on the basis of the data in the archive of the interpretant. The interpretant is embodied in the brains or nervous systems of the organisms. The human brain is by far the most elegant interpretant developed ever and as such an object lesson about how a potentiality may become actual when time goes and the circumstances become favorable.

Yet a short summary of what we have learnt about the interplay between the objects and the representamens. Everything we can grasp or sense is actually a representamen. The greater part of the entities in the universe includes evolutionary representations. Theoretically, it is possible to think the whole existence is ultimately an expression of the potential. The apothegm in the second part of the 'Faust' by Goethe, "Alles Vergängliche ist nur ein Gleichnis" (All that is transient is but a likeness), would agree with such kind of a reality, assuming that the adjectives 'vergänglich' and 'transient' were synonymous with the adjective 'actual.' The distinction between potential and actual is far from clear-cut. Above, an intentional meaning and its utterance illustrated it. The meaning in relation to the utterance exemplifies the difference between the two. The actual or the spoken or written utterance has a physical shape and is thus concrete. The meaning intentions reflect the variety of possibilities, are largely subconscious and suggestive of the potential in all their vagueness. However, meanings are condensed in conceptual structures and belong as such to the representamens. Semiotic objects are abstract intentional potentials, no distinct shapes and impossible to be defined. The statement tries to demonstrate the process character of meanings.

The interpretant acts – as mentioned – on the boundary between the potential and the actual. It runs its task by applying certain theoretical principles to the process. Semiosis is a system and systems and chaos theories can be used to demonstrate the functional cycle of it. It forms complex systems like weather. The detailed knowledge of the weather on Earth at a certain moment of time is so far beyond grasp, though the system of weather as such and the interdependent variables of it are well known. Today, the 'butterfly effect' is almost a household phrase. It encapsulates the notion of 'sensitive dependence on initial conditions' in chaos theory. The term is coined by Edward Norton Lorenz, an American mathematician and meteorologist, who in 1972 delivered a talk at a meteorological conference entitled "Does the flap of a butterfly's wings in Brazil set off a tornado in Texas?" The provocative coinage has survived ever since. The system of semiosis is like weather. Like any other dynamical systems it sensitively depends on its initial conditions and approximately recurs towards the conditions over and over again. From the viewpoint of semiosis, there is no cardinal difference between the whole and the parts; the whole, of course, is more determining. The regulation of the interpretant equally concerns both. The same principles apply to individual cases and the system in general.

FUNDAMENTALS IN THE INTERPRETANT

The physical constants are an expression of the principles in accord of which the universe is made up. Our hypothesis about a continuous all-embracing cosmological evolution with life as the ultimate in it is based on the rationality of the evolutionary principles. These basic rules of evolution are only boundary conditions but underlie also the natural selection in the evolution of life. The border conditions don't predetermine everything. On the contrary, they allow a certain randomness and unpredictability of emergence.

In my view, the rational ingredients of the interpretant include the governance principles of general intentionality in evolution and the logical, mathematical and linguistic determinants. Intentionality is tentatively dealt with above. Intentionality is an inherent part of semiosis and can be, at the most general level, defined as being about something or representing something: the representamen cannot be imagined without being about any object. In this adaptation of the Peircean model, all objects represent something potential and potentials tends to be actualized in representamens. As stated above, the adjective 'potential' is largely synonymous with the 'intentional.'

Intentionality is in key in the speech act theory indirectly based on the *Philosophical Investigations* by Wittgenstein from 1999 (1953). Intentionality is inseparably bound to speech; it is impossible to think that any utterance might be unintentional. Thus, for instance, children like to speak pig Latin without meaning anything literal with it but there is an intention in their speech, nevertheless – a desire for speaking foreign language. The function cycle between an organism and its umwelt is always intentional. In this framework, we prefer the term 'semiotic acts' to the term 'speech acts.' The natural processes in all evolution might be regarded as semiotic acts with a general built-in intentionality in them, although we don't know what kinds of goals general intentions really tend to. From a systems theoretical point of view, the basic intentionality of nature might be its tendency to more complex evolutionary forms.

Mathematicians often wonder about how in principle quite simple formulas may have an amazingly wide range of applications. In systems theory, Mandelbrot sets or fractals are used to mirror especially the forms of life. A simple equation appears to be a key to an uncanny metamorphosis. The underlying mathematical factors of evolution form a cardinal part of the rational guiding principles of the interpretant. As I have assumed above, the interpretant as such can be considered a potential object tending to realize in the actual representations of life, especially in the brains and nervous systems and in mind. The explanation of mathematics as an embodiment of mind (see Lakoff & Núñez 2000) correlates with the assumption.

The mathematical generator is a sibling of the logical and linguistic principles in the interpretant. The tradition of the discussion dealing with the connection between mathematics, logic and language goes back to Classical Greece, flourishes in Scholasticism of the Middle Ages and reaches its modern apex in the era from the 1870s through today, from the German mathematician Gottlob Frege's contribution to the recent post-Wittgensteinian debate. Frege started his design of 'Begriffs-schrift,' the writing system aimed at explicating the conceptual relations that underlie in the natural languages, to find a logical metalanguage for describing the relations between numbers in arithmetic. Frege's point of departure often seems self-evident to us. He wished to deny Kant's dictum (Kant 1998: A51/B75), "Without sensibility, no object would be given to us" and claimed that numbers like 0 and 1 are objects but that they "can't be given to us in sensation" (Frege 1997: 121 (101). From the viewpoint of our analysis of the interpretant, numbers like any other mathematical sets and the rules governing them are abstract objects. They include in the theoretical apparatus underlying the evolution. Frege's trial of using first-order logic as a metalanguage of arithmetic failed. Logicians of today, however, hold that second-order logic might be used as such a metalanguage (SEP 2006). All in all, Frege's contribution has showed that mathematics and logic are close relatives.

The connection of language and logic is more apparent. Frege has gone down in the history of semantics of language. Alike to Peirce, his model of semantic analysis is also triadic (Sowa 2000). Peirce's model applies to language, as well. As for the semantics of language, Peirce is practically left in the wilderness – an illustrative example of the scholarly camps with their prejudices prevalent still today and of the walls between them.

I have worked for about a decade since the mid-1970s to design a model for semantic analysis of verbs. The stage was closed in 1986 when a book was published (Jämsä 1986, see also Jämsä 1997). The study concerned the 150 most frequent Finnish verbs. Finnish, like any other language, has ways of its own kind to convey meanings. A later application of the model of analysis to English has shown that the basic model is repeated in it and probably in all other languages, too, but the lower-order qualifiers are used in English differently from Finnish – a highly predictable state of affairs because the languages belong to different phyla; Finnish is a Finno-Ugrian and English an Indo-European language. Languages (like Finnish and English) form an illustrative example of how evolution works. Languages differentiate from their original inner environments, their systemic predispositions, like organisms differentiate from their protoforms in the natural selection. Though evolution possibly involves emergence and the traces to the origin may grow dim, the basic point of departure remains, nonetheless.

My model of semantic analysis is easily applicable to semiotics. In the model, the meanings of verbs are enlarged upon nuclear phrases with the predicates in focus. The point of departure takes to relations, usually composed of three elements: the things, called in this context 'arguments' (representing subjects and objects), and the relator (representing predicates) between them. The logical formula xRy corresponds to the model. On the basic level, the meanings are described in strings of relations. The basic meanings are represented in the following typical strings (the arguments not appearing defined but only in letters): x CAUS (y PROC (y STAT z)) ~ x CAUS (PROC (x STAT y)); x PROC (x STAT y) ~ PROC (x STAT y); x STAT y. Noteworthy: arguments may vary differently. The symbol CAUS stands for 'causative.' The argument on the left of it is an agent that utilizes its energy to get something accomplished. The symbol PROC stands for "processive" and usually refers to a change while the symbol STAT stands for "stative" and refers to a state between x and y.

The string x CAUS (x PROC (x STAT y)) can be exemplified by a sentence like "The heavy rains have raised the level of water." The argument x, the heavy rains, have got an effect (x CAUS) on the change of the argument y, the level of water, so that it has a quality of z, reaching a level higher than before. The string like PROC (x STAT y) describes the sentence "A storm arose," for instance. The symbol PROC refers to a change resulting in the expression in parentheses, x STAT y, where a storm, x, has the quality of y, being existent. The sentence "About a fourth of the area of Finland lies inside the Arctic Circle" demonstrates the string x STAT y; x or about a fourth of the area of Finland has the quality of y or of being inside the Arctic Circle. In Finnish, sentences may have both a definite and indefinite variant of interpretation. For example, the sentence "Kissat syövät hiiriä" may have a definite interpretation (like in the English sentence. "The cats are eating mice") or an indefinite interpretation referring to a permanent quality of cats that they eat mice. In the indefinite case, English uses the sentence "Cats eat mice." Not only Finnish but also German and Swedish and probably most other languages don't know a particular predicate form to differ the definite from the indefinite.

The strings introduced represent only the skeleton of the system. In the final version of the system of the semantic description, there are 141 hierarchically arranged symbols that qualitatively define the main symbol STAT. The basic system refers to the physical space-time with matter and energy. The causative component of the system implies an entity, a source of energy that causes a process or a change resulting in a state, in which the qualitative representation, the qualifier, will be different. The physical model is widely applied to psychological or metaphorical contexts, in which the sources of energy are not physically but metaphorically actual. In the clause like "We have made a decision that," the x, we, is maybe no actual source of energy but is reflected as such and interpreted also as a causative argument (agent).

Semiotics has six components: material, morphological, syntactic, textual, semantic and pragmatic. The material component corresponds to the phonological one in language and contains the chemical (but also physical) elements in semiosis. The morphological component aggregates the combinations of the elements into meaningful units. The syntactic component arranges the morphological combinations into structures constituting propositions, i.e., structures with meaning. The textual component balances the sentences and the larger textual units, ultimately the discourse as a whole. Propositional meanings are represented in the semantic component. The system it follows is introduced above. The pragmatic component reflects the metalevel, the addresser's cognitive, emotional and attitudinal participation with the text. To go on with the paralleling of world to language, semiosis might be regarded as an expression of semiotic grammar. The codex of the semantic rules introduced above might be called the 'semantic grammar.'

The triad of energy, process and state forms the kernel of my model of semantics. Language is not the pattern the world mirrors; on the contrary, the physicochemical universe reflects in language. The theory of modeling is one of the key doctrines in semiotics. The notion of primary modeling was developed by the Moscow–Tartu School in the early 1960s (see Zaliznjak, Ivanov & Toporov 1988; Lotman 1988a,b). Sebeok and Danesi's book describes primary, secondary and tertiary modeling in more detail than anyone else so far. In the case of primary modeling they summon up the seminal work in the 1960s and 1970s stating that there exist two kinds of primary modeling, which they call 'osmosis' and 'mimesis' (Sebeok & Danesi 2000: 45). They say the conclusion is based on "biosemiotic research" subsequent to the Moscow–Tartu School but don't make any reference. The bifurcation into osmosis and mimesis takes place on the basis of intentionality; mimesis refers to the intentional making of forms in a simulative manner, osmosis is natural simulation. From the viewpoint of my model of semiotics, the division fails. Both natural and artificial simulation is intentional in it. Anyway, it comes down to primary modeling, to a model and the copies of it and the copies are icons of the model. The semantic grammar I have designed is modeled, as stated in the initial sentence of this paragraph, by the cosmological and especially the physicochemical essentials.

There is also interdependence between logic and language in my semantic grammar. The system of semantic description introduced is not, of course, a system of logic though it represents the logical basics underlying language. The mother of the word 'logic,' the Greek 'logos,' derives from the verb 'legein' (to say). 'Logos' is the result of the act of saying, the utterance or sentence. In linguistics, words and sentences (and utterances) are considered to be composed of the minimal meaningful units called 'morphemes.' The English translation of the sentence of Heraclitus, "Logos governs all things," comprises the morphemes 'logos,' '0' (singular), '0' (nominative, the subject or theme of the sentence), 'govern-' (the stem of the verb), '-s' (the sign of the 3rd person, which indicates the theme of the sentence is not the addresser or the addressee but somebody or something outside), 'all' (a quantifier), '0' (indifferent in relation to the grammatical number), '0' (an adjunct qualifying the following head), 'thing' (the stem of the noun), '-s' (plural), '0' (accusative, the object of the noun phrase before). The morpheme inventory is tentative but will give a rough idea of what morphemes are about. Besides, the interdependency of the morphemes should be described by syntactical means, which are familiar especially of the analysis model, usually conducted in tree diagrams, by Chomsky and his colleague grammarians. Syntax represents one kind of logic, too.

'Logos' is a deep metaphor. It reflects the division of language into two: the physical and sensible spoken or written discourse and the mental content in the background of the discourse. The thesis of Heraclitus, "Logos governs all things," (see Heidegger 1985: 71) might be applied also to the world model put forth here. The duality of physical matter such as it appears in the actual concrete signs and of spiritual substance such as it appears in the thought that underlies the physical discourse, deeply represents also the evolutionary framework of this essay. I propound – at least sketchily – a theory that the double articulated (or double-coded) language embodies the great principles that the universe realizes in its interplay of material shapes and dynamical forces. In parallel with double coding, the world divides in two: in cosmological, physical, chemical and biological manifestations that can be observed and in the rational rules that govern the manifestations and cannot be immediately observed but only intuited or inferred.

Popper, a great admirer of Pre-Socratics, particularly Heraclitus and Parmenides, translates 'logos' into "the true word" (in German, "das wahre Wort") (Popper 2005: 46). The commonest translation of 'logos' is 'word.' To understand the translation, we have to remember that 'word' (or many equivalents of it in other languages, like the Finnish 'sana') did not originally have the same precise meaning as it has today but referred then to a thought and its utterance. Heraclitus is famous for seeing the world being "in flux." Out of the four elements recognized in the early Classical Greece, he preferred Fire. It has yet a particular meaning as Heidegger emphasizes; it burns out the non-essentials and sheds light on the essentials, the basic structure of the world, its logos, described by Heraclitus in the phrase "hen kai pan" (one and all). In the middle of the continuous change and death and disaster, Heraclitus saw the 'logos,' or the reason and rationality, to represent something that remains, the underlying principle permanent in the illusory ever-changing world.

Departing from the claim that "the logos has man" and not the reverse, Heidegger (1979: 295) interprets Heraclitus by saying that the 'logos' unifies beings, brings them together and to their proper identity, "their original *Versammlung*." Thus, Heidegger emphasizes by implication that the human condition is a reflection of the natural condition and the latter embraces all other creatures, too, and everything. Language and the cognizant semiosis is based on a discovery, in the original sense of the word.

The following quote of Heraclitus (2006) illustrates his view of 'Logos.' Besides, it is a polemic address that draws us aside from the highway. In my view, the quote is, though, noteworthy also here as a document, which shows that from the point of view of everyman's philosophy, nothing has changed in the course of thousands of years: "Men have no comprehension of the Logos, as I've described it, just as much after they hear about it as they did before they heard about it. Even though all things occur according to the Logos, men seem to have no experience whatsoever, even when they experience the words and deeds which I use to explain physis, of how the Logos applies to each thing, and what it is. The rest of mankind are just as unconscious of what they do while awake as they are of what they do while they sleep."

THE CODES MAPPING SEMIOSIS

The popularity of the term 'code' in biology originally comes from those little heroic stories about code breaking at World War II. Marshall Warren Nirenberg, together with Har Gobind Khorana and Robert W. Holley, was awarded the Nobel Prize in 1968 for describing the genetic code and how it operates in protein synthesis. Nirenberg, Khorana and Holley explained the function of the genetic code and thus saw Watson and Crick's and Maurice Wilkins's (and some significant others') determination of the structure of DNA through to the finish. Watson, Crick and Wilkins had started their work in the early 1950s and shared the Nobel Prize in 1962.

The physicist Erwin Schrödinger (1992: coined the term 'code,' actually 'code-script,' for the molecular 'texture' in his lecture series in 1943. Schrödinger postulated that the design of life probably followed a code that could be found among the molecules of the body. He was an expert on subatomic systems and believed that the riddle of life was written in an inherited code at a very small scale similar to that he was familiar with in physics. Schrödinger challenged his colleagues in biology to find the code.

Mathematics, logic and language exemplify the rationale of the interpretant. They together with many other systems of rules underlie all processes of nature and all evolution. As far as they function regularly, they can be called 'codes.' Codes are semiotic systems based on some kind of 'conditioning' – to use a classical renowned term for mechanic behaviorist psychology. I discount the term and the tradition of research in the background but anyway; it exemplifies the interdependence of learnability and memory in the interpretant. In their typically fossilized repetition, they are like natural laws, cold examples of the Dawkins's metaphor of the blind watchmaker as the lord of the universe. Yet the idea of the interpretant shuns functioning like a combustion engine and would immediately try to correct the mistake if it occurred. In theory of semiotics, codes are latecomers. They appear in Sebeok and Danesi (2000: 37). The term 'code' is defined in the glossary of the book as a "system of signifying elements which can be deployed to represent types of phenomena in specific ways" (p. 191). Barbieri (2003: 94) defines the code "as a set of rules that establish *a correspondence between two independent worlds* [the italics included in the quote]."

Partly to illustrate that codes never work as sheer automatons do I will next use language to show how the language code manifests. Language mirrors the interaction of matter and energy in its morphological structure, in the morphemes of it. The main interpretant is composed of grammar. It regulates how morphemes can be combined, how the utterances (like written or spoken sentences) may be formed, etc. Morphemes, "free" ones that can be used singly (like 'too') and "bound" ones that are used only connected with the "free" morphemes to constitute words (like 'underpinnings'), make up the morpheme inventory or dictionary. The language code has the base structure of grammar and dictionary. The base structure is a theoretical model that never occurs in concrete reality. Nowadays, people on Earth use about 6,600 different languages; a great part of people uses only few of them. Each language has a grammar and a dictionary of its own and develops further in the framework of the particular predispositions concerning it. The variety of different languages offers a good example of how the rules of evolution may come true and be culturally conveyed. The products of culture and nature are usually juxtaposed against each other. That illusion, for instance, distorts the relationship between language faculty and speech of language and has given rise to the raging debate about the question whether language is innate or not. In reality, cultures don't shape up in isolation from nature.

According to the interpretation of mine, codes are composed of rules and objects and in principle, the semiosis in each code is fulfilled automatically. The language with definite rules of grammar and with a dictionary, in which every word has an established meaning, exemplifies a strictly code-like use of language. Fortunately, the cultural evolution modifies the language. In practice, no one speaks her or his native language uncompromisingly. The colloquial speech of anybody somehow always differs from the speech of someone else, especially in different areas and social groups of the language community. The variety of the norms for language use is in a continuous flux. Words break from traditional contexts, in particular when utterances are representations of a metaphoric or humorous double-speak as they very often do, and the connections between the meanings and their morphemes are enriched.

Languages are human systems of acoustic communication. Other species have systems of communication of their own. Anyway, they have the basic structure of grammar and dictionary (sign inventory) in common with humankind. The so-called 'animal' communication systems are more code-like compared to human languages but evolutionary variation does appear in the ethological semiotic codes, too. Human dialects are bound to communities living in regions that, at least originally, may have been separated from other ones and thus resemble animals evolving in isolated islands. Regional dialects abound among the humans. The same trait occurs also in animal languages. Nightingales sing differently in different parts of the world, for instance in Poland and Finland. When living in Indiana, in the US, I noticed on my way back home from work in the afternoon how crows used to assemble for a session in the tulip trees near by. In my ears, the croaking was so loud, gravelly and penetrating that I thought the fellow creatures in Finland would badly startle at the conversation habits of the Indiana crows, probably at some foreign meaningful units or morphemes of squawks, too. The animal and vegetable "languages" are, of course, typically much simpler than human languages. Anyway, they are complex enough to make possible the social life in the umwelt they live. The abundance of a sign inventory is a very inadequate criterion of the qualitative level of communicative semiosis in certain species. For instance, ants and termites are believed to have only about half a dozen morphemes but the semiotic grammar of their species-specific interpretant is able to take control over their labyrinthine society with a high-tech nests construction and a sophisticated cooperation at war and 'love'. Besides, in comparison to human languages, animal communication has the basic indexical utterances in common with human communication. The acoustic tones are always fine-tuned to the context they are expressions of. The shouts of crows when they are coming together are an expression of happiness while the voices are fully different in jeopardy – when they, for instance, try to take flight as fast as possible. Likewise, the human tones tell different stories when two bosom friends convene, especially in comparison with the situation in which the same people are in open quarrel.

Marcello Barbieri, the editor of this book, has applied organic codes into biosemiotics and in particular into the genetic code. The work from 2003 is a new edition of the book published two years earlier. The subtitle of the 2001 edition was a bit bolder, "The Birth of Semantic Biology." Kalevi Kull (2001) gave his deeply sympathetic comments on the book and showed how close its semantic framework actually was to the paradigm of biosemiotics. Apparently, Barbieri was not familiar with biosemiotics. In 1981, as his first article on the topic was published, only few people knew about the trials of applying semiotics to biology so far. The term 'biosemiotics' was probably coined only at the turn of the 1990s or a bit earlier. In many cases, the term is too narrow. Therefore, a more extensive term 'semiotics of nature' would be better advised (see Hoffmeyer & Emmeche 1991a,b).

Barbieri's contribution is highly original. No wonder that his work was largely ignored for about twenty years. In the silence, the only remarkable sympathizers were René Thom who wrote the foreword to Barbieri's book in 1985 and Karl Popper who clearly understood the originality and significance of Barbieri's work and spurred the author in two private letters. It seems to me that Barbieri had come across his great ideas particularly in the course of the 1970s by industry, insight and serendipity, under the protection of genii. His theoretical model is, like that of Peirce, a triad. Instead of object, representamen and interpretant, Barbieri speaks about genotype, phenotype and ribotype or energy, information and meaning. However, he does not mention Peirce and from all accounts he does not, even know him. Barbieri deals with the 'semantics' of life. The key to the origin of life is in the ribotype (RNA). As if in passing, he refers to his triad. His semantic component constitutes not only the RNA of the ribotype but also the information in general. In fact, it is a close cousin of the Peircean interpretant. Barbieri applies information theory and in the spirit of it, divorces meaning from information. The meanings occur in certain contexts while information is composed of the permanent features underlying the meanings (cf. my above account of semiotic relations).

The definition of the ribotype resembles the interpretation of the Peircean interpretant in the sense put forth here. Barbieri (2003: 209–215) assumes, for instance, that body plans are equivalent to the ribotype. The information needed for evolution is recorded in the interpretant; it is one of the key principles also in my description of the Peircean triad (cf. to my assumption of the brain and mind as expressions of the interpretant). Barbieri's 'genotype' is close to the Peircean 'object' and 'phenotype' to the Peircean 'representamen.' My description of the triad is deeply influenced by Barbieri but his biological account is one of the richest in its kind and cannot be compared with a linguist's trials to a semiotic overview.

Barbieri (2003: 220) defines his semantic theory of life as follows, "The cell is an epigenetic system [continuously tending to a more complexity, TJ] of three fundamental categories (genotype, phenotype and ribotype) which contains at least one organic memory (the genome) and and at least one organic code (the genetic code)." The relationship between meaning and its material equivalents are dealt with closer in the next and last section of this essay. The explanation of biological processes in terms of semantics may go along two ways, morphologic-syntactically or pragmatically. The morphologic-syntactic way demands for dividing the life processes in morpheme-like units: symbols or processes that carry through the semantic or semiotic directions in question. The pragmatic stance again departs from the assumption that the processes of life are some kind of utterances. Pragmatically, they remain composed of morphemes and syntactic units (phrases and sentences), but the morphemes and syntactic chains are of no cardinal importance in utterances. Cardinal is from the pragmatic point of view the intentionality of the meaning that governs the morphologic-syntactic discourse and changes it into a process for a certain function.

As well known, DNA is a two-stranded molecule composed of 'A' (adenosine), 'T' (thymidine'), 'C' (cytidine) and 'G' (guanosine) with different coding sequences. The object–representamen coupling introduced in our semiotic model is realized by the template or antisense strand with anticodons and the coding or sense strand with codons. The interpretant of the model is represented in transfer RNAs (tRNAs). They decode the sense strands of the molecules by interacting with ribosome-bound messenger RNAs (mRNAs) that contain the coding sequences. Surprisingly, the majority of the polynucleotide molecules does not take part in coding and is therefore called the 'junk-DNA.' From the stance of information theory, the noncoding regions have statistically more 'linguistic' properties than the coding ones, occurring especially in regard to Zipf's law and Shannon's redundancy rule (Mantegna & al. 1994). Zipf's law concerns the probability of occurrence of words in language: a few of words – like 'and,' 'the' and 'I' – occur very often while many others are more rarely used. On the basis of it, there are frequency dictionaries that don't introduce words in alphabetical but frequency order. Thanks to redundancy, there is always a lot of excess in discourses conveying messages. The excess makes the role of the receiver easier while the poverty of redundancy makes it more difficult, respectively. Unfortunately, the linguistic evidence value of Mantegna and his colleagues' claim is rather insignificant. Both arguments tell more about the general pragmatic implications of speech.

Barbieri (2003: 184) gives to understand that the noncoding regions of DNA may, in fact, use a system of coding but in a way unknown so far. He compares the "junk-DNA" to a foreign language. The comparison seems apt. The decoding of the 'junk' will possibly be a long step forward. As a linguist, I dare to try to inspire the discussion by comparing the coding DNA to the subject and predicate of a sentence and the noncoding region of DNA to a yet unknown model of adverbial and object complements of the same sentence. Anyway, DNA primarily stands for the syntactic component in semiosis.

DOUBLE CODING

So far, DNA is usually considered a metaphor of language. Barbieri (2002) denies that his organic codes were linguistic metaphors. On the contrary, they resemble a first-order language. In this, he is not alone. In line with Barbieri, Sungchul draws a parallel between DNA and language. As he puts it, there are two paradigms used in the research of biological processes: the PC paradigm with its belief that physics and chemistry are sufficient to account for life and the PCS paradigm with the assumption that semiotics, linguistics included, is needed for that. Sungchul regards John von Neumann as the father of the PCS paradigm and credits him with the foundational insight that self-replication would have been impossible without 'symbolic' self-representation. von Neumann's doctrine was developed further – as Sungchul points out – by Howard H. Pattee (1995) in his theory of matter–symbol complementarity The theory has been under construction since the late 1960s. Pattee's model follows in principle the morphological double articulation of linguistics, in the sense of the duality of morphemes introduced above. Jesper Hoffmeyer and Claus Emmeche have coined the term 'code duality' (1991; for a fresher interpretation see also Hoffmeyer 2006[2002]). André Martinet (1960) dubbed the linguistic term 'double articulation.' Because it is ambiguous, largely due to its use in different fields of research, I have proposed in the section title above the term 'double coding' and meant by that the simple division of the semiotic discourse into form and meaning. Factually, the division is the very same as in linguistic morphology. The codes operative in living systems, and thus in code duality (or double coding), are indeed codes in the sense of modern semiotics, where a code may be defined as "a semiotic resource—a meaning potential—that enable certain kinds of meanings to be made (in language, in the ways we dress, in our eating rituals, in the visual media, and so on) while others are not, or at least in that code" (Thibault 1998).

Sungchul acts as an exponent of the view that cell language or cellese and human language or humanese are isomorphic. Cell language has, as he highlights, 10 to 13 design features of human language (such as Hockett and Lyons have characterized it) in common. I won't go deeper into the details. Anyway, Sungchul's address is worth all attention. Nevertheless, there are some linguistic points of analysis that would need a more thoroughgoing additional argumentation. From the viewpoint of the interpretation of the Peircean model in this essay, a close kinship between language and language-like functions of cells is anticipated. The information potentials in the interpretant work on the same conditions and into the same direction. No one has claimed that language and the organic codes were, in fact, one and the same but that they follow the same primary model rather. Sungchul's characterization of isomorphism is thus well grounded.

Language has undergone a long history of evolution. The development of language is focused, though, on the era of the last half million years, probably on the very last dozens of millennia. Most experts on language evolution stress the specifics in the formation of their research object. Thus, for instance, Ray Jackendoff (1999) holds that the emergence of current language is preceded by nine partially ordered steps, each of them contributing to the precision and variety of expression.

Hoffmeyer and Emmeche's contribution to the basic theory of semiotics (1991b) is of exceptional historical importance; it is the first explicit analysis of signs and their strings in life processes. The introduction that follows is based on a personal interpretation and may differ from that of the authors. The notion of code duality comprises the coin of semiosis at both heads and tails. On one side, semiosis is digital and composed of information ultimately based on 0–1 opposition; on the other side, it is analog and represented, for instance, in the myriad figures we encounter in our daily life. The indexes of things are but one kingdom of the analog semiosis. Another must be out there in the meanings. As stated above, the meanings occur in definite contexts while the rules governing them reside in the indefinite information of the interpretant. The basic problem in biosemiotics is about how the code duality realizes in practice: What are the members of the semiotic process that interlink the different parts of analog semiosis to the digital one, and vice versa? Particularly problematic is the bond, say, between a certain meaning and the molecular process triggered and steered by it. Thus, Pattee (2001) assumes the biosemiotic system is not in balance and harmony and has coined the terms 'semiotic closure' and 'epistemic cut' to encapsulate the inherent deficiency.

The linguistic term 'morpheme' refers to the duality of meaning and form at its simplest. There is no form without meaning. Language, isomorphic at least to DNA as Sungchul assumed it, represents the general semiotic model between sign and meaning. At the level of language, we might put the verisimilitude of the principle to the test. We can quote for instance James Joyce's *Finnegans Wake* from 1939, "As we there are where are we are we there UNDE ET UBI. from tomtittot to teetootomtotalitarian. Tea tea too oo" (an electronic edition, 1.2:1, 1-2, p. 260) and write down sets of letters in random order, "qwertui" and "wfgro." There have been panels of experts trying to translate *Finnegans Wake* into foreign languages but with relatively poor fortunes. Everyone will give the quote his individual interpretation, though he possibly does not want to do it and in that case is not even aware of any sense of his. As for the sets of letters, I have briefed the interpreters and most readers may take it for granted that the sets don't mean anything. Factually, they do have a meaning, though – a meaning that they don't have any meaning, that is to say they have a 'zero meaning.' But inevitably, we cannot be released from our built-in principle of giving sense to everything. The hypothesis that everything has a meaning and can be understood is a very strong idiosyncrasy and would plausibly appear in the context of the nonsense strings of letters as 'possible' meanings in the spirit of what might be called 'the principle of a sensible interpretation.' It resides in the background of religion and science, too, as a presupposition that anything means something to be found out. Subjectively, religion is a shortcut and science a roundabout way to the meanings that are not opened straight away.

The twofold parsing comes from the model that obviously governs the entire universe and the iconic, diagrammatic semiotic model of it. In particular, the parsing principle stems from the dialectic between object and representamen, the meaning potentials and the actual entities. The interpretant of the model shall reconcile the conflict between them.

All signs we think of have a 'physical' shape. In case they are abstract objects they are dealt with as if they had one. The meanings given to the shapes are psychological products and regarded to be 'nonphysical' or 'spiritual.' Essential of our species is the signification. That's why, it is justified to define the member of humankind as 'homo significans' or 'homo semioticus.' As biosemiotics postulates, all organisms live in the umwelts they are habituated to interpret. Fundamentally, the semiotic principles of interpretation are based on nature.

ASYMMETRIES IN BIOSEMIOSIS

Before taking a closer look at the semiotic closures and epistemic cuts, let's return yet to the genetic code. The genetic code is, in principle, well known and serves therefore for a model example of semiosis in nature. The morpheme model is, of course, applicable into genetic processes. The morphological form–meaning coupling is realized, for the first, on the level 1 where a certain meaning of the set of the alternative meanings takes the role of the trigger and for the second, the combination of DNA molecules that matches with the actual meaning will unfold. Thus, single molecules have the same role as the phonemes do in the phonological component of language. The above levels are meant to be merely illustrative. Theoretically, they can be severed but function in an inseparable coupling in practice. To go on with our language isomorphism, morphemes are the smallest meaningful units but that combination of morphemes that has a real contextual meaning occurs only in clauses and sentences. Sentence meanings are propositions. The forms of sentences come from the combinations of the morphemes and the arrangement of morphemes in clauses and sentences is determined by the syntactic rules. The syntactic component has an important role also in the genetic code and the biological processes in general. The transfer and messenger RNAs seem to run the syntax of the genetic code. If the genetic code represented the noun phrase or the subject then there must be a verb phrase, too, with an object. The verb phrase is represented, for instance, in a bodily process catalyzed by a certain genetic sequence.

Decisive for biosemiosis is what I would call the 'textual component.' Kalevi Kull (2002) has with reason highlighted the precedence of the text – the 'biotext,' as he labels it – over the sign. There is no deeper polarity between the two, however. Naming the sign processes 'texts' underscores the presence of the semiotic grammar, which is so closely alike to that of language. The biotexts in the bodies of organisms differ some from language texts because the propositions they are composed of are often simultaneous and the repetition makes them more redundant. The term 'text' is factually more relevant and gives a better understanding of the unbelievable complexity of the bodily text. Our knowledge of the biotext is imperfect, with many lacunae, so far. The bodily text compares to the notions of text and utterance in speech act theory. In it, 'illocutionary force' is the basic intentional meaning what a text or utterance has. From this point of view, the illocutionary force of a biotext is to keep a certain body alive. The statement brings us into the domain of the pragmatic semiotic component. Survival is the highest pragmatic meaning. The co-action of the pragmatic components with the other ones is of particular interest. Assuming that body and mind constitute a first-order subject (interpretant) there must be a pragmatic second-order subject, too. It looks the first-order subject like in the mirror and the autoimmune system of the pragmatic subject, for instance, tries to protect the first-order subject from infection and disease.

In the history of biophysical semiotics, Howard H. Pattee is one of the key figures. He was one of the first to realize that symbolic semiosis of life processes intrinsically conflicts with the dynamical side of the same processes (see in particular Pattee 2001). Pattee reminds of John von Neumann's statement from 1955 (p. 419),

> "– we must always divide the world into two parts, the one being the observed system, the other the observer. In the former, we can follow up all physical processes (in principle at least) arbitrarily precisely. In the latter, this is meaningless. The boundary between the two is arbitrary to a very large extent...but this does not change the fact that in each method of description the boundary must be placed somewhere, if the method is not to proceed vacuously, i.e., if a comparison with experiment is to be possible."

von Neumann's and Pattee's theoretical argument lies in quantum mechanics where – as well known – the observer always has a certain effect on the observed. Niels Bohr propounded the complementarity principle in 1928. Electro-magnetic phenomena in waveforms and as streams of particles (electrons or photons) cannot be measured at the same time. At the subatomic level, the measuring instruments influence the action of the objects. The theoretical and the context-dependent knowledge are not right or wrong but should be seen as complementary. Today, we know that for instance the behavior of electro-magnetic waves and large-scale waves (like water waves) is incompatible and a true picture can be received only by investigating both.

Pattee (1995) uses the term 'matter–symbol distinction.' It was von Neumann (1966) who first recognized the complementarity in a semiotic sense. In the context of DNA, for instance, the matter–symbol complementarity is principally realized at two levels, at the material one where dynamical laws of molecules represent the same what sounds and letters do in language and at the symbolic one where functions have the center stage. From the point of view of this essay, Pattee's symbolic level is close to the meanings. Pattee has an exceptionally deep and versatile conception of symbols (in a broad sense different from that in semiotics). He seems not to be fully aware of the linguistic morphological tradition, not even of the notion of morpheme. However, he has a clear insight of how only the material form – as I have interpreted his account – can let out the functional (in this essay: semantic or meaningful) potentials. And he includes in the functional implications, for instance, survival value, goals, significance, meaning and self-awareness. Semantic closure requires, he thinks with reason, complementary models of the material and symbolic aspects of the organism. Something important remains missing, however. Pattee does not develop in more detail such a semiotic model that would entirely meet his conditions.

Semiotic closure is self-reference – the subject or the 'self' may have and often does have indeed himself as an object. Only in the context of the self, organisms may develop to the demands of their umwelt or 'Mitwelt.' Pattee also uses the term 'epistemic cut,' assuming that the cut divorces the organisms from the world. He mentions (2001) how language probably abstracts away the epistemic cuts. Hoffmeyer (2000) rebuts the interpretation and explains in the framework of code duality that not only the 'symbolic' domain but also the one Pattee calls 'dynamical' are in fact semiotic. The dynamical processes of molecules embody the indexical or iconic modes of semiosis. Hoffmeyer is right: from a Peircean view, the so-called 'reality' is composed of analog signs, indexes and icons. It is a different thing and of secondary importance that these signs can be double-coded as well. From the point of view of the epistemic cut, it is crucial that there is no divide between an organism and its Umwelt and Mitwelt (about 'Mitwelt' see Witzany 2005). Hoffmeyer's stance accords with that proposed here. There seems to be no cut between our mind and the physical constants we live in. Antti Revonsuo (2006) lines up with some kind of 'phenomenological' (phenomenal) externalism. Nearly all the qualities of experience, even those of dreams, have a phenomenal location and extension; in other words, our experience refers to events that take place in a space-time and have spacial dimensions. Thus, our experiences show we are inseparably bound to the physical world. Spacial properties constitute the omnipresent background of our inner semiosis.

In a sense, the notion of epistemic cut reveals the intrinsic imbalance peculiar to all semiosis. Juri Lotman coined the term 'semiosphere' referring by that above all to culture. In the course of time, he was not any more as certain on the boundary between the semiosphere and nature as he had been before. As Boguslaw Zylko (2001: 393–394) points out Lotman's antinaturalism changed in his last years into a view of nature and culture as complementarities. He questioned the boundary between the spheres, "The boundaries are blurred, and defining each concrete fact as belonging to either the cultural or the noncultural sphere involves a high degree of relativity" (a quote by Zylko from Lotman's book, 1992, "Kultura I vzryv" [Culture and explosion]). The natural world comes into culture by means of language, more precisely, through naming.

I hypothesize that Lotman's semiosphere covers all semiosis and borrow the term 'bipolar asymmetry' from Lotman to illustrate the relationship especially between object and representamen in my adaptation of the Peircean model. Lotman's description of semiosphere is different; only the idea of asymmetry is applicable. In Lotman (1990), asymmetry characterizes the dialectic between two things: an intelligence (or a language or a text) and an other peculiar to semiosis, particularly to the generation of new information. The bipolar asymmetry refers to two aspects making up any semiotic phenomenon. The two asymmetric poles represent entities, neither of which can be fully translated into the other.

Asymmetry and intranslatability characterize things at the poles of objects and representamens. The intrinsic dissimilarity between the two could be called also 'epistemic cut,' in a sense very close to that in Pattee. At the physical level, the principle of complementarity of Bohr seems to realize in the domain of objects and representamens. Quantum mechanics and classical physics complete each other. So far, we don't have a so-called 'theory of everything' to bind both together. Empirically, we know, though, that we can get a true picture of physical events only by applying both models to them. It is an indirect evidence of semiosis and the dominant role of the interpretant. The interpretant integrates the asymmetric items into a whole and functions so neatly that there seems to be nothing left from the conflicts. Pattee suspects that language abstracts the epistemic cuts away. In fact, the great fusion seems to take place by semiosis and particularly by the interpretant instead. If the quantum mechanics and large-scale physics asymmetry really has that kind of a role in semiosis it supports the semiotic model based on the initial physicochemical stage of cosmological evolution.

Object and representamen constitute an asymmetric (dissymmetric) pair in many dimensions. Ingolf Schmid-Tannwald (2006) highlights that the dominant model of man actually distorts the true essence of the biological nature of our species. It does not take into account "that the biological reproduction in man is taking place within the unique and common reality of two partners." From a semiotic point of view, this implies that to understand the partnership we have to find out for it a new common denominator. Semiotically, the interpersonal reality of the mate partners is to be described by two interconnected circular sign-processes that bind two individual systems to one supra-system.

The principle of duality passes through all the sexual and asexual reproduction. The same duality concerns not only two single individuals but also the system of paired halves of one and the same individual. Especially from a biological point of view, the demands of sexual reproduction often cause anxiety because people searching for a mating partner and even after having found him or her reside in a permanent state of a feeling of insufficiency. It is due to the asymmetry principle of semiosis and occurs everywhere in an individual's consciousness. Individuals apart from mating partners and the society they belong to are sheer abstractions. The object–representamen dissymmetry – a stable semiosic tension and anticipation – features everybody's experience of the world leaving us in the middle of inexplicable feeling of the heaviness of life. Heidegger, the master especially of Sartre, holds there is no pregiven human essence but humans make them up in the course of their lives. Contrary to Heidegger and Sartre, our hypothesis of the intrinsic semiosic asymmetry refers to another kind of interpretation. Thus, in my mind, Heidegger's 'existentialia:' like Being-in-the-world ('Dasein'), Emotional tone ('Stimmung'), Anxiety ('Angst') and Care or Concern ('Sorge') are ontological elements preconditioning all human existence and conducting and determining it (cf. Guignon 1996). The primary feeling of life-world ('Lebenswelt,' Husserl; 'Firstness', Peirce) similarly to the Heideggerian existentialia probably comes too from the principle of dissymmetry in semiosis.

The dividing line between object and representamen can be seen in several such contexts, in which the polarity of duality occurs in pairs. Amino acids and sugars, the building blocks of the biological world, are distinctively left- or right-handed – possessing a quality known as chirality (the structural characteristic of a molecule that makes it impossible to superimpose it on its mirror image); scientists have been puzzling to answer how and why. Another divide coupled up runs through the corpus callosum, which severs the brain into two hemispheres, both of them having their typical functions. The interplay between the hemispheres is fine-tuned and stunning; only the whole is able to nurture the task of the brain (or, semiotically) that of the interpretant. Moreover, the dissymmetric coupling is very ordinary in human and other cellular bodies. Humans have two legs, two arms, two breasts, two ears, two eyes, two temples, two rows of teeth, two nostrils, etc. Semiosis produces body plans and due to asymmetry between object and representamen, the perfect symmetry is extremely rare in paired organs and body parts.

Like Revonsuo (see above) gives to understand, external phenomenalism (the term 'phenomenology,' due to the rich and long philosophical tradition of phenomenology, is quite misleading) is ultimately rooted in the physical condition. Semiosis started within the Big Bang and bears since the primary specifications tied up with the circumstances at the cosmological dawn. Cardinal epistemic cuts originate from the very beginning. Thus, even the human mind-interpretant cannot grasp the void preceding the initial explosion and inflation. Cosmologists claim it is absurd to ask around the pre-state of Big Bang because there was not out anything and everything, the existence and all attributes of it, like space and time, were generated in the 'catastrophe.' Though, there might have been a long pre-evolution finally leading to the great evolutionary burst. The assumption of such a pre-evolution does not conflict with our 'default settings' in semiosis; the invalidation of the old "Ex nihilo nihil fit" (Nothing comes from nothing) principle does. In agreement with the 'horror vacui' (fear of empty space), there are several other boundaries alike in our thought. Thus, it contradicts our thinking if the notion of 'infinite' number series contained a certain ultimate number; our physical semiosic existence presumes we must be in a physical space-time, even in that of numbers, but not in a void beyond the 'boundary.' Death stands for our journey in space-time. We can think of the breakdown of our life story in a space-time but the void after the last breath is definitely beyond our grasp. Our death demonstrates the same semiotic boundary that we cannot cross – stepping into nothing, the void.

REFERENCES

Adams, F. 2002. *Origins of Existence. How Life Emerged in the Universe*. New York: Free Press.

—. 2004. *Leben im Universum*. Eine deutsche Übersetzung des 'Origins of Existence' von Hainer Kober. München: Deutsche Verlags-Anstalt.

Adams, F. & Laughlin, G. 1999. *The Five Ages of the Universe – Inside the Physics of Eternity*. New York: Touchstone.

Barbieri, M. 1981. The ribotype theory of the origin of life. *Journal of Theoretical Biology* 91: 545–601.

—. 1985. *The Semantic Theory of Evolution*. London: Harwood.

—. 2002. Organic codes: metaphors or realities? *Sign Systems Studies*, 30.2: 743–754.

—. 2003. *The Organic Codes. An Introduction to Semantic Biology*. Cambridge: Cambridge University Press.

Bar-Hillel, Y. 1964. *Language and Information*. Reading, MA.: Addison-Wesley.

Barrow, J.D. & Tipler, F.J. 1986; *The Anthropic Cosmological Principle*. Oxford: Clarendon Press.

Chapelle and his research group. 2002. Chapelle, F.H., O'Neill, K., Bradley, P.M., Methe, B.A., Ciufo, S.A., Knobel, L.L., Lovley, D.R. A hydrogen-based subsurface microbial community dominated by methanogens. *Nature*, vol. 415, pp. 312–315.

CP = *Collected Papers of C. S. Peirce*, ed. by Hartshorne, C., Weiss, P. & Burks, A. 1931–58. 8 vols. Cambridge, MA: Harvard University Press.

de Gruyter. Kant, I. 1998. *Critique of Pure Reason*. The original [A], *Kritik der reinen Vernunft*, published in 1781, the second edition (B) in 1787. Transl. by P. Guyer and A. Wood. Cambridge: Cambridge University Press.

Dyson, F. 1979. Time without end: physics and biology in an open universe *Reviews of Modern Physics*, vol. 51, no. 3. Available online at the address http://www.aleph.se/Trans/Global/Omega/dyson.txt.

—. 1985. *The Origins of Life*. New York: Cambridge University Press.

Eco, U. 1976. *Theory of Semiotics*. Bloomington: Indiana University Press.

Ehrenfreund, P. & Menten, K. 2003. From molecular clouds to the origin of life. Homeck, G. & Baumstark-Khan, C. (eds). *Astrobiology: The Quest for the Conditions of Life*, pp. 7–24. Berlin: Springer.

Frege, G. 1997. *The Foundations of Arithmetic: A Logico-Mathematical Enquiry into the Concept of Number*. The Original: *Die Grundlagen der Arithmetik: eine logisch-mathematische Untersuchung über den Begriff der Zahl* from 1884. Transl. by Michael Beaney. Beaney, M. (ed.). *The Frege Reader*, pp. 84–129. Oxford: Blackwell.

Guignon, C.B. 1996. Heidegger, Martin. *The Cambridge Dictionary of Philosophy*, pp. 317–319. Reprinting. Cambridge: Cambridge University Press.

Harold, F.M. 2003. *The Way of the Cell: Molecules, Organisms, and the Order of Life*. New York: Oxford University Press.

Heidegger, M. 1979. *Heraklit.* Gesamtausgabe, vol. 55. Frankfurt: Klostermann.

—. 1985. *Early Greek Thinking*. 'Logos' (Heraclitus, Fragment B 50). New York: HarperCollins.

Heraclitus. 2006. Fragment 1, quoted in Sextus Empiricus at the address http://www.wsu.edu/~dee/GREECE/HERAC.HTM, retrieved on March 12.

Hintikka, J. 1968. The varieties of information and scientific explanation. Van Rootselaar, B. & Staal, J.F. *Logic, Methodology, and Philosophy of Science*, pp. 311–331. Amsterdam: North Holland.

Hoffmeyer, J. 2000. Code duality and epistemic cut. *Annals of the New York Academy of Sciences*, 901: 175–86.

Hoffmeyer, J. 2006 (2002). Code duality revisited. Retrieved on March 30 from the address http://www.library.utoronto.ca.

Hoffmeyer, J. & Emmeche, C. 1991a. From language to nature - the semiotic metaphor in biology. *Semiotica* 84 (1/2): 1–42.

—. 1991b. Code-duality and the semiotics of nature. Anderson, M. & Merrell, F. (eds.): *On Semiotic Modeling*, pp. 117–166. Berlin & New York: Mouton de Gruyter.

Inagaki & al. 2006. Inagaki, F., Nunoura, T., Nakagawa, S., Teske, A., Lever, M., Lauer, A., Suzuki, M., Takai, K., Delwiche, M., Colwell, F.S., Nealson, K.H., Horikoshi, K., D'Hondt, S. & Jørgensen, B.B.. Biogeographical distribution and diversity of microbes in methane hydrate-bearing deep marine sediments on the Pacific Ocean Margin. *PNAS (Proceeding of the National Academy of Sciences of the United States of America)*, vol. 103, no. 8, pp. 2815–2820.

Jackendoff, R. 1999. Possible stages in the evolution of language capacity. *Trends in Cognitive Science*, 3 (7), pp. 272–279.

Jämsä, T. 1986. *Suomen kielen yleisimpien verbien semantiikkaa* ('On semantics of the most frequent Finnish verbs'). Acta Universitatis Ouluensis: Humaniora, 12.

—. 1997. Semiotics of relations. Rauch, I. & Carr, G.F. (eds.). *Semiotics around the World: Synthesis in Diversity*. Proceedings of the Fifth Congress of the International Association for Semiotic Studies, Berkeley, 1994, 2 vols. Vol. 2, sect. XII (Semiotics and metatheory). Berlin & New York: Mouton.

Kull, K. (ed.). 2001. Jakob von Uexküll: A paradigm for biology and semiotics. Special issue of *Semiotica*, vol. 134 – 1/4.

—. 2002. A sign is not alive — a text is. *Sign System Studies* 30.1, pp. 327–336.

—. 2006 (2001). [Comments on Barbieri's *The Semantic Codes*]. Retrieved on March 15 from the address http://www.biologiateorica.it /organiccodes/comments/kull.htm.

Lakoff, G. & Núñez, R.E. 2000. *Where Mathematics Comes From: How the Embodied Mind Brings Mathematics into Being*. New York: Basic Books.

Lotman, Ju.M. 1988a (1974). Primary and secondary communication-modeling systems. Lucid, D.P. (ed.). *Soviet Semiotics*, pp. 95–98. Baltimore & London: The Johns Hopkins University Press.

—. 1988b (1970). Two models of communication. Lucid, D.P. (ed.). *Soviet Semiotics*, pp. 99–101. Baltimore & London: The Johns Hopkins University Press.

—. 1990. *Universe of the Mind: A Semiotic Theory of Culture*, ed. and transl. by A. Shukman. Bloomington, IN: Indiana University Press.

Luis, G. & Kumar, A.S. 2006. The red rain phenomenon of Kerala and its possible extraterrestrial origin. A preprint accepted for publication in *Astrophysics and Space Sciences*. Retrieved on March 7 from the address www.arxiv.org/abs/astro-ph/0601022.

Mantegna, R.N., Buldyrev, S.V., Goldberger, A.L., Havlin, S., Peng, C.K., Simons, M. and Stanley, H.E. 1994. Linguistic features of noncoding DNA sequences. *Physical Review Letters*, vol. 73 (23), pp. 3169–3172.

Martinet, A. 1960. *Eléments de linguistique générale*. Colin: Paris.

Maturana, H. & Varela, F. 1973. *Autopoiesis and Cognition: the Realization of the Living*. Cohen, R.S. & Wartofsky, M.W. (eds.). Boston Studies in the Philosophy of Science, 42. Dordecht: Reidel.

Muir, H. 2006. It's raining aliens. *New Scientist*, 4 March 2006, pp. 34–37.

Pattee, H.H. 1995. Evolving Self-Reference: Matter, Symbols, and Semantic Closure. Evolving self-reference: matter, symbols, and semantic closure. Rocha, L. (ed.) *Self-Reference in Biological and Cognitive Systems*. Special issue of *Communication and Cognition*, vol. 12, nos 1–2, pp. 9–27.

—. 2001. The physics of symbols: bridging the epistemic cut. *Biosystems*, vol. 60, pp. 5–21.

Popper, K.R. 2005 (1998). *Die Welt des Parmenides. Der Ursprung des europäischen Denkens.* Hrg. von Arne F. Petersen unter Mitarbeit von Jørgen Mejer und übersetzt von Sibylle Wieland und Dieter Dunkel. Das Original: *The World of Parmenides – Essays on the Presocratic Enlightenment.* München: Piper.

Pueschel, R.F. 2006. Methanogenesis, mesopheric clouds, and global habitability. Retrieved on March 3 from the address http://geo.arc.nasa.gov/sgp/aero/aerocloud3.html.

Revonsuo, A. 2006. *Inner Presence: Consciousness as a Biological Phenomenon*. Cambridge, MA: MIT Press.

Schmid-Tannwald, I. 2006. Towards a more comprehensive scientific model of man. An abstract of a paper to be delivered at the 6th Gathering of Biosemiotics in Salzburg, July.

Schrödinger, E. 1992 *What Is Life? With Mind and Matter and Autobiographical Sketches*. Cambridge: Cambridge University Press.

Sebeok, T.A. 1968. *Animal Communication*. Bloomington: Indiana University Press.

—. 1986. *The Semiotic Sphere*. New York: Plenum.

Sebeok, T.A. & Danesi, M. 2000. *The Forms of Meaning. Modeling Systems Theory and Semiotic Analysis*. Berlin & New York: Mouton de Gruyter.

SEP 2006 = Frege. Stanford Encyclopedia of Philosophy online. Retrieved on March 10 from the address http://plato.stanford.edu/entries/frege/#logic.

Shinitzky and his research group 2006 = Scolnik, Y., Portnaya. I., Cogan, U., Tall, S., Haimovitz, R., Fridkin, M., Elitzur, A.C., Deamer, D.W., & Shinitzky. M. Is water the answer to nature's handedness? *Physical Chemistry, Chemical Physics*, 2006, vol. 8, p. 333–.

Sowa, J.F. 2000. Ontology, Metadata, and Semiotics. Ganter, B. & Mineau, G.W. (eds). *Conceptual Structures: Logical, Linguistic, and Computational Issues*, pp. 55–81. Berlin: Springer.

Stjernholm Madsen, A. 2006 (1999). A theory of organizational consciousness. Retrieved on Jan. 30 from the address http://www.strategic-innovation.dk/ Engelsk/DeepCons.html.

Sungchul, J. 1999. The Linguistics of DNA: Words, Sentences, Grammar, Phonetics, and Semantics. *Annals of the New York Academy of Sciences*, 870:411–417.

Tegmark, M. 2004. Parallel Universes. Barrow, J.D. & Davies, P.W.C. & Harper, C.L. (eds.). *Science and Ultimate Reality: From Quantum to Cosmos*, honoring John Wheeler's 90th Birth Day, pp. 459–491. Cambridge: Cambridge University Press.

Tembrock, G. 1971. *Biokommunikation*. Berlin: Akademie-Verlag.

Thibault, P. J. 1998. Code. Bouissac, P. (ed.), *Encyclopedia of Semiotics*. pp.125–129. Oxford: Oxford University Press.

v. Neumann, J. 1955. *The Mathematical Foundations of Quantum Mechanics*. Princeton: Princeton University Press.

—. 1966. *The Theory of Self-Reproducing Automata*. Ed. and compl. by A.W. Burks. Urbana, IL: University of Illinois Press.

v. Uexküll, J. 1909. *Umwelt und Innenwelt der Tiere*. Berlin: Springer.

—.1920. *Theoretische Biologie*. Berlin: Paetel.

—. 1928. *Theoretische Biologie*. 2. gänzl. neu bearb. Aufl. Berlin: Springer.

—. 1940. Bedeutungslehre (= Bios, Abhandlungen zur theoretischen Biologie und ihrer Geschichte sowie zur Philosophie der organischen Naturwissenschaften, Bd. 10). Leipzig: Barth.

—. 1982. The theory of meaning. *Semiotica* 42(1), pp. 25–82.

Wittgenstein, L. 1999 (1953). *Philosophical Investigations*. 3rd ed.Upper Saddle River, NJ: Prentice Hall.

—. 2001 (1922). *Tractatus Logico-Philosophicus*. 2nd ed. Oxford: Routledge.

—. 1964. *The Blue and Brown Books*. Oxford: Blackwell.

Witzany, G. 2005. From Umwelt to Mitwelt. Natural laws versus rule-governed sign-mediated interactions (rsi's). Paper at the 5th Gathering of Biosemiotics at Urbino.

WMAP 2006 = [The finding based on data from NASA's WMAP satellite]. Big Bang unfolded in the blink of an eye. *USA Today*, 3/16/2006.

Zaliznjak, A.A., Ivanov, V.V. & Toporov, N. 1988 (1962). Structural typological study of semiotic modeling systems. Lucid, D.P. (ed.). *Soviet Semiotics*, pp. 47–58. Baltimore & London: The Johns Hopkins University Press.

Zylko, B. 2001. Culture and Semiotics: Notes on Lotman's Conception of Culture. *New Literary History*, vol. 32, no 2: pp. 391–408.

CHAPTER 3

HAS BIOSEMIOTICS COME OF AGE?*+ AND POSTSCRIPT+

MARCELLO BARBIERI

Dipartimento di Morfologia ed Embriologia, Via Fossato di Mortara 64, 44100 Ferrara, Italy, brr@unife.it

Abstract: The 2001 Special Issue of *Semiotica* has been dedicated to celebrating Jakob von Uexküll as a founding father of biosemiotics. The two main points of the volume – the making of biosemiotics and the recovery of Jakob von Uexküll from oblivion – come out with clarity and force, and are definitely a success. The volume is also an excellent example of interdisciplinarity, with contributions from history, philosophy, linguistics, biology, art, literature and computer science that integrate each other with admirable ease. There is however a third message of the special issue that is less agreeable. It is the message that biosemiotics has been the crowning achievement of the tradition that goes back to Goethe, von Baer, Driesch and von Uexküll, and many contributors did not hide their preferences for neo-vitalism and anti-darwinism. The author of the review welcomes the project of introducing meaning in biology but points out that neo-vitalism is not the best approach. The existence of organic codes and organic meaning in nature are scientific problems that can and should be investigated with the classical method of science, i.e. with the mechanistic approach of model building. This led the reviewer to conclude that biosemiotics had not yet come of age in 2001. In the Postscript of 2005, however, the same reviewer acknowledges that in a few years the situation has rapidly changed. Biosemiotics has become a pluralistic field of research that no longer excludes the mechanistic method, and today it is a vibrant young science where all approaches to the problem of biological meaning are investigated without preconditions

Keywords: Biosemiotics, signs, codes, meaning, vitalism, mechanism

---

* Review of Semiotica 134 – 1/4 (2001). Special Issue. Jakob von Uexküll: A paradigm for biology and semiotics Review article by Marcello Barbieri © published in Semiotica, 2002, 139, 283–295.
+ Chinese translation by Juipi Chien published in Chung Wai Literary Monthly, 2005, 403, 11–26.

M. Barbieri (ed.), Introduction to Biosemiotics, 101–113.

INTRODUCTION

This very special issue of *Semiotica* is dedicated to celebrating the coming of age of biosemiotics, the study of living systems from a semiotic perspective, by celebrating Jakob von Uexküll as the once forgotten chief architect of the new discipline. The recovery of *a "neglected figure in the history of semiotic inquiry"*, and the proclamation of a new scientific domain, are certainly worthwhile academic pursuits, but the general reader may feel, at first, that such specialized items do not concern him. In this case, however, he would be wrong, because underneath the academic niceties one can feel, in almost all papers of this volume, a powerful, pulsating new vision about the fundamentals of life: a new theory of signification and biological meaning.

In the very opening paper, Kalevi Kull, the guest editor, puts the cards on the table in no uncertain terms: *"Sign science and life science are coextensive"*, *"semiotics is biology and biology is semiotics"* (p.3). And the message is promptly reinforced with a quotation from Umiker-Sebeok: *"If, according to semiobiological theory, all living things are signs, and signs are living things, then life* qua *signs must be seen as constantly evolving according to certain general rules, for 'symbols grow'"*. The concept is hammered out again in the following paper by Thomas Sebeok: *"Because there can be no semiosis without interpretability – surely life's cardinal propensity – semiosis presupposes the axiomatic identity of the semiosphere with the biosphere"* (p.68).

The reader is warned. It is not a small academic niche which is discussed here, but an entirely new conception of biology: *life as semiosis*. And more is to come. The third paper, by Frederik Stjernfelt, begins with the announcement that *"Jakob von Uexküll's theoretical biology is a main contribution to the developmental, or epigenetic, trend in the biology of recent centuries, a lineage involving scholars like Goethe, Saint-Hilaire, von Baer, d'Arcy Thompson, Spemann, Driesch, Waddington, Brian Goodwin, René Thom and Stuart Kauffman"* (p.79). This lineage has been the historical antagonist of the 'mechanistic' approach of Galileo, Descartes, Newton, Lamarck, Darwin, Mendel, James Watson, Francis Crick and Jacques Monod, an approach which has produced what is still the main paradigm of modern biology. This special issue, in short, not only presents a revolutionary idea of biology, but also announces that such a revolution comes from the heirs of the historical opposition to mechanism. And this is no isolated announcement. 41 distinguished academics from 15 different countries have produced a 828-page volume with papers on history, philosophy, theoretical biology, ecology, linguistics, arts, literature and computer science, and all come, by varying degrees, to similar general conclusions. The volume owes in fact its remarkable overall unity to this ideal convergence, and there is no doubt that its aim is to strike at the very heart of the life sciences.

In such a situation, a reviewer can hardly avoid getting involved, but first I will try to describe, as impartially as I possibly can, the three main points of this book: (1) the making of biosemiotics, (2) the redeployment of Jakob von Uexküll, and (3) the endorsement of a non-mechanistic science of life.

THE MAKING OF BIOSEMIOTICS

Semiotics, *the science of signs*, belongs to the humanities, because it has always been taken for granted that signs, or symbols, are quintessential cultural products. The idea that man evolved from animals implies of course that culture has biological roots, but this has never undermined its independence from 'the brute forces of nature'. For the development of mathematics, for example, it is irrelevant whether animals are able to count or not, and similar considerations appeared to apply to the humanities in general and to semiotics in particular. Even the discovery of animal psychology and the development of ethology did not much change the status of semiotics, since primitive forms of communication seemed to have very little to say about the complex architecture of human semiosis. It was therefore very bold, and very risky, of Thomas Sebeok, to suggests, in the early 1960s, that human semiotics *needs* to be complemented by animal semiotics (or *zoosemiotics*, as he called it in 1963) in order to find its proper place, and its real nature, within the larger framework of general semiotics.

Sebeok knew of course that this idea would not have stood a chance unless he could back it up with some experimental data, and so he started looking around and digging in various gardens, particularly in psychology, medicine and molecular biology. And the hunt paid off. In his contribution to this volume, Sebeok presents a streamlined account of his life-long chase, and declares that he got the crucial experimental clues from three men: Heini Heniger (1908–1992), Giorgio prodi (1929–1988), and, above all, Jakob von Uexküll (1864–1944). The first two he met personally, while the encounter with Jakob came from reading, in 1976, the original German edition of *Theoretische Biologie* (1928). This was apparently a kind of 'fulguration on the road to Damascus', so much so that Sebeok decided to end his individual search and start an active campaign of proselytization. In August 1977 he delivered his now historical speech on 'Neglected figures' at a congress in Vienna, and from that time on the making of biosemiotics became a collective enterprise that Sebeok led with all the academic and editorial power he could muster. In this campaign he was quickly joined by Thure von Uexküll (Jakob's son), who could argue, from within his own professional field, that medicine has been a semiotic discipline since antiquity, because it has always been concerned with the interpretation of clues.

Sebeok's own idea of zoosemiotics was generalized in the process. Originally he had only taken animals into consideration because of the commonly held idea that semiosis requires a nervous system, but in 1981 Martin Krampen showed that plants too engage in vegetable semiosis (phytosemiotics). In 1988, Giorgio Prodi spoke of *signs and codes in immunology*, thus extending his 1977 idea of the natural semiosis of cells, and Sorin Sonea proposed that semiosis goes on even in the bacterial world, which can be regarded as a true global organism. The word 'zoosemiotics' was clearly inadequate, and Sebeok decided to replace it officially with 'biosemiotics', a term proposed by Juri Stepanov in 1971, but which appeared for the first time (with a restricted meaning) in 1961, when Friedrich Rothschild used it to indicate a semiotic approach to psychology.

The making of biosemiotics gathered further momentum in the 1990s, with the joining in of a younger generation and the increasing participation of biologists. Among these, Jesper Hoffmeyer and Claus Emmeche, who founded a Biosemiotic Group in Copenhagen, and Kalevi Kull, who became director of the Jakob von Uexküll Center in Tartu, Estonia. Biosemiotics has thus become a well established interdisciplinary field, and perhaps it is fair to say that its development was formally completed in 2001, when the first Gathering *exclusively* dedicated to biosemiotics took place in Copenhagen (by which time the official Directory listed 61 biosemioticians from all countries of the world).

The making of biosemiotics has been therefore a 40-year-long affair (1961–2001), which can be divided into two phases: the first (1961–1977) was a period of uncoordinated attempts, often of utterly isolated initiatives, while the second (1977–2001) was a period in which individual ideas could fall on a more receptive ground and contribute, under the discreet supervision of Thomas Sebeok, to the collective growth of the field.

THE REDEPLOYMENT OF JAKOB VON UEXKÜLL

In 1905, Jakob von Uexküll published a book on the physiology of marine animals (*Leitfaden in das Studium der experimentellen Biologie der Wassertiere*) where he described a neuromuscular cycle (later a functional cycle) which can be regarded an early formulation of the feedback principle. In a second book, published in 1909 (*Umwelt und Innenwelt der Tiere*), he adopted the word *Umwelt* (coined by a poet in 1800) to indicate the subjective world of an organism (the combination of its *perceptual* world with its *operational*, or *motor*, world). These two concepts – functional cycle and Umwelt – formed the basis of his major book, *Theoretische Biologie* (1920 and 1928), and are to this day his true biological heritage.

The functional cycle was further developed by Charles Sherrington into the negative feedback principle of the reflex arc, while the Umwelt idea had an influence on some philosophers (Ernst Cassirer and Martin Heidegger, for example), and was instrumental for Konrad Lorenz's development of ethology. All of which explains why, in biology's history books, Jakob von Uexküll is mentioned as a precursor (some say a co-founder) of both animal cybernetics and ethology. And that is probably what he would have continued to be remembered for, had it not been for Thomas Sebeok's 'discovery' that Uexküll's greatest merit was his contribution to *semiotics*.

As a matter of fact, Uexküll himself categorically denied (in a letter to a linguist friend) that he ever had anything to do with semiotics, and Sebeok did not actually call him a semiotician, but only a *cryptosemiotician* (*"einer der grössten Kyptosemiotiker seiner Zeit"*). Be that as it may, the fact is that in this special issue of *Semiotica* Uexküll is extensively redeployed as a chief architect of the modern discipline of biosemiotics, and so we can only ask ourselves if such a claim can be justified.

First however let us notice that the redeployment operation had to face the fact that Uexküll's philosophy of life was, to put it lightly, somewhat dated. He was not only a staunch anti-Darwinist, but also a feeble evolutionist (verging on creationism), had strong sympathies for Hans Driesch's vitalism, and often resorted to musical metaphors to explain the perfection of the living world. To the credit of this issue's contributors, these points have not been swept under the carpet, and this makes it easier to leave them in the background. It is only the main idea, after all, which must be good. And Umwelt is indeed an excellent idea. More than that. It is a deep concept, not at all easy to grasp, but once grasped is breathtaking. Surely one of the most original concepts of twentieth-century biology.

Uexküll started from Kant's idea that we do not know the world as it is (the thing in itself), but only what our mind reconstructs from the inputs of the senses, and the mind does not start from scratch in this enterprise but from inborn ideas, that is, from 'inner forms of perception'. The mind is therefore two things in one: an inner mind which provides the basic tools, and an outer mind which builds the world of appearances around us. Uexküll was deeply committed to this brand of idealism, and even when he had to take some liberties in order to apply it to biology, he never thought he was being unfaithful to his master. But liberties he had to take, and he took them.

The first was to recognize that the body takes an active part in the production of mental objects. This is why Uexküll did not speak of inner and outer minds, but of inner and outer subjective worlds: *Innenwelt* and *Umwelt*. In any organism there is something which remains private, and something else which is projected to the outside to become the world of appearances. This is Umwelt: the mental bubble that we perceive as our surrounding world.

The second liberty that Uexküll had to take was the recognition that animals have nervous systems similar to ours, and so they too must have Umwelts. Which is more or less what we mean today when we say that animals too have minds. But together with similarities there are also differences between the nervous systems of our fellow animals, and so their Umwelts are not alike. As a matter of fact, every species must have its own Umwelt because it reacts in a distinctive way to the same signals from the physical world. The concept of space is a good example here. We are convinced that we live in a three-dimensional world, but this is because the semicircular canals of our inner ear are at right angles and allow us to perceive three different directions. The same is probably true for all animals which have semicircular canals, but not for those which are deprived of them. Many animals therefore have a perception of space which is totally different from ours, and we cannot even imagine what it must be like living in a different space.

Uexküll's greatest insight, however, was probably his third amendment to Kant. He was drawn to it by the fact that animals can play, cheat, threaten, court and act (and now even dream), all of which suggests that they can react to the same stimulus in many different ways. Which in turn means that animals are *interpreters*, not just receivers, of signals. Interpreting implies the ability to transform signals into signs by giving meaning to them, and so we have before us all three basic elements of semiosis: object, interpreter and sign. Uexküll however used the word 'cue' instead of 'sign', and this is why he was not aware that he was dealing with semiotics. But 'cue' can well be regarded as a 'cryptosign', and so Sebeok was right after all in defining Uexküll a 'cryptosemiotician'. Which brings us to the conclusion that it was indeed right and proper to celebrate him as a chief architect of the modern field of biosemiotics.

THE ENDORSEMENT OF A NON-MECHANISTIC BIOLOGY

Most contributors to this special issue of *Semiotica* have openly admitted that some of Uexküll's ideas are no longer tenable, but no one has objected to his opposition to mechanism. And this for the very good reason that they too share, more or less wholeheartedly, this attitude. The endorsement of a non-mechanistic approach to life is indeed a constant underlying theme of this volume, to the point that one is almost unwittingly driven to the conclusion that biosemiotics must be incompatible with mechanism. The most outspoken expression of this stance comes from Claus Emmeche, and so it is his paper that I am referring to, but only in respect to the points that are shared by most biosemioticians. Emmeche starts by debunking one of the most common cliché of our times: the idea of *"twentieth-century biology as a fight between vitalism and mechanicism that finally was won by mechanicists"*. He points out that *"the resolution of the debate was not a mechanistic stance but a sort of historical compromise"* that has been called *organicism* (p.657).

This is very thoughtful and, sadly, very true. The fight has indeed been between three ideologies, and the winner, today, is not mechanism but the 'third road' that has *improperly* been called 'organicism'. I say improperly because the debate has been between three theoretical stances that according to logic correspond to *mechanism*, *antimechanism* and *non-mechanism*. Antimechanism is virtually synonymous with vitalism, and in this case the two names are interchangeable (antimechanism is only slightly more general). Organicism and non-mechanism, instead, are not equivalent at all, and in this case we will see that the distinction is important because it is a matter of substance, not of terminology. In order to develop this point, however, we first need to make clear what mechanism actually is.

One of the expressions that best catches the spirit of mechanism is John Maynard Smith's statement that *"we understand biological phenomena only when we have invented machines with similar properties"*. In fact, 'understanding' something means explaining it with a model that we are familiar with, and a machine gives us an immediate sense of familiarity. When we see it working before our eyes, we instinctively feel that we 'know' it. Actually, we do not even need to build a machine to get this feeling. A description is enough, and so a 'machine' is often just a 'machine-model', or an *algorithm*. One of the most famous machines of all times was built by Turing with just pencil and paper. A machine model, furthermore, is not necessarily a set of mathematical equations. Natural selection, for example, is a mechanistic model which is entirely expressed in words. The important point is that the model has the *logic* of a machine (i.e. that it delivers the same sense of familiarity that we get from a real functioning machine). Mechanism, in short, is the view that scientific knowledge is obtained by building machine-like models of what we observe in nature.

Let us now examine 'organicism', the dominant paradigm of modern biology. The standard view, faithfully reported by Emmeche, is that organicism is qualified by being *"non-vitalist, non-reductionist, and emergentist"*, and so let us see where the contrast with mechanism comes from. Needless to say, mechanism too is non-vitalist, and so the difference seems to come from non-reductionism and emergence. But a machine is a machine not when it is reduced to pieces, but precisely when it is put together into a functioning whole, which means that mechanism is *quintessentially* a non-reductionist approach. As for emergence, there is absolutely nothing in mechanism that prevents it. Take a machine that brings in hydrogen from one side and oxygen from another and there you can witness the emergence of water. In brief, it is not only misleading but downright wrong to say that the combination of 'non-vitalism+non-reductionism+emergence' amounts to something different from mechanism, because those three properties actually *belong* to mechanism.

The real problem is: why do so many biologists believe in such a muddled conclusion? This is where the difference between organicism and non-mechanism becomes a matter of substance. What most contemporary biologists share is actually *non-mechanism*, i.e. the feeling that mechanism is not enough, that there must be something else in life, and this is a profoundly respectable view. The point is that one cannot make a science out of doubts, and so some people (Ernst Mayr first in line) had this splendid idea of taking three outstanding properties of mechanism and saying that together they form 'organicism', the new philosophy of life that liberates biology from mechanism.

Claus Emmeche does not comment on organicism as such, but he must have felt uncomfortable with it, because he quickly distinguishes between *mainstream organicism* (the official version) and *qualitative organicism*, an interpretation that comes out of this volume as the view which is shared by most biosemioticians. Qualitative organicism (that in my opinion should be called qualitative biology) is the most serious attempt produced so far to put some substance into the empty container of non-mechanism without resorting to the tricks of official organicism. Emmeche describes it in this way: *"Qualitative organicism is concerned with qualities which are not only of the famous category of 'primary' qualities (roughly corresponding to the scientifically measurable quanta), but also concerned with the 'secondary' qualities of color, taste, sound, feeling, etc."* (p.658)... *"It is obvious that the Umwelt notion is of central importance to the development of a coherent theory of the qualitative experiential world of the organism, a task present day biology must face, instead of continuing to ignore a huge phenomenal realm of the living world – the experiential world of animal appetites, desires, feelings, sensations, etc."* (p.659).

Here we are then. So far biosemiotics has been the discipline which has discovered that animals are interpreters, or semiotic agents; now we are told that mechanism is not competent to study this new world. Only a qualitative science can do that. Why? Because mechanism cannot explain *meaning*, that's why.

One is not supposed to object to this piece of wisdom, but mechanism has proved to have an unsuspected resilience in the past, and has changed its skin many times in the face of adversity. The first mechanistic model of the body was the clock-machine, then came the steam-engine-machine, and lately the computer-machine. Which is equivalent to saying that mechanism introduced in biology first *mechanical energy*, then *chemical energy*, and finally *information*. Are we really sure that it cannot introduce meaning? This special issue of *Semiotica* seems to be saying that only a qualitative approach can cope with biological meaning, but we really cannot rule out a mechanistic approach. Our only option, I am afraid, is to discuss both views.

THE BIOSEMIOTIC APPROACH TO MEANING

Perhaps the most instructive and surely the most enjoyable part of Uexküll's work is his description of real-life cases of animal behaviour, cases that he reports with the light touch of the consummate expert and a masterly attention to details. It is in fact this evidence which convinces us that animals are indeed interpreters of the world, and not pre-programmed puppets. And that is really all we need to know in order to conclude that animals are 'subjects' (i.e. semiotic agents). We would need something else if we wanted to know *how* they do their semiosis, but in order to realize that they do it, the description is enough. In this case, seeing is believing.

Uexküll's biology is therefore first and foremost a descriptive science, not an explanatory one, a point which is underlined in the excellent paper contributed by Tuomo Jämsa: *"Uexküll does not expressly aim at creating an articulated theory of sign and meaning but at a description of the functions of the organisms in terms of signs and meaning"* (p.493). On many occasions it may appear that Uexküll is not just describing animal behaviour but also explaining it, and his theory of functional cycles may look like a mechanistic model, but it's nothing of the kind. A mechanistic model would say that animals receive signals from the physical world and react to them in a subjective way, which means that somehow and somewhere the signals are transformed into signs and the signs are processed by an interpreter, but all this partitioning is thoroughly alien to Uexküll. As a dyed-in-the-wool idealist, he regards signifying and interpreting as subjective operations which are *consubstantial* to any signal processing act.

In the physical world there are inanimate objects without 'qualities', but life is like a world where a sun is shining, and in such a world you cannot have objects without shadows, the shadows being the qualities which are cast into the subjective world from the quantities of the physical world. In a truly idealistic philosophy, the positions are actually reversed: the qualities, or ideas, being the real bodies, and living organisms being their shadows. This 'swapping of the roles' was operated by Plato in his famous *myth of the cave*, and has stuck in our culture ever since.

When Uexküll speaks of perception, he may seem to be putting signals first and signs second, which would be like an attempt to 'naturalize' Kant and Plato, but that is just an impression. The idea that perception starts with objective signals from the outside which are transformed into subjective signs by the organism is a mechanistic way of looking at things that Uexküll regards as pitifully naive. For him, the objects 'out there' are in fact mental entities created by the subject and projected to the outside, and so there is never a divorce between signals and signs. Quantities and qualities, objects and subjects, bodies and shadows can never be taken apart when the sun of life is shining.

A similar view of the world was also proposed by Charles Peirce, the acknowledged father of semiotics. He stated clearly that semiosis requires three basic elements – object, interpreter and sign – which are *preconditional* and therefore *primitive* entities. They must be present simultaneously in order to have semiosis, and since they are the starting point they cannot be reduced any further. Again, they are consubstantial agents of semiosis, just as Uexküll treated them.

From this special issue of *Semiotica* it seems that most biosemioticians accept both Uexküll's approach and Peirce's scheme, and therefore it is fair to say that the theoretical framework of biosemiotics is based on the ideas of Immanuel Kant, Charles Peirce and Jakob von Uexküll. In this framework, objects, interpreters and signs are primitive entities, consubstantial protagonists of any semiotic act. But a semiotic act is always associated with a signaling process, and signaling is everywhere in life, so semiosis too must be everywhere. This is why we are told *that "semiotics is biology and biology is semiotics","the semiosphere is coextensive with the biosphere","the basic unit of life is the sign, not the molecule".*

If semiosis is everywhere, life is immersed in it like in Hegel's famous night where all cows are black. When something is everywhere and primitive you don't need anything else. You just accept it and contemplate the omnipresence of meaning in life. That is the comfort that philosophy and biosemiotics can give us, and thank you very much for it.

But mechanistic science is not like that. Stubborn, nosy old mechanism wants to open up even primitive boxes and look inside them: Where is the interpreter? What is it made of? How is the interpreting done? What is a sign? Where does meaning come from? What is meaning?

A MECHANISTIC APPROACH TO MEANING

If you are a mechanist and you want to study the problem of biological meaning, you don't start with the assumption that meaning is everywhere, but the other way round. You say that you are not going to recognize the existence of anything which may be called 'meaning' unless you fall flat in its face. The evidence provided by animal behaviour is convincing enough, but animals are complicated things, and are also late-comers in the history of life, so they are not the best to start with. It seems much more sensible to begin from the beginning (i.e. from molecules), and work our way up.

Down there, at the molecular level, there are all sorts of transactions going on, and we can scrutinize them one by one to find out what makes them tick. Take for example the breakdown of sugar. At every step there is a transformation of energy with the assistance of catalysts, and that is really all there is to it, from beginning to end. In the replication of DNA molecules, instead, we encounter something else. Again we find energy exchanges and catalysts at each step, but here the end-products can be very different even when the energies and the catalysts are exactly the same. The difference is due to the linear order of the building-blocks (the nucleotides), an order that amounts to no less than a new physical quantity which has been called *information.*

If we extend our scrutiny, we realize that countless biochemical reactions can be divided, like the above examples, into two great categories: the class where all that takes place is transformations of energy, and the class where the physical quantities that count are two, energy and information. With just these two quantities we account for so many molecular transactions that there seems to be no need for anything else. Except that Nature, as usual, seems to like coming up with surprises. In our case the unexpected takes the form of protein synthesis, because by using only energy and information there is no way we can explain the fact that a chain of nucleotides is always transformed into a unique chain of amino acids. It is like obtaining the letters of the alphabet from combinations of dots and dashes. There simply is no necessary connection between those entities, and the only way to establish one is by inventing it (i.e. by creating a code). A Morse code in the case of the alphabet, and a genetic code in the case of protein synthesis.

The Morse code is built by a human being (the interpreter) who attaches a meaning to arbitrary combinations of dots and dashes (the signs) by setting them in correspondence with the letters of the alphabet (the objects). In protein synthesis, combinations of three nucleotides (the signs) are set in correspondence with amino acids (the objects) by special compounds known as *adaptors*, i.e. *by molecules that perform two independent recognition processes* (one in the nucleotide world and one in the amino acid world). In this case the agent implementing the code (the interpreter) is the entire set of adaptor molecules. Protein synthesis, in short, is a true semiotic act because the three defining protagonists of semiosis are all present, the only difference with cultural semiosis being that the interpreter is not outside the system but very much inside it. Can we generalize this experimental result? Indeed yes, we can. Any time we discover that the link between two organic worlds requires not only catalysts but also adaptors, we are very likely to be in the presence of an organic code, and therefore of organic meaning.

All this seems to be accountable with the step-by-step procedures of mechanism, but qualitative organicists apparently do not agree. Semiosis is not explicable by mechanism, they say, and a good empirical proof of this is that semiotic machines do not exist. Nor it is helpful to suggest that they may exist in the future, because the issue is not the technicalities of implementation of such machines, but the very possibility of their ontological existence. We seem to be in a no-win situation here, because it is true that all mechanistic models proposed so far, including the heterogeneous approaches which are collectively known as 'non-linear methods', have nothing to do with codes and meaning.

Clock-mechanism, steam-engine-mechanism, computer-mechanism, and now even nonlinear-mechanism, have solved many biological problems, but have not even touched the problem of meaning, which remains 'in principle' outside them, like an alien creature. In a recent book however (Barbieri, 2001), I have described a mathematical model of epigenesis where codes have an *internal* role, and this does make a difference. It tells us that there is a whole new class of models, and therefore a new type of mechanism (we may call it *epigenetic mechanism*), where codes are not alien creatures but components of the system that must necessarily be present for the system to work. It is not the solution, yet, but it does look like the foot in the door.

Mainstream organicists and qualitative organicists may well be responding with enthusiasm to the *"United against mechanism?"* rallying question that Malte Herwig is launching from this special issue (p.569), but I am not. The best chances to solve the new problems of life are still likely to come from where all our solutions have always come in the past: from good, rational, old-fashioned machine-like models.

CONCLUSION

This special issue of *Semiotica* is truly a rare book. It gives an overall view of a fascinating new field in its adolescence, when it has just completed the difficult process of growth and it is preparing to enter the creative phase of maturity. The volume is also an excellent example of a fresh way of doing academic research, where interdisciplinarity is not just an empty word but a convincing reality. The contributions from history, philosophy, linguistics, biology, art, literature and computer science integrate each other with an ease that makes one wonder why on earth the same approach has not been applied to all other fields.

The first two main points of this special issue – the making of biosemiotics and the recovery of Jakob von Uexküll from oblivion – come out with clarity and force, and are definitely a success. Normally this would have been more than enough for one book, but unfortunately there is also a third less happy theme that is developed throughout the volume. The endorsement of non-mechanism, or qualitative organicism, is in my opinion the first serious mistake of the young field of biosemiotics. Indeed it is the one drawback that can prevent it from growing into a true science. I must conclude therefore that biosemiotics has not yet come of age, but I do hope that this criticism is taken for what it is: a diagnosis that is supposed not to hurt but to help.

POSTSCRIPT (AUGUST 2005)

I sent the above review to Thomas Sebeok in August 2001, saying that I had not been able to write an impartial report and therefore that I would not be surprised if he turned it down. Sebeok, however, accepted it, and I received a biography form and a copyright transfer form to fill out. That gave me the idea to test his determination so I answered that he was welcome to publish the paper but that I needed to keep the copyright to myself for a forthcoming book. Since he had been taken ill, it was his wife Jean who replied and wrote that *"he has made some rare exceptions to the copyright rule when necessary, and he would be willing to do so in this case"*. That convinced me that Sebeok *wanted* to use the review in order to send a message to the biosemiotic community. It was an implicit message, of course, but to me it was something like this: *neo-vitalism and anti-darwinism are not forbidden in biosemiotics, but they are not compulsory either. A good, old-fashioned mechanistic approach to the problem of meaning could not be ruled out, and people who were proposing it should be listened to.*

Sebeok died a few months later, on December 21, 2001, and that indirect message was probably his last contribution to biosemiotics. Personally, I took it as an invitation to join the biosemiotic community and to argue in favour of a mechanistic approach to the problem of meaning from within that community. I decided to give it a try and asked to take part in the second Gathering in Biosemiotics that was going to take place in Tartu, Estonia, in June 2002. Since then I have been to all subsequent Gatherings and I have never regretted it, even if my proposals were received with some incredulity. The idea of a mechanistic model of epigenesis (Barbieri, 2003) was dismissed in no uncertain terms (Markoš et al., 2003). The concept that biological meaning has the same status as a fundamental physical quantity (Barbieri, 2003b; 2004) raised a few comments and then was politely ignored. Same story for the two ideas that I have been proposing since 1981: the idea that the cell is a trinity of genotype, phenotype and ribotype, and the idea that evolution took place with two distinct mechanisms, i.e. by natural selection and by natural conventions (Barbieri, 1981; 1985). To me, however, the important point was not the reception of those ideas. It was the discovery that the problems of biosemiotics were being discussed without the constraint of *ideological* principles. Gone were the triumphant tones and the neo-vitalistic declarations of the Special Issue. The reality was the feeling that nothing had been settled yet, that everything was on the move, that the exploration of the new continent of meaning had just begun.

The decisive change came in 2004, at the fourth Gathering organized by Anton Markoš in Prague. Jesper Hoffmeyer, Claus Emmeche, Kalevi Kull, Anton Markoš and myself met in a pub and decided that what was uniting us – the introduction of meaning in biology – was far more important than our divisions, and we should make that visible. Up until then, I had been referring to the science of biological meaning as *semantic biology*, or *biosemantics*, whereas Markoš had been calling it *biohermeneuthics*, but we accepted to give up those favourite names of ours and to adopt the term *biosemiotics* that Sebeok had been campaigning for with so much passion and vigour.

We also decided to make the problem of meaning visible by producing a new Journal specifically dedicated to biosemiotics. That, in my opinion, is when biosemiotics came of age. It happened when people decided to work together not because they had the same ideas but because they accepted to put their differences aside in the interest of a greater goal.

REFERENCES

Barbieri, Marcello (1981). The Ribotype Theory on the Origin of Life. *Journal of Theoretical Biology*, 91, 545–601.

Barbieri, Marcello (1985). *The Semantic Theory of Evolution*. Harwood Academic Publishers, London and New York.

Barbieri, Marcello (2001). *The Organic Codes. The birth of semantic biology*. Pequod, Ancona.
Barbieri, Marcello (2003). *The Organic Codes. An Introduction to Semantic Biology*. Cambridge University Press, Cambridge, UK.
Barbieri, Marcello (2003b). Biology with Information and Meaning. *History and Philosophy of the Life Sciences*, 25, 243–254.
Barbieri, Marcello (2004). The Definitions of Information and Meaning. Two possible boundaries between physics and biology. *Rivista di Biologia-Biology Forum*, 97, 91–110.
Krampen, Martin (1981). Phytosemiotics. *Semiotica*, 36 (3/4), 187–209.
Markoš, Anton, Eduard Gaidoš, Lászlό Hajnal and Fatima Cvrčková (2003). An epigenetic machine. Review of Marcello Barbieri: "The Organic Codes - An Introduction to Semantic Biology", Cambridge University Press, 2003. *Sign Systems Studies*, 31(2), 605–616.
Prodi, Giorgio (1977). *Le basi materiali della significazione*. Bompiani, Milano.
Prodi, Giorgio (1988). Material bases of signification. *Semiotica*, 69 (3/4), 191–241.
Rothschild, Friedrich S. (1961). The Psychology of the Self. Conference held at the New York Academy of Sciences.
Rothschild, Friedrich S. (1962) Laws of symbolic mediation in the dynamics of self and personality. *Annals of the New York Academy of Sciences*, 96, 774–784.
Sebeok, Thomas A. (1963). Communication among social bees; porpoises and sonar; man and dolphin. *Language*, 39, 448–466.
Sonea, Sorin (1988). The global organism: A new view of bacteria. *The Sciences*, 28 (4), 38–45.
Stepanov, Juri S. (1971). *Semiotika*. Nauka, Moscow.
Uexküll, Jakob von (1905). *Leitfaden in das Studium der experimentellen Biologie der Wassertiere*. J.F. Bergmann, Wiesbaden.
Uexküll, Jakob von (1909). *Umwelt und Innenwelt der Tiere*. Julius Springer, Berlin.
Uexküll, Jakob von (1920). *Theoretische Biologie*. Paetel, Berlin.
Uexküll, Jakob von (1928). *Theoretische Biologie*. 2te Auflage. Julius Springer, Berlin.

CHAPTER 4

THE NECESSITY OF BIOSEMIOTICS: MATTER-SYMBOL COMPLEMENTARITY

H. H. PATTEE*

Department of Systems Science and Industrial Engineering, Thomas J. Watson School of Engineering and Applied Science, State University of New York at Binghamton, Binghamton, NY 13902–6000

Abstract: Biosemiotics distinguishes life from inanimate matter by its dependence on material construction controlled by coded symbolic information. This irreducible primitive distinction between matter and symbol is necessary for open-ended evolvability and the origin of life as we know it. This type of subject/object distinction is reestablished at many levels throughout all of evolution. In physics this becomes the distinction between material laws and symbolic measurements and models; in philosophy this is the distinction between brain and mind. These are all emergent epistemic distinctions, not ontological dualisms. The origin of life requires understanding the origin of this symbolic control and how inanimate molecules become functional messages. I discuss the necessary physical conditions that would allow such evolvable symbolic control of matter to arise

Keywords: self-replication, von Neumann, symbolic control, semantic information, measurement, constraints, epistemic cut, protein folding

LIFE DEPENDS ON SEMIOTIC CONTROLS

We easily agree with Einstein that a Beethoven symphony cannot be appreciated as only "a graph of air pressures," although in principle it has such a physical description. In the same way we understand Bohr that, "You don't explain a tea party by quantum mechanics." On the other hand, it is not so easy to understand why you cannot adequately explain genetics with biochemistry or enzyme catalysis with quantum mechanics. Because we believe no events at tea parties, in genes, or in enzymes violate any physical laws we might assume that their descriptions differ

* Present address: 1611 Cold Spring Road, Williamstown, MA 01267. hpattee@adelphia.net

M. Barbieri (ed.), Introduction to Biosemiotics, 115–132.

only in their degrees of complexity. What biosemiotics illustrates is that symbolic controls are categorically different from laws and that they are irreducible to physical laws even though their material vehicles obey the laws and have a correct physical description.

What we need to understand is that physical laws are universal and must apply to all conceivable systems. Therefore laws are empirically moot with respect to any particular system until its particular initial conditions are specified. This requires information, and physical laws cannot specify this information. In physics jargon symbol systems are special types of initial condition called boundary conditions or constraints (Polanyi, 1968; Pattee, 1972). Consequently an adequate explanation of any living organism requires more than a detailed lawful physical description or merely the confirmation that the laws of nature are always inerrantly followed. One must explain how informational constraint structures locally control the universal physical laws so as to propagate and evolve.

All living organisms exist by virtue of hierarchies of control by informational constraints. This is the case at all levels, from the genes, to development, to sensorimotor controls, to abstract thinking, and to our technical artifacts. Symbol systems are rate-independent informational constraints that control rate-dependent dynamics by means of coding systems.

To understand what this implies one must first recognize that physical laws are universal and objective. This means that the fundamental principled requirement for a law of nature is that it is as independent as possible of all conceivable individual organisms and observers. Consequently, physical laws are based on invariance and symmetry principles that guarantee the irrelevance and impotence of any observer, organism, or mechanism to affect the laws. In other words, *physics focuses on all those universal regularities of nature over which organisms and observers have no control.* Physical laws are universal and inexorable. By contrast, the study of *biology focuses on those specific events over which the organisms and observers have local control.* Beginning with the organism's specific catalytic rate control by enzymes, evolution progresses by elaborating and testing many types of controls at many hierarchical levels. Over the course of evolution organisms have gradually increased their ability to control their internal and external environments on which their survival depends.

Survival is the ultimate function of controls, but unfortunately controls do not assure survival. In the case of humans the brain through the freedom of language and the prostheses of technology has developed controls with such Promethean powers that the art of artificial control may turn out to have exceeded what is adaptive as a survival strategy for our species.

THE RELATION BETWEEN PHYSICAL LAWS AND CONTROL CONSTRAINTS

There is a clearly problem of language here that creates confusion. What does it mean to say that universal inexorable physical laws over which organisms can have no control are in fact controlled by individual organisms? The answer requires understanding a distinction fundamental to all physical theory, the primitive separation of the laws themselves that are universal and inexorable, and initial conditions and constraints that are local and controllable, and that must be measured to have any effect. Eugene Wigner calls this principled distinction between laws and initial conditions "Newton's greatest discovery." That is because it is an epistemic necessity that is essential for all conceivable physical laws, not just Newton's laws.

Briefly, the idea is this. The universe and all systems within it are assumed to run according to universal laws whether or not observers or life exist. The mathematical descriptions of these laws are interpreted by ontological concepts of space, time, matter and energy but the laws themselves do not include the epistemological concepts of measurement and control events. However, measurement is essential if we want to predict any consequence of laws on a specific observable system. There must be measurement of initial conditions and the measurement process requires local control constraints of a measuring device or instrument.

Physical laws and initial conditions are therefore necessary irreducibly complementary categories. That is, neither can be reduced to, or derived from, the other. Measurement instruments and control constraints are special, usually complex, structures with initial conditions that are largely arbitrary. In practice measurements and controls are distinguished from the action of physical laws by how time and energy enter their descriptions. Fundamental physical laws are time and energy dependent in a mathematically rigorous sense. That is, the equations describing these laws require the concept of time-derivatives or rates of change of the states and energies of the system[1]. Also, the fundamental microscopic laws are time reversible. This physical time, sometimes called "real time," and the rates described by time derivatives are intrinsic to natural laws and are not controllable, although they may be different when measured by different observers in relative motion.

The concept of *control* of rates does not apply to universal laws but only to local structural constraints. The classical example of both rate control and time measurement is a clock. By contrast to the real-time of laws, clock-time depends on some form of local structure or constraint. We speak of clocks *measuring* time intervals but, unlike laws, clocks do not have an intrinsic rate independent of how we measure it. Also, unlike microscopic laws, measurement and control are irreversible concepts. Clocks function only by measuring local periodic structures such as a pendulum with an escapement or counter. Of course the pendulum swings according to laws, but its period depends on its length, and that is entirely arbitrary boundary condition. Escapements, whether mechanical, electronic, or chemical, can be said to control the rate at which energy "escapes" or is dissipated from the driving source, and these constraints are also arbitrary[2]. Some form of measurement is a necessary component of any functional control process.

The point I want to emphasize here is that we say a clock is a control constraint only by virtue of its locally "escaping" the inexorable time, rate, and energy dynamics of physical laws. In other words, the laws exist in time but cannot make measurements of time. Within wide limits imposed by natural laws, a clock keeps its own arbitrary time and runs at its own arbitrary rate. *This concept of local "escape" is important because life depends on it.* Enzymes control the rates of specific chemical dynamics in all of life allowing local organisms to locally escape the universal rates we associate with unconstrained physical laws. The existence of an isolated catalyst that controls rates of reaction is not considered as functional. Function, as I use the term, applies only coordinated controls initiated by organisms or executed by their artifacts.

RATE-INDEPENDENT CONSTRAINTS; SYMBOL SYSTEMS

Biosemiotics recognizes many levels of control. Rate control, as in enzyme catalysis, is only the first level of control constraints. Symbolic constraints are a type of record that requires higher levels of organization. Ernst Mayr has often pointed out that biology is not explained by natural laws because life requires the concept of an adaptive evolutionary *history*, an actual record in the organism that while obeying natural laws cannot be derived from these laws. Records are a special form of constraint that can "instruct" controls. Polanyi (1968) classifies these structures in physical terms as special types of boundary conditions that he aptly describes as "harnessing" the laws.

The word "history" has two profoundly different usages. The looser meaning is simply an implied ontological sequence of events, as in the history of the universe, or geological history, where there is no explicit record other than the actual events or structures themselves. The more specific meaning of history as used by Mayr requires a separate record of events. This latter specific meaning that is essential for evolution implies an epistemic record that is a representation or description distinct from the events that it records. In all known living systems, the genes are such historical records of innumerable adaptive natural selection processes. The relative simplicity of the record itself (the DNA) is deceptive. What is important is that for a record to have any function or meaning requires complex coding, reading and interpreting mechanisms. Along with measurement and control the concepts of biological information and instruction are not a part of physical theory in so far as they are functional concepts. That is, we cannot identify a molecule as informational unless we can identify how it is interpreted by the organism and how it functions in the organism[3]. The question remains, how does symbolic information actually get control of physical systems when it appears to be a separate category?

PHYSICAL LAWS CANNOT ADDRESS THIS QUESTION

This matter-symbol separation has been called the epistemic cut (e.g., Pauli, 1994). This is simply another statement of Newton's categorical separation of laws and initial conditions. Why is this fundamental in physics? As I stated earlier, the laws are universal and do not depend on the state of the observer (symmetry principles) while the initial conditions apply to the state of a particular system and the state of the observer that measures them. What does calling the matter-symbol problem "epistemological" do for us? Epistemology by its very meaning presupposes a separation of the world into the knower and the known or the controller and the controlled. That is, if we can speak of knowledge *about* something, then the knowledge representation, the knowledge vehicle, cannot be in the same category of what it is about.

The dynamics of physical laws do not allow alternatives paths between states and therefore the concept of information, which is defined by the number of alternative states, does not apply to the laws themselves. A measurement, in contrast, is an act of acquiring information about the state of a specific system. Two other explicit distinctions are that the microscopic laws are universal and reversible (time-symmetric) while measurement is local and irreversible. There is still no question that the measuring device must obey the laws. Nevertheless, *the results* of measurement, the timeless semantic information, cannot be usefully described by these time-dependent reversible laws (e.g., von Neumann, 1955).

THE EPISTEMIC PROCESS IN BIOLOGY

The problem is that physics greatly simplifies the matter/symbol relation by more or less arbitrarily making an epistemic cut. This avoids as far as possible the influence of measurement on the state of the measured system. Whenever an attempt is made to include the measuring device in the system it becomes the notorious "measurement problem" on which there is still no consensus.

The origin of life must address the question: How did this separation, this epistemic cut, originate? As Hoffmeyer (2000) has pointed out, the assumed sharp conceptual epistemic cut between these highly evolved categories of knower and known makes it difficult to imagine how life began and how these two categories separated at primitive levels. The epistemic cut is often treated as a conceptual discontinuity. Indeed it is difficult to imagine a "gradual cut." How does a reversible dynamics gradually become an irreversible thermodynamics? How does the concept of counting discrete units transform into the concept of a continuum (Zeno's paradox of motion)? How does a paradigm shift from classical determinism to quantum indeterminism occur gradually?

The problem arises acutely with the genetic code. A partial code does not work, and a simple code that continuously works as it evolves is hard to imagine. In fact, this is a universal problem in evolution and even in creative thought. How does a complex functioning set of constraints originate when no subset of the constraints appears to maintain the function? At least in the case of thought we can trace some of the history, but in the origin of life we have no adequate history. Even in the case of creative thought, so much goes on in the subconscious mind that the historical trace has large gaps.

The problem is that conceptually the epistemic cut divides the world in two, and the central problem is how the two worlds are connected. As C. S. Peirce has emphasized, all symbol systems are necessarily triadic systems, and the epistemic cut itself is actually a complex process. It corresponds to the *interpretation* that relates the symbol to its referent. In the cell this is an enormously complex process of transcription, translation, synthesis, folding, distribution, and selective control of many proteins. How this coordinated interpreting system originated is the central problem of the origin of life.

SYMBOLIC CONTROL IS NECESSARY FOR EVOLVABILITY

The categorical distinctions between matter described by physics and matter functioning as symbol are different at each level of biological organization. The distinction needs to be made clearly at each hierarchical level or conceptual and terminological confusion will result. It is easy to distinguish symbols at highly evolved levels like symphonies and tea parties. The words on this page are clearly symbols. Their material embodiment is arbitrary. The font is not relevant, nor do we pay attention to their material embodiment, whether they are displayed on a liquid crystal screen, a cathode ray tube, or ink on paper. Even the language we are using is arbitrary.

It is not so easy to see that the DNA of genes is also an arbitrary embodiment of a record because it happens to be the only one we know from life on earth. However, within the fields of exobiology and artificial life studies the arbitrariness of DNA is generally assumed. Many other copolymer strings or even bit strings in a computer could be interpreted or translated by a suitable coding mechanism to synthesize the same proteins as a DNA sequence.

Why is this arbitrariness of symbols essential for open-ended evolution? The most obvious property of highly evolved symbol systems such as natural language and mathematics is their enormous open-ended variety that is not limited in any significant way by physical laws. This independence is also illustrated by the fact that, unlike physical laws, the function and meaning of symbols is not dependent on the rate at which they are written or read. A mathematical proof does not depend on how long it took to produce or to read. The same it true of a work of literature. In other words, the basic observables of physical laws, space, time, matter, energy, and rates of change, have no significance for the semantic information of symbol systems. The symbolic expressions of physical laws are "about the laws" but the mathematical symbols that describe the laws do not appear to be restricted by the laws. It is just this arbitrariness that allows organisms freedom to harness laws. The necessity of symbols for open-ended evolution was first discussed by von Neumann (1966) in his lecture on the logic of self-replication.

VON NEUMANN'S DESCRIPTION AND CONSTRUCTION

Von Neumann was the first to argue that the two categories, *symbolic description* and *material construction*, are essential for self-replication that is capable of open-ended evolution. His argument was entirely abstract and by no means logically complete. It explicitly abstracted away matter, energy and all physical laws. I will first elaborate on von Neumann's logic and then I will take up the necessary *physical* conditions to realize this logic, or what he thought, "may be the more important half" of the problem.

Von Neumann's logical argument for the necessity of symbols as distinct from dynamics in self-replication was informal and largely intuitive. Nevertheless, if you understand his argument you will find it hard to think how evolvable self-replicating units could work any other way. The motivation for his argument was to understand the "threshold of complication" that would allow systems to evolve increasing complexity rather than wearing out or decaying. His logic is all the more remarkable because it correctly predicted how cells actually replicate before the discovery of the mechanisms of genetic description, coding and protein synthesis. Von Neumann began by observing that the medium of communication that feeds a material automaton is completely different than the automaton itself or its output. This was his recognition that symbols are a different category than matter. He also recognized that this was important for general-purpose computers, what is called the software-hardware distinction.

Von Neumann emphasized the "completely decisive property of complexity, that there exists a critical size below which the process of synthesis is degenerative, but above which the phenomenon of synthesis, if properly arranged, can become explosive." He was thinking of biological evolution and its open-ended variety. The essential condition here is that the individual self that is being replicated must be only one of an indefinite number of different potential selves all of which can be replicated by the same process. This raised two questions: (1) what defines the set of all possible individual selves that potentially can be replicated? And (2) how do you describe or represent the individual *self* that is being replicated?

Logic will get you only so far with these questions. For example, the concept of replication implies assembling or constructing a new individual that is like another. Von Neumann realized that how this construction can be done will depend on the nature of the available parts and on how the parts are to be assembled. He saw that if the parts were too elementary, like atoms, then both the description and construction would be a long and complicated process, while if the parts were too complex, like robots or rabbits, then there would be no real problem. He called this the "parts problem" and abstracted away the matter and energy of real construction by defining some functional operations on parts, like recognizing, moving, cutting, joining, etc., that are to be symbolically represented. There is a great amount of arbitrariness in these choices of parts and operations, but as we shall see, the basic logical separation of symbolic description and material construction does not depend on these choices.

The more fundamental question is how you make sure the replicated individual is like the original. How do you construct a copy of an organized structure made up of parts from a reservoir of these parts? There are two approaches. One is to identify the original parts themselves by *inspection* and then assemble the corresponding parts to form the copy. The other approach is to use a *description* of the original that when interpreted amounts to instructions enabling the assembly of the parts in the copy. Note that the concepts of inspection and description require an epistemic cut that separates the object being inspected or described and the record of the inspection or description. Both of these methods have advantages and disadvantages that go beyond logic and depend on the physical nature of space, time, and the nature of the parts. Von Neumann using heuristic reasoning found that taking advantage of both approaches gives the most promising results, and in fact we now know that both approaches are used in all living systems in the way that von Neumann proposed.

VON NEUMANN'S LOGIC OF SELF-REPLICATION

Following these intuitions, von Neumann began simply by postulating the existence of both symbolic and material components in the forms of a *description* and a *constructor*. The constructor would both interpret and construct what was described using parts from a reservoir. The constructor was universal with respect to an open-ended set of descriptions one of which he assumed could be the description of the constructor itself. In his notation, A was the material constructor and $\varphi(A)$ was the description of the constructor. If the description $\varphi(A)$ was fed to the constructor A, then A would construct a copy of itself, A′. We can symbolize this as $\varphi(A) \rightarrow A = A'$. This is not self-replication because the description $\varphi(A)$ has not been replicated. One might at first think that to copy the description we would simply feed the constructor a description of the description, $\varphi(\varphi(A))$, but this leads to an infinite regress since that description must also be copied, and so on.

This leads to the crucial recognition that *a symbolic description, whatever form it may take, has a physical structure that is independent of its interpretation.* In other words, to *read* the description means to *interpret* the description. To *copy* the description means *not to interpret* the description but to copy only its physical structure. Since the description is quiescent, copying can be done by inspection or by some template process. The constructor is defined to only interpret the description, so it is necessary to add another component, B, called the copier and its description $\varphi(B)$. We then can write $\varphi(A+B) \rightarrow (A+B) = \varphi(A'+B') \rightarrow (A'+B')$. This is almost self-replication except it is ambiguous. What is missing is how the new descriptions and constructions are related. Von Neumann "solved" this logically by creating a new control component, C, that takes care of housekeeping details such as inserting the new description into the new hardware constructor and separating the offspring from the parent. This component, C, amounts to what is called the operating system of a computer that takes care of the software-hardware relationship.

Von Neumann's logic and computer analogies are by no means a clear solution to the material semantics of cells. In the cell we know that the control required for cell division is a very complex process that is not yet fully understood. But the essential evolutionary consequence of von Neumann's logic is that now any additional description, D, of some new structure or function when added to this basic description will be constructed and incorporated into all future generations:

$$\varphi(A+B+C+D) \rightarrow (A+B+C)$$
$$= \varphi(A'+B'+C'+D') \rightarrow (A'+B'+C'+D')$$

This is as far as von Neumann's logic takes us. The main point of his logic is that open-ended evolution requires more than a complex time-dependent dynamics and complex chemical reactions. There must be a time-independent passive memory that by means of a coded description controls the dynamical rates of specific constructions or chemical syntheses. What I will now take up are the physical requirements that would allow such a complicated symbol-matter logical scheme to actually work in a reasonably effective way. I repeat that I am not solving the origin problem. Von Neumann himself had no clue. He thought, "That such complex aggregations should occur in the world at all is a mystery of the first magnitude." In my view, the place to look for clues is in the actual physical requirements of symbol systems where we may imagine simpler systems than we find in today's highly evolved organisms that satisfy these requirements.

VON NEUMANN'S "MORE IMPORTANT" QUESTION

Von Neumann was fully aware that logic alone was not adequate to explain cells. He warned: "By axiomatizing automata in this manner one has thrown half the problem out the window and it may be the more important half. One does not ask the most intriguing, exciting and important questions of why the molecules or aggregates that in nature really occur…are the sorts of thing they are, why they are essentially very large molecules in some cases, but large aggregations in other cases."

Von Neumann's use of *inspection* and *description* are really generalizations of highly evolved cognitive activities that need to be more precisely defined in the context of the simplest replicating unit. Copying by inspection means using physical interaction with the object directly without the use of symbols, codes, translation, or interpretation. Casting from a mold and template matching are such direct processes, as in base pairing in copying nucleic acids and the binding of a substrate by an enzyme. I should emphasize here that the physical interaction of base pairing and substrate binding are not in themselves functional or semiotic processes. It is only by virtue of their roles in the overall process of self-replication that they are interpreted as functional. Such material matchings might be interpreted in Peirce's terms as iconic signs.

A description, on the other hand, requires more complicated physical interactions that couple the description to what it stands for, its referent. This interaction in the context of self-replication can be called a code or an interpretation, and because the code constraints are themselves constructed from a description they are not determined by physical necessity. It is implicit in the concept of a code that it must apply to more than one description. In fact, to allow evolution the code must apply to an open set of potential descriptions. Again I emphasize that only by virtue of its potential function for an individual's survival can this be distinguished as a semiotic process. This chemical arbitrariness in the coding enzymes Jaques Monod (1971) calls the "principle of gratuity." It is also this construction from a description that Barbieri (2004) calls "artifact-making," a distinguishing characteristic of life. It is because of this freedom or lack of physical necessity that genetic symbol systems and the novelties of evolution have no adequate physical explanation even though they can in principle be correctly described by physical laws in every detail.

It is not clear that von Neumann saw this point since he was concerned with the logic, not the physics. However, he did argue that a description had the advantage of being quiescent, relatively time-independent, and free of the dynamics of the system it describes. It could then be copied by direct inspection. On the other hand, copying a dynamic system by direct inspection in real time would run into a problem with the parts continually changing in time. How would the system choose what state should be copied in that case? He also suggested that a complete and detailed inspection, including inspecting the inspection components themselves, would probably lead to logical antinomies of self-reference. He did not elaborate on this, but he may have been thinking of the measurement process in physics where he showed elsewhere that measuring the initial conditions of the measuring device itself leads to an infinite regress. Only by choosing at some point to make the distinction between the system being measured and the measuring device, i.e., an epistemic cut, can this regress be terminated (von Neumann, 1955).

PHYSICAL REQUIREMENTS FOR EFFICIENT MEMORY

The physical conditions necessary for memory storage are relatively simple to state as contrasted to the conditions for writing and reading of memory. The first condition is that there exist many inherently equiprobable constraint structures with adequate stability. Equiprobable means that the structures are energy degenerate or the energy of each state is the same. These states need not be exactly the same energy as long as the energy differences do not significantly affect the setting of the state by writing or the communication of the state by reading. One-dimensional copolymers and linear symbol strings are the simplest common physical structures satisfying these conditions. Such relatively time-independent memory structures function as long-term, high capacity storage.

Memory structures can also exist physically in one, two, three dimensions, or in n-dimensional networks but explicit syntax for access must be supplied. The advantages of the linear sequence memory, like nucleic acids and Turing machine tapes, and language text are (1) open-endedness or extendable capacity, (2) uniformity and simplicity of writing and reading, including ease of random access, (3) universal coding for all sequences, (4) relative isolation from the dynamics that it controls because of coding or the interpretation process. In the context of the origin of life, copolymer chains are the simplest abiogenic structures that have the necessary stability and potential memory capacity. The disadvantages of linearity are (1) lack of parallel processing or associative access, (2) low density of information storage, and (3) the necessity for an explicit code to couple one-dimensional energy degenerate sequences to the energy-dependent three-dimensional dynamics.

One can also define analog memory and codes as in analog computation. Analogs need not involve discrete symbols. This has been suggested by Hoffmeyer and Emmeche (1991), Juarrero (1998), Hoffmeyer (1998) and Barbieri (2003) in contrast to discrete or digital memories and codes. The problem with analogs is that they are all special purpose structures like individual molecular messengers that have limited informational capacity and that have no common code or interpreting process, as do genetic sequences. An autocatalytic or metabolic network may be interpreted as containing an implicit informational dynamics, but lacking an explicit passive memory structure and code it is difficult to imagine any open-ended evolvability. On the other hand, as Hoffmeyer (2000) suggests, some form of implicit analog codes may have existed as precursors of the explicit discrete codes of present life.

PHYSICAL REQUIREMENTS FOR CODING AND CONSTRUCTION

In even the simplest existing cells the steps from the symbolic base sequence in DNA to a functioning enzyme are too complex to have originated without simpler intermediate stages. However, to control construction or synthesis, even the simplest one-dimensional discrete-state memory storage that exists by virtue degenerate energy states, must somehow control the rates of specific dynamical interactions. This means that the linear degeneracy must be broken. This must be done by new interactions between the linear storage elements. In present cells this is a complex process that requires several steps. First, the DNA sequence is transcribed to messenger RNA by template copying. Next the coding enzymes and transfer RNAs translate the base triplet code to the corresponding amino acids that are then joined in sequence by the messenger RNA and ribosome machinery. Finally, the one-dimensional sequence folds into a functioning enzyme. In this process there are cases of descriptions and constructions by both template inspection and coded descriptive translations.

The discovery of enzymatic RNA made it possible to imagine a much simpler translation process in which RNA can function both as a constructing enzyme and as a symbolic description of an enzyme. By description I mean a passive structure that can be copied by template inspection, and by construction I mean a dynamic catalytic process that joins molecules by strong, covalent bonds. The main point is that this double function is only possible by virtue of the two configurations of RNA, the passive one-dimensional sequence memory and the folded three-dimensional active ribozyme.

THE PHYSICAL REQUIREMENTS FOR FOLDING AND FUNCTION

Folding transformations are the most fundamental semiotic processes in all living systems. Folding is fundamental because it is the process that transforms the passive symbolic gene sequences into the dynamic rate-control of enzymes. Folding transforms what are essentially rate-independent syntactically coded sequences into rate-dependent functional controls. Protein folding is a highly parallel process with so many degrees of freedom that is difficult to model even on supercomputers. Physically to describe folding in any structure requires two types of bonds, strong bonds that preserve the passive topological structure of what is folded, and weaker bonds that acting together hold the active folded structure in place.

This physical requirement follows from the logical definition of "folding." For example, to fold a sheet of paper means forming a three-dimensional shape without changing the two-dimensional topology of the sheet by tearing or gluing. As long as the strong-bond topological sequence structure is energy degenerate it can serves as an informational constraint or a passive memory. Folding removes this degeneracy by allowing new weak bond interactions between the elements resulting in an active enzyme. A *physical description* of protein folding is an energy minimization process or a relaxation of many weak bond interactions under the constraints of the strong bonds holding the linear sequence together (e.g., Frauenfelder and Wolynes, 1994).

How should we describe the semiotics of this process? I want to distinguish the physics and the semiotics. First, I defined a condition for symbolic information storage as a physically indeterminate (energy degenerate) structure. I assumed that all symbol vehicles obey physical laws and have, in principle, a physical description, but as I explained, that does not imply that symbol structures are physically determined. Quite the contrary is the case. Such a degenerate sequence structure can have an immense number of physically indeterminate sequences. Therefore the interpretation or function of any such semiotic or informational sequence is literally metaphysical (beyond physics).

The actual folding process, on the other hand, is an entirely physical process of minimizing the energy under the semiotic constraints of the sequence. In other words, the strong-bonded sequence can be called informational because it is one of many physically equivalent alternative sequences, while the folding dynamics itself is not informational because no new information is added in the process of minimizing the energy. (There are special cases where folding information may be added from scaffolding molecules.)

THE SEMIOTIC CLOSURE REQUIREMENT FOR "SELF"

How do we define the individual system that is interpreting the information? We need an objective criterion for what "self" is doing the interpreting and replicating, because there are innumerable energy degenerate structures that are not descriptions and many catalytic events that are not functional. What additional conditions are required to satisfy a *physical* implementation of the *logical* "self" that reads and interprets descriptions and constructs and assembles parts in von Neumann's formal self-replication.

The essential logical requirement for self-replication that von Neumann described is that all the components that implement description, translation, and construction are themselves described, translated and constructed within the "self" that is being replicated. This amounts to a *logical closure* that defines a "self." Physically this requires elaboration. There is more to the strong and weak bond requirement than the ability of the weak bonds to cause the strong bonds to fold into a functioning enzyme. The strong bonds also stabilize the passive memory and the integrity of the primary structure of enzymes. The weak bonds bind the enzyme to its substrate and control the rate of catalyzed strong bond formation. In effect, the strong bonds form the skeleton for both descriptive and constructive molecules while the coordinated organization of weak bonds define the shapes necessary to control the strong bonds, both the strong bond folding and individual strong bond formation or breaking.

These are the *physical* conditions required to implement von Neumann's logical closure. I have called this *semantic closure*, but Luis Rocha (2001) has more accurately called it *semiotic closure* because its realization also includes the syntax and pragmatic physical control processes. This complex interrelationship of strong and weak bonds is the minimum physical requirement that allows the realization of von Neumann's quiescent symbolic description and dynamic material construction. Of course the actual physical forces come in more than two strengths and evolution has refined structures at many hierarchical levels using different types of forces. Many types of strong and weak bonds enter into the complex process of folding (e.g., Wolynes, et al., 1995).

EVOLUTION REQUIRES POPULATION DISTRIBUTIONS

Based on the concept of semiotic closure, I would define an interpreter as a semiotically closed localized (bounded) system that survives or self-reproduces in an open environment by virtue of its memory-stored constructions and controls. That distinguishes interpreters from inanimate physical systems that evolve dynamically simply because they follow the memoryless state-determined laws of nature. I believe that this minimal concept of interpreter is consistent with Ghiselin's (1997) more elaborate definition of an "individual" that also applies to higher levels, like species. However, just as there are no single symbols that have meaning, so there are no single interpreters capable of efficient evolution.

Symbols exist only in the context of codes and interpreters. Symbols are recognized in an individual interpreting system just because they function in propagating the system. But we cannot stop there. We immediately see that "propagating a system" is ambiguous. The individual interpreter is not enough. The whole idea of evolution by variation and natural selection depends on a *population* of individuals that can differ in their heritable memories. This leads directly to the central issue of evolution: what kinds of symbolic descriptions, control constraints and material constructions promote survival of *populations*? Of course there is no predictable answer to this question except the course of evolution itself. All we can do is look carefully at what is actually going on in existing organisms, and see if we can discover some answers to von Neumann's question of why the molecules are the sorts of thing they are. I will mention some properties of memory, codes, symbolic control, and material construction that studies suggest promote efficient evolutionary search and natural selection.

REQUIREMENTS FOR EFFICIENT SEARCH AND SELECTION

After asking this question von Neumann remarked that it was "a very peculiar range" for the parts since they were many orders of magnitude larger than the physically elementary particles. He did not discuss this except to suggest that the size had to do with the reliability of control since in automata there is a direct correlation between number and size of parts and reliability. A certain level of reliability is certainly one requirement in order to prevent error catastrophe, but another way to look at the question is in terms of function. How small could an enzyme be and accurately bind a substrate and catalyze a specific single bond. It would have to be a large enough structure to establish a shape with the necessary specificity to recognize a substrate by folding up a linear chain. Simple models suggest that of the order of 100 amino acids is necessary.

This size creates two fundamental problems. The first problem is that the number of copolymer sequences of such lengths is immense, well beyond actual enumeration. One of the oldest, non-religious arguments against Darwinian evolution is the apparent improbability of chance mutations producing any successful protein, let alone a species. This is still an argument used by "intelligent design" advocates. This argument is based on the assumption of the sparseness of functional sequences and the immensity of the search space. The weakness of this argument is that the actual probabilities of the events in question are largely unknown.

Formulated in biosemiotic terms, to address this problem we need to know what fraction of the innumerable potential symbol strings in a genetic memory has some meaning or function when expressed by a population of individual interpreters. We need to know how the enormous space of sequences maps into the space of biological functions. The second classical problem is that functions appear to be discretely separated. That is, one function does not smoothly transform into another function. This results in the so-called trapping problem on a function or fitness landscape.

Both these problems have been studied extensively, greatly assisted by the use of computational models. Of course, there are no pure theoretical answers. Some basic empirical knowledge is required of the actual polymers that form the memory sequence space, the nature of codes that map to protein sequences, the nature of folding, and the nature of the constructive or controlling enzymes. The auspicious discovery of molecular genetics was that many mutations are neutral with respect to function and fitness (Kimura, 1983). Along with the redundancy in the genetic code, this neutrality permits searches over a wide region near a function optimum or a local fitness peak thereby alleviating the trapping problem. Trapping is also greatly reduced by the large number of saddle regions that increases with the dimensionality of the memory sequence space (e.g., Kanerva, 1988). This lends weight to the concept of quasispecies and the advantages of mutation rates near the error threshold (Eigen, 1971; Eigen and Schuster, 1979).

This search problem has been studied extensively for the simple RNA worlds of sequences and their folding (e.g., Schuster, et al., 1994; Schuster, 1998; Crutchfield and Schuster, 2003). Again the mapping of passive memory sequences to shapes that could function as enzymes appears to be highly redundant with many sequences resulting the same three-dimensional shape. Furthermore, these sequences are distributed more or less uniformly over the entire sequence space. This means that a random search need not find just one needle in a haystack, but only one of many needles uniformly distributed over the whole haystack. That is, wherever a random search begins in sequence space, it appears likely that a description of a useful molecule will be found nearby.

ANALOGIES AND DISANALOGIES OF GENETICS WITH NATURAL LANGUAGE

Biosemiotics is the study of all forms of signification and communication. It recognizes that life is distinguished from the nonliving world by its dependence on signs and symbols. However, of the innumerable examples of pattern recognition, recording, signaling, and communication throughout all levels of living organizations only two clear examples of open-ended, creative language systems exist, the genetic language and natural languages. The similarities of genetic sequences and natural language have struck linguists as well as biologists and physicists (e.g., Jakobson, 1970). These two languages can be characterized by (1) a small, fixed alphabet, (2) one-dimensional expressions in discrete sequences, (3) an immense sequences space with no significant restriction or bias from physical laws, (4) expressions not limited in what they can potentially describe by what currently exists, (5) the interpretation of sequences, their function or meaning is complex requiring highly parallel processing. In the case of genetic sequences, the essential step is folding in which many strong constraints and weaker forces act in parallel. In the case of the brain, millions of neurons are involved in interpreting even the simplest expressions (Pattee, 1980).

Natural language structure also illustrates the strong and weak bond principle, not with a hierarchy of physical forces but with a hierarchy of rules. The lexical rules are the most rigid beginning with the alphabet and the words in the lexicon. The grammar rules are weaker than the lexical rules in the sense that syntax cannot control or modify the alphabet or the dictionary. The semantics of the text does not generally alter syntax. We usually assume our writing will not change the basic meanings of words or the grammar rules depending on what we write. Similarly the sequence or meaning of the code's base triplets is not changed by the functions of enzymes they describe. Notably however, both languages have evolved exceptions to these rules, the genetic system with special editing enzymes, reverse transcription and cutting and splicing, and natural language with freedom to invent metaphors, add new words, and to violate grammar rules with figures of speech.

Of course there are enormous differences between these languages both in their embodiments, their stability, and in their range of meanings which one would certainly expect considering they originated only at the very beginning and the end of the evolutionary time scale. The genetic language began with the origin of life, and it took 4 billion years of evolution to create brains with the capability to create natural languages. The genetic language can be called highly successful in creating adaptive functions that have kept life going over this enormous time span.

As I suggested at the beginning of this paper, whether language will turn out to be a long-term evolutionary success is not at all obvious. We often refer to natural language as the defining characteristic of human intelligence. The power of language has dominated history and shaped all our cultures. Human language has not existed for more than 100,000 years and it is quite possible that it has become too persuasive for generating myths and wishful thinking that avoid basic survival necessities for the species. Also, the technology that depends on language now allows us to design genetic messages that satisfy immediate human desires rather than long-term survival of the species. Assuming humans survive the dangers of natural language and technology, one wonders what higher level of languages might evolve in 100,000 years. If humans do not survive natural language and technology, one wonders what alternative biosemiotic structures might evolve in its place.

NOTES

Sections of this paper are edited and updated selections from H. H. Pattee, The Physics and Metaphysics of Biosemiotics, *Journal of Biosemiotics* 1(1), 223–238 (2005).

[1] This statement applies to the relatively narrow range of time and energy domains within which living organisms have been found to exist on earth. Fundamental particle and cosmological theories are far outside these domains, although the possible relevance of these theories to other conceivable forms of life is an open question.

[2] Natural periodic motions like the rotation of the earth and the emission frequencies of atoms also serve as a reference for clocks, but without arbitrary and often elaborate dissipative constraints the function of any clock, that is, the measurement of time, does not occur. The word control is also sometimes used in a more general sense to describe parameters in physical systems where no function or measurement is involved.

[3] Physicists and engineers often use information in a structural rather than functional sense because of its formal relation to the entropy of a system. *Structural information* is defined in communication theory (e.g., Shannon and Weaver, 1949). Also in quantum processes one may think of structural information being transferred from the quantum system to the observing system (e.g., Zurek, 1990). I am restricting my usage to *semantic information* that functions in the survival of biological organisms and populations.

REFERENCES

Barbieri, M. (2003). *The Organic codes. An Introduction to Semantic Biology.* Cambridge University Press, Cambridge, UK.

Barbieri, M. (2004). Life is "artifact-making,"*J. Biosemiotics*, 1(1), 81–101.

Crutchfield, J. P. and Schuster, P., Eds. (2003). *Evolutionary Dynamics: Exploring the Interplay of Selection, Accident, Neutrality, and Function.* Oxford University Press, New York.

Eigen, M. (1971). Selforganization of matter and the evolution of biological macromolecules, *Naturwissenschaften*, 58, 465–523.

Eigen, M. and Schuster, P. (1979). *The Hypercycle – A Principle of Natural Self-organization.* Springer-Verlag, Berlin.

Frauenfelder H. and Wolynes, P. G. (1994). Biomolecules: Where the physics of complexity and simplicity meet, *Physics Today* February 1994, pp. 58–64.

Ghiselin, M. T. (1997). *Metaphysics and the Origin of Species.* University of New York Press, Albany, New York.
Hoffmeyer, J. and Emmeche, C. (1991). Code duality and the semiotics of nature. In M. Anderson and F. Merrell, eds., *On Semiotic Modeling.* Mouton de Gruyter, New York.
Hoffmeyer, J. (2000). "Code-duality and the epistemic cut". In Jerry Chandler and Gertrudis Van de Vijver, Eds., *Closure. Emergent Organizations and their Dynamics.* Annals of the New York Academy of Sciences, Vol. 901. pp. 175–186.
Hoffmeyer, J. (1998). The unfolding semiosphere. In G. V. de Vijver, S. Salthe, and M. Delpos, Eds., *Biological and Epistemological Perspectives on Selection and Self-Organization*, Kluwer, Dordrecht.
Jakobson, R. (1970). Linguistics. *In Main Trends of Research in the Social and Human Sciences* I, Paris-The Hague: Mouton-UNESCO, p. 438.
Juarrero, Alicia (1998). Causality as constraint. In *Evolutionary Systems. Biological and Epistemological Perspectives on Selection and Self-Organization*, G. V. de Vijver, S. Salthe, and M. Delpos, Eds., Kluwer, Dordrecht.
Kanerva, P. (1988). *Sparse Distributed Memory*, MIT Press, Cambridge, MA.
Kimura, M. (1983). *The Neutral Theory of Molecular Evolution.* Cambridge University Press, Cambridge, UK.
Mayr, Ernst (1982). *The Growth of Biological Thought.* The Belknap Press of Harvard University Press, Cambridge, MA.
Monod, Jacques (1971).*Chance and Necessity*, Knopf, New York.
Pattee, H. H. (1969). How does a molecule become a message? *Developmental Biology Supplement*, 3, 1–16.
Pattee, H. H. (1972). Laws and constraints, symbols and languages. In Waddington, C. H. Ed., Toward a Theoretical Biology 4. Essays, Edinburgh University Press, Edinburgh, pp. 248–258.
Pattee, H. H. (1973). The physical basis and origin of hierarchical control, in *Hierarchy Theory: The Challenge of Complex Systems*, H. Pattee, Ed., George Braziller, New York, pp. 73–108.
Pattee, H. H. (1980). Clues from molecular symbol systems. In *Signed and Spoken Languages: Biological Constraints on Linguistic Forms*, Bellugi, U. and Studdert-Kennedy, M. Eds. Dahlem Konferenzen, Verlag Chemie GmbH, pp. 261–273.
Pattee, H. H. (2001a). The physics of symbols: bridging the epistemic cut, *BioSystems*, 60, 5–21.
Pattee, H. H. (2001b). Irreducible and complementary semiotic forms, *Semiotica*, 134(1–4), 341–358.
Polanyi, M. 1968. Life's irreducible structure. *Science* 160, 1308–1312.
Planck, Max (1960). *Survey of Physical theory.* Dover, New York, p. 64 and p. 66.
Pauli, W. (1994) The philosophical significance of the idea of complementarity, in *Writings on Physics and Philosophy*, C. P. Enz and K. von Meyenn, eds., Springer-Verlag, Berlin, 1994, pp. 35–48 (see p. 41). First published under the title "Die philosophische Bedeutung der Idee der Komplementarität" in *Experientia* 6(Heft 2), pp. 72–75, 1950.
Rocha, Luis (2001). Evolution with material symbol systems, *BioSystems*, 60(1–3), 95–121.
Schuster, P, Fontana, W., Stadler, P. F., and Hofacker, I. L. (1994). From sequences to shapes and back: a case study in RNA secondary structures, *Proc. Roy. Soc. Lond.* B 255, 279–284.
Schuster,P. (1998). Evolution in an RNA world. In M. G. Ord and L. A. Stocken, Eds., *Foundations of Modern Biochemistry, Vol. IV: More Landmarks in Biochemistry.* JAI Press, Stamford, CT, pp. 159–198.
Shannon, C. and Weaver, W. (1949). *The Mathematical Theory of Communication.* Univ. of Illinois Press, Urbana, IL.
Simon, H. A., (1962). The architecture of complexity: hierarchic systems, *Proc. Am. Philosophical Soc.* 106 (Dec. 1962), 467–482. Reprinted in Simon, H. A., *The Sciences of the Artificial*, MIT Press, Cambridge, MA, 1996, pp. 183–216.
Von Neumann, J. (1966). *Theory of Self-reproducing Automata.* Edited and completed by A. W. Burks, University of Illinois Press, Urbana and London, pp. 74–87 and pp. 121–123.

Von Neumann, J. (1955). *Mathematical Foundations of Quantum Mechanics*. Princeton University Press, Princeton, NJ, pp. 418–421.

Wolynes, P. G., Onuchic, J. N., and Thirumalai, D. (1995). Navigating the folding routes, *Science*, 267, 1.

Zurek, W. H. 1990. *Complexity, Entropy, and the Physics of Information*. Addison-Wesley, Redwood City, CA.

CHAPTER 5

WHAT IS THE SCOPE OF BIOSEMIOTICS? INFORMATION IN LIVING SYSTEMS

STANLEY N. SALTHE

Visiting Scientist in Biological Sciences, Binghamton University, Natural Systems, 42 Laurel Bank Avenue Deposit, New York 13754 ssalthe@binghamton.edu

Abstract: Because biology involves so many scalar levels and entrains so many aspects of Nature, the notion that it is essentially a direct consequence of genetic information is questioned. After discussing some general points about information and semiosis, the hegemony of genetics is rejected on the grounds that: basic cellular phenomena are generic for micellar systems, and that the overall pattern of development is generic for dissipative structures, and that supramolecular information emerges during development, and that closely similar forms and functions can emerge at the scale of organisms and above in distantly related genealogical lineages. During the course of the paper I suggest that various epigenetic systems may be semiotic entities

Keywords: Convergent evolution, ecological equivalence, extended phenotype, genetics, habitus, hierarchies, origin of life, species, structures, systems of interpretance

INTRODUCTION

We may well ask 'where *is* biology located'? The problem here is that biology as a discourse ranges over several levels of scale (thereby subsuming dynamics over several orders of magnitude), from populations – even species and ecosystems! – to macromolecules within cells (Hoffmeyer, 1996, Salthe, 1998). This may be contrasted with the commonly held notion within the discourse that biology is essentially a product of genetic information.

Using the specification hierarchy of integrative levels (Salthe, 1993, 2002a) to organize our thinking about Nature, we have, for example:

{physical dynamics {chemical recognition/reactions {biological forms {sociopolitical arrangements}}}}

M. Barbieri (ed.), Introduction to Biosemiotics, 133–148.

This hierarchy is interpreted as {lower integrative level {higher integrative level}}, with a lower level giving rise to the next higher, which in turn integrates (contextualizes, regulates, interprets) all those lower than it. The lowest integrative level where biology is known to have a recognizable footprint is in the chemical (i.e., material) realm, where, for example, it generates accumulations of chemicals against diffusion gradients. It is generally considered that we have the basis of biology when differences between individuals of kinds of molecules (in biology, macromolecules) come to 'make a difference' (Bateson, 1971) in the results of local chemical reactions. In contrast, at the purely chemical integrative level, small differences between molecules of the same kind would have no significant consequences globally upon chemical reactions.

We may ask whether, in view of the fact that what characterizes biology most deeply is the presence of molecular level information held in the genetic system, would it not be reasonable to suppose that biology is fundamentally nothing more than the ramified consequences of a highly specified kind of chemistry? Unless we subscribe rigorously to a bottom-up ideology, biology's range (reach or footprint) over so many scalar levels (Salthe, 1985), as in:

[species [population [organism [cell [macromolecule]]]]]

would seem to argue against this. Only if all of biology could be completely explained as the direct result of effects generated by proteins could it reasonably be taken to be just an elaboration of chemistry. As a hint of where I will go in this paper, note that, even though the chemical integrative level gives (and must also originally have given) rise to biological phenomena, these in turn integrate chemical processes under biological regulation – by way, for example, of deploying substrates and depleting end products in various patterns within particular regions. Indeed, one could say that biology harnesses (Polanyi, 1968, see also Ulanowicz, 2004) chemistry to its own ends.

INFORMATION AND SEMIOSIS

As a preliminary, I will delimit the concept of 'information' as I conceive it. From information theory we know that information is engendered by a decrease in uncertainty, or by a reduction in the variety of possibilities. More important in the present context, materially, (i.e., thermodynamically), information acts as any constraint (restrictive or enabling or both) on entropy production – which is to say, on anything at all that might happen in the natural world (Salthe, 2003b). It is in this connection that information can become associated with meanings[1], thereby becoming a semiotic concept, given that its presence or effects as constraint makes a difference to some system of interpretance (sensu Salthe, 1998, see also Spinks, 1983[2]; and see the Appendix). Now, here we have implicitly circled back again to the information theoretic view because it would necessarily have to be a system of interpretance that could evaluate whether a decrease in uncertainty relevant to its interests has occurred. Failing that, there could be no decrease in uncertainty, since that would require a system of habits and expectations to be in place to begin with. Furthermore, I note that in the context of complex systems, complication requires some classification scheme to diminish the number of significant states that a system must recognize. If we examine the number of possible states a complicated system might access [see, e.g., the analysis of the immune response in Calvano et al. (2005) or of protein interaction networks in Rual et al. (2005)], it becomes clear that only certain states in a very large number of possible ones could have significance for any finite system. This requires a standing reduction in the number of meaningful states by way of classifying recognized differences in only a few of the many degrees of freedom afforded by complexity, which few then carry the information used by, and meaningful to, the system. A functional complex system must, therefore, be able to classify its own states as well as those of its surroundings, and so must necessarily have information..

INFORMATION AND HISTORY

We can proceed further using Pattee's (1977, 1982) distinction between dynamics and informational constraints. Informational constraints are configurations that modulate entropy production. As a radically out- of- equilibrium, presumably isolated system, the main business of Nature must be to return to thermodynamic equilibrium (Salthe, 2002b, 2003a, 2004a) by way of dissipating all energy gradients (Schneider and Kay, 1994). Underwritten fundamentally by gravitation, organizations, forms and masses, as well as matter itself, have all erected barriers to a quick passage of the world to the 'blessed' equilibrial state. Every form is potentially a constraint on energy dissipation, instituting friction on the dissipative process, generally slowing it down (Lotka, 1922). Such informational constraints, in their role of informing natural processes, are prominent among the objects of semiosis (which I take to be the construction and interpretation of meaning as mediated by signs[2], and the Appendix). A very simple abiotic example of informational constraint would be a ridge in the pathway of running water, imposing friction upon its flow.

Constraining configurations like this are generally the results of historical contingencies – and so information generally is instituted, and therefore meanings mediated, by the effects of history, at all scales. These effects tend to be preserved because the material world is sticky and easily marked, this being supplemented in biological systems by the process of preservation of genetic marks by natural selection. Eventually new marks may obliterate older ones (think of craters on the moon), but usually not completely. Erasing information (instead of just modulating it) is in fact extremely difficult in the material world, usually consuming considerable effort and energy to get to where every trace has been deleted. In biology this can take many generations of natural selection gradually diminishing the representation of some genetic information in the gene pool of a population, after which even more activities must transpire before it will be eliminated completely by way of genetic drift.

Here we may note that natural science has been almost completely concerned with ahistorical possibilities (e.g., the typical function of mitochondria in a cell, or the average mode of generation of storms by way of energy gradient dissipation, or the normal course of development of a kind of embryo or in the main sequence of stars), in which history is obliterated statistically, whereby variety of values is decreased down to just a few statistical moments, establishing average values as the normal data of science. But in the context of the Big Bang even the universal constants of Nature could be thought of as the results of history. I should acknowledge as important exceptions to the nomothetic focus of natural science, that we do have, e.g., cosmology, historical geology, and evolutionary biology, which seek as part of their practice to pinpoint when and in what sequence various particular events occurred. However, even here statistics are often used as tools to establish likelihoods of precedence. And so, historically acquired information as such (unassimilated by statistics), has had only a small place among the facts of natural science.

In biology, genetic information imposes constraints on chemical dynamics, the results of which go on to inform the activities and generation of dynamic micro- to mesoscopic forms – living cells. This information then continues to inform cellular behavior to the effect, among other things, of generating emergent macroscopic forms, like layers of mucus, syncytia, organisms, colonies, and their populations. These in turn participate in megascopic ecosystemic activities organized around energy flows emanating from the dissipation of energy gradients. Thus, using the scale hierarchy format (Salthe, 1985), we have for example: [ecosystem [population [organism [cell [macromolecule]]]]], the levels here, [megascopic [macroscopic [mesoscopic [microscopic]]]], being separated by average dynamical rate differences of at least about an order of magnitude (Salthe, 2004b) which keeps the levels functionally separated at different scales.

I note here in passing that, important as genetic information is at the molecular level, there is as yet no understanding of how the genetic system could have evolved (see some thoughts in, e.g., Guimaraes, 2006). This remains at present the major unsolved mystery in biology. Here we will merely accept this microscopic level of information as being present, and proceed from there, even when discussing, below, some aspects of the origin of life.

REPRESENTATION IN DNA

Biology assumes its ecological role – increasing the variety of energy gradients being dissipated, as well as increasing the thoroughness of some of that dissipation to heat energy (Salthe, 2004a, 2005a) – when different historically generated 'versions' of the informational macromolecules (mutants) become associated with dissipative structures in different types of locales, or with different energy sources. These associations, which are mediated in biology by the diverse adaptations of organisms and free living cells, establish formal causes of meaning (i.e., what?, where?, how?), and are the sources of biosemioses spanning several levels of scale. Thus, a kind of organism or colony finds itself successfully reproducing its genetic information in a given habitat in a given biome, living in a given Umwelt, working a given ecological niche (Salthe, 2001). This genetic information, whose interpretation by the involved cells was in part responsible for the success, is in that way reproduced as well. Biosemiosis results in increases in mutual information between locales and their non-constitutive occupants over time (Collier, 1998), a process that is understood in biology today as mostly the results of natural selection. In that model, initially adequate accidental relationships between biological systems and their environments become gradually refined into better adaptations – or, indeed, the biological system may just manage to hold on when faced with the exigencies of environmental change (viz., the 'continual deterioration of the environment' of Van Valen, 1973 – see Salthe, 2005b).

So macroscopic to megascopic environmental associations are what afford meanings for the microscopic sequence information in nucleic acids, as interpreted by mesoscopic systems of interpretance (cells, organisms) successfully engaged in macroscopic niche transactions. (See the Appendix for restatement in more detail.)

These semiotic associations are conceptually related to Dawkins' (1982) "extended phenotype" concept, wherein a nucleic acid sequence is held to be capable of representing the combined informational constraints held in environmental configurations extending from cells outward even to as large a scale as climatic weather patterns. The genetic information consulted by cells reflects aspects of those cells, and, since they are doing well in an organism or colony, it reflects to a degree aspects of these as well. Since that organism or colony is successful in a given habitat in a given biome, these too are to some degree reflected in that information – which could therefore be expected to be different in a different biome.

But careful consideration of these relations shows that the simple notion of an actual transition of meaning across levels of scale, characterized as these levels are by dynamics of very different rates, cannot really be the case. Meanings represented in DNA sequences reflect relations that overall can be modeled as the scale hierarchy: [climatic region [biome [population [organism [cell [macromolecule]]]]]], interpreted as [higher level [lower level]], with increasing span of numbers of individuals as we go down to lower levels. Meaning cannot normally transit directly through these levels, but cascades from level to level, and is transformed (transduced) at each one. The information present at any given level of scale is made possible by informational configurations at the next lower level, and contextualized by configurations at the next higher level, which have permitted it to thrive (Salthe, 1985). This means that configurations at every level function immediately as informational constraints in respect only to the next lower level. So a pattern of weather, say, would have – as such – no meaning for a cell within an organism, because that cell relies for its effective information on cues only from conditions within the organism itself. The organism, however, can respond to changes in weather. So, e.g., a hurricane, as such, cannot touch a cell within an organism even if it destroys the organism. So meanings in this system are transformed over a cascade of downward constraint relations. For example, the scouring pressure of a waterfall is 'too far away' in scale to carry meaning for an algal cell growing in a colony on its rocks. The layer of slime secreted by the colony of cells is here interposed between a cell's DNA and the rushing water. In this case the DNA would hold information relative only to some glycoproteins or mucopolysaccharides, with no direct reference at all to flowing water.

So, accumulated information in the DNA of, say, an algal cell at any moment might be parsed, for example, as: [winter [ice [many neighbors [metabolic pattern X [increased molecular stability]]]]]. It is true that genes coding for proteins would be 'judged' with respect to all these levels simultaneously. Frequent failure of a given kind of protein to function adequately for a cause traceable to some fluctuation beyond normal at any one of these levels, even if the protein's function would have been adequate relative to conditions at all the other levels, could result in death of the cell. Now, this, however, would not be a direct transit of information across all the intervening levels because arrangements at any of these other levels might have been able to compensate for the fluctuation so that the function in question would not have been so severely challenged by an effect of that fluctuation. Thus, information at any genetic locus is potentially contextualized by all the information in the genome, and – crucially – configurations at other loci, some referring to configurations or activities at different levels of scale, could compensate for stress bearing upon the product of a given genetic locus, thereby preventing the action of selection. Since meanings reflected in a given gene product are contextualized by meanings reflected in many others, which could either support or problematize its function, no locus can be said to hold, by itself, information about any biological activity.

THE ORIGIN OF LIFE

Materially, ecosystems were originally just abiotic energy flow landscapes holding dispersed energy gradients, minerals and other potential resources (Ehrensvärd, 1960, Odum, 1971). These systems fostered chemosyntheses globally powered by the sun, as well as locally in some places by vulcanism. These were powerful energy sources, the tremendous magnitude of whose energy flows would have prevented development beyond relatively simple proto-living systems because they would disrupt more delicate micro- and macroscopic forms almost as soon as they formed. Solar intermittence was therefore necessary to provide a dark period for molecular and supramolecular folding into least free energy configurations with enhanced stability. And the presence of shaded ledges could also have been the sites of further complication, as well as any other more temperate locales that would have been reached by turbulent dispersal away from submerged outlets of terrestrial heat energy which had driven concatenations of chemosyntheses (Corliss, 1988).

The origin of life – that is, the emergence of replicability of linear macromolecules that can betoken meaning – needs (whatever its mechanisms were) to have been fostered by macroscopic dissipative structures mediating the required energy flows and resources. Some of these dissipative structures plausibly later became co-opted into the biological system proper, becoming in that process more defined and complicated. Eventually they would function as the very focused systems of interpretance that we know as organisms, colonies and symbiotic associations. These would continue to be dependent upon their local megascopic ecologies, to which they would have become especially adapted for energy, materials and waste sinks. It should not be thought that the microscopic structure and functioning of the cell itself was a particular product of biology, without abiotic precursors. Sydney Fox's extensive work on proteinoid microspheres (reviewed in 1988) showed that these could perform many of the functions of living cells – maintenance of internal difference, including charge separation between inside and outside, division, fusion, growth, and other functions. Once formed, they are quite stable, but do show the effects of aging. As well, when artificially fossilized, they do not differ from some Precambrian microfossils (Francis et al, 1978). The point I am getting at here is that many cell functions are generic for enclosed microscopic physicochemical systems like micelles, and do not depend, as such, upon genetic information (Salthe, 1972).

SUPRAGENETIC INFORMATION

Species

Once an array of genetic information is being replicated, copies of it can get transported to other areas, some perchance with similar enough ecosystemic conformations to allow tentative colonization. Successful colonization by an informed system is mediated by meanings evoked from its own configuration/conformations (which are generated in contemporary biological systems by translation of DNA 'messages' during development) if the system happens to find itself in a supportive environment. Successful colonization of different habitats isolated from each other would allow the preservation of different genetic arrays, eventually generating different biological species, each related to, and eventually adapted to, a different habitat. Species can be said to hold ownership of the biological information embodied in the nucleic acid arrays held within cells and organisms, even though their scale is very much larger than that of the actual informational molecules. We know this because local extinctions of populations do not necessarily eliminate a species' information in these locales, as there may be other populations of the same species elsewhere which can supply immigrants for repopulation (Eldredge, 1989). Furthermore, no single population, any more than any individual organism or cell, will have all the genetic variants belonging to a species. So the species is the actual storehouse – and indeed the owner – of genetic information, deploying it by way of mediating dispersal of its organisms or their propagules. This is managed either by the shape and location of its areography (Rapoport, 1982), or by way of organismic and propagule dispersal abilities coded for in its genes. So, even if one assumes that the most characteristic biological information exists in nucleic acids, it must be admitted that it is deployed through the agency of species. Since this involves at least some material constraints independent of genetic information, as in the areographic pattern of a species occupation of space, or in migration routes learned by its organisms, so species have a separate existence as systems of interpretance (somewhat, perhaps, in the way that a bank has an existence separate from the money it manipulates).

As an exercise in pansemiotics, we might consider species as systems of interpretance in more detail. We first need to cast our thoughts upward in scale, so that, as observers, we see them as if smaller than ourselves. We would then see a species' areographic range looking rather like a carelessly assembled amoeba, with parts of its pseudopods dissociated from the main body, which itself does not occupy space fully, but rather with greater or lesser aggregations of the organisms or cells that form its 'atoms'. There are as well gaps between its populations, which are its reproductive organs. These are connected by way of gene flow carried by immigrants from one to another, which, if our observation scale is large enough, we would see scooting back and forth between them. Now, considering Peirce's triadic formulation of semiosis (see the Appendix), we must contrive to find a species' interpretants, the signs it 'attends', and the objects it is relating to. The external objects of importance to it are various environmental properties. The signs it is attuned to are the environmental affordances that it is particularly capable of constructing out of environmental raw materials, should the latter be present. Such an affordance might be, say, soils of a given pH with just the right amount of shade during a given season and time of day. Its interpretants are emigrant populations established by what seem to be wayward propagules or organisms. Note that Peircean semiotics is an abstract system of relations that is capable of being overlaid upon any complex enough material system. Resistance to this procedure would likely be driven by refusal to open up observational scale, motivated by pragmatic rather than philosophical concerns.

Convergent Evolution

With this issue we have broached the general problem of where biology exists more explicitly. A major phenomenon to support my perspective here is convergent evolution (Conway Morris, 2004, Willey, 1911). This is the situation whereby quite distantly related biological lineages generate similar anatomical organs, organisms, or even whole ways of life from different ancestors that were not similar in these respects, and whose genetic information therefore could not have been the same. Perhaps the most famous example is the independent evolution of essentially the same kind of eye in both vertebrates and cephalopods (squids, octopi and cuttlefish). Being only very distantly related, fishes and squids, for example, do not generally resemble each other closely other than by being streamlined for swift motion in water – this being a very general adaptation easily conceived as the separate results of natural selection in different lineages. Moving swiftly through the dense medium of water requires this shape of any organization that takes up that way of progression, and is such a general requirement that almost any starting point could be molded in this direction – even people have invented submarine boats! Another such example would be the evolution of woody trees in many different plant lineages. Such very general similarities need not be reckoned to show convergent evolution.

The focusable eye, however, is quite a different matter. It is a complicated arrangement of cornea, lens and retina, capable of focusing light from different distances. The key point for us here is that the genetic information involved in this eye must be completely different in the two lineages under consideration. The alpha-crystalline proteins in the vitrius humor of the lens were found to have very different conformation when comparing those from mammals and squids immunologically. This raises a key point here. Genetic information is today accepted as the basis of biological form, yet here we see that such form can exist independent of the information in that basis. It is well to note here also that evolution has produced a number of completely different kinds of eyes among animals, and so it is not the case that an eye, for some unknown lawful, nomothetic reason, must always have the same form. So it is as if the focusable eye, as an independent form, is enlisting different informational tokens to code for it in different biological lineages, thus acting like a 'deep structure' in Nature.

An interesting example of a kind of evolutionary convergence is the iterative evolution of the same suite of multiple body forms (pikes, panfish, eels, etc.) in three fish faunas that replaced each other after major periods of extinction throughout the fossil record – the paleoniscoids were followed by holosteans, who were replaced by teleosts. These faunas independently evolved these same forms starting in each case from a more typical fish form, and since they were not closely related, the genetic basis of these forms in the different lineages must have been quite different. The teleosts, who are still with us, and so known in more detail, have evolved many other kinds of 'bizarre' forms which, since we don't know the earlier faunas nearly as well, we cannot say whether they were represented in them. A favorite example of mine to illustrate evolutionary convergence is the comparison between one kind of bizarre fish – seahorses – with chameleons, because these two are basically so different in underlying structure, and in habitat as well. And, of course, they are not closely related genetically. These two vertebrates are similar in size and in a bushwhacking mode of predation, lurking or creeping slowly along on vegetation, then striking swiftly. Both are cryptically colored, both have slender prehensile tails, and both have independently moving eyes as well. Furthermore, seahorses bear live young, and so do some chameleons. Such examples of evolutionary convergence are particularly notable in light of the fact that evolutionary biologists today use only Darwin's descent- with- modification model of evolution, conceptually delivering only a diverging process of evolutionary change. In any case, we see that organismic forms at the macroscopic level have a potential existence independent of the particular microscopic configurations of gene arrays.

Habitus

Another way to problematize gene hegemony in biology is to consider exactly where a species' form might be said to exist. Darwinian biology has established the fact of individual variability in details of form, linking this with concurrent variability among individual genomes. Yet individuals of given species can generally be identified as members of that species with little trouble. Each is individually different, yes, but in some larger sense all are the same. The concept of 'habitus' is useful here. For example, the North American sugar maple tree has a well-known egg-shaped form in middle age when free standing, yet the branching pattern producing that form differs in every individual. This overall form is even preserved when two or three individuals grow so close to each other that they share in producing it. The branching pattern of growth is certainly directly influenced by genetic information, as can readily be seen by comparing the twigs of different species (say, ashes contrasted with maples). Even individuals of one species can have clearly visible branching pattern differences; individual sugar maples can be consistently more, or less, densely twiggy for example. As well, the branching pattern responds easily to environmental perturbations, and yet somehow that overall egg shape of this kind of maple is regulating overall growth of these trees. Mushrooms are interesting in this way as well, it having been noted that the mycelial hyphae making up the mushroom cap are haphazardly tangled together in a way that does not prefigure, or in any way relate to, the species-specific shape of the cap (Bonner, 1952). It is as though the mycelia were poured upward into a mold. So a species has a form which it imposes upon its individual participants, but which cannot be said to be imposed directly by way of a chain of information originating in microscopic genetic transcripts, which could be said to directly produce, for example, the process of branching of twigs. No doubt a geneticist could suppose that some sort of cellular interpretation of environmental cues allows this kind of regulation. If so, it needs to be demonstrated.

At a larger scalar level there is a related phenomenon – the migratory routes of various animals. These are quite stable, and cannot plausibly be represented in the genome, but are learned by individuals from their parental generation. The ability to learn would be, of course, a genetic predisposition, but the actual routes themselves exist only as used. This may be another kind of information that could be said to be "owned" by species, or at least by populations.

Genetic Discourse

I will note here that genetic information shows up in biology discourse as differences between individuals, or species; to the extent that these are similar, genetic information is not invoked, and is discursively irrelevant. Once an inheritable difference is discovered, it is true that searching the DNA will likely uncover a difference therein that can be understood as the underlying informational distinction that can be preserved by inheritance in a population. Yet it remains the case that in the absence of phenotypic difference there is no reason to invoke genetic information – and of course there might be genetic differences that are phenotypically neutral as well. This latter point obviates the argument that phenotypic difference is no longer required to find genetic difference given modern gene sequencing technology. Without correlation to phenotypic difference, however, genetic difference is semiotically meaningless – 'neutral' (the exception here is their discursive use in taxonomy). The import of this would be that, for example, the presence of four limbs in all tetrapods is not a fact related to genetics. Discursively, genes code only for differences, while materially they provide the means for recording those differences. It might even be said that genes create differences, in the sense of stabilizing them.

It has recently been argued that, besides genes, organisms inherit many other 'resources for development' (Griffith and Gray, 1994). First and foremost we should note here the biological system itself, either a daughter cell or a propagule, which clearly are systems of interpretance with respect to the genetic information. No genetic information exists, or is passed on, independently of them. Even viruses have some non-nucleic acid parts. But there is more. Consider local mycorrhizal relationships among plants, or beaver dams and pack rat nests, both of which are inherited within a primary family. And there are trails in the mountains that have been used by deer for many generations. Plants inherit environments that have been favorably modified by their immediate progenitors, as when certain trees and shrubs in dry habitats produce very flammable leaves which accumulate around them to the point where they eventually ignite, burning out competitors that are not, like them, capable of surviving the flames. Then there is the extended family in many mammals and birds – flocks, packs and herds of related individuals, functioning as units. It might be said that these resources could not be exploited except by systems having a particular array of genes, but it is clear that these examples show that biology is much more than a molecular phenomenon. Molecular information in genes functions as switches and placeholders in a much larger material system, itself maintaining non-genetic informational constraints.

What appears to be the case is that macroscopic biological forms, produced after all by nothing more than physico-chemical processes, were implicit – even immanent – in Nature prior to being co-opted by genetic systems by way of a "genetic takeover" (Cairns-Smith, 1982, see also Odum, 1971).

Ecological Equivalence

Robust biological forms independent of genetics exist even at the megascopic level of the biome. There are a number of well-known examples. The pine barrens vegetation is very similar up and down the sandy coasts of eastern North America, even though most of the species existing in the separate regions north and south are not the same. In this case, for example, two very distantly related lineages have produced a striking plant with a curly wooly body that lives draped over tree branches without contact with soil. In the north this is the old man's beard, a lichen, while down south it is the flowering plant (a bromeliad), Spanish moss. A famous example of biome level convergence is the existence of nearly identical Mediterranean vegetation forms in Australia and California, as well as in the Andes Mountains, all regions having similar climatic regimes. We must conclude that potentially accessible forms (deep structures in the structuralist sense) exist at many levels of scale in the world, and can be represented indirectly in coded form in the microscopic gene arrays in living cells by quite different arrangements. That is to say, we have evidence for stable meanings at all of these levels – ecosystemic and organismic as well as molecular. And so organisms, as such, and biomes, appear to be genuinely semiosic entities, and inform us that biology is not essentially only a microscopic, molecular phenomenon, even if genes are essential to understanding it. Here is an interesting quote from Guenther Stent (1981) on this point:

> Consider the establishment of ecological communities upon colonization of islands or the growth of secondary forests. Both of these examples are regular phenomena in the sense that a more or less predictable ecological structure arises via a stereotypic pattern of intermediate steps, in which the relative abundances of various types of flora and fauna follow a well-defined sequence. The regularity of these phenomena is obviously not the consequence of an ecological program encoded in the genomes of the participating taxa.

I should not close this section without noting the extensive mycorrhizal connections between plants in a flora, mediated by various fungi linking plants through their root hairs. These connections suggest the possibility of supraorganismic individuality, as it is known that substances are passed from plant to plant by this route.

Development

Reaching even further afield from genomes, it is interesting to see that processes often held to be biological (and therefore presumably under regulation by genes) are actually found throughout dynamical material systems. A case in point is the pattern of development during ontogeny. When examined using very general informational and thermodynamic criteria, it can be seen that all dissipative structures, living and abiotic, follow the same general pattern, which can be described with four rules, as follows (Salthe, 1993, see also Ulanowicz, 1997):

(1) There is an asymptotic increase in size, in information content, in orderliness, and as well in gross energy throughput (power).
(2) There is an initial increase up to a peak, followed by a gradual decrease, in mass specific power, these three phases (peak included) being constructible as the developmental stages, immaturity, maturity, and senescence. Dissipative systems thus exist as higher level developmental trajectories, (immature → mature → senescent), ending naturally in failure followed by recycling. Because of the development of senescence, it is often said that dissipative structures move during their existence toward a mass specific minimum entropy production regime, but few natural ones get anywhere near that before getting recycled.
(3) There is an increase in internal stability, involving increases in stereotypic behavior which produce loss of flexibility, demonstrating for an outside observer an increasing predictability of internally generated activity.
(4) Consequent upon the increasing rigidity devolving from (3), there is a decrease in stability to perturbations.

Since this pattern is found in ecosystems and even in abiotic dissipative structures like tornadoes and hurricanes, it cannot be the case that this important developmental trajectory was an invention of biology, requiring genetic information to be passed on to future generations. Rather this canonical pattern is more like a law of matter, 'inherited' by biological systems from their abiotic precursors as a condition for existence as dissipative structures

CONCLUSION

I believe I have demonstrated that there is more information bearing upon biological systems than just that carried in their genetic arrays, and that we must therefore suppose that biology is semiotically more than the consequences of cells interpreting genetic information.

For one thing, the basic microscopic forms and functions of cells appear to be generic for micellar structures, and so these chemical level properties were apparently just co-opted and stabilized by living systems. Therefore, no special information regarding these structures (like the bimolecular leaflet form of the cell membrane) needs to be supposed to be carried in genes. As well, biological systems have inherited the canonical developmental system (Salthe, 1993) that appears to be generic for dissipative structures. Beyond these, and more particularly biological, are forms assumed by organisms as revealed by a species' habitus, which emerges from processes mediated by gene products but seemingly cannot be directly derived from them. This implies some sort of emergent, supramolecular information. Then we have forms involved in phenomena of ecological equivalence, particularly as revealed by convergent evolution, which do not seem to have any connection to genetic information at all, and are the best observations in biology for suggesting the possibility of deep structures in Nature. This view is further enhanced by examples of similarities of whole vegetations at a much greater than organismic scale.

From these considerations, it seems most reasonable to assume that the major role of genetic information in biology has been to stabilize spontaneously emergent material forms and to provide access to structural attractors, which is to say, to harness informational constraints present generally in Nature. (Subsequent elaboration of structure was of course facilitated by genes.) Biological processes appear to have led to the emergence, therefore, of semiotic systems of interpretance at several scalar levels, and so, while biology's key invention – the genetic system – has been the basis of its success, this success was attained only by conspiring with informational properties at large in Nature.

ACKNOWLEDGEMENTS

I thank Marcello Barbieri, Gary Fuhrman, Edwina Taborsky and Bob Ulanowicz for helpful comments

NOTES

[1] I elaborate here on what I intend by "meaning" and "sign". Meaning is a degree of matching of indications from an external object's counterstructures (von Uexküll, 1926) with a system of interpretance's forms or categories, increasing that system's sensitivity and readiness relative to that object, as embodied in its constructed interpretants (see Appendix). A sign is the focus of mediation of meaning. A system of interpretance constructs a system of signs – its Umwelt (von Uexküll, 1926, Salthe, 2001) – from its lexicon of meanings. A pansemiotician would suggest that these relations can be generalized throughout Nature.

[2] A system of interpretance is a locale capable of relating to another (called the 'object') by way of generating interpretants, via the process of semiosis (see figure in the Appendix), some of which interpretants then facilitate interaction.

BIBLIOGRAPHY

Bateson, G., 1971. The cybernetics of "Self": A theory of alcoholism. *Psychiatry* 34: 1–18.

Bonner, J. T., 1952. *Morphogenesis: An Essay on Development.* Princeton: Princeton University Press.

Cairns-Smith, A.G., 1982. *Genetic Takeover and the Mineral Origins of Life.* Cambridge, UK: Cambridge University Press.

Calvano, S.E., et al, 2005. A network-based analysis of systemic inflammation in humans. *Nature* 437: 1032–1037.

Collier, J.D., 1998. Information increase in biological systems: how does adaptation fit? in G. Van de Vijver, S.N. Salthe and M. Delpos (eds.) *Evolutionary Systems: Biological and Epistemological Perspectives on Selection and Self-organization.* Dordecht: Kluwer Academic.

Conway Morris, S., 2004. *Life's Solution: Inevitable Humans in a Lonely Universe.* Cambridge, UK: Cambridge University Press.

Corliss, J.B.,1988. Hydrothermal energy flow on planetary bodies and the creation of living systems. In Marx, G. (ed.), *Bioastronomy – The Next Steps.* Dordrecht: D. Reidel. pp. 195–200.

Dawkins, R. 1982. *The Extended Phenotype.* San Francisco: Freeman.

Ehrensvärd, C., 1960. *Life: Origin and Development.* Chicago: University of Chicago Press.

Eldredge, N., 1989. *Macroevolutionary Dynamics: Species, Niches, And Adaptive Peaks.* New York: McGraw-Hill.

Fox, S., 1988. *The Emergence of Life: Darwinian Evolution From the Inside.* New York: Basic Books.

Francis, S., Margulis, L. and E.S. Barghoorn, 1978. On the experimental silification of microorganisms II. On the time of appearance of eucaryotic organisms in the fossil record. *Precambrian Research* 6: 65–100.

Griffith, P.E. and D. Gray, 1994. Developmental systems and evolutionary explanation. *Journal of Philosophy* 91: 277–304.

Guimaraes, R. C., 2006. Stability in the genetic connections: The code and the nucleoprotein associations. *Journal of Biosemiotics* 2: in press.

Hoffmeyer, J., 1996. *Signs of Meaning in the Universe.* Bloomington, IN: Indiana University Press.

Lotka, A.J., 1922. Contribution to the energetics of evolution. *Proceedings of the National Academy of Science.* 8: 147–151.

Odum, H.T., 1971. *Environment, Power and Society.* New York: John Wiley and Sons.

Pattee, H.H., 1977. Dynamic and linguistic modes of complex systems. *International Journal of General Systems.* 3: 259–266.

Pattee, H.H., 1982. Cell psychology: an evolutionary approach to the symbol-matter problem. *Cognition and Brain Theory* 5: 325–341.

Polanyi, M., 1968. Life's irreducible structure. *Science* 160: 1308–1312.

Rapoport, E.H., 1982. *Areography: Geographical Strategies of Species.* New York: Pergamon.

Rual, J.-F. et al, 2005. Towards a proteome-scale map of human protein-protein interaction network. *Nature* 437: 1173–1178.

Salthe, S.N., 1972. *Evolutionary Biology*. New York: Holt, Rinehart and Winston.
Salthe, S.N., 1985. *Evolving Hierarchical Systems: Their Structure And Representation*. New York: Columbia University Press.
Salthe, S.N., 1993. *Development And Evolution: Complexity and Change in Biology*. Cambridge, MA: MIT Press.
Salthe, S.N., 1998a. Naturalizing semiotics: an extended review of J. Hoffmeyer, 1996, *Signs of Meaning in the Universe*. *Semiotica* 120: 389–394.
Salthe, S.N., 1998b. Semiosis as development. In J. Albus and A. Meystel (eds.) *Proceedings of the 1998 IEEE ISIC/CRA/ISAS, On Intelligent Systems*. pp. 730–735. Gaithersberg: IEEE Press.
Salthe, S.N., 2001. *Theoretical Biology* as an anticipatory text: the relevance of Uexküll to current issues in evolutionary systems. *Semiotica* 134: 359–380.
Salthe, S.N., 2002a. Summary of the principles of hierarchy theory. *General Systems Bulletin* 31: 13–17.
Salthe, S.N., 2002b. An exercise in the natural philosophy of ecology. *Ecological Modelling* 158: 167–179.
Salthe, S.N., 2003a. Infodynamics, a developmental framework for ecology/economics. *Conservation Ecology* 7. www.consecol.org/vol7/iss3/art3.
Salthe, S.N., 2003b. Entropy: what does it really mean? *General Systems Bulletin* 32: 5–12.
Salthe, S.N., 2004a. The natural philosophy of ecology: developmental systems ecology. *Ecological Complexity* 2: 1–19.
Salthe, S.N., 2004b. The origin of new levels in dynamical hierarchies. *Entropy* 2004, 6[3]: 327–343. http://www.mdpi.net/entropy/list04.htm
Salthe, S.N., 2005a. Energy and semiotics: the Second Law and the origin of life. *Cosmos and History* 1: 128–145.
Salthe, S.N., 2005b. Semiotics in biology: inside neoDarwinism. *Journal of Biosemiotics* 2: in press.
Schneider, E.D. and Kay, J.J., 1994. Life as a manifestation of the Second Law of thermodynamics. *Mathematical and Computer Modelling* 19:25–48.
Spinks, C.W., 1983. Peirce's demon abduction: or how to harm the truth out of a quark. *American Journal of Semiotics* 2: 195–209.
Stent, G., 1981. Strength and weakness of the genetic approach to the development of the nervous system. *Annual Review of Neuroscience* 4: 16–19.
Uexküll, J. von, 1926. *Theoretical Biology*. London: Harcourt, Brace.
Ulanowicz, R.E., 1997. *Ecology, the Ascendent Perspective*. New York: Columbia University Press.
Ulanowicz, R.E., 2004. New perspectives through brackish water ecology. *Hydrobiologia* 514: 3–12.
Van Valen, L., 1973. A new evolutionary law. *Evolutionary Theory* 1: 1–30.
Willey, A., 1911. *Convergence in Evolution*. New York: E.P. Dutton.

APPENDIX

This figure shows a system of interpretance engaged in semiosis, as understood in the triadic formulation of C.S. Peirce. I have supplied terms that would be used in human discourse in order to facilitate understanding. However, I intend that these relations should be generalizable to any systems in Nature, providing that they have the requisite complexity for the mapping. As human systems are more specified, or just better known than others, this would usually require simplification of the diagram. Interaction – between some of the system's interpretants and the object – is not represented here.

In biology and in general, macroscopic to megascopic environmental associations generate the objects of semiosis, and so are what afford meanings for the microscopic sequence information in nucleic acids, as interpreted by mesoscopic

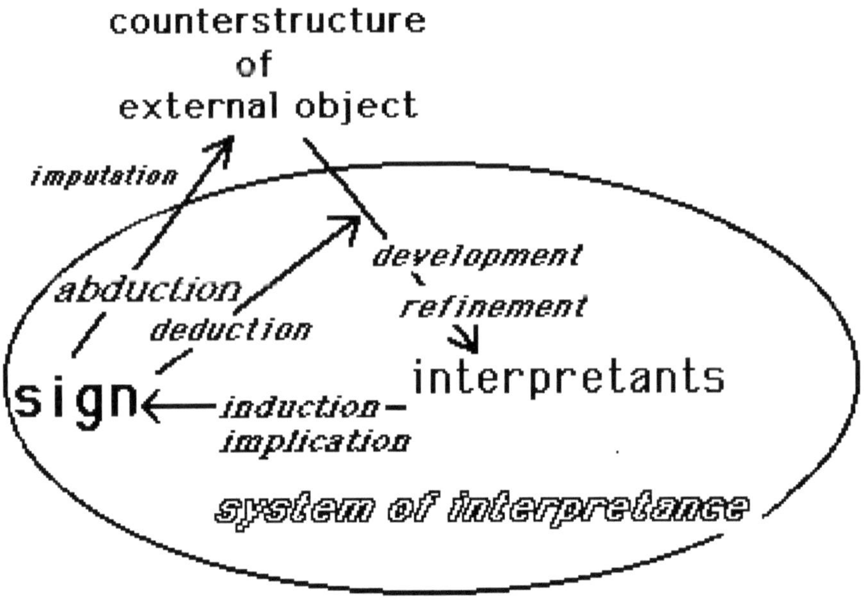

systems of interpretance (cells, organisms) successfully engaged in macroscopic niche transactions. Biological interpretants range from induced enzymes in cells to organismic behavioral reactions. Signs in these two systems are constructed as perceptions based on sensations. In cells these involve various transport processes mediated by chaperones. It should be noticed that sensation and perception are themselves (systems of) interpretants.

CHAPTER 6

SEMIOTIC SCAFFOLDING OF LIVING SYSTEMS*

JESPER HOFFMEYER

Department of Biological Chemistry, University of Copenhagen, Denmark jhoffmeyer@mail.dk

Abstract: The apparently purposeful nature of living systems is obtained through a sophisticated network of semiotic controls whereby biochemical, physiological and behavioral processes become tuned to the needs of the system. The operation of these semiotic controls takes place and is enabled across a diversity of levels. Such semiotic controls may be distinguished from ordinary deterministic control mechanisms through an inbuilt anticipatory capacity based on a distinct kind of causation that I call here "semiotic causation" to denote the bringing about of changes under the guidance of interpretation in a local context. Anticipation through the skilled interpretation of indicators of temporal relations in the context of a particular survival project (or life strategy) guides organismic behavior towards local ends. This network of semiotic controls establishes an enormously complex semiotic scaffolding for living systems. Semiotic scaffolding safeguards the optimal performance of organisms through semiotic interaction with cue elements which are characteristically present in dynamic situations. At the cellular level, semiotic scaffolding assures the proper integration of the digital coding system (the genome) into the myriad of analogical coding systems operative across the membranes of cells and cell organelles

Keywords: Biosemiotics, Emergence, Scaffolding, Anticipation, Sign, Evolution

LIFE AND MEASURING: BASICS OF A SEMIOTIC ONTOLOGY

The vision of nature as an intelligible place has nourished confidence in the scientific project ever since the times of the Enlightenment. One prominent source for this belief was in Thomas Aquinas' teaching in the 13th century which strongly emphasized the inner connection between the two great books, the book of God,

* Parts of this paper have appeared as: "From Thing to Relation: On Bateson's Bioanthropology" In *Journal of Biosemiotics* 3 (in preparation).

M. Barbieri (ed.), Introduction to Biosemiotics, 149–166.

i.e., the Bible, and the book of nature. The will of God manifested itself in his creation as well as in the Bible and therefore reading the "Book of Nature" was a necessary supplement to reading of "The Book of God". That God in his benevolence would not have created nature as an unruly and lawless place seemed obvious to most Christian thinkers (note 1).

Orderliness does indeed seem to be a widespread property of our universe, but it should be noticed that: (1) such orderliness need not embrace all phenomena in nature, and (2) it need not have been instituted in our world from the beginning, but may as well have appeared in the world through an emergent process. Strangely enough, however, many modern scientists do not seem embarrassed by the obvious Christian metaphysical heritage of science, whereas the idea of emergence, i.e. the idea that the orderliness we observe in nature has itself emerged through processes which are not yet effectively understood, is often seen as a smuggling in of supernatural intervention through the backdoor. It is hard to see, however, why the belief in an orderly universe as instituted from the beginning (by a benevolent God?) should be seen as a less supernatural explanation than the belief in orderliness as something arising by its own "force" in an unruly and largely random universe. Both ideas are dependent on ontological presuppositions which cannot themselves be ultimately proven.

The ontology of an emergent universe was explored in the evolutionary cosmology of the American chemist and philosopher Charles Sanders Peirce, who posited the idea that our universe has an inherent tendency to "take habits" (note 2). Taken in its broadest or most primitive sense, the Peircean idea of "habit taking" can be seen as an act of interpretation, i.e. the formation of a mediating link between one regularity and another, as when a bell is interpreted by a Pavlovian dog to mean food. Habituation, in other words is *semiosis* (sign activity) in its most general sense, and seeing habit taking as a general property of our universe immediately lets us reconcile our cosmology with the fact that semiotic creatures (such as ourselves) exist on planet Earth – or, in other words, that there are creatures in this world capable of "making sense" of their environment, i.e. measuring it and making choices based on such measurements.

That measuring processes constitute a central aspect of life processes in general has been emphasized throughout the work of Howard Pattee (Pattee 1977; Pattee 1997). Convergently, Stuart Kauffman has recently also discussed natural measuring processes in the context of the "non-ergodicity" of our universe. That the universe is non-ergodic implies that the universe never has had the time it would have needed, should its present state of affairs be in any way representative of its in-built possibilities (Kauffman 2000). The persistent movement of the universe into the next possible state, or the "adjacent possible" as Kauffman calls it, precludes its ever reaching a state that depends on statistical likelihood. Instead the universe is historical, for "history enters when the space of the possible that might have been explored is larger, or vastly larger, than what has actually occurred" (p. 152).

Kauffman is fully aware that the "burgeoning order of the universe" cannot be reduced to matter alone, to entropy (or the negation of entropy for that matter), to information, or to anything that simple. The propagation of organization and the subsequent growing diversification of the world is enabled in Kauffman's terminology by *autonomous agents* and, as we shall see, these agents are in a deep sense semiotic creatures. For an autonomous agent may be defined quite rigorously as an "autocatalytic system able to reproduce and able to perform one or more thermodynamic work cycles" and in his earlier work, Kauffman has shown that such agents will be expected to self-organize given the kind of system our Earth belongs to (Kauffman 1993).

However, in *Investigations,* Kauffman explicitly observes that such a definition leads to the more intractable questions of "measuring" and "recognition". For if work is defined as "the constrained release of energy" where will such "constraints" come from? Minimally, it will take work to produce them, and this is not all, for: "autonomous agents also do often detect and measure and record displacements of external systems from equilibrium that can be used to extract work, then do extract work, propagating work and constraint construction, from their environment" (Kauffman 2000, 110). Now, since a measurement always implies an act of interpretation, this immediately brings us to the heart of biosemiotics.

Because the ability to measure is the exclusive property of living systems; measurements do not take place in ordinary chemical systems. By way of illustration, let us consider the chemotactic behavior of the E. coli cell. In naturally ocurring systems, nutrients will not normally be homogenously distributed in space but will typically be present in certain localities and absent in others. This is the logic behind the appearance of the chemotactic capacity in early evolution. E coli cells are certainly primitive creatures, but they are, in fact, capable of measuring the concentration of nutrients they encounter while swimming, and to register any change in the concentration they might come upon. An eventual change in the concentration of an edible amino acid will – at least when more profitable nutrient sources are absent – cause the bacterium to swim upstream toward the source of the amino acid.

This behavior depends upon a sophisticated interaction of some fifty different proteins that co-operate in executing a *comparison of measurements* taken at two successive points in time – as well as in mediating the result of this comparison to the many aggregates of proteins spread along the surface of the cell that are responsible for flagellar movements. The collective effect of this co-operative effort is the establishment of a scaffolding mechanism assuring that the bacterium moves towards the best available nutrient source – or eventually, if no such nutrients are available, that it changes its movements into a random search behavior, i.e., tumbling around itself without a definite direction.

What goes on in this measuring sequence is the formation of an interpretant (here: the change in flagellar movement) that is related to something exterior to the cell (the distribution of nutrients in the outside environment) in a way that reflects the historically- and evolutionarily- acquired integration between the sensory system of the cell to its motoric faculties. In other words, the reason why an interpretant is formed here and now is that the cell through its evolutionary ancestry has evolved this particular mechanism for a mediation between its sensoric capacity (e.g., the receptors at its surface) and its needs (the regularly assured movement towards nutrients). History thus not only matters to the cell, but literally operates inside the cell through the structural couplings – or semiotic scaffolds – that it has served to build into the system. And this is exactly what distinguishes living systems from non-living systems: the presence in the former of historically created semiotic interaction mechanisms which have no counterpart in the latter.

This mechanism is so different from anything taking place in the non-living sector of nature, that it deserves to be distinguished as constituting a special kind of causation, different from but dependent upon traditional efficient causation. I have coined the term *semiotic causation* for this kind causation of *bringing about effects through interpretation* (note 3), as when, for example, bacterial movements are caused through a process of interpretation based on the historically defined needs of a sensitive system (Hoffmeyer 2005).

Of course, such "semiotic causation" cannot *in itself* execute its effects, but must always operate through the mechanisms of material efficient causation. The relation between the two kinds of causation is like the relationship between a court of law and a sheriff, says Peirce: "*Law*, without *force* to carry it out, would be a court without a sheriff; and all its dicta would be vaporings" (*CP* 1. 213 (note 4)). Or to paraphrase Kant: *Semiotic causation without efficient causation is helpless, but efficient causation without semiotic causation is blind.* Semiotic causation is the term for that system of relations that gives direction to the flows of metabolic energy through a living system and thus to the behavior of an organism (Santaella-Braga 1999).

Biosemiotics deals with habit taking in this precise sense, as exhibited by living cells and the interactive patterns in which they take part, i.e., as organisms or as supra-individual entities. But whether or not such habituation is a meaningful concept in the inorganic world, as Peircean cosmology requires, is a metaphysical question that is not of direct concern to biosemiotics. For it is an empirical scientific fact that the equivalent of measuring processes do undoubtedly take place in every living system, and this basic semiotic activity alone amply justifies the study of living systems as semiotic entities.

THE SCAFFOLDING OF LIFE PROCESSES

Life depends on the fine tuned co-ordination of an astronomical number of biochemical reactions taking place inside and across different kinds of membranous structures (Hoffmeyer 1998; Hoffmeyer 1999). The total area occupied by cell membranes in the human body, for example, has been calculated as one third of a squared kilometer (Hoffmeyer 2000). Moreover, the area of membranes filling up the internal space of cells, i.e. the membranes around the mitochondria, endosplasmatic reticulum, Golgi apparatus, and the many smaller cell organelles, are probably several orders of magnitude larger than the area of the outer cell membrane itself, resulting in a total area of membranes in the human body of perhaps $30\,km^2$.

These membranes generally are fluid structures that need to be upheld actively at the expense of metabolic energy. Thus, nothing in this complex arrangement has any *inherent* stability, so the maintenance of a living system requires a very intricate system of dynamic interactions. And while this system is of course *powered* by metabolic energy – *it must be controlled by semiotic means*. This is to say that local processes must subserve the needs of global functions, and this result can only be obtained through communicative activity connecting distant parts and different functional domains of the body (or organism) to each other.

For even inside the single cell – and a human body consists of some 50,000 billion single cells – the task of communicative coordination is not a simple one. A multitude of pathways for signal transduction are involved in each and every task, and one major problem to avoid is the possibility of so-called signal transduction "cross-talk" interfering with the transfer of messages, i.e., the prevention of signal molecules destined for one distinct pathway becoming interpreted as relevant by other pathways (see Bruni 2003; and this volume for a discussion of the intricate semiotics of signal transduction).

The semiotic coordination of the processes described here makes up the branch of biosemiotics called *endosemiotics*, i.e. the semiotics of processes taking place inside the organism. *Exosemiotics*, on the other, hand is the term used for biosemiotic processes going on between organisms, both between and within species, as well as for the semiotic processes connected with the interpretation of abiotic markers in the environment, as when migratory birds make use of stellar configurations in order to find their way. That these endo- and exo- prefixes have thus come to refer to the two sides of the borderline around bodies, is an terminological distinction only and should not be taken to signify any privileged role in biosemiotics for either side of the interface, or boundary. In fact, semiotics is in principle *always* connected with some kind of inside-outside interaction.

Thus, through the totality of life processes in the world, a *semiosphere* is created that envelops the earth in much the same way the atmosphere, hydrosphere or biosphere envelops the planet (Hoffmeyer 1996; Hoffmeyer 1997) (note 5). This semiosphere truly is a sphere like the atmosphere, the hydrosphere or the biosphere, in that it penetrates these spheres for living organisms and consists in communication: sounds, odors, movements, colors, electric fields, waves of any kind, chemical signals, touch, etc.

The concept of the semiosphere adds a semiotic dimension to the more well-known concept of the biosphere, emphasizing the need to see life as belonging to a shared universe of sign activity through which cells, organisms and species all over the planet interact in ways that we still hardly understand. And yet every single species (including humans) has only limited access to this semiosphere, because each species' capacity for sensing and interpreting potential cues in its surroundings, i.e. its *interpretance* (note 6), has evolved to fit a particular ecological niche. Put in the terminology of Jacob von Uexküll, each species is confined to its own limited Umwelt, or "internal model" with which individuals of a species constructs an understanding of its surroundings.

Moreover, the semiosphere poses constraints and boundary conditions upon the Umwelts of various species populations, since each are forced to occupy specific *semiotic niches,* which is to say that each will have to master different sets of visual, acoustic, olfactory, tactile and chemical signs in order to survive in the semiosphere. It is thus entirely possible that the semiotic demands made upon species' populations are often a decisive challenge to their success. If this is so, then ecosystem dynamics, for example, shall have to include a proper understanding of the semiotic networks operative in ecosystems.

The network of semiotic interactions by which individual cells, organisms, populations, or ecological units are controlling their activities can thus be seen as scaffolding devices assuring that an organism's activities become tuned to that organism's needs. And just as the scaffold raised to erect a building will largely delimit what kind of building is raised, so too do the semiotic controls on biological activities delimit when and how such fine-tuned activity should take place. To conceptualize and analyze the myriad of semiotic scaffolding mechanisms operative at and across different levels in natural systems is the core subject matter of biosemiotics.

THE CONCEPT OF SCAFFOLDING

Semiotic scaffolding operates by assuring performance through semiotic interaction with cue elements that are characteristically present in dynamic situations such as the catching of prey, invading host organisms, or mating. The significance of dynamic scaffolding in the human sphere has been pointed out already by the Russian psychologist Lev Vygotsky, who was probably the first to emphasize the importance during child development of scaffolding, i.e. experiences with external supporting structures (including linguistic ones). At crucial developmental moments, adults help give the child the experience of successful actions that child alone would not be able to produce (Vygotsky 1986). Some obvious examples include physically supporting the first few faltering steps of a near-walker, or supporting a baby in water to allow for swimming movements.

A striking case of a linguistic scaffolding is when a child is "talked through" a tricky challenge by a more experienced agent and thereby succeeds in solving a problem which was otherwise beyond its abilities (such as learning to tie his or her shoelaces). Later, when the adult is absent, the child may often conduct a similar dialogue with herself – in which case the speech sounds serve as an external memory-scaffold to guide the difficult activity and to avoid errors. In such cases "the role of language is to guide and shape our own behavior – it is a tool for structuring and controlling action, not merely a medium of information transfer between agents" (Clark 1997, 195).

In turn, the concept of scaffolding was later taken up and further developed within the fast-growing segment of robotics research concerned with so-called autonomous agents (e.g., Brooks 1993; Hendriks-Jansen 1996; Clark 1997). Autonomous agents are mobile robots ("mobots") capable of functioning in messy and unpredictable real-world settings such following as close as possible alongside a wall in a crowded office. "The New Robotics revolution" writes Andy Clark "rejects a fundamental part of the classical image of mind. It rejects the image of a central planner that is privy to all the information available anywhere in the system... The problem with the central planner [model] is that it is profoundly impractical... The reason is that the incoming sensory information must be converted into a single symbolic code so that such a planner can deal with it" (Clark 1997, 21).

Instead, autonomous agents operate on the principle that Hendriks-Jahnsen calls "interactive emergence": "Patterns of activity whose high-level structure cannot be reduced to specific sequences of movements may emerge from the interactions between simple reflexes and the particular environment to which they are adapted... The emergent behavior of the system as a whole is the result of various autonomous activities interacting with each other and with the environment, and not a centralized system making decisions based on internally represented courses of action or goals" (Hendriks-Jansen 1996, 8–9).

Significantly, as seen from a biosemiotic point of view, Hendriks-Jansen emphasizes that "Interactive situated behavior cannot be explained in terms of a deductive or generative law. It requires a *historical* explanation because there can be no rules to predict the sorts of behavior that might emerge" (p.9). Addressing the question of similar scaffolding mechanisms in biological creatures, Clark has suggested a "007-principle": "In general, evolved creatures will neither store nor process information in costly ways when they can use the structure of the environment and their operations upon it as a convenient stand-in for the information-processing operations concerned. That is, know only as much as you need to know to get the job done" (Clark 1997, 46).

Both Clark and Hendriks-Jansen are writing interchangeably about autonomous agents and biological creatures in ways which are likely to meet objections from many biologists. Thus, in the case of a wall-following robot, a human constructor pre-selects the "situatedness" of the given activity. But how does an organism select its own "situatedness"? Situated activity presupposes a kind of fitting between environment and organism, so Hendriks-Jansen contents himself by pointing to "natural selection" as *the* explanation for interactive emergence in the organic realm.

But the anticipatory or teleological capacity of natural selection is not as self-evident as most scientists seem to suppose. For natural selection not only presupposes superfecundity – i.e., that more individuals are born than can possibly survive, but more generally it presupposes the operation in organisms of a "strive" for survival. But a strive already implies a telos, something of the kind philosophers call 'aboutness' or intentionality, (although *intentionality* in this case does not imply thoughts and consciousness).

From the very beginning, even the simplest prokaryotic (bacteria-like) life forms take an interest in their surroundings with regard to finding solutions to survival problems such as how to feed, how to escape predation, and how to reproduce. None of these strivings are explainable through schemes of simple efficient causation, for they all presupposes some kind of "orientation" from the system towards the environment and towards the *future*. Thus, the inherent teleological nature of livings systems cannot be "explained" by natural selection because natural selection wouldn't work without it. And this "teleological property" of living beings was, of course, exactly the reason why Kant reached the conclusion that life evades scientific analysis. The Kantian conception of science was purely physicalist and he did not doubt that functionalist explanations in biology – e.g, the perennially beloved "machine metaphor" – violate the physicalist ideal to which biology still often subscribes – thus unwittingly (one must suppose) admitting logical contradictions into its core body.

And this is why the semiotic dimension of natural processes must be drawn into any dynamic analysis of such systems. Rather than talking about interactive emergence, as Hendriks-Jansen suggested, I would recommend using instead the concept of *semiotic emergence*. The primary mechanism behind semiotic emergence is semiotic scaffolding, the key to nature's tendency to take habits in the biological realm.

SCAFFOLDING AND EMERGENCE

The emergence of new scaffolding devices (unknowingly) function like stepping stones in a river, leading evolutionary processes forward one step at a time and – in average – farther away from the bank at each step. In themselves, such semiotic scaffolding patterns may take many forms and rely on many different principles, but the core property of a semiotic scaffold remains that of focusing the energy flow (behavior) of the concerned system or subsystem upon a rigidly limited repertoire of possibilities, or in guiding the system's behavior to realize a definite sequence of events. A receptor molecule at the surface of a cell may be tuned to open a neighboring channel when, and only when, being hit by a small set of possible domains on protein surfaces, just as the offspring of a bird may be tuned to learn only one or a very narrow band of sequences of sounds. When the proper cue arrives, the receptor opens the channel or the young bird learns the species' song. The receptor may be misled, however – as when an HIV virus iconically mimics the surface domains of one of the organism's own proteins – and the bird may be fooled e.g. if it's been deposited in the nest of a foster bird without further contact with adult birds of its own species (cf. Soler and Soler 1999). Semiotic scaffolding mechanisms depends on acts of *interpretation* (understood in the most encompassing, and not merely human anthropomorphic sense) and interpretation always runs the risk of being wrong.

For illustration, let us consider the case of infertility in the so-called eyeless mutant of the axolotl, as discussed by Leo Buss (Buss 1987). Under normal conditions, the amphibian eye will be produced by chemical interactions between the newly formed optic vesicle and the embryonic ectoderm layer. A chemical inducer produced by the optic vesicle is used for the scaffolding of this interaction. What happens in the *eyeless* mutant of the axolotl is that this step is disturbed because the ectoderm of the mutant does not respond properly to the inducer, so that no eye will be formed and the mutant develops blind. This however, is not the only problem this poor creature has to cope with – for it also lacks the capacity for leaving offspring. This is because the eyeless mutant develops a *secondary* deficiency in the region of the brain called hypothalamus, which will only be properly developed through induction via signals that are sent to it from the eye. In the *eyeless* individual, no eye exists to direct the development of the hypothalamus, and thus the hypothalamus therefore cannot produce gonadotropin hormones – and in the absence of these hormones, the individual becomes sterile.

The deficiencies of the eyeless mutant clearly illustrates the tinkering ways in which ontogeny has become scaffolded by evolution. There is presumably no other reason why the development of hypothalamus should depend on the presence of a functional eye than the eventual formation of the eye in a location that happens to be anatomically close to that region of the brain where hypothalamus is normally developed in this lineage. Making the development of hypothalamus dependent upon the prior formation of an eye effectively assures that hypothalamus will become constructed at the exact right moment in embryogeny. And this is precisely the situation that went wrong, of course, in the eyeless mutant – but rare mutants are statistically of little concern in evolution. *Rather,* the axolotl eye just happened to be in the neighborhood of the nascent hypothalamus-region in normal individuals and it is most likely for no other reason than this that evolution managed to exapt the eye for a secondary role as an ontogenetic switch for the initiation of proper development of a hypothalamus. As Buss says: 'Ontogeny must re-enact the interactions which gave rise to it' (Buss 1987, 97). In the terminology of this paper, ontogeny is safeguarded by myriads of semiotic scaffolds that depend on one another in long chains of successive steps.

In other words, all that was needed to assure that proper induction would take place was that some factor could be counted on as a reliable cue for the onset of hypothalamus development. We must suppose that the sensitivity of cells in the hypothalamus region to induction from the eye has been molded by natural selection, but very probably a number of other constituents might equally as well have become its target. Natural selection thus is responsible for producing the safe channeling (under normal conditions) of an inductive relation between eye and brain development, but the choice of this *particular* relation (between the development of those parts of the hypothalamus that will later enable gonadotropin production and the presence during embryological development of a rudimentary eye) as a theme for the semiotic scaffolding of hypothalamus development was probably more or less random.

In the same way, natural selection has safeguarded the inductive relation between optic vesicle and ectoderm layer at the location where the eye is supposed to form, but the fact that *this* relation became the focus for the selection process didn't necessarily itself offer any "selective advantage." Yet once this relation had become safeguarded through the strengthening influence of natural selection, it did thenceforth, however, offer a reliable cue for the successive construction of yet further semiotic scaffolding. In this way, ontogeny may be seen as being based upon a highly integrated web of historically coordinated semiotic scaffolding devices that guide the construction of the embryo safely through a procedure involving millions of possible choices taken by cell lines all along the developmental patchwork of the ontogenic process.

The emergence of this pattern of subtle scaffolding devices through evolution is, of course, in a certain sense the outcome of natural selection. But it should also be noticed that an important aspect of this process is the capacity – or talent one might say – of individual cells and cell assemblies to change their internal settings in integrated waves (e. g. signal transduction cascades) under the influence of external (or new) molecular cues. The semiotic logic of localized dynamic biochemistry in a given embryonic tissue thus would tend to tell us as much or more about the actual "causality" behind semiotic emergence than do explanations in terms of "natural selection" alone.

SPAM

According to neo-Darwinian understanding, the gradual fixation of favorable mutations in DNA molecules due to natural selection is *the* mechanistic backbone for change in organic evolution. One reason for the overwhelming success of this understanding may be that genomes are in fact the most fundamental scaffolding devices for the ontogenetic production of organisms, and if you monitor the construction of a scaffold you are sure to catch also an important temporal aspect of the dynamic processes scaffolded. If, for instance, you monitor the ongoing addition of new notes to a musical score you will certainly get some insight into the tempo and speed of the composing process, and even a deaf person might become an expert in this. Too, there can be no doubt that the genomic structure is the most rigid and conservative scaffold for the evolutionary process, and monitoring the changes in gene frequencies down through generations will, of course, present you with a timetable which is tightly coupled to true evolutionary change.

For ease of illustration, let us consider a well-known case of linguistic scaffolding taken from the history of the development of the Internet, namely the term *spam.* This word has become customary all over the world for referring to the violation of privacy through invasion of other people's electronic mailboxes by non-invited advertising mail. Yet only a few people outside of the English speaking world would know that this new meaning of the word "spam" derives from a very specific situation – in particular, a single *Monty Python's Flying Circus* skit in which a group of Vikings sing a chorus of "SPAM, SPAM, SPAM, SPAM. . .lovely SPAM, wonderful SPAM. . ." in an increasing crescendo, drowning out nearby conversation, much in the way that unsolicited e-mail "drowns out" other email on the Internet. Additionally, the song's lyrics themselves costitute an endless repetition of worthless or annoying text similar to the e-mail variety of "spam."

Now, technically, the term "spam" is a so-called telescope word formed by the contraction of the phonemic extremities of "sp(iced) (h)am", i.e. *spam.* But although this term has been intentionally introduced and functions now, of course, in its present significative role – we can be quite sure that neither the original inventors of the spiced ham brand name SPAM, nor the cast members of *Monty Python*, had any idea of the particular world wide usage awaiting it. Rather, he metaphoric transformation of "spam" from the *Monty Python* setting to the Internet vocabulary was a non-necessary and *creative* act.

Moreover, this new relation ("spam" = electronic junk mail) only stuck because it happened to hit an unfilled locus in linguistic space, nominalizing a non-verbalized general experience in modern society. Yet through this metaphorical transformation onto the experiential plane of the Internet, the term now has become an linguistic actant in its own right, generating a range of new habits – e.g., we can engage in making rules for Internet services to "eliminate spam" or discuss and execute punishments towards "spam-sinners." And we may eventually expect even further conceptualizations to develop on the top the original concept. In this way, new terms can themselves be "scaffolding devices" for cultural development.

The point is that the coining of the term spam was a *creative* response to the *novel needs* of a new cultural situation. As such, it was not a result of conscious deliberations; instead, it just happened to grow spontaneously out of already existing linguistic resources by a sort of tacit interactive consensus.

It is tempting to see the appearance of the term "spam" as a prototype case for the origin of new digitally coded signs in nature, i.e. in evolution. Thus the digitalization of the *Monty Python* sketch in one simple phonetic sequence served to *scaffold* a complex social experience by making it an easy general resource for communication. And this kind of semiotic scaffolding, I suggest, is exactly what digitally coded messages such as genes are in general good for.

For new genes may often be formed very much through the same kind of scaffolding conversions that we have seen to be instrumental in furthering the inclusion of new words in a language. Thus, in the case of the new term "spam" the decisive point was the conjunction of a pointed meaning (submitted by *Monty Python*) and a social need (created by spam sinners). Likewise, in the biological realm, we can suppose that gene duplication accompanied by the hitch-hiking down through generations of one of the copies of non-essential or masked genetic material (prone to all kinds of non-lethal mutations) would assure the availability of a rich resource base for potential future genes. The decisive cause for the birth of a new functional gene would be a lucky conjunction of two events: (1) an already existing non-functional gene might acquire a new "meaning" through integration into a functional (transcribed) part of the genome, and (2) this gene-product would hit an unfilled gap in the "semiotic needs" of the cell or the embryo.

In this way, a new gene may become a scaffolding mechanism supporting a new kind of interaction by imbuing some kind of semiotic advantage upon its bearer – and this is what I mean by the term *semiotic scaffolding.* By entering the realm of digitality, the new semiotic functionality becomes available not only to the cells of the organism carrying it, but also to future generations as well (and, if we allow for horizontal gene transfer, possibly even to unrelated organisms).

Digitality in the life sphere thus provides for the sharing (or objectivity) of ideas (functions) and thereby also assures their conservation over time. But this very function is itself dependent on the relative inertness of the genetic material and its very indirect and highly sophisticated way of interfering with the worldliness of cellular life. Genes, like human words, do not *directly* cause change in the world around them (i.e., we do not believe in spells), but do so only when some body interprets them. And just as words serve to support human activity and communication, genes support cellular activity and communication. Genes and words are both marvelous semiotic scaffolding tools.

ANTICIPATION AND BRAINS

Scaffolding mechanisms in general depend upon the ability to anticipate and prepare for important situations and events in the life cycles of the concerned entities. To scaffold life processes through genomic control mechanisms obviously becomes difficult when organismic life cycles become more complicated or when animals engage in complex social processes. Genomic scaffolding necessarily operates through the controlled assembling of protein resources which are then released in sophisticated temporal patterns reflecting the upcoming needs of the organism. Such mechanisms do work sufficiently well so long as the behavioral repertoires of animals are limited to instinctually triggered responses to foreseeable events.

But large-brained animals such as birds and mammals are generally dependent upon not just instinctual reflexes, but the processes of *learning* – and while such processes are assisted by genetically assured behavioral preferences, the whole advantage of learning ability must be the inherent element of flexibility inferred on behavior by the learning process – and thus the transfer of behavioral control from the genomic level to the cerebral level. This introduces the need for even newer sets of scaffolding mechanisms – and foremost among these are the diversity of control mechanisms exhibited by the neuro-endochrinological apparatus.

An amusing example of the semiotics of neuro-endochrinological scaffolding has been observed in the cooing behavior of ring doves (*Streptopelia risoria*). Before a female ring dove lays her eggs, she and her mate go through a series of courtship displays. As courtship proceeds, hormonal changes in the female trigger the growth of follicles in her ovaries, each of which eventually bursts to release an egg. Now it has been shown that if a female dove is operationally hindered in making the so-called "nest-coo" she will not be able to ovulate, even despite the enthusiastic courting by males. Yet in control experiments, tape recordings of nest-coos were played to females with no males present. Now follicles thus immediately began to grow.

The conclusion seems simple: Female doves are not cooing, ultimately, at the males – they are, in fact, cooing at their own ovaries to trigger the release of eggs. And since ovaries are not supposed to posses means for meaningful absorption of sound, this mechanism must be operating through the brain of the animal. And what the experiment shows is that the brain does not tell the ovaries to make eggs until after it has interpreted the sounds emanating from its own throat.

Why such a strange mechanism has actually evolved can only be guessed at, but perhaps courtship ritual and thus cooing behavior is in fact more safely correlated with the actual time of mating than a purely endogenously-based release system would have been. The obvious, although speculative, explanation would be that the cooing behavior measures the state of a relation between two birds and two sexes, which is likely to be superior metric for reproductive success than would be a simple measure of the hormonal state of the female organism itself.

The advantage of preparing for future events or situations – i.e. *anticipation* – is, of course, the main reason for the evolution of *semiotic causation*. The animal that flees at the moment that it senses the presence of smoke obviously runs a better chance of leaving offspring than does an animal that doesn't respond until the heat is actually felt. The most important tool for surviving is thus anticipation and organisms are involved in anticipatory action and relations all the time. They must decide when to grow and when to withhold growth, when to move, when to hide, when to sing and so on, and this way of adjusting one's own behavior depends on a capacity to predict the future at least to some limited extent.

For instance: "Is it likely the sun will shine or not?" "Is it likely that little flies will pass by if I make my web here?" "Will the predator be fooled away from the nest if I pretend to have a broken wing?" etc. In most cases, it will be the instinctual reflex system of the animal, rather than the brain, that makes this kind of prediction – but the underlying logic is the same: an animal profits from its ability (whether acquired through phylogeny or through ontogeny) to identify trustworthy regularities in the surroundings and to orient its own actions accordingly.

Now most – if not all – such trustworthy regularities are *relations*. For instance: the relation between the amount of daylight and the approaching summertime that tells the beech when to burst into leafs; or the play of sun and shadows which tells the spider where to construct its web; or the relation between the clumsy movements and an easy catch that tells the predator which individual prey animal to select – and that thus also tells the bird how to fool the predator away from its nest.

In the first two of these examples (the beech and the spider), a certain organismic activity is released as a response to pure (non-semiotic) natural relations, so-called *categorical relations* – whereas in the third example the bird produces a fake categorical relation (clumsy behavior as expectedly related to easiness of catch) and then takes advantage of the semiotic relation established by the predator when it lets itself be fooled by a false sign. In this case, in other words, the bird fools the predator because it somehow (genetically or ontogenetically) 'knows' how the predator is going to (mis)interpret the seeming categorical relation. Observe, too, however, that in this case the predator may not always be fooled – and this shows us that we are not here dealing with just material forces (*efficient* causality) – but also and more importantly with *semiotic* causality whose consequences are not strictly deterministic: e.g., the predator may misinterpret the sign (the faked clumsy behavior), but it also may not.

Anticipation is thus essentially a semiotic activity in which a sign is interpreted as a relation between something occurring now and something expected to occur later, such as a dark cloud alarming us to an upcoming thunderstorm. And from its very first beginnings in Augustine's writings in the fourth century, a *sign* is conceived as something awakening us to infer something else. For Augustine, a *signum* (sign) "is anything perceived which makes something besides itself come into awareness" (quoted from Deely 2001, 221). And while Augustine's definition is too narrow in its focus on perception, since elements of awareness may well be signs also without being perceived, yet he did point to the core of the matter when he defined a "thing" as: "what has so far not been made use of to signify something" (p. 221). This implies that "things" may well be "signs" but that *they need not be so*. It also implies that the essence of the sign is its *formal relational character* of evoking an awareness of something which it is not itself – and thereby implying the full Peircean triad of sign, object and interpretant (here: the altered awareness). The evoking of such a referential triad is, of course, by no means exclusive to the workings of human awareness but is rather, as was later realized, a purely *logical* relation to be established in *any* system capable of autonomous anticipatory activity.

Moreover, just as predictability must precede prediction, a system of useful dyadic relations must first have been realized on planet Earth before the emergence of biological life. Only then could more sophisticated systems survive based on a the capacity for anticipation – i.e., for bringing themselves in relation to the pre-established set of dyadic relations under the formation of true triadic or semiotic relations. And while the underlying system of dyadic relations may well be understood in terms of the things related, the emergence of true triadic semiosis in the shape of living beings and their activities established a new kind of causality peculiar to this new form of *relative being* – causalities which are far too sophisticated to be accurately grasped through the simple dynamics of dyadic relations between inanimate things.

SEMIOTICS AND RELATIVE BEING

Alhough most biologists do in some sense recognize that communicative processes are part of natural systems, many instinctively conceptualize these processes *only* in terms of the biochemical and genetic processes involved, and that are proposed to result in such communicative behaviors. To talk of messages or semiosis, they feel, just blurs our minds – and this is the reductionist credo ruling almost every department of biology throughout the world. And so the simple question asked from these quarters when confronted with "biosemiotics" normally is: What's all this fuss about?

What it is all about, I think, is quite a simple thing: namely, the reality of *relative being*. Relative being is a strangely obvious thing which is nevertheless generally dismissed by science as not really "real". For example, Jupiter has a number of moons circling around it, but the *relation* between the moons and the planet is not seen as anything "real in itself" in that doing so doesn't add anything to a strict analysis of the properties of the individual celestial bodies themselves. The simple genitive case seems neatly to exhaust the whole relation: the moons are indeed Jupiter's. And it is of course true that, in principle, a "relation" could be drawn between *any* two physical objects in the world, and in all but a very few cases, such relations would turn out to be absolutely uninteresting whether seen from the point of view of science, or from the point of view of ordinary people's everyday life.

However: not all relations are of this inconsequential kind, and to give an example of 'relative being' which cannot easily be dismissed as "ficticious" let me suggest the relation of *parenthood*. For all we know, King Frederik the Ninth of Denmark was the father of Queen Margrethe the Second. But His Majesty passed away a long time ago, and we have no doubt that Margrethe will likewise pass away, too, at some point in the future. Yet, due to royal destiny, their *relation* will in all likelihood persist for a *very* long time as the relation between two real entities that it is: i.e., the relation of parenthood, of this particular father to this particular daughter. *Thus*, this kind of 'relative being' seems to have a reality of its own which *cannot* be reduced to the individual persons that substantiate the relation. Such relations have been called *ontological relations* in that they are "real" (i.e., actually existing) functional factors of the actually existing world (Deely 1990; Deely 1994; Deely 2001).

But are there ontological relations in nature? One of the first to answer this question in the affirmative was the anthropologist and biologist Gregory Bateson (Bateson 1972; Bateson 1987). According to Bateson, the reality of ontological relations is exactly what distinguishes life from non-life. For relations in the prebiotic sphere have also sometimes been thought of as being ontological, as for instance in the case with astrology. But since no likely mechanism whereby relations between planets (say a conjunction between Mars and Venus as seen from Earth), could possibly influence the destiny of individuals or nations on Earth has ever been established, such a belief is generally (and correctly) rejected by scientists as superstition. For we have absolutely no warrant for believing that those relations have any distant causal effects on the world *qua* relations. In this case – as in the prebiotic world in general – it makes more sense to talk about "related things" rather than about *relations* – and maybe the general unwillingness of science to accept relations as ontologically real owes much of its strenght to the ancient (and now strangely revived) struggles science had to fight against dogmatic beliefs connected to mystical or religious persuasions.

When we turn to the investigation of animate nature, however, *relations* tend to become considerably more important than autonomous "things". The human shoulder, for instance, is a ball-and-socket joint that enables a person to raise, twist, bend, and move the arms forward, to the sides and behind. The head of the upper arm bone (*humerus*) is the ball and a circular depression (*glenoid*) in the shoulder bone (scapula) is the socket. A soft-tissue rim (*labrum*) surrounds and deepens the socket. The head of the upper arm bone is coated with a smooth, durable covering (*articular cartilage*) and the joint has a thin, inner lining (*synovium*) for smooth movement, while the surrounding muscles and tendons provide stability and support.

Here, then, are a whole assembly of *relations* which are all remarkably adjusted to each other. The primary functional relation, of course, is that between the shape of the ball of the arm bone and the contour of the shoulder socket, and we can assume that this relation has indeed been functionally modulated by natural selection, all along the way from its likely evolutionary origin as the appendages, or fins, in fish. Clearly these relations are of quite another kind than the relations pertaining to the planetary system.

The relation, in fact, is so central to the function of the animal that one can hardly imagine the one bone changing without entailing a corresponding change occurring in the other bone (or in some other relation within the system). Or, if such a unilateral change should happen due to an unfortunate mutation, the resulting individual would surely be functionally deficient and leave little or no offspring. Conversely, if a mutation should occur that affected both bones in a coordinated way, *conserving* their internal relation – the resulting individual might perhaps manage quite well in the evolutionary competition. In this case, the relation *as such* does indeed seem as real and perhaps even more important to the system than the individual bones making up the relation. And this state of affairs may well be the rule rather than the exception in the realm of the biological world.

I conclude that not only is it absurd to deny the reality of *relative being*, because it is *relative being* rather than things (i.e., individual creatures or populations) that evolution persistently optimizes – and by denying this, one is prevented from developing a proper scientific understanding of both biosemiosis and of purposefulness. For semiosis is all about bringing oneself *in* relation *to* a relation. And from the beginning of life, organisms have based their survival on this capacity for anticipation – i.e. for interpreting events or structural configurations as *signs* for one thing and another. For as Peirce saw, the proper and most fundamental definition of a sign is something "which is in a relation to its object on the one hand and to an interpretant on the other, in such a way as to bring the interpretant into a relation to the object, corresponding to its own relation to the object" (*CP* 8. 322).

The fact that signs are often false or that relations (at least for humans) are more often than not imaginary does not preclude signs or relations from having causal significance. When the predator hunts the bird with a clumsy behavior its actions are equally real whether the bird actually does have a broken wing or not. If the bird is only pretending that its wing is broken, then the predator will most likely not catch it – but the movements of the predator were not for that reason any less caused by the (misleading) interpretant formed in its brain. By accepting the reality of relative being – and thus of semiotic causation – we not only open up an explanatory space for a reconciliation of human semiotic existence (such as is manifested in the writing a scientific paper) with that of organic existence in general, but we are also immediately brought to see the semiosphere as an emergent process nourished by the interpretative interaction of countless organisms and cells – or in other words, by *biosemiosis.*

NOTES

[1] An exception, perhaps, would be the early nominalism followers of Sir William of Ockham, who would not accept any limits to the freedom of God and who, by implication, would not exclude the possibility that God might have created a totally unintelligible nature. Such a view would hardly have allowed for science to develop, but eventually it was replaced by the ideas expressed by philosophers such as Voltaire, Rousseau and the other Enlightenment thinkers.

[2] "... if the laws of nature are results of evolution, this evolution must proceed according to some principle; and this principle will itself be of the nature of a law. But it must be such a law that it can evolve or develop itself. Not that if absolutely absent it would create itself perhaps, but such that it would strengthen itself, and looking back into the past we should be looking back through times in which its strength was less than any given strength, and so that at the limit of the infinitely distant past it should vanish altogether. Then the problem was to imagine any kind of a law or tendency which would thus have a tendency to strengthen itself. Evidently it must be a tendency toward generalization, – a generalizing tendency. But any fundamental universal tendency ought to manifest itself in nature. Where shall we look for it? We could not expect to find it in such phenomena as gravitation where the evolution has so nearly approached its ultimate limit, that nothing even simulating irregularity can be found in it. But we must search for this generalizing tendency rather in such departments of nature where we find plasticity and evolution still at work. The most plastic of all things is the human mind, and next after that comes the organic world, the world of protoplasm. Now the generalizing tendency is the great law of mind, the law of association, the law of habit taking. We also find in all active protoplasm a tendency to take habits. Hence I was led to the hypothesis that the laws of the universe have been formed under a universal tendency of all things toward generalization and habit-taking" (CP 7. 515, see note 4).

[3] An interpretation is a sign process as seen from the point of view of the person or living system engaging in it. Sign processes – or semiosis – are processes whereby something refers to something else, as when an animal is seized by alarm upon the smell of smoke. The smoke in this case acts as a sign vehicle that provokes the formation of an interpretant in the animal, i.e., a sense of danger causing it to flee. A sign then consists in a triadic relation of a sign vehicle, an object (here danger) and an interpretant. According to Peirce: "A sign ... is an object which is in a relation to its object on the one hand and to an interpretant on the other, in such a way as to bring the interpretant into a relation to the object, corresponding to its own relation to the object" (*CP* 8. 322)

[4] The designation *CP* abbreviates (Peirce 1931–1935) and (Peirce 1958). The abbreviation followed by volume and paragraph numbers with a period between follows the standard CP reference form.

[5] The concept of *the semiosphere* was originally introduced by the Russian-Estonian semiotician Yuri Lotman (Lotman 1990) who explicitly used it in analogy with Vernadsky's concept of the *biosphere*. In Lotman's writings however, the semiosphere remained a concept primarily connected to cultural processes: "The unit of semiosis, the smallest functioning mechanism, is not the separate language but the whole semiotic space in question. This is the term we term *semiosphere*. The semiosphere is the result of and the condition for the development of culture; we justify our term by analogy with the biosphere, as Vernadsky defined it, namely the totality and the organic whole of living nature and also the continuation of life" (*ibid.* 125) (Vernadsky 1926; Vernadsky 1945). Additionally, one might claim that Vernadsky's concept of the *biosphere* does indeed cover the meaning the I have given here to the term *semiosphere,* but the concept of biosphere has not survived in the sense given to it by Vernadsky, but the latter is now used simply as "the ecosystems comprising the entire earth and the living organisms that inhabit it" (Webster's Encyclopedic Unabridged Dictionary 1996). For further details on the origin of these terms, see (Sebeok 2001). Semiotician and historian John Deely approves of my use of the term *semiosphere*, however, and suggests the term "signosphere as a term more appropriate for the narrower designation of semiosphere in Lotman's sense, leaving the broader coinage to Hoffmeyer's credit" (Deely 2001, 629).

[6] *Interpretance* may be defined as the capacity of a system to respond to signs through the formation of 'meaningful' interpretants. High interpretance allows a system to "read" many sorts of "cues" in the surroundings. High-level interpretance means that the system will form interpretants in response to complex cues, which might not be noticed or even be noticeable by low-level agents.

LITERATURE

Bateson, Gregory (1972). *Steps to an Ecology of Mind*. New York, Ballantine Books.

Bateson, Gregory and Mary Catherine Bateson (1987). *Angels Fear. Towards an Epistemology of the Sacred*. New York, Macmillan.

Brooks, R. (1993). "A robot that Walks: Emergent Behaviors from a Carefully Evolved Network". In *A robot that Walks: Emergent Behaviors from a Carefully Evolved Network*. R. Beer and et. al., Academic Press.

Bruni, Luis Emilio (2003). A Sign-Theoretic Approach to Biotechnology. *Dissertation. Institute of Molecular Biology,* University of Copenhagen. Copenhagen.

Buss, Leo (1987). *The Evolution of Individuality*. Princeton, Princeton University Press.

Clark, Andy (1997). *Being There. Putting Brain, Body, and World Together Again*. Cambridge, MA, MIT Press, A Bradford Book.

Deely, John (1990). *Basics of Semiotics*. Bloomington, Indiana University Press.

Deely, John (1994). "How does Semiosis Effect Renvoi?" *The American Journal of Semiotics* 11 (1/2): 11–61.

Deely, John (2001). *Four Ages of Understanding. The First Postmodern Survey of Philosophy from Ancient Times to the Turn of the Twenty-first Century*. Toronto, Toronto University Press.

Hendriks-Jansen, Horst (1996). *Catching Ourselves in the Act. Situated Activity, Interactive Emergence, and Human Thought*. Cambridge, MA, MIT Press.

Hoffmeyer, Jesper (1996). *Signs of Meaning in the Universe*. Bloomington, IN, Indiana University Press.

Hoffmeyer, Jesper (1997). *The Global Semiosphere*. Semiotics Around the World. Proceedings of the Fifth Congress of the International Association for Semiotic Studies, Berkeley 1994, Berlin/New York: Mouton de Gruyter.

Hoffmeyer, Jesper (1998). "Surfaces Inside Surfaces. On the Origin of Agency and Life." *Cybernetics & Human Knowing* 5(1): 33–42.

Hoffmeyer, Jesper (1999). "Order out of Indeterminacy." *Semiotica* 127 (Biosemiotica II, Special issue on Biosemiotics): 321–343.

Hoffmeyer, Jesper (2000). "The Biology of Signification." *Perspectives in Biology and Medicine* 43(2): 252–268.

Hoffmeyer, Jesper (2005). *Biosemiotik. En afhandling om livets tegn og tegnenes liv*. København, Ries Forlag.

Kauffman, Stuart A. (1993). *Origins of Order: Self-Organization and Selection in Evolution*. New York/Oxford, Oxford University Press.

Kauffman, Stuart A. (2000). *Investigations*. Oxford/New York, Oxford University Press.

Lotman, Yuri M. (1990). *Universe of the Mind. A Semiotic Theory of Culture*. London, I. B. Taurus and Co.

Pattee, Howard H. (1977). "Dynamic and Linguistic Modes of Complex Systems." *International Journal for General Systems* 3: 259–266.

Pattee, Howard H. (1997). "The Physics of Symbols and The Evolution of Semiotic Controls". In *The Physics of Symbols and The Evolution of Semiotic Controls*. Redwood City, CA, Addison-Wesley.

Peirce, Charles Sanders (1931–1935). *The Collected Papers of Charles Sanders Peirce, Vols. I–VI*. Cambridge, MA, Harvard University Press.

Peirce, Charles Sanders (1958). *The Collected Papers of Charles Sanders Peirce, Vols. VII–VIII*. Cambridge, MA, Harvard University Press.

Santaella-Braga, Lucia (1999). "A New Causality for the Understanding of the Living. Biosemiotics." *Semiotica* 127 (1/4): 497–519.

Sebeok, Thomas A. (2001). "Biosemiotics. Its, Roots, Proliferation, and Prospects." *Semiotica* 134 (1/4): 61–78.

Soler, Manuel and Juan José Soler (1999). "Innate Versus Learned Recognition of Conspecifics in Great Spotted Cuckoos Clamator glandarius." *Animal Cognition* 2: 97–102.

Vernadsky, Vladimir I. (1926). *Biosfera*. Leningrad, Nauka (French version: Paris 1929).

Vernadsky, Vladimir I. (1945). "The Biosphere and the Noösphere." *American Scientist* 33: 1–12.

Vygotsky, L. S. (1986). *Thought and language*. Cambridge, MA, MIT Press.

CHAPTER 7

BIOSEMIOTICS AND BIOPHYSICS — THE FUNDAMENTAL APPROACHES TO THE STUDY OF LIFE

KALEVI KULL
Department of Semiotics, University of Tartu, Estonia

Abstract: The importance, scope, and goals of semiotics can be compared to the ones of physics. These represent two principal ways of approaching the world scientifically. Physics is a study of quantities, whereas semiotics is a study of diversity. Physics is about natural laws, while semiotics is about code processes. Semiotic models can describe features that are beyond the reach of physical models due to the more restricted methodological requirements of the latter. The "measuring devices" of semiotics are alive — which is a *sine qua non* for the presence of meanings. Thus, the two principal ways to scientifically approach living systems are biophysics and biosemiotics. Accordingly, semiotic (including biosemiotic) systems can be studied both physically (e.g., using statistical methods) and semiotically (e.g., focusing on the uniqueness of the system). The principle of code plurality as a generalization of the code duality principle is formulated

Physical or natural-scientific methodology sets certain limits to the acceptable ways of acquiring knowledge. The more alive the object of study, the more restrictive are these limits. Therefore, there exists the space for another methodology – the semiotic methodology that can study the qualitative diversity and meaningfulness of the world of life.

THE DEVELOPMENT (OR SPECIATION) THAT HAS RESULTED IN BIOSEMIOTICS

An analysis of the early development of the approach that is nowadays called biosemiotics shows that it has emerged from several trends and branches concerned with the study of life processes. There are three main approaches.[1]

1. The *Holistic* approach in biology treats biological phenomena and organic forms as functional parts of organic wholes. Relatedness to the wholes is thus opposed

M. Barbieri (ed.), Introduction to Biosemiotics, 167–177.

to reductionism that explains the structures on the basis of constituent elements of these structures.

2. The *Structuralist* approach, or study of the specifically biological rules has often demonstrated the autonomy of these rules which means that these rules cannot be deduced from general physical or chemical laws. Biological rules, or organic codes, are historically and locally generated, thus providing a certain natural analogy to the codes of cultures.
3. The *Communicational* approach, including the study of animal communication, follows the models of communication worked out as models of sign communication. Particularly, the general semiotic theory of Charles Morris (1971) includes the communication of non-human animals as a form of sign communication. A more systematic review of the forms of animal communication has been compiled by Thomas Sebeok (1977).

In theoretical biology, at least since Karl Ernst von Baer's "*Welche Auffassung der lebenden Natur ist die richige*?" (Baer 1864), a trend (or paradigm) has developed which seeks the construction of an extended biology, one that could deal with both subjectivity and meaning. One may include in this trend: Nicolas Rashevsky's idea of relational biology, Brian Goodwin's idea of organisms as hypothesis testing systems (Goodwin 1972: 267), Howard Pattee's idea of self-interpretation as the basis of life (Pattee 1972: 249), Robert Rosen's analysis of the dualism between quantitative and qualitative (Rosen 1991).

A special place in the history of biosemiotics belongs to Jakob von Uexküll. In his *Theoretische Biologie* (1928) and particularly in *Bedeutungslehre* (1940) he developed an approach that could encompass the kind of subjectivity that every organism possesses in its unique relationship to its world (Deely 2004; Rüting 2004; Kull 2001).

More recently there has been research on the parallels of non-verbal communication in animals, which led to the formulation of zoosemiotics by Thomas A. Sebeok since 1963 (Sebeok 1972; 1990).

Another addition to the field, endosemiotics, developed from the study of the communicative nature of many intraorganismic relationships (T. Uexküll et al. 1993), led to the placing of a semiotic threshold[2] at the border of life (Anderson *et al.* 1984; Sebeok 2001). The latter would mean that the sphere of sign processes, or semioses, corresponds with the sphere of living processes. Thus, semiosphere — the notion introduced by Juri Lotman — coincides with the biosphere (Lotman 1990; 2005; Kotov, Kull 2006). Or, as it has been put by T. A. Sebeok — life and semiosis are one and the same (Sebeok 2002).

Biosemiotics is deeply rooted both in (the holistic, epigenetic, organicist) biology, and in semiotics proper. Of the semioticians, whose work and concepts have been particularly important for biology, one should mention Charles S. Peirce (Santaella 1999), Charles Morris (Petrilli 1999), Roman Jakobson (Shintani 1999), Juri Lotman (Kull 1999c), and again Thomas A. Sebeok (Deely 2005b; Kull 2003).

The term 'biosemiotics' probably appeared for the first time in an article by Friedrich S. Rothschild (1962).[3] Books under this title started to appear since 1992

(Sebeok, Umiker-Sebeok, 1992), and international annual meetings ('Gatherings in Biosemiotics') have been organized regularly since 2001 (however, the history of international biosemiotic meetings dates back at least to 1978). The 'International Society for Biosemiotic Studies'[4] was established in 2005, and the *Journal of Biosemiotics* began in the same year.

This convergence of sources has created a dynamic approach that tends to reshape biology — this can be seen in a series of texts written in the past decade, including books (Hoffmeyer 1996; Deacon 1997; Sebeok 2000a; Emmeche et al. 2002; Markos 2002; Barbieri 2003; Weber 2003; Schult 2004; Witzany 2006; etc.), special issues of journals (e.g., *Semiotica* 127(1/4), 1999; 134(1/4), 2001; *Sign Systems Studies* 30(1), 2002; 32(1/2), 2004; *Cybernetics and Human Knowing* 10(1), 2003; etc.), and a large number of single articles.

It can be argued that the whole domain of biology — the study of living systems — belongs to the domain of semiotics. However, living systems are simultaneously a special type of physical systems (the self-organizing, locally autocatalytic, complex dynamic systems) — which means that the whole domain of biology simultaneously belongs to the domain of physics. This paradoxical relationship between semiotic and physical approaches also poses the greatest methodological problem in the life sciences.

SEMIOTICS AND PHYSICS

Roland Posner, in his Presidential address in Semiotics Congress, 2000, in Dresden, has characterized "semiotics as the physics of the XXI century" (Posner 2001). This underlines that semiotics is not only an alternative, but also similar to physics in its role, universality, and theoretical depth. Such a comparison between physics and semiotics is not new.

John Locke, in *An Essay Concerning Human Understanding* (1690), argued that all human knowledge can be divided into three major sciences, into three 'sorts of sciences' — ethics, physics, and semiotics.[5] As he said, "This is the first and most general division of the objects of our understanding". Leaving aside the analysis of the domain of ethics, our task here is to describe the divide between physics and semiotics, and by doing so help clarify the basic differences between the physical (or natural-scientific) and semiotic biology.

Both physics and semiotics have expressed their ambition to be able to deal with everything in the world. However, the world either as a whole, or as a space is described quite differently by these two transdisciplinary sciences.

When speaking about the semiosphere as the space of meaning-generation — or semiosphere as a continuum of life — it would be helpful to distinguish it from space that is not (part of the) semiosphere. For instance — atmosphere is obviously not semiosphere. Similarly, whenever a semiotician uses the expression "purely physical" we should realise that he or she is in the majority of cases not referring to something belonging to the semiosphere. Thus it is reasonable to ask

Table 1. Interrelation of semiotic/non-semiotic methods and semiotic/non-semiotic things as generating a principal classification of sciences

Things \ Methodologies	Non-semiotic (detextualised) approach	Semiotic (textualised) approach
Non-semiotic (not alive)	physics *s. str.*	Ecosemiotics *s. str.*, semiotics of environment
		Primary semiotic threshold
Semiotic (alive)	biology *s. str.* — physical or natural-scientific biology	Biosemiotics, including endosemiotics, zoosemiotics, etc.
		Secondary semiotic threshold
Semiotic (lingual)	sociology *s. str.*	Semiotics of culture

what is the difference between physical space (physical world) and semiotic space (or semiosphere).

It is useful to remember that 'semiotic' means both an approach and an object — likewise, 'physical' means both an approach and an object. In addition to the study of semiotic processes or semioses (i.e., semiotics *s. str.*, including semiotics of culture and biosemiotics), there also exists a semiotic study of the environment that is not necessarily a living one or semiotic *per se* (this environment is studied, e.g., by ecosemiotics, or semiotics of environment), which means the textualization (or rather signification) of everything, independent of its nature (Nöth 2001). Besides the study of non-semiotic (or study of everything as "meaningless", as in physics), there also exists a non-semiotic approach to the living, i.e. a physical study of semiosis-consisting or self-defining objects (examples of this approach include large part of biology, and the natural-scientific study of society) (Table 1).

This classification follows from the nature of semiosis that locally multiplies the reality. Consequently, semiosphere can be viewed as the region of multiple realities. However, the region and phenomena of multiple realities can be described as all belonging to one single reality (as in the physical approach). Furthermore, the regions of single reality can be projected into the multiple one via the description process itself (like semiotics does). Therefore, four groups of sciences can be distinguished in this respect (Table 2).

Table 2. Projections of realities from two types of world (of one or several realities) into two types of models (of descriptions in a single or multiple languages) as a basis for classification of sciences

World \ Models	Non-semiotic models	Semiotic models
Non-semiotic (world of no semiosis)	Single reality into Single reality	Single reality into Multiple realities
Semiotic (world of semioses)	Multiple realities into Single reality	Multiple realities into Multiple realities

These four basic types of modeling differ also in the sense of simplification. One can claim that any modeling simplifies, however, in a more general sense, physics of *physis* (upper left in the Table 2) and semiotics of *semiosis* (bottom right in the Table 2, both biosemiotics and semiotics of culture in Table 1) tend not to make dimensional simplification, whereas physics of semiosis (bottom left, both natural-scientific biology and sociology in Table 1) does, and semiotics of non-semiosis (upper right — 'physiosemiotics') adds some dimensions when textualizing non-signs.[6]

The two principal types of inquiries, or classes of sciences — *physica et semiotica* — provide two distinct types of descriptions. A brief comparison of these two points of view is presented in Table 3.

Both physics and semiotics have expressed their ambition to study everything in the world, or at least their ability to cover everything. These can be seen as different types of sciences, approaches, or points of view. In principle, any phenomenon can be studied both physically and semiotically.

From Table 3 it appears to be quite clear that the difference between physics and semiotics, when studying seemingly one and the same thing, is rooted in their methodology. Physics and semiotics, are just two different ways of seeing the same world, each with their own separate point of view[9] — and set of methods. They differ on how the world is perceived; in the case of physics there exists just

Table 3. Relationships between the two types of transdisciplinary sciences — physical, and semiotic.

	Physics	Semiotics
Study fields, e.g.	natural sciences	sciences of meaning
	study of quantities	study of qualitative diversity
	physical ecology	semiotic ecology
	biophysics	biosemiotics
Objects (models) of study	physical space	semiotic space, semiosphere
	non-textual or detextualised	textual or textualised
	things and interactions	signs and semioses[7]
	laws	codes, habits
	transformations	translations, interpretations
	quantities	qualitative diversity
	multiple objects	*unique* objects
	world as non-living	living world
Features of objects (models)	commensurability	incommensurability
	context-independence	context-dependence
	errorless nature	Fallibilism
Methods of study	measurements	qualitative methods
	experimental	experiential
	from outside	from inside
	by independent researcher	participatory study
	reductionism	holism, mimesis[8]
	statistical tests	comparisons
Truth, reality	*single*	*plural*

a single physical world that is studied via repetitiveness (both of its elements and of the measurements), whereas in the case of semiotics there exist many individual (subjective) worlds each requiring different approaches. Semiotics therefore covers an area (multiple or plural realities of life) which is inaccessible to physics, because by its very definition, physics requires repeatability and abhors uniqueness.

Hence, we may study the physics of an organism, and alternatively, we may study the semiotics of an organism. The former is about many things (its mechanics, dynamics, chemistry), but not about meanings. The latter is the study in terms of semiotic space, and consequently emerging meanings can be studied. Likewise, Yoshimi Kawade (1999: 370) has distinguished between 'physicobiology' and 'semiobiology'.

It is important to note that both — physics and semiotics — make predictions. However, the methods of making these predictions are principally different. The physical types of predictions are quantitative — either deterministic, probabilistic, or statistical. Semiotic predictions however are qualitative ones. For instance, when studying a text that is currently in the process of being written, it is possible to make a scientific prediction about the next word to appear. In the case of a physical approach, the prediction would use correlations between adjacent words in the language, and it will be possible to calculate the statistical probabilities for the next word. A semiotic approach, on the other hand, would look at the possible meanings of the expression, and provide a prediction about the next word on a purely qualitative basis.

When speaking about the environment and ecology in this context, one realises that ecology is clearly twofold. There is an ecology that has been developed as a natural science, i.e., according to the Modernist model of science — a field of quantitative research of environment with organic systems in it, without any intrinsic value or meaning in itself. This can be seen as a branch of physics, or biophysics. And there is an ecology that includes meaning and value — ecophilosophy, biosemiotics, semiotic ecology — which is a branch of semiotics. The latter has been characterized as belonging to the post-Modern approach (Deely 2005a; M. Lotman 2002).

The environment as a physical concept, it should be emphasized, is not the same as semiosphere. But it is however permissible to view the ecosphere as a semiotic concept. According to biosemiotics, the concept of the semiosphere is synonymous with ecosphere. Therefore, this is a concept that can deal with environmental problems without the nature-culture opposition; instead, these problems can be formulated in terms of specific features of sign systems.

Here we should also mention *umwelt* — a concept introduced by Jakob von Uexküll — and closely allied to the notion of semiosphere. We may redefine it: *umwelt* is a personal semiotic space. Thus — *Umweltforschung*, or *umwelt*-research is a semiotic study. From the above we can see that it is conceivable to have parallel approaches to the environment, one in which meaning is generated, the other the more traditional quantitative approach.

Of course — the issue is more complicated, because one can distinguish between physical and semiotic *things*, physical and semiotic *methods or approaches*, and physical and semiotic *models and knowledge.*

Since the semiosphere, being a concept or model belongs to semiotic knowledge, then evidently we can speak about semiosphere everywhere semiotic knowledge extends. Conversely, everything physical is capable of being textualised, and physical models can be seen (interpreted) as special cases of semiotic ones.

If we look at the level of methods, it turns out that the physical method is restricted – it is incapable of discovering meaning or meaningfulness. In order to detect meaning, we need a semiotic approach. One needs a "living device", because meaning is exclusive to living beings. Thus — semiosphere is a creation or a construct of the semiotic method.

The world of signs is formed by those who are capable of making differences. The power of distinction-making is, in a way, also a method. Only those who use several codes, several sign systems, etc., can be a part of the semiotic world, the semiosphere.

The acceptance of two fundamental methodologies — physical monism and semiotic (realist) pluralism — can be described as the methodological duality. Because these, indeed, are complementary to each other, they can co-exist. Their co-existence, however, requires a "bridge", and this, again, can be provided by semiotics — as claimed, e.g., by Yates (1985), Perron et al. (2000), Sebeok (2000b).

THE PRINCIPLE OF CODE PLURALITY

Mind is never single, in the sense of how the physical world is viewed — there are always many minds. Even in ones own mind, there exists a multitude of mind-creating semioses, since mind is in effect an intersection of codes.

Hoffmeyer and Emmeche (1991) have cogently argued that life can only be understood as including two codes — the digital and the analogue. Or — *at least* two codes, as stated by Juri Lotman (1990; 1992).

In one of his articles, Claus Emmeche (2004) speaks about the code plurality, in the sense that there are many codes in any living organism:

> A biological notion of *function* is not a part of physics, while it is crucial for all biology. Biofunctionality is not possible unless a living system is self-organizing in a very specific way, based upon a memory of how to make components of the system that meet the requirement of a functional (autopoietic and homeostatic) metabolism of high specificity. For Earthly creatures this principle is instantiated as *a code-plurality* between a 'digital' genetic code of DNA, a dynamic regulatory code of RNA (and other factors as well), and a dynamic mode of metabolism involving molecular recognition networks of proteins and other components. (Emmeche 2004: 120)

This feature — of the necessary co-existence of many codes — however, is far more general. As it characterizes all life processes, it also characterizes any sign process.

Consequently, we can formulate this as the *principle of code plurality* — which states that any text, any sign, any semiosis assumes the co-existence of several codes, of many codes.

Semiosphere is the web of interpretations and reinterpretations. Every text is written by many, and read by many. Any piece of biosphere, any site in culture is reshaped and recognized by many.

Likewise to the semiosphere, one can characterize the organism or organic "body as a web of semioses" (T. von Uexküll et al. 1993: 9). Further, any organic whole — for instance, biological species as a self-defining category — is not real, strictly speaking, from the physical point of view. Still, biological species are real, however, in a certain fuzzy way. This "fuzziness" is an apt description of its indeterminate reality, because the species (any biological species, as well as any organic, perceptual etc. category) is a creation of multiple recognition-representation processes of the organisms (or the cells, etc.) that together make the species — a multi-real entity.

The models of communication that characterize the communication process as a dialogue, a movement of messages between two, consequently, oversimplify the situation, because any communication is communication between many. Accordingly, the principle of code plurality implies the fundamental communication model of many-to-many.

Semiosis is the process which both creates the realities and connects them. Since, according to the semiotic approach, "what is important in biology is not how we see the systems which are interacting, but how they see each other" (Rosen et al. 1979: 87), therefore there are always many realities, many descriptions. This can be derived from the multitude of umwelten, or from the co-existence of both synchronic and diachronic semiosis, or from the multitude of the levels of self-representation.

Robert Rosen has noted that, "so long as we persist in equating mechanism with objectivity, and hence with science, the mind-brain problem and, even more, the life-organism problem are inherently outside the reach of that science" (Rosen 1999: 94). The enlargement of objectivity that would lead to the solution of these problems, cannot be less radical than the abandonment of the assumption of the unity of the reality and the acceptance of realist pluralism. *Meaning is plurality.* The difficulty of the mind-body problem has been very much rooted — and known since Aristotle — in the assumption about a certain primary equivalence of the minds or mental experiences. A version of such equivalence is represented by dualism. The similarity of minds, however, is secondary, it is a result and construct of their communication. This is why the concept of sign processes, semioses, resolves the mind-body problem.

BIOSEMIOTICS AND BIOPHYSICS: INSTEAD OF CONCLUSION

Robert Rosen (1999: 105) has said that "Life poses the most serious kinds of challenges to physics itself". His approach has demonstrated a way in which the understanding of life processes requires the assumptions that create the models beyond physical ones.

In order to understand the scope, role, and tasks of biosemiotics, it is obviously necessary to describe its framework, to compare it with other approaches.

Biosemiotics, on the one hand, is semiotic methodology as applied to the study of living systems. On the other hand, biosemiotics means the study of living systems that interprets these as sign systems, or communicative structures, and involves the description and analysis of various organic codes (e.g., epigenetic, genetic, behavioural codes, including intracellular, intercellular, and interorganismic codes). Important examples include the immunological code, and ecological code (e.g., Levich, Lovyagin 1977). Biosemiotics comprises of a rich set of processes and phenomena, categories and categorization that are achievable via the multiple descriptions using different means and languages of description.

The conclusion of our analysis demonstrates that biosemiotics stands simultaneously in three different relationships with biophysics (a) biosemiotics is opposed to biophysics, (b) biosemiotics is more general than biophysics, (c) biosemiotics is more specific than biophysics.

Thus, this adds an additional meaning to the formulation expressed by Jesper Hoffmeyer (1997): "Rather than understanding biology as a separate layer 'between' physics and semiotics, we should then see biology as a science of the interface in which these two sciences meets, an interface in which we study the origin and evolution of sign processes, semiosis."

If biophysical modeling is the highest achievement of modernist biology, then biosemiotics must be the biology of the post-modern era.

ACKNOWLEDGEMENTS

I thank Claus Emmeche, Peeter Torop, Silvi Salupere, Riste Keskpaik, Kaie Kotov and Andres Luure for helpful comments and corrections and Stephen Pain for copy-editing the manuscript. Tartu University Press has published an earlier version of this chapter as "Semiotics and physics". The work is related to ETF 6669.

NOTES

1 A more detailed history of biosemiotics is described in Favareau (2006) and Kull (1999a); also in *Journal of Biosemiotics* 1(1): 1–25 (2005).

2 The term 'semiotic threshold' has been introduced by Umberto *Eco* (1979). He wrote: 'By *natural boundaries* I mean principally those beyond which a semiotic approach cannot go; for there is non-semiotic territory since there are phenomena that cannot be taken as sign-functions' (Eco 1979: 6).

3 Though, in the form of 'biosemiotic'. About Rothschild's work, see Anderson (2003), Kull (1999b).

4 See its homepage at http://www.biosemiotics.org/.

5 See Deely 2001: 593ff.

6 It is also important to distinguish between the textualization (thus semiotization) of non-living world, and the semiotics of the study of the non-living world, e.g., semiotics of physics, or semiotics of chemistry (cf., e.g., Hoffmann 1995; Pirner 2002; Yates, Beaugrande 1990).

7 Or *objects*, in the sense of Deely 2005.

8 The role of mimesis as a study method apposed to reductionism has been described by Rosen 1999; 116ff.

9 Cf. Deely 2005: 12ff.

REFERENCES

Anderson, Myrdene; Deely, John; Krampen, Martin; Ransdell, Joseph; Sebeok, Thomas A.; Uexküll, Thure von 1984. A semiotic perspective on the sciences: Steps towards a new paradign. *Semiotica* 52(1/2), 7–47.

Baer, Karl Ernst von 1864. *Reden gehalten in wissenschaftlichen Versammlungen und kleinere Aufsätze vermischten Inhalts*. Erster Theil. St. Petersburg: H.Schmitzdorff.

Barbieri, Marcello 2003. *The Organic Codes*. Cambridge: Cambridge University Press.

Deacon, Terrence 1997. *The Symbolic Species*. London: Penguin.

Deely, John 2001. *Four Ages of Understanding*. Toronto: University of Toronto Press.

— 2004. Semiotics and Jakob von Uexkull's concept of umwelt. *Sign Systems Studies* 32(1/2): 11–34.

— 2005a. *Basics of Semiotics*. 4th ed. (Tartu Semiotics Library 4.) Tartu: Tartu University Press.

— 2005b. *Thomas Albert Sebeok and Semiotics*. Sofia: Southeast European Center for Semiotic Studies.

Eco, Umberto 1979. *A Theory of Semiotics*. Bloomington: Indiana University Press.

Emmeche, Claus 2004. A-life, organism, and body: The semiotics of emergent levels. In: Bedeau, Mark; Husbands, Phil; Hutton, Tim; Kumar, Sanjev; Hideaki, Suzuki (eds.), *Workshop and Tutorial Proceedings: Ninth International Conference on the Simulation and Synthesis of Living Systems (Alife IX)*. Boston, 117–124.

Emmeche, Claus; Kull, Kalevi; Stjernfelt, Frederik 2002. *Reading Hoffmeyer, Rethinking Biology*. (Tartu Semiotics Library 3.) Tartu: Tartu University Press.

Favareau, Donald 2006. The Evolutionary History of Biosemiotics. This volume, P. 1–67.

Goodwin, Brian C. 1972. Biology and meaning. In: Waddington, Conrad Hal (ed.), *Towards a Theoretical Biology. Vol. 4: Essays*. Edinburgh: Edinburgh University Press, 259–275.

Hoffmann, Roald 1995. *The Same and Not the Same*. New York: Columbia University Press.

Hoffmeyer, Jesper 1996. *Signs of Meaning in the Universe*. Bloomington: Indiana University Press.

— 1997. Biosemiotics: Towards a new synthesis in biology. *European Journal for Semiotic Studies* 9(2): 355–376.

Hoffmeyer, Jesper; Emmeche, Claus 1991. Code-duality and the semiotics of nature. In: Anderson, Myrdene; Merrell, Floyd (eds.), *On Semiotic Modeling*. Berlin: Mouton de Gruyter, 117–166.

Kawade, Yoshimi 1999. The two foci of biology: Matter and sign. *Semiotica* 127(1/4): 369–384.

Kotov, Kaie; Kull, Kalevi 2006. Semiosphere versus biosphere. In: Brown, Keith (ed.), *Encyclopedia of Language and Linguistics*, 2nd Edition, vol. 11: 194–198. Oxford: Elsevier.

Kull, Kalevi 1998. On semiosis, Umwelt, and semiosphere. *Semiotica* 120(3/4): 299–310.

— 1999a. Biosemiotics in the twentieth century: A view from biology. *Semiotica* 127(1/4): 385–414.

— 1999b. On the history of joining *bio* with *semio*: F. S. Rothschild and the biosemiotic rules. *Sign Systems Studies* 27: 128–138.

— 1999c. Towards biosemiotics with Yuri Lotman. *Semiotica* 127(1/4): 115–131.

— 2001. Jakob von Uexküll: An introduction. *Semiotica* 134(1/4): 1–59.

— 2003. Thomas A. Sebeok and biology: Building biosemiotics. *Cybernetics and Human Knowing* 10(1): 47–60.

Levich, Alexander P.; Lovyagin, S. M. 1977. Suschestvuet li ekologicheskij kod? In: *Metodologicheskie i teoreticheskie aspekty biologii*. Petrozavodsk: Institut Biologii Karelskogo filiala, 71–74.

Lotman, Juri M. 1990. *Universe of the Mind: A Semiotic Theory of Culture*. London: I.B.Tauris.

— 1992. *Kul'tura i vzryv* [Culture and Explosion]. Moskva: Gnozis.

— 2005 [1984]. On the semiosphere. *Sign Systems Studies* 33(1): 215–239.

Lotman, Mihhail 2002. Umwelt and semiosphere. *Sign Systems Studies* 30(1): 33–40.

Markos, Anton 2002. *Readers of the Book of Life: Contextualizing Developmental Evolutionary Biology*. Oxford: Oxford University Press.

Morris, Charles 1971 [1946]. *Signs, Language, and Behavior*. New York: Braziller.

Nöth, Winfried 2001. Ecosemiotics and the semiotics of nature. *Sign Systems Studies* 29(1): 71–81.

Pattee, Howard H. 1972. Laws and constraints, symbols and languages. In: Waddington, Conrad Hal (ed.), *Towards a Theoretical Biology. Vol. 4: Essays*. Edinburgh: Edinburgh University Press, 248–258.

Perron, Paul; Sbrocchi, Leonard G.; Colilli, Paul; Danesi, Marcel (eds.) 2000. *Semiotics as Bridge between the Humanities and the Sciences*. New York: Legas.
Petrilli, Susan 1999. Charles Morris's biosemiotics. *Semiotica* 127(1/4): 67–102.
Pirner, Hans-Jürgen 2002. The semiotics of "postmodern" physics. In: Ferrari, Massimo; Stamatescu, Ion-Olimpiu (eds.), *Symbol and Physical Knowledge*. Berlin: Springer, 211–229.
Posner, Roland 2001a: Semiotics: The Physics of the 21st century. Presidential Address at the 7th Congress of the International Association for Semiotic Studies (IASS). In: Schmitz, Walter; Froeschle, Ulrich (eds.), *Sign Processes in Complex Systems. Proceedings of the 7th Congress of the IASS, Dresden, October 6–11, 1999*. Dresden: WEB-Verlag, p. 40.
Rosen, Robert 1991. *Life Itself: A Comprehensive Inquiry into the Nature, Origin, and Fabrication of Life*. New York: Columbia University Press.
— 1999. *Essays on Life Itself*. New York: Columbia University Press.
Rosen, Robert; Pattee, Howard H.; Somorjai, R. L. 1979. A symposium in theoretical biology. In: Buckley, Paul; Peat, F. David (eds.), *A Question of Physics: Conversations in Physics and Biology*. Toronto: University of Toronto Press, 84–123.
Rothschild, Friedrich S. 1962. Laws of symbolic mediation in the dynamics of self and personality. *Annals of New York Academy of Sciences* 96: 774–784.
Rüting, Torsten 2004. History and significance of Jakob von Uexküll and of his institute in Hamburg. *Sign Systems Studies* 32(1/2): 35–72.
Schult, Joachim (ed.) 2004. *Biosemiotik — praktische Anwendung und Konsequenzen für die Einzelwissenschaften*. [Studien zur Theorie der Biologie 6.] Berlin: Verlag für Wissenschaft und Bildung.
Santaella, Lucia 1999. Peirce and biology. *Semiotica* 127(1/4): 5–21.
Sebeok, Thomas A. 1972. *Perspectives in Zoosemiotics*. The Hague: Mouton.
— (ed.) 1977. *How Animals Communicate*. Bloomington: Indiana University Press.
— 1990. *Essays in Zoosemiotics*. Toronto: Toronto Semiotic Circle.
— 2000a. *Essays in Semiotics — I: Life Signs*. New York: Legas.
— 2000b. Semiotics as bridge between humanities and sciences. In: Perron, Paul; Sbrocchi, Leonard G.; Colilli, Paul; Danesi, Marcel (eds.), *Semiotics as Bridge between the Humanities and the Sciences*. New York: Legas, 76–100.
— 2001. Biosemiotics. In: Cobley, Paul (ed.), *The Routledge Companion to Semiotics and Linguistics*. London: Routledge, 163–164.
— 2002. La semiosfera come biosfera. *Athanor* 5: 11–18.
Sebeok, Thomas A.; Umiker-Sebeok, Jean (eds.) 1992. *Biosemiotics: The Semiotic Web 1991*. Berlin: Mouton de Gruyter.
Shintani, Laura 1999. Roman Jakobson and biology: 'A system of systems'. *Semiotica* 127(1/4): 103–113.
Uexküll, Jakob von 1928 [1920]. *Theoretische Biologie*. 2te Aufl. Berlin: J.Springer.
— 1940. *Bedeutungslehre*. Leipzig: J.A.Barth.
— 1982 [1940]. The theory of meaning. *Semiotica* 42(1): 25–82.
Uexküll, Thure von; Geigges, Werner; Herrmann, Jörg M. 1993. Endosemiosis. *Semiotica* 96(1/2): 5–51.
Weber, Andreas 2003. *Natur als Bedeutung: Versuch einer semiotischen Theorie des Lebendigen*. Würzburg: Königshausen & Neumann.
Witzany, Günther 2006. *The Logos of the Bios 1: Contributions to the Foundation of a Three-Leveled Biosemiotics*. Tartu: UMWEB.
Yates, F. Eugene 1985. Semiotics as a bridge between information (biology) and dynamics (physics). *Recherches Sémiotiques/Semiotic Inquiry* 5: 347–360.
Yates, F. Eugene; Beaugrande, Robert de 1990. Physics and semiotics. In: Koch, Walter A. (ed.), *Semiotics and the Sciences*. Bochum: Brockmeyer, 318–351.

CHAPTER 8

IS THE CELL A SEMIOTIC SYSTEM?

MARCELLO BARBIERI
Dipartimento di Morfologia ed Embriologia, Via Fossato di Mortara 64, 44100 Ferrara, Italy, brr@unife.it

Abstract: Semiotics is the study of signs and initially it was thought to be concerned only with the products of culture. Mental phenomena, however, exist also in animals, and cultural semiotics came to be regarded as a special case of biological semiotics, or *biosemiotics*, a science that started by studying semiotic phenomena in animals and then it was gradually extended to other living creatures. Eventually, the discovery of the genetic code suggested that the cell itself has a semiotic structure and the goal of biosemiotics became the idea that all living creatures are semiotic systems. This conclusion, however, is valid only if we accept that the genetic code is a *real* code, but an influential school of thought, known as *physicalism*, has apparently convinced many people that it is only a *metaphor*, a mere linguistic expression that we use in order to avoid long periphrases. The argument is that the genetic code would be real only if it was associated with the production of *meaning*, but modern science does not deal with meaning and is bound therefore to relegate the genetic code among the metaphorical entities.

In this paper it is shown that there is no need to avoid the issue of meaning and to deny the reality of the genetic code. On the contrary, it is shown that organic meaning can be defined with operative procedures and belongs to a new class of fundamental natural entities that are as objective and reproducible as the physical quantities. It is also shown that the presence of molecular adaptors gives us an objective criterion for recognizing the existence of organic codes in Nature, and that criterion proves that the genetic code has all the qualifying features of a real code. It also proves that the genetic code is not alone in the cellular world, and that many other organic codes appeared in the history of life, especially in eukaryotic cells.

The conclusion that the cell is a semiotic system, in short, is based on the experimental evidence provided by the adaptors, but also requires a new theoretical framework where concepts like sign, meaning and code are not put aside as metaphorical entities but are defined by operative procedures and are recognized as fundamental components of the living world

Keywords: Semiotics, biosemiotics, information, meaning, organic codes, ribotype, physicalism

M. Barbieri (ed.), Introduction to Biosemiotics, 179–207.

INTRODUCTION

Semiotics is the study of signs, and *biological semiotics*, or *biosemiotics*, is a new field of research that started by studying semiotic phenomena in animals and then was gradually extended to other organisms. Eventually, the goal of biosemiotics became the idea that all living creatures are semiotic systems and that semiosis is not a side-effect but a fundamental process of life. So far, however, biosemiotics has made little impact in the scientific world, and is largely regarded as a philosophical approach rather than a science.

There are many reasons for this, but the most important, in my opinion, is the fact that *biosemiotics has not yet proved that the cell is a semiotic system*. The cell is the unity of life and there is no chance that biosemiotics can become a science if it does not prove that signs exist in the cell, at the molecular level. This is the first problem of biosemiotics: *can we prove that the cell is a semiotic system?*

The starting point, in this enterprise, is a definition of semiosis, and to this purpose it is natural to turn to the classical model that Charles Peirce developed in semiotics and that Thomas Sebeok adopted in biosemiotics. According to this model, a semiotic system is a triad of *"sign, object and interpretant"*, and interpretation is regarded as the *sine qua non* condition of semiosis. Sebeok insisted that *"there can be no semiosis without interpretability"* and that interpretation is *"a necessary and sufficient condition for something to be a semiosis"* (Sebeok, 2001). Thanks to Sebeok, the Peirce model of semiosis has become extremely popular and today it is adopted by most biosemioticians.

There is no doubt that processes of interpretation take place almost everywhere in the living world, and the Peirce model applies therefore to an impressive range of biological phenomena. There is however one outstanding exception to that rule. The exception is the genetic code, a set of conventions that does not depend on interpretation because the rules of the genetic code are virtually the same in all organisms and in all circumstances, and that has been true for almost the entire history of life on Earth. This suggests that interpretation was not necessary for the origin of the genetic code and, more in general, that the origin of semiosis was not the origin of interpretation.

But can we separate semiosis from interpretation? Here it will be shown that semiosis can indeed be defined without any reference to interpretation because its qualifying feature is coding and coding requires a codemaker, not an interpreter. What is most important, however, is that a model of semiosis that does not depend on interpretation gives us precisely what we have been looking for in biosemiotics. It allows us to prove that the cell is a semiotic system.

In order to reach that conclusion, however, it will be necessary to go through a series of stages. To start with, it has to be shown that semiosis was instrumental not only to the origin but also to the evolution of the cell. Another issue that must be dealt with is the paradigm of *physicalism*, the claim that the origin and the evolution of life can be explained in principle only by physical quantities, because that implies that signs, meanings and codes are redundant concepts. A third challenge is how to account for interpretation, a phenomenon that cannot be ignored because it is almost ubiquitous in the living world.

In order to deal with these problems, the paper has been divided into four Parts. Part 1 will show that the origin of semiosis was a precondition for the origin of life. Part 2 will show that a whole stream of organic codes came into existence during the evolution of the cell. Part 3 will face the claim of physicalism and expose its weakness. Part 4 will argue that interpretation can be accounted for by the evolution of semiotic systems. Altogether, we will have to overcome five major obstacles, and only at that point we will be able to conclude that the cell is a true semiotic system.

PART 1 – SEMIOSIS AND THE ORIGIN OF LIFE

A New Model of Semiosis

Semiotics is usually referred to as *the study of signs* but this definition is too restrictive because signs are always associated with other entities. A sign, to start with, is always linked to a *meaning*. As living beings, we have a built-in drive to make sense of the world, to give meanings to things, and when we give a meaning to something, that something becomes a sign for us. Sign and meaning, in other words, cannot be taken apart because they are the two sides of the same coin. We cannot have a sign without a meaning because a sign must signify something, and we cannot have a meaning without a sign because a meaning must be the meaning of something. Semiotics, therefore, is not just the study of signs; it is the study of signs and meanings together. The result is that a system of signs, i.e. a *semiotic system*, is always made of two distinct worlds: a world of objects that we call *signs* and a world of objects that represent their *meanings*.

The link between sign and meaning, in turn, calls attention to a third entity, i.e. to their *relationship*. A sign is a sign only when it stands for something that is *other than itself*, and this *otherness* implies at least some degree of *independence*. It means that there is no deterministic relationship between sign and meaning. Different languages, for example, give different names to the same object precisely because there is no necessary connection between names and objects. A semiotic system, therefore, is not merely a combination of two distinct worlds. It is *a combination of two worlds between which there is no necessary link*, and this has an extraordinary consequence. It implies that a bridge between the two worlds can be established only by *arbitrary* rules. A link between signs and meanings, in other words, can be produced only by *conventions* or *codes*. This is what qualifies the semiotic systems, what makes them different from everything else: *a semiotic system is a system made of two independent worlds that are connected by the conventional rules of a code.* A semiotic system, in conclusion, is necessarily made of *three* distinct entities, and is represented by the triad *"signs, meanings, code"*.

Here at last we have a definition where it is mentioned explicitly that a code is an essential component of a semiotic system. It is the rules of a code that create a correspondence between signs and meanings, and we can say therefore that an act of semiosis is always an act of coding, i.e. it is always a convention. More precisely, we can say that an elementary act of semiosis is a triad of *"sign, meaning and convention"*, whereas a semiotic system as a whole is a set of signs and meanings linked by the conventions of a code.

Signs, meanings and conventions, however, do not come into existence of their own. There is always an "agent" that produces them, and that agent can be referred to as a *codemaker* because it is always an act of coding that gives origin to semiosis. In the case of culture, for example, the codemaker is the mind, since it is the mind that produces the mental objects that we call signs and meanings and the conventions that link them together. We come in this way to a general conclusion that can be referred to as "the code model" of semiosis: *a semiotic system is a triad of signs, meanings and code that are all produced by the same agent, i.e. by the same codemaker.*

This conclusion is highly relevant to biology because it tells us that the existence of the genetic code in every cell is a strong indication that the cell is a semiotic system. This would be one of the greatest generalizations of science because it would mean that semiosis exists in the whole organic world and not just in culture. It would also mean that the first semiotic structures appeared on Earth at a very early stage of evolution and were instrumental for the origin of the genetic code. The idea that the cell is a semiotic system, in short, raises fundamental questions about the nature of life and invites us to take a new look at the problem of its origin.

The Molecules of Life

Modern biology is based on three extraordinary experimental facts: (1) the discovery that most biological structures and functions are ultimately due to *proteins*, i.e. to strings of amino acids; (2) the discovery that the hereditary instructions for making proteins are carried by strings of nucleotides called *genes*; and (3) the discovery that gene are translated into proteins by a universal set of rules which has become known as *the genetic code*.

These discoveries have confirmed that genes and proteins are the key molecules of life, but have also revealed something totally unexpected about them. They have shown that genes and proteins differ from all other molecules not because of their size, shape or chemical composition, but because they are *produced* in a totally different way. In the inorganic world, the structure of molecules is determined by the bonds that exist between their atoms, i.e. by *internal* factors. In living systems, instead, genes are built by molecular machines which physically stick their nucleotides together following the order of a template which is *external* to the growing molecule. In a similar way, proteins are made by molecular machines which bind amino acids in the order prescribed by an external template of nucleotides.

Genes and proteins, in short, are assembled by molecular robots on the basis of outside instructions. They are *manufactured* molecules, as different from ordinary molecules as *artificial* objects are from *natural* ones. Indeed, if we accept the commonsense view that molecules are natural when their structure is determined *from within*, and artificial when it is determined *from without*, then genes and proteins can truly be referred to as *artificial molecules*, as *artifacts made by molecular machines*. This in turn implies that all biological objects are artifacts, and we arrive at the general conclusion that *life is artifact-making*.

Molecular biology, in short, has revealed the existence of a totally unexpected gulf between life and non-life. The great divide is not between organic and inorganic structures. It is between structures which are built *from within* and structures which are built *from without*. Between molecules which are made by spontaneous assemblies and molecules which are manufactured by molecular machines. All of which has two outstanding implications. The first is that there must have been a point, in the early history of the Earth, when molecular machines appeared on our planet and started producing *manufactured genes*. The second is that there must also have been a point in history when molecular machines appeared on our planet and started producing *manufactured proteins*. Those two great events had to take place long before the first cells could appear, and it was they that set in motion the odyssey of evolution. They were the real beginning of life on Earth.

The Origin of Genes

The discovery that genes and proteins are *manufactured* molecules has direct implications for the origin of life, because it tells us that primitive molecular machines came into existence long before the origin of the first cells. The simplest molecular machines we can think of are molecules that could join other molecules together by chemical bonds, and for this reason we may call them *bondmakers*. Some could form bonds between amino acids, some between nucleotides, others between sugars, and so on. It has been shown, for example, that short pieces of ribosomal RNA have the ability to form peptide bonds, so it is possible that the first bondmakers were RNA molecules of small or medium-size molecular weights. Among the various types of bondmakers, furthermore, some developed the ability to join nucleotides together in the order provided by a *template*. Those bondmakers started *making copies* of nucleic acids, so we can call them *copymakers*. We don't know when they appeared on the primitive Earth but at some stage they did and that was a real turning point, because it set in motion an extraordinary sequence of events.

The copying of a template is the elementary act of gene duplication, the very first step toward the phenomenon of *heredity*. When a process of copying is repeated indefinitely, furthermore, another phenomenon comes into being. Copying mistakes become inevitable, and in a world of limited resources not all changes can be implemented, which means that a process of selection is bound to take place. Molecular copying, in short, leads to *heredity*, and the indefinite repetition of molecular copying leads to *natural selection*. That is how natural selection came into existence. Molecular copying started it and molecular copying has perpetuated it ever since.

In the history of life, molecular copying came into being when the first copymakers appeared on the primitive Earth and started making copies of nucleic acids. This implies that *natural* nucleic acids had already been formed by spontaneous reactions on our planet, but that was no guarantee of evolution. Only the copying of genes could ensure their survival and have long-term effects, so it was really the arrival of copymaking that set in motion the extraordinary chain of processes that we call evolution. The first Major Transition of the history of life (Maynard Smith and Szathmáry, 1995) is generally described as the origin of genes, but it seems more accurate to say that it was the origin of molecular *copying*, or the origin of *copymakers*, the first molecular machines that started multiplying nucleic acids by making copies of them.

The Origin of Proteins

Proteins truly are the stuff of life. They are the key building blocks of all living structures, as well as the engines of countless reactions that go on within those structures. For all their extraordinary versatility, however, there is one thing they cannot do. Unlike genes, they cannot be their own templates. It is simply not possible to make proteins by copying other proteins. The transition from natural to manufactured molecules, therefore, was relatively simple for genes but much more complex for proteins. Manufactured genes could be made simply by copying natural genes, and all that was required to that purpose were molecules which had a polymerase-like activity. Manufactured proteins, instead, could not be made by copying, and yet the information to make them had to come from molecules that can be copied, because only those molecules can be inherited. The information for manufacturing proteins, therefore, had to come from genes, so it was necessary to bring together a carrier of genetic information (a messenger RNA), a peptide-bondmaker (a piece of ribosomal RNA) and molecules that could carry both nucleotides and amino acids (the transfer RNAs). The first protein-makers, in short, had to bring together three different types of molecules (messenger, ribosomal and transfer RNAs), and were therefore much more complex than copymakers.

The outstanding feature of the protein-makers, however, was not the number of components. It was the ability to ensure a one-to-one correspondence between genes and proteins, because without it there would be no biological specificity, and without specificity there would be no heredity and no reproduction. Life as we know it simply would not exist without a one-to-one correspondence between genes and proteins.

Such a correspondence would be automatically ensured if the bridge between genes and proteins could have been determined by *stereochemistry*, as one of the earliest models suggested, but that is not what happens in Nature. The bridge is always provided by molecules of transfer RNA, first called *adaptors*, that have two recognition sites: one for a group of three nucleotides (a *codon*) and another for an amino acid. In this case, a one-to-one correspondence could still be guaranteed automatically if one recognition site could determine the other, but again that is not what happens. The two recognition sites of the adaptors are physically separated in space and are chemically independent. There simply is no necessary link between codons and amino acids, and a one-to-one correspondence between them can only be the result of conventional rules. Only a real code, in short, could guarantee biological specificity, and this means that the evolution of the translation apparatus had to go hand in hand with the evolution of the genetic code.

Protein synthesis arose therefore from the integration of two different processes, and the final machine was a *code-and-template-dependent-peptide-maker*, or, more simply, a *codemaker*. The second Major Transition of the history of life (Maynard Smith and Szathmáry, 1995) is generally described as the origin of proteins, but it would be more accurate to say that it was the origin of *codemaking*, or the origin of *codemakers*, the first molecular machines that discovered molecular coding and started populating the Earth with codified proteins.

The Cell as a Trinity

The idea that life is based on genes and proteins is often expressed by saying that every living system is a duality of genotype and phenotype. This model was proposed by Wilhelm Johannsen in 1909, but was accepted only in the 1940s and 50s, when molecular biology discovered that genes are chemically different from proteins, and, above all, when it became clear that genes carry *linear information* whereas proteins function by their *three-dimensional structures.* The genotype-phenotype duality is therefore a dichotomy that divides not only two different biological functions (heredity and metabolism), but also two different physical quantities (information and energy). It is the simplest and the most general way of defining a living system, and has become the founding paradigm of modern biology, the scheme which has transformed the *energy-based* biology of the 19th century into the *information-based* biology of the 20th.

In the 1950s and 60s, however, the study of protein synthesis revealed that genes and proteins are not formed spontaneously in the cell but are manufactured by a system of molecular machines based on RNAs. In 1981, the components of this manufacturing system were called *ribosoids* and the system itself was given the collective name of *ribotype* (Barbieri 1981, 1985). The cell was described in this way as a structure made of genes, proteins and ribosoids, i.e. as a trinity of genotype, phenotype and ribotype.

This model is based on the idea that it was molecular machines made of ribosoids that gave origin to the first cells. Spontaneous genes and spontaneous proteins did appear on the primitive Earth but could not give origin to cells because they do not have biological specificity. They gave origin to copymakers and codemakers and it was these molecular machines that evolved into the first cells.

The RNAs and the proteins that appeared spontaneously on the primitive Earth produced a wide variety of ribosoids, some of which were synthetizing ribosoids whereas others were ribogenes and others were riboproteins (or ribozymes). The systems formed by combinations of all these molecules, therefore, had a ribotype, a ribogenotype and a ribophenotype. Eventually, evolution replaced the ribogenes with genes and the riboproteins with proteins but the synthetising ribosoids of the ribotype have never been replaced. They are the oldest phylogenetic molecules that exist on Earth and they firmly remain at the heart of every living cell. Genes, proteins and ribosoids are all manufactured molecules, but only the ribosoids can become makers of those molecules. From a functional point of view, in other words, the ribotype is the maker of genotype and phenotype, and has therefore a logical and a historical priority over them.

It is an experimental fact, at any rate, that every cell contains a system of RNAs and ribonucleoproteins that makes proteins according to the rules of a code, and that can be described therefore as a "code-and-template-dependent-protein-maker", i.e. as a "codemaker". That is the third party that makes of every living cell a trinity of genotype, phenotype and ribotype. The genotype is the seat of heredity, the phenotype is the seat of metabolism and the ribotype is the seat of the genetic code, the codemaker of the cell.

The Defining Feature of Signs and Meanings

A semiotic system is made of signs, meanings and code, and we know that there is a code in the apparatus of protein synthesis. But what about the other two entities? In addition to the genetic code, can we say that there are also signs and meanings in protein synthesis? Can these entities exist at the molecular level? In order to answer this question, let us examine the traditional signs and meanings of our cultural world and see if they have a qualifying feature that can be extended to all systems.

The signs and meanings that we are familiar with are normally associated with objects or events of the physical world, but they do not coincide with those entities. They are always *mental representations*. A sign, for example, can be the mental representation of the word "tree", and its meaning can be the mental image of a tree, in which case the sign is a mental sound and its meaning is a mental image. This shows that signs and meanings are distinct mental objects, and in principle therefore they can be separated. We have also seen, however, that they cannot be taken apart because they are the two faces of the same coin, so we have a problem. Can we separate signs and meanings or not?

The mental image of an object is normally associated with different sounds in different languages, and this clearly shows that mental sounds and mental images can be separated. When they are separated, however, they no longer function as signs and meanings. To a non-English speaker, for example, a sound like "tree" may have no linguistic meaning and in this case it would be just a sound, not a sign. A mental object, in short, becomes a sign only when the mind *treats it as a sign*. There is no contradiction therefore in saying that signs and meanings are distinct mental objects and that they cannot be taken apart, because when they are taken apart they simply stop functioning like signs and meanings.

It is the mind, and more precisely the mental codemaker, that brings signs and meanings into existence. A codemaker, furthermore, is necessary not only to bring them into existence, but also to keep them in existence, because they do not have a life of their own. Signs and meanings exist only when a codemaker creates a correspondence between two distinct entities, and only for as long as the codemaker keeps producing that correspondence. The codemaker is the agent, whereas signs and meanings are the instruments of semiosis. They are totally dependent on codemaking, and it is precisely this total dependence that gives them their defining feature: *signs and meanings are codemaker-dependent entities.*

This is the qualifying feature that we were looking for because it applies to all systems where there is a code and a codemaker. In order to say that signs and meanings exist in protein synthesis, therefore, we must prove that there are codemaker-dependent entities in that process and that the genetic code is a correspondence between those entities.

Organic Signs and Organic Meanings

The signs and meanings of our mental world have properties that do not seem to exist in molecules. A mental sign, or a mental meaning, is never an *intrinsic* property of a mental object. It is something that the mind can give to a mental object and that the mind can take away from it. Mental signs and mental meanings simply do not exist without a codemaker and outside a codemaking process.

In protein synthesis, however, things appear to be different. The translation apparatus is the agent, and therefore the codemaker of the code that creates a correspondence between genes and proteins, but these molecules appear to have only "objective" chemical properties, not the "codemaker-dependent" properties that qualify signs and meanings. A messenger RNA, for example, appears to be a unique and objective sequence of codons, but let us take a closer look.

A messenger RNA is certainly a unique and objective chain of *nucleotides* but in no way it is a unique sequence of *codons* because different codemakers could scan it in different ways. If the nucleotides were scanned two-by-two, for example, the sequence of codons would be totally different. A chain of nucleotides is a potential carrier of many sequences of codons and it is always the codemaker that determines which sequence is going to be realized because it is the codemaker that *defines* the codons. A linear sequence of codons, in short, does not exist without a codemaker and outside a codemaking process. It is totally dependent on codemaking, and is therefore a "codemaker-dependent" entity, exactly as a sign is.

In the same way, the linear sequence of amino acids that is produced by a translation apparatus is also "codemaker-dependent", because only a codemaker can produce it. Any spontaneous assembly of amino acids would produce three-dimensional arrangements, not linear chains, and above all it would not arrange the amino acids in a specific sequence. Proteins, on the other hand, cannot be copied so they cannot be produced by copymakers. A specific protein can be produced only by a codemaker, but different codemakers would arrange the same amino acids in different ways, which shows that the linear sequence of a protein is only one of the many possible "meanings" that could be given to a string of nucleotides.

The sequence of a gene and the sequence of a protein, in conclusion, may look like "objective" properties of these molecules, but they are not. They are "codemaker-dependent" entities because they do not exist without a codemaking process, and because they would be different if the codemaker had a different conformation. The sequences of genes and proteins, in short, have the essential characteristics that define signs and meanings, and we can say therefore that they "are" organic signs and organic meanings.

More precisely, we can say that "an organic sign is the sequence *used* by a codemaker in a coding process", and that "an organic meaning is the sequence *produced* by a codemaker in a coding process". All we need to keep in mind is that *signs and meanings are mental entities when the codemaker is the mind, but they are organic entities when the codemaker is made of organic molecules.*

We come in this way to the conclusion that the apparatus of protein synthesis contains all the three essential components of semiosis (signs, meanings and code) and is therefore a real semiotic system. To this conclusion we can now add the fact that the RNAs of that system are the oldest phylogenetic molecules in the world, and we come to another major conclusion. We realize that the apparatus of protein synthesis was the very first semiotic system that appeared in the history of life. The origin of protein synthesis, in other words, was also the origin of semiosis.

PART 2 – THE ORGANIC CODES

How Many Organic Codes?

According to modern biology, the genetic code is the only code that exists in the organic world, whereas the world of culture is populated by a virtually unlimited number of codes. We know, furthermore, that the genetic code came into being with the origin of life, whereas the cultural codes arrived almost four billion years later, with cultural evolution. This appears to suggest that organic evolution went on for four billion years, almost the entire history of life on Earth, without producing any other organic code after the first one, while cultural evolution has produced an astonishing number of cultural codes in just a few thousand years. If this were true, there would be an impressive asymmetry between the codes of the organic world and those of the mental world. But is the asymmetry real? Are we sure that the genetic code is the only organic code of life? Luckily, this is a problem that we can deal with, because if other organic codes exist in Nature we should be able to find them by the classic experimental method of science, just as we have found the genetic code.

The very first step, in this enterprise, is to underline the difference that exists between *copying* and *coding*, a difference that is particularly evident in *transcription* and *translation*. In transcription, an RNA sequence is assembled from the linear information of a DNA sequence, and in this case a normal biological catalyst (an RNA polymerase) is sufficient, because each step requires a single recognition process. In translation, instead, two independent recognition processes must be performed at each step, and the system that performs the reactions (the ribosome) needs special molecules, first called *adaptors* and then *transfer RNAs*, in order to associate codons to amino acids according to the rules of the genetic code. Without a code, in fact, a codon could be associated with different amino acids and *biological specificity*, the most precious of life's properties, would be lost.

These concepts can easily be generalized. We are used to think that biochemical processes are all *catalyzed* reactions, but the difference that exists between *copying* and *coding* tells us that we should distinguish very sharply between *catalyzed* and *codified* reactions. The catalyzed reactions are processes (like transcription) that require only one recognition process at each step. The codified reactions, instead, require two independent recognition processes at each step and a set of coding rules. The catalyzed reactions, in other words, require *catalysts*, while the codified reactions require *adaptors*, i.e. catalysts plus a code.

Any organic code is a set of rules that establish a correspondence between two independent worlds, and this necessarily requires molecular structures that act like *adaptors*, i.e. that perform two independent recognition processes. The adaptors are required because the two worlds would no longer be independent if there were a necessary link between them, and a set of rules is required in order to guarantee the specificity of the correspondence. In any organic code, in short, we should find three major features:

(1) A correspondence between two independent worlds.
(2) A system of molecular adaptors.
(3) A set of rules that guarantee biological specificity.

We conclude that the key molecules of the organic codes are the adaptors. They are the molecular *fingerprints* of the codes, and their presence in a biological process is a sure sign that that process is based on a code. This gives us an *objective criterion* for the search of organic codes, and their existence in Nature becomes therefore, first and foremost, an experimental problem.

The Splicing Codes

One of the greatest surprises of molecular biology was the discovery that the primary transcripts of the genes are often transformed into messenger RNAs by removing some RNA strings (called *introns*) and by joining together the remaining pieces (the *exons*). The result is a true assembly, because exons are assembled into messengers, and we need therefore to find out if it is a *catalyzed* assembly (like transcription) or a *codified* assembly (like translation). In the first case the cutting-and-sealing operations, collectively known as *splicing*, would require only a *catalyst* (comparable to RNA-polymerase), whereas in the second case they would need a catalyst and a set of *adaptors* (comparable to ribosome and tRNAs).

This suggests immediately that splicing is a codified process because it is implemented by structures that are very similar to those of protein synthesis. The splicing systems, known as *spliceosomes*, are huge molecular machines like ribosomes, and employ small molecular structures, known as *snRNAs* or *snurps*, which are very much comparable to tRNAs. The similarity, however, goes much deeper than that, because the snRNAs have properties that fully qualify them as *adaptors*. They bring together, in a single molecule, two independent recognition processes, one for the beginning and one for the end of each intron, thus creating a specific correspondence between the world of the primary transcripts and the world of messengers.

The two recognition steps are independent not only because there is a physical distance between them, but above all because the first step could be associated with different types of the second one, as demonstrated by the cases of *alternative splicing*. The choice of the beginning and of the end of an intron, furthermore, is the operation that actually defines the introns and gives them a *meaning*. Without a complete set of such operations, primary transcripts could be transformed arbitrarily into messenger RNAs, and there would be no biological specificity whatsoever.

In RNA splicing, in conclusion, we find the three basic characteristics of all codes: (1) a correspondence between two independent worlds, (2) the presence of molecular adaptors, and (3) a set of rules that guarantee biological specificity. We conclude therefore that the processing of RNA transcripts into messengers is truly a codified process based on adaptors, and takes place with rules that can rightly be given the name of *splicing codes* (Barbieri, 1998; 2003).

The Signal Transduction Codes

Cells react to a wide variety of physical and chemical stimuli from the environment, and in general their reactions consist in the expression of specific genes. We need therefore to understand how the environment interacts with the genes, and the turning point, in this field, came from the discovery that the external signals (known as *first messengers*) never reach the genes. They are invariably transformed into a different world of internal signals (called *second messengers*) and only these, or their derivatives, reach the genes. In most cases, the molecules of the external signals do not even enter the cell and are captured by specific receptors of the cell membrane, but even those that do enter (some hormones) must interact with intracellular receptors in order to influence the genes (Sutherland, 1972).

The transfer of information from environment to genes takes place therefore in two distinct steps: one from first to second messengers, which is called *signal transduction*, and a second path from second messengers to genes which is known as *signal integration*. The surprising thing about signal transduction is that there are hundreds of first messengers (hormones, growth factors, neurotransmitters, etc.) whereas the known second messengers are only four (cyclic AMP, calcium ions, inositol trisphosphate and diacylglycerol) (Alberts *et al.*, 1994).

First and second messengers, in other words, belong to two very different worlds, and this suggests immediately that signal transduction may be based on organic codes. This is reinforced by the discovery that there is no necessary connection between first and second messengers, because it has been proved that the same first messengers can activate different types of second messengers, and that different first messengers can act on the same type of second messengers.

The experimental data, in brief, prove that external signals do not have any instructive effect. Cells use them to *interpret* the world, not to yield to it. Such a conclusion amounts to saying that signal transduction is based on organic codes, and this is in fact the only plausible explanation of the data, but of course we would also like a direct proof. As we have seen, the signature of an organic code is the presence of adaptors, and the molecules of signal transduction have indeed the typical characteristics of the adaptors. The transduction system consists of at least three types of molecules: a *receptor* for the first messengers, an *amplifier* for the second messengers and a *mediator* in betwen (Berridge, 1985). The system performs two independent recognition processes, one for the first and the other for the second messenger, and the two steps are connected by the bridge of the mediator. The connection however could be implemented in countless different ways since any first messenger can be coupled with any second messenger, and this makes it imperative to have a code in order to guarantee biological specificity.

In signal transduction, in short, we find all three characteristics of the codes: (1) a correspondence between two independent worlds, (2) a system of adaptors that give meanings to molecular structures, and (3) a collective set of rules that guarantee biological specificity. The effects that external signals have on cells, in conclusion, do not depend on the energy or the information that they carry, but on the *meaning* that cells give them with rules that we can rightly refer to as *signal transduction codes* (Barbieri, 1998; 2003).

The Compartment Codes

Eukaryotic cells not only produce molecules of countless different types but manage to deliver them to different destinations with astonishing precision, and this gives us the problem of understanding how they manage to cope with such an immensely intricate traffic. The first step in the solution of this mystery came with the discovery that the Golgi apparatus is involved not only in the biochemical modification of innumerable molecules but also in the choice of their geographical destination. But the truly remarkable thing is that all this is achieved with an extremely simple mechanism. More precisely, the Golgi apparatus delivers an astonishing number of molecules to their destinations with only three types of vesicles. One type has labels for the transport of proteins outside the cell and another for their delivery to the cell interior, whereas the vesicles of the third type carry no destination label, and are programmed, *by default*, to reach the plasma membrane. As we can see, the solution is extraordinarily efficient. With a single mechanism and only two types of labels, the cell delivers a great amount of proteins to their destinations, and also manages to continually renew its plasma membrane.

The Golgi apparatus, however, is a transit place only for a fraction of the cell proteins. The synthesis of all eukaryotic proteins begins in the soluble part of the cytoplasm (the *cytosol*), together with that of a signal that specifies their geographical destination. The piece of the amino acid chain that emerges first from the ribosome (the so-called *peptide leader*) can contain a sequence that the cell interprets as an *export signal to the endoplasmic reticulum.* If such a signal is present, the ribosome binds itself to the reticulum and delivers the protein into its *lumen*. If not, the synthesis continues on free ribosomes, and the proteins are shed into the cytosol. Of these, however, only a fraction remains there, because the amino acid chain can carry, in its interior, one or more signals which specify other destinations, such as the *nucleus*, the *mitochondria*, and other cell compartments. Proteins, in conclusion, carry with them the signals of their geographical destination, and even the absence of such signals has a meaning, because it implies that the protein is destined to remain in the cytosol.

The crucial point is that there is *no necessary correspondence* between protein signals and geographical destinations. The export-to-the-nucleus signals, for example, could have been used for other compartments, or could have been totally different. They and all the other geographical signals are purely conventional labels, like the names that we give to streets, to cities, to airports and to holiday resorts. The existence of eukaryotic compartments, in other words, is based on natural conventions, and to their rules of correspondence we can legitimately give the name of *compartment codes* (Barbieri, 2003).

The Cytoskeleton Codes

A cytoskeleton is absolutely essential for typical eukaryotic processes such as phagocytosis, mitosis, meiosis, ameboid movement, organelle assembly and three-dimensional organization of the cell, i.e., for all those features that make eukaryotic cells so radically different from bacteria. The actual cytoskeleton, in reality, is an integrated system of three different cytoskeletons made of filaments (*microfilaments*, *microtubules* and *intermediate filaments*) each of which gives a specific contribution to the three-dimensional form of the cell and to its mobility.

The driving force of the cytosleton is a very unusual mechanism that biologists have decided to call *dynamic instability.* The cytoskeletal filaments – especially microtubules and microfilaments – are in a state of continuous flux where monomers are added to one end and taken away at the other, and the filament is growing or shortening according to which end is having the fastest run. But what is really most surprising is that all this requires *energy*, which means that the cell is investing enormous amounts of energy not in building a structure but *in making it unstable!*

In order to understand the logic of dynamic instability, we need to keep in mind that cytoskeletal filaments are unstable only when their ends are not attached to special molecules that have the ability to anchor them. Every microtubule, for example, starts from an organizing center (the *centrosome*), and the extremity which is attached to this structure is perfectly stable, whereas the other extremity can grow longer or shorter, and becomes stable only when it encouters an anchoring molecule in the cytoplasm. If such an anchor is not found, the whole microtubule is rapidly dismantled and another is launched in another direction, thus allowing the cytoskeleton to explore all cytoplasm's space in a short time.

Dynamic instability, in other words, is a mechanism that allows the cytoskeleton to build structures with an *exploratory strategy*, and the power of this strategy can be evaluated by considering how many different forms it can give rise to. The answer is astonishing: the number of different structures that cytoskeletons can create is *potentially unlimited*. It is the anchoring molecules (that strangely enough biologists call *accessory proteins*) that ultimately determine the three-dimensional forms of the cells and the movements that they can perform, and there could be endless varieties of anchoring molecules. The best proof of this enormous versatility is the fact that the cytoskeleton was invented by unicellular eukaryotes but was later exploited by metazoa to build completely new structures such as the axons of neurons, the myofibrils of muscles, the mobile mouths of macrophages, the tentacles of killer lymphocytes and countless other specializations.

Dynamic instability, in conclusion, is a means of creating an endless stream of cell types with only one common structure and with the choice of a few anchoring molecules. But this is possible only because there is *no necessary relationship* between the common structure of the cytoskeleton and the cellular structures that the cytoskeleton is working on. The anchoring molecules (or accessory proteins) are true *adaptors* that perform two independent recognition processes: microtubules on one side and different cellular structures on the other side. The resulting correspondence is based therefore on *arbitrary* rules, on true natural conventions that we can refer to as *the cytoskeleton codes* (Barbieri, 2003).

The Sequence Codes

In the 1980s and 90s, Edward Trifonov started a life-long campaign in favour of the idea that the nucleotide sequences of the genomes carry several messages simultaneously, and not just the message revealed by the classic triplet code. According to Trifonov, in other words, the genetic code is not alone, and there are many other codes in the nucleotide sequences of living organisms. This conclusion rests upon Trifonov's definition that *a code is any sequence pattern that can have a biological function*, or *codes are messages carried by sequences*, or *a code is any pattern in a sequence which corresponds to one or another specific biological function* (Trifonov, 1989; 1996; 1999).

The plurality of codes described by Trifonov is a result of his particular definition of a code, but it is not necessarily limited by that, and could well be compatible with different definitions. The splicing code, for example, is a code not only according to his criterion, but also according to the operative definition that a code is a set of rules of correspondence implemented by adaptors. This suggests that Trifonov's conclusions may have a general validity, and at least some of his sequence codes may turned out to be true organic codes. For the time being, however, let us acknowledge the fact that according to Trifonov's definition there are at least eight sequence codes in the genomes of living creatures, in addition to the classic triplet code (Trifonov, 1996):

(1) The *transcription codes* include promoters and terminators, and are rather universal, though different in prokaryotes and in eukaryotes.
(2) The *gene splicing code* for the processing of nuclear pre-mRNA is largely undeciphered. Its main components are obligatory GU- and AG-ends of introns, as well as rather conserved consensus sequence features around the ends.
(3) The *translation pausing code*, for the regulation of translation, is encoded by clusters of rare triplets for which the aminoacyl-tRNAs are in limited supply.
(4) The *DNA structure code*, or *DNA shape code*, is a sequence-dependent local shape of DNA which is a crucial component of the protein-DNA recognition.
(5) The *chromatin code* describes those sequence features that direct the histone octamer's binding to DNA and the formation of nucleosomes.
(6) The *translation framing code* is overlapping with the triplet code (Trifonov, 1987), and ensures the correct reading frame during translation.
(7) The *modulation code* is about the repeating sequences and regulates the number of repeats as an adjustable variable to modulate expression of the nearby gene.
(8) The *genome segmentation code* is one of the emerging new codes, and is due to fact that the genomes appear to be built of rather standard size units.

A Stream of Codes

Most of the papers which have been published on biological codes do not make any reference to their definition. This is the case, for example, of the reports which have described and discussed a truly remarkable wealth of experimental data on the *Adhesive Code* (Redies and Takeichi, 1996; Shapiro and Colman, 1999), on the *Sugar Code* (Gabius, 2000; Gabius et al., 2002), and on the *Histone Code* (Strahl and Allis, 2000; Jenuwein and Allis, 2001; Turner, 2000; 2002; Gamble and Freedman, 2002; Richards and Elgin, 2002).

The practice of studying something without precisely defining it is fairly common in many sciences, and biology is no exception. The paradigmatic example is life itself, a phenomenon that we keep studying even if nobody seems to agree on its definition. Another instructive case is the concept of species, for which there is no definition that is universally valid and yet this does not prevent biologists from doing experiments, obtaining results and making sensible predictions on countless species of living creatures. Precise definitions, in short, are not always essential, but in some cases they are, and this is one of them. More precisely, we should be aware that an operative definition of organic codes in terms of adaptors would have provided a crucial guideline in at least two important cases.

(1) One is the research on new biological codes such as the Adhesive Code, the Sugar Code and the Histone Code. The problem here is that the experimental data *suggest* the existence of organic codes but do not *prove* it. And yet the results could have been conclusive because they are all compatible with the existence of true adaptors. On the face of the evidence, for example, it is most

likely that lectins are the adaptors of the Sugar Code and that cadherins are the adaptors of the Adhesive Code. If that had been proved, there would be no doubt that we are in the presence of true organic codes. But people did not use a definition of codes based on adaptors, so they did not look for adaptors.

(2) The second case is that of the classical research on signal transduction. Here the amount of experimental data is so enormous to be beyond description, and yet there is a remarkable paradox in this field. The only logical explanation of the facts is that signal transduction is based on organic codes and yet the word "code" has never been mentioned, so people have never looked for coding rules. The evidence has actually proved that signal transducers have the experimental characteristics of true adaptors, and yet the word "adaptors" has never been mentioned with reference to a code. This habit could well go on indefinitely by inertia, and only a precise definition of organic codes can convince people that an alternative (and much more convincing) explanation of the facts already exists.

It has been the existence of adaptors which has proved the reality of the Genetic Code, and the same is going to be true for the Signal Transduction Codes, for the Adhesive Code, for the Sugar Code and for the Histone Code. An operative definition based on adaptors, furthermore, is the only scientific instrument that can allow us to prove the existence of other organic codes in Nature. And when we really start looking for them, we may well discover that so far we have only scratched the surface. That there is a long golden stream of organic codes out there.

PART 3 – INFORMATION AND MEANING

The Claim of Physicalism

In 1953, Watson and Crick proposed that the linear sequence of nucleotides represents the *information* carried by a gene. A few years later, the mechanism of protein synthesis was discovered and it was found that the sequence of nucleotides in genes determines the sequence of amino acids in proteins, with a process that amounts to a transfer of linear information from genes to proteins. In both types of molecules, therefore, *biological information* was identified with, and defined by, the specific sequence of their subunits.

The concept of biological information threw a completely new light of the century old mystery of inheritance (*"heredity is the transmission of information"*) and quickly transformed the whole of biology from an energy-based into an information-based science. Despite this outstanding success, however, it has been repeatedly pointed out that biological information, or *biological specificity* (as some prefer to call it), cannot be measured and cannot therefore be regarded as a physical quantity, neither a fundamental nor a derived one. This is further underlined by the fact that there is another type of information which can be measured and which is a true physical quantity. It is the information of a message that Shannon defined in 1948 with an entropy-like formula, and which can be referred to as *physical information*.

The crucial point is that Shannon's information does not depend on the sequence of subunits, while biological information is defined precisely by that sequence. Physical information, in other words, has nothing to do with *specificity*, while biological information has everything to do with it. The two concepts are literally worlds apart, and this reinforces the conclusion that biological information is not, and cannot be, a physical quantity. So what is it? A similar problem arises with the genetic code. The rules of a code cannot be measured and a code therefore cannot be a combination of physical quantities. So what is it?

According to an influential school of thought, biological information and the genetic code are simply *metaphors*. They are linguistic constructions that we use in order to avoid long periphrases when we talk about living systems, but no more than that. They are like those computer programs that allow us to write our instructions in English, thus saving us the trouble to write them with the binary digits of the machine language. Ultimately, however, there are only binary digits in the machine language of the computer, and in the same way, it is argued, there are only physical quantities at the most fundamental level of Nature.

This conclusion, known as *physicalism*, or *the physicalist thesis*, has been proposed in various ways by a number of scientists and philosophers (Chargaff, 1963; Sarkar, 1996; 2000; Mahner and Bunge, 1997; Griffiths and Knight, 1998; Griffith, 2001, Boniolo, 2003). It is probably one of the most deeply dividing issues of modern science. Many biologists are convinced that biological information and the genetic code are real and fundamental components of life, but physicalists insist that they are real only in a very superficial sense and that there is nothing fundamental about them because *they must be* reducible, in principle, to physical quantities.

It has to be pointed out that the physicalist thesis *could,* in theory, be true. In fact it would be rigorously true if genes and proteins were made by spontaneous assemblies because these processes are fully described by physical quantities. The point however is precisely that genes and proteins are *not* spontaneous molecules. They are *molecular artifacts* because they are manufactured by molecular machines, and the physicalist thesis collapses because there is a *fundamental* difference between spontaneous objects and artifacts. Spontaneous objects can be completely accounted for by physical quantities, whereas artifacts require additional entities like sequences and codes.

This makes us understand how important it is to realize that genes and proteins are molecular artifacts, and that artifacts cannot be accounted for by physical quantities *alone*. The issue is not so much the physicalist thesis as the nature of life itself. The claim of physicalism is that there is no substantial divide between inanimate matter and life. The experimental reality is that genes and proteins are manufactured by molecular machines with processes of copying and coding that require sequences and codes. The very existence of life, in short, depends upon entities which are objective and reproducible but which cannot be measured. They appear to be a completely new type of fundamental natural entities, and in the next sections we are going to see that that is precisely what they are.

Organic Information

In both genes and proteins, as we have seen, *biological* (or *organic*) *information* has been defined by the specific sequence of their subunits. This definition however is not entirely satisfactory because it gives the impression that information is a *static* property, something that molecules have simply because they have a sequence. In reality, there are countless molecules which have a sequence but only in a few cases this becomes information. That happens only when copymakers use it as a guideline for copying. Even copymakers, however, do not account, by themselves, for information. Copymakers can stick subunits together and produce sequences, but without a template they would produce only *random* sequences, not specific ones. Sequences alone or copymakers alone, in other words, have nothing to do with information. It is only when a sequence provides a guideline to a copymaker that it becomes *information* for it. It is only during an act of copying, in other words, that information comes into existence.

This tells us that organic information is not just the specific sequence of a molecule, but *the specific sequence produced by a copying process*. This definition underlines the fact that information is not a thing or a property, but the result of a process. It is, more precisely, an "operative" definition, because *information is defined by the process that brings it into existence*. There simply is no difference between saying that molecule B is a copy of molecule A, and saying that molecule B carries the same information as molecule A. We realize in this way that organic information is as real as the copying process that generates it, but we still do not know what kind of entity it is. How does it fit into our description of Nature?

According to a long tradition, natural entities are divided into *quantities* and *qualities*. Quantities can be measured and are objective, while qualities are subjective and cannot be measured. In the case of organic information, however, this scheme breaks down. Organic information is not a quantity because a specific sequence cannot be measured. But it is not a quality either, because linear specificity is a feature that we find in organic molecules, and is therefore an *objective* feature of the world, not a subjective one.

A scheme based on quantities and qualities alone, in short, is not enough to describe the world. In addition to quantities (*objective and measurable*) and qualities (*subjective non-measurable)* we must recognize the existence in Nature of a third type of entities (*objective but not measurable*). Information is one of them, and we can also give it a suitable name. Since it can be described only by *naming* its sequence, we can say that organic information is a *nominable* entity, or that it belongs to the class of the nominable entities of Nature (Barbieri, 2003b; 2004).

We conclude that organic information is a new type of natural entity, but we also conclude that it belongs to the same class of *objective* entities that contains all physical quantities. Therefore, *it has the same scientific "status" as physical quantities*. This however gives us a new problem, because there are two distinct classes of physical quantities: a small group of *fundamental* quantities (space, time, mass, charge and temperature) and a much larger group of *derived* quantities. This distinction applies to all objective entities, so we need to understand whether organic information belongs to the first or to the second group.

Luckily, this problem has a straightforward solution because the sequences of genes and proteins have two very special characteristics. One is that *a change to a biological sequence may produce a sequence which has entirely new properties.* This means that although a biological sequence can be said to have "components", it is at the same time a single indivisible whole. The second outstanding feature is that *from the knowledge of n elements of a biological sequence it is impossible to predict the element (n+1).* This is equivalent to saying that *a specific sequence cannot be described by anything simpler than itself*, so it cannot be a derived entity.

We conclude that organic information has the same scientific status as the physical quantities, because it is an objective and reproducible entity. But we also conclude that it does not have the status of a *derived* physical quantity because it cannot be expressed by anything simpler than itself. This means that organic information has the same scientific status as the fundamental physical quantities, i.e. that it is a *fundamental* (or *irreducible*) entity of Nature.

Organic Meaning

A code is a set of rules which establish a correspondence between the objects of two independent worlds. The Morse code, for example, is a correspondence between groups of dots and dashes with the letters of the alphabet, and in the same way the genetic code is a correspondence between groups of nucleotides and amino acids. Let us notice now that establishing a correspondence between, say, object 1 and object 2, is equivalent to saying that object 2 is the *meaning* of object 1. In the Morse code, for example, the rule that "dot-dash" corresponds to letter "A", is equivalent to saying that letter "A" is the meaning of "dot-dash". In the code of the English language, the mental object of the *sound* "apple" is associated to the mental object of the *fruit* 'apple', and this is equivalent to saying that that fruit is the meaning of that sound.

By the same token, the rule of the genetic code that a group of three nucleotides (a codon) corresponds to an amino acid is equivalent to saying that that amino acid is the organic meaning of that codon. Anywhere there is a code, be it in the mental or in the organic world, there is meaning. We can say, therefore, *that meaning is an object which is related to another object by a code*, and that organic meaning exists wherever an organic code exists (Barbieri, 2003).

It must be underlined that meaning is always a *codemaker-dependent* entity, because it is always brought into existence by a codemaker during a coding process. This suggests that we should distinguish between "meaning" and "meaning-vehicle". The sequence of a protein is the "meaning" produced by a codemaker in protein synthesis, whereas the protein itself is a "meaning-vehicle", or a "meaning-carrier". In the same way, the sequence of a messenger-RNA is the "sign" used by the codemaker in protein synthesis, whereas the RNA-molecule itself is a "sign-vehicle" or a "sign-carrier".

The existence of meaning in the organic world may seem strange, at first, but in reality it is no more strange than the existence of codes because they are the two sides of the same coin. Saying that a code establishes a correspondence between two objects is equivalent to saying that one object is the meaning of the other, so we cannot have codes without meaning or meaning without codes. All we need to keep in mind is that *meaning is a mental entity when the code is between mental objects, but it is an organic entity when the code is between organic molecules.*

Modern biology has readily accepted the concept of information but has carefully avoided the concept of meaning, and yet organic information and organic meaning are both the result of natural processes. Just as it is an act of *copying* that creates organic information, so it is an act of *coding* that creates organic meaning. Copying and coding are the processes; copymakers and codemakers are their agents; organic information and organic meaning are their results. But the parallel goes even further than that. We have seen that organic information *cannot be measured*, and the same is true for organic meaning. We have seen that organic information is an *objective* entity, because it is defined by the same sequence for any number of observers, and that is also true for organic meaning, which is defined by coding rules that are the same for all observers. Finally, we have seen that organic information is an *irreducible* entity, because it cannot be described by anything simpler than its sequence, and the same is true for organic meaning, which cannot be defined by anything simpler than its coding rules.

Organic information and organic meaning, in short, belong to the same class of entities because they have the same general characteristics: they both are *objective-but-non-measurable* entities, they both are *irreducible*, or *fundamental*, entities of Nature, and since we can describe them only by naming their components, they both are *nominable* entities (Barbieri, 2003b; 2004). Finally, let us underline that they are the twin pillars of life because organic information comes from the copying process that produces genes, while organic meaning comes from the coding process that generates proteins.

Operative Definitions

Physical quantities have three fundamental properties: (1) they are *objective*, (2) they are *reproducible*, and (3) they are *defined by operative procedures*. This last property is particularly important because it has provided the solution to one of the most controversial issues of physics. The critical point was the theoretical possibility that the entity which is measured may not be the same entity which has been defined. This led to the idea that there should be no difference between what is measured and what is defined, i.e., to the concept of *operative* (or *operational*) definition: *a physical quantity is defined by the operations that are carried out in order to measure it.*

It was this operational approach that solved the definition problem in physics, and it is worth noticing that we can easily generalize it. Instead of saying that *a natural entity is defined by the operations that measure it*, we can say that a *natural entity is defined by the operations that evaluate it in an objective and reproducible way.* The advantage of this generalized approach is that it applies to *all* objective entities, so it can be used not only in physics, but in biology as well. To this purpose, we only need to notice that *a measurement* is an objective and reproducible description of a physical quantity, just as *the naming of a specific sequence* is an objective and reproducible description of organic information, and just as *the naming of a coded object* is an objective and reproducible description of organic meaning.

Whereas the physical quantities are evaluated *by measuring*, our biological entities are evaluated *by naming their components*, but in both cases the entities in question *are defined by the operations that evaluate them*, and this is the essence of the operative approach. We can add that organic information and organic meaning cab also be defined by the processes of copying and coding that bring them into existence, and that too amounts to an operative definition (Barbieri, 2003b; 2004).

We conclude that organic information and organic meaning can be defined by generalized operative procedures that are as reliable as the operative procedures of physics. This means that the definitions of information and meaning should no longer be at the mercy of endless debates on terminology as they have been in the past. The operative definitions are scientific tools which are justified by their own prescriptions, so there is no point in asking if they are right or wrong. All we can ask of them is whether they contribute or not to our description and to our understanding of Nature.

At this point, we can summarize all the above concepts by saying that *in protein synthesis*:

(1) Organic information is the sequence used by a copymaker during a copying process.

(2) An organic sign is the sequence used by a codemaker during a coding process.

(3) An organic meaning is the sequence *produced* by a codemaker during a coding process.

(4) Organic information, organic signs and organic meaning are neither *quantities* nor *qualities*. They are a new kind of natural entities which are referred to as *nominable* entities.

(5) Organic information, organic signs and organic meaning have the same scientific status as physical quantities because they are *objective* and *reproducible* entities which can be defined by operative procedures.

(6) Organic information, organic signs and organic meaning have the same scientific status as *fundamental* physical quantities because they cannot be reduced to, or derived from, simpler entities.

PART 4 – THE MODELS OF BIOSEMIOTICS

Three Models of Semiosis

The existence of the genetic code suggests that the cell is a semiotic system, and this in turns implies that all living organisms are semiotic systems. But what is a semiotic system? Today the answers to this question are still provided by the two classical models of cultural semiotics. The model proposed by Ferdinand de Saussure, who defined a semiotic system as a duality of *"signifier and signified"*, or *"sign and meaning"*, and the model of Charles Peirce, who pointed out that interpretation is an essential component of semiosis and defined a semiotic system as a trinity of *"sign, object and interpretant"*.

In 1974 Marcel Florkin argued that "signifier and signified" are equivalent to "genotype and phenotype" and proposed that *molecular biosemiotics* should be based on the dualistic model of Saussure. A biological system, according to Florkin, is a duality of genotype and phenotype because it is entirely accounted for by genes and proteins. Florkin was aware that the bridge between genes and proteins is provided by RNAs, but was convinced that the connections made by the RNAs are entirely determined by chemistry and not by conventional rules. In this case there would be no autonomous third party between genes and proteins, and the genotype-phenotype duality would be an exact description of a living system. The model proposed by Florkin, in short, assumes that the genetic code is determined by chemistry and that means that the cell is not a real semiotic system.

In the 1960s Thomas Sebeok started a lifelong campaign in favor of the idea that language must have biological roots, and gave the name of *zoosemiotics*, and later of *biosemiotics*, to the new science of biological semiotics. He realized however that any semiosis requires a third party between signifier and signified, and adopted the triadic scheme of Peirce where the third party is an "interpreter". He insisted that *"there can be no semiosis without interpretability"* and that interpretations is *"a necessary and sufficient condition for something to be a semiosis"* (Sebeok, 2001). In this framework the cell is a semiotic system only if the genetic code is based on some kind of interpretation. The model of Saussure-Florkin, in short, assumes that the genetic code is not a real code, whereas the model of Peirce-Sebeok assumes that the code is based on interpretation.

A third model, however, does exist and was suggested by the theory that the cell is a trinity of genotype, phenotype and ribotype (Barbieri, 1981, 1985), where the ribotype is the ribonucleoprotein system of the cell and represents its "codemaker", i.e. the seat of the genetic code. This amounts to saying that the cell contains a "codemaker" but not an "interpreter", because the genetic code does not depend on interpretation. The simplest semiotic system, in other words, is a trinity made of *"sign, meaning and code"*.

We have, in conclusion, three different definitions of a semiotic system and therefore three different models of the cell. The model of Saussure-Florkin describes the cell as a duality of genotype and phenotype. The model of Peirce-Sebeok describes it as a trinity of genotype, phenotype and *interpreter*. The ribotype model describes it as a trinity of genotype, phenotype and *codemaker*.

It may appear that the concepts of interpreter and codemaker are very close, if not equivalent, but this is only because semiosis appears to be inseparable from interpretation. This is true in the world of culture, but it is not true in protein synthesis and would not help us to understand the origin of semiosis at the molecular level.

The time has come therefore to separate semiosis from interpretation and to study them as two distinct processes that appeared at different times in the history of life.

The Problem of Interpretation

Biologists have quickly accepted the idea of molecular information but not the idea of molecular meaning. Oddly enough, one of the most popular arguments against the existence of meaning at the *molecular* level has come from the study of animal communication, the very science that argued for the existence of meaning at the *animal* level. A signal that reaches an animal may set in motion an automatic response, and what we observe in these cases is the unfolding of a deterministic sequence of reactions. In other cases, however, there is an intermediate phase between signal and response, a phase where the signal is *interpreted* and the response changes accordingly. These are the cases where it is said that meaning appears, because interpretation is assumed to be the process that gives a meaning to a signal. Hence the conclusion that *meaning is always the result of an interpretation process.* No interpretation, no meaning (Markoš, 2002).

In the case of protein synthesis, we know that codons are translated into amino acids according to the rules of the genetic code, but we also know that these reactions take place in a totally automatic way, with no room for interpretation. Hence the conclusion that there is no addition of meaning here, because protein synthesis is a deterministic chain of biochemical reactions. This argument is still popular and yet it is flawed because the automatism that we observe in protein synthesis is also observed in many cultural processes where we know that meaning does exist. The very act of speaking, for example, is based on the automatic application of prefixed rules. Language itself would not be possible if the meaning of its words had to be negotiated every time they are spoken. Once the basic rules have been fixed in the initial stage of learning, they are no longer changed and the code of a language becomes as deterministic as the genetic code.

An elementary act of semiosis, the process that gives a meaning to a sign, is always a convention, and it is always a codemaker therefore that produces meaning. There are cases, however, when the same sign can be given different meanings, and it is in these cases that interpretation appears. We say that interpretation takes place when, for example, a sign A is given the meaning X in context 1, the meaning Y in context 2 and the meaning Z in context 3. Interpretation, in other words, is a *context-dependent* semiosis, and requires a context-dependent code. This gives the *impression* that the generation of meaning comes from the interpreting process, but the truth is rather different. The interpretation process is necessary to make a choice among a plurality of rules, not to create the meaning of those rules. Meaning, in short, is always the result of a code, but sometimes the code is context-dependent, and in these cases it is associated with a process of interpretation.

A context-dependent code can be determined by genes, but the number of coding rules that can be stored in a genome is bound to be limited, and this favored the evolution of a new type of semiosis. A semiosis where some coding rules are acquired by *learning*. This is particularly evident in the case of animal behavior. It is likely that the most primitive behaviors were genetically determined, but the number of hard-wired responses could not increase indefinitely, and animals became more and more dependent on processes of learning in order to increase their behavioral repertoire.

Learning how to respond to a signal, on the other hand, means learning *how to interpret* that signal. Rather than memorizing an unlimited number of incoming stimuli, it was far more convenient to learn a few basic rules and let an interpretation phase decide which meaning a signal has in any given context. A process of learning, in short, amounts to the construction of a behavioral code whose rules are *context-dependent*. At the same time, learning requires a memory where the results of experience are accumulated, and in these cases interpretation is not only a *context-dependent* but also a *memory-dependent* process. This amounts to a new type of semiosis that can be referred to as *hermeneutics*, a term which is commonly used for the processes of cultural interpretation.

Our culture is so totally dependent on context-dependent and memory-dependent processes that we can hardly imagine codes and conventions without them. This is why we find it so difficult to separate semiosis from interpretation, and yet we must do that because they truly are two distinct processes.

The Evolution of Semiosis

The genetic code was the first code of the history of life and the apparatus of protein synthesis was the first semiotic system that appeared on Earth. But what happened afterwards? According to modern biology, the only other codes that exist in the world are the mental codes of culture which appeared almost four billion years after the origin of life. In this case there would have been no evolution of semiosis for almost the entire history of life. The evidence, however, suggests that many other organic codes came into being during that history. This is because the existence of organic codes is revealed by the presence of molecular adaptors and these molecules have been found in many cellular processes such as splicing, signal transduction, compartment formation, cytoskeleton assembly, histone assembly and cell movement.

The origin of new organic codes took place particularly in eukaryotic cells, and accounts for the great innovations that appeared in those cells. Any new organic code added a new degree of complexity because it allowed the cell to perform a new set of codified reactions, i.e. a new semiotic process. Life remained at the level of single cells for almost three billion years and in single cells only the origin of new organic codes could produce more complex types of semiotic systems.

The addition of new organic codes continued to be a mechanism of semiotic evolution even in multicellular organisms because it has been shown that molecular adaptors exist in all major processes of embryonic development like cell-to-cell interactions, cell migration and cell differentiation. In multicellular systems, however, there were also the conditions that favoured the appearance of a new type of semiosis. The complexity of the genome could not increase indefinitely and was setting a limit to the number of codes that could be programmed in the genes. This led to the evolution of codes that could be acquired by *learning*, and this in turn required a new type of semiosis based on interpretation.

The evolution of semiosis took place therefore in two different ways: (1) by the addition of new organic codes, and (2) by the invention of semiotic systems that were capable of interpretation, i.e. of *hermeneutic* systems. The origin of semiosis (the *semiotic threshold*) and the origin of interpretation (the *hermeneutic threshold*) were separated by an extremely long period of evolution, because interpretation is dependent on context, memory and learning, and probably evolved only in multicellular systems. The history of semiosis, in short, was a process that started with context-free codes and produced codes that were more and more context-dependent. Today, our cultural codes are so heavily dependent on context that we can hardly imagine semiosis without interpretation, and yet these are distinct processes and we need to keep them apart if we want to understand their origin and their evolution in the history of life.

Conclusions

A genetic code exists in every cell, but modern biology has not accepted the idea that the cell is a semiotic system. The main reason is that there are five major obstacles against this idea, and all of them must be overcome in order to realize that the cell has a semiotic structure.

(1) The first obstacle is the claim of physicalism, the idea that sequences and codes are useful metaphors but not fundamental entities because they can be reduced, in principle, to physical quantities. This objection has been overcome by pointing out that only the formation of spontaneous molecules can be completely described by physical quantities, whereas molecular *artifacts* like genes and proteins require new fundamental entities. It has also been shown that organic information, organic signs and organic meaning are indeed a new type of fundamental entities that can be defined by operative definitions and that are as objective and reproducible as physical quantities.

(2) The second obstacle is the idea that semiosis is always the result of a process of interpretation. If this were true, the genetic code would be no proof of semiosis because it does not depend on interpretation. In order to overcome this obstacle, it has been necessary to show that the qualifying feature of semiosis is coding not interpretation, and that coding requires only adaptors and codemakers, not interpreters.

(3) The third obstacle is the paradigm that the cell is a duality of genotype and phenotype, whereas a semiotic system is necessarily made of three distinct categories. The duality of genotype and phenotype would be a valid model if the cell had evolved from spontaneous genes and spontaneous proteins, but this could not have happened because spontaneous processes do not produce specific sequences. Only manufactured molecules can have biological specificity. The

"agents" that gave origin to the cell, therefore, were the molecular machines that manufactured genes and proteins by copying and coding, and these machines form the ribotype, the "third party" that still exists in every cell and that represent its codemaker, the seat of the genetic code.

(4) The fourth obstacle is the belief that the cell does not have all the essential components of a semiotic system because signs and meanings do not seem to exist at the molecular level. This objection has been overcome by showing that signs and meanings are *codemaker-dependent* entities and that the sequences of RNAs and proteins that appear in protein synthesis have precisely the codemaking-dependent characteristics that define signs and meanings.

(5) The fifth obstacle is the idea that there are only two types of codes in the living world: the genetic code that appeared at the origin of life and the cultural codes that came at the end of evolution. If this were true, we would have to conclude that the cell did not produce any other code for almost four billion years, virtually the entire history of life on Earth, and it would be legitimate to doubt that it is a true semiotic system. This obstacle has been overcome by showing that the genetic code was only the first of a long series of organic codes that appeared during cellular evolution and that gave origin to more complex types of cells. That means that semiosis was instrumental not only to the origin but also to the subsequent evolution of life, all the way up to the origin of language and culture.

So far, the above five obstacles have effectively prevented modern biology from accepting the idea that semiosis is essential to organic life, but here we have seen that they do not stand up to scrutiny. This makes us realize that signs, meanings and codes exist not only in the mental world but also at the molecular level, and we are finally able to conclude that the cell is a genuine semiotic system.

REFERENCES

Alberts, B., Bray, D., Lewis, J., Raff, M., Roberts, K., & Watson, J.D. (1994). *Molecular Biology of the Cell.3rd edition.* Garland, New York.

Barbieri M. (1981). The Ribotype Theory on the Origin of Life. *Journal of Theoretical Biology*, 91, 545–601.

Barbieri M. (1985). *The Semantic Theory of Evolution.* Harwood Academic Publishers, London and New York.

Barbieri M. (1998). The Organic Codes. The basic mechanism of macroevolution. *Rivista di Biologia-Biology Forum*, 91, 481–514.

Barbieri M. (2003). *The Organic Codes. An Introduction to Semantic Biology.* Cambridge University Press, Cambridge, UK.

Barbieri M. (2003b). Biology with Information and Meaning. *History and Philosophy of the Life Sciences*, 25, 243–254.

Barbieri M. (2004). The Definitions of Information and Meaning. Two possible boundaries between physics and biology. *Rivista di Biologia-Biology Forum*, 97, 91–110.

Berridge, M. (1985). The molecular basis of communication within the cell. *Scientific American*, 253, 142–152.

Boniolo G. (2003). Biology without Information. *History and Philosophy of the Life Sciences*, 25, 255–273.

Chargaff, E. (1963). *Essays on Nucleic Acids.* Elsevier, Amsterdam.

Florkin, Marcel (1974). Concepts of molecular biosemiotics and molecular evolution. In *Comprehensive Biochemistry,* vol.29 part A (Comparative Biochemistry, Molecular Evolution), Marcel Florkin and Elmer H. Stotz (eds.), Elsevier, Amsterdam, 1–124.
Gabius H.-J. (2000). Biological Information Transfer Beyond the Genetic Code: The Sugar Code. *Naturwissenschaften*, 87, 108–121.
Gabius H.-J, André S., Kaltner H., and Siebert H.-C. (2002). The sugar code: functional lectinomics. *Biochimica et Biophysica Acta*, 1572, 165–177.
Gamble M.J. and Freedman L.P. (2002). A coactivator code for transcription. *Trends in Biochemical Sciences*, 27 (4), 165–167.
Griffith, P.E. (2001). Genetic Information: A Metaphor in Search of a Theory. *Philosophy of Science*, 68, 394–412.
Griffith, P.E. and Knight, R.D. (1998). What is the developmental challenge? *Philosophy of Science*, 65, 276–288.
Jenuwein T. and Allis D. (2001). Translating the Histone Code. *Science*, 293, 1074–1080.
Johannsen, W. (1909). *Elemente der exacten Erblichkeitslehre.* Gustav Fisher, Jena.
Mahner, M. and Bunge, M. (1997). *Foundations of Biophilosophy.* Springer Verlag, Berlin.
Markoš, A. (2002). *Readers of the Book of Life: Conceptualizing Developmental Evolutionary Biology.* Oxford University Press, Oxford.
Maynard Smith, J. & Szathmáry E. 1995. *The Major Transitions in Evolution.* Oxford University Press, Oxford.
Pattee, H. H. (1969). The physical basis of coding and reliability in biological evolution. In C. H. Waddington (ed.) *Toward a Theoretical Biology* Vol. 1, Edinburgh Univ. Press, 1969, pp. 67–93.
Pattee, H. H. (1972). Laws and constraints, symbols and languages. In C. H. Waddington (ed.) *Towards a Theoretical Biology* Vol. 4,, Edinburgh Univ. Press, 1972, pp. 248–258.
Pattee, H. H. (2001). The physics of symbols: bridging the epistemic cut, *BioSystems*, 60, 5–21.
Peirce, Charles S. (1931–1958) *Collected papers of Charles Sanders Peirce.* Harvard University Press, Cambridge Massachusetts.
Prodi, G. (1988). Material bases of signification. *Semiotica*, 69 (3/4), 191–241.
Readies C. and Takeichi M. (1996). Cadherines in the developing central nervous system: an adhesive code for segmental and functional subdivisions. *Developmental Biology*, 180, 413–423.
Richards E.J. and Elgin S.C.R. (2002). Epigenetic Codes for Heterochromatin Formation and Silencing: Rounding up the Usual Suspects. *Cell.* 108, 489–500.
Sarkar, S. (1996). Biological Information. A Skeptical Look at some Central Dogmas of Molecular Biology. In S. Sarkar (ed.) *The Philosophy and History of Biology.* Kluwer Academic Publishers, Dordrecht, 187–231.
Sarkar, S. (2000). Information in Genetics and Developmental Biology. *Philosophy of Science*, 67, 208–213.
Saussure, Ferdinand de (1916). *Cours de linguistique générale.* Payot, Paris.
Sebeok, T. A. (1963). Communication among social bees; porpoises and sonar; man and dolphin. *Language*, 39, 448–466.
Sebeok, T. A. (1972). *Perspectives in Zoosemiotics.* Mouton, The Hague.
Sebeok, T. A. (1986). The doctrine of signs. *Journal of Social and Biological Structures* 9, 345–352.
Sebeok, T. A. (2001). Biosemiotics: Its roots, proliferation, and prospects. *Semiotica*, 134, 68.
Shannon, C. E. (1948). A mathematical theory of communication. *Bell Systems Technical Journal*, 27, 379–424, 623–656.
Shapiro L. and Colman D.R. (1999). The Diversity of Cadherins and Implications for a Synaptic Adhesive Code in the CNS. *Neuron*, 23, 427–430.
Strahl B.D. and Allis D. (2000). The language of covalent histone modifications. *Nature*, 403, 41–45.
Sutherland, E.W. (1972). Studies on the mechanism of hormone action. *Science*, 177,401–408.
Trifonov E.N. (1987). Translation framing code and frame-monitoring mechanism as suggested by the analysis of mRNA and 16s rRNA nucleotide sequence. *Journal of Molecular Biology*, 194, 643–652.
Trifonov E.N. (1989). The multiple codes of nucleotide sequences. *Bulletin of Mathematical Biology*, 51, 417–432.

Trifonov E.N. (1996). Interfering contexts of regulatory sequence elements. *Cabios*, 12, 423–429.
Trifonov E.N. (1999). Elucidating Sequence Codes: Three Codes for Evolution. *Annals of the New York Academy of Sciences*, 870, 330–338.
Turner, B.M. (2000). Histone acetylation and an epigenetic code. *BioEssay*, 22, 836–845.
Turner, B.M. (2002). Cellular memory and the Histone Code. *Cell*, 111, 285–291.
Watson, J.D. and Crick, F.H.C. (1953). Genetical Implications of the Structure of Deoxyribose Nucleic Acid. *Nature*, 171, 964–967.
Woese, C.R. (2000). Interpreting the universal phylogenetic tree. *Proceedings of the National Academy of Science USA*, 97, 8392–8396.

CHAPTER 9

COMPUTING CODES VERSUS INTERPRETING LIFE

Two alternative ways of synthesizing biological knowledge through semiotics

STEFAN ARTMANN

Friedrich-Schiller-University, Institute of Philosophy, Zwätzengasse 9, 07737 Jena, Germany
stefan.artmann@uni-jena.de
Ludwig-Maximilians-University, Human Science Center, Goethestraße 31/I, 80336 Munich

Seeing deep correspondences in seemingly unrelated things is the essence of science and is vital in mathematics and philosophy as well.

William D. Hamilton

Abstract: One of the most important aims of biosemiotics is to provide a conceptual framework for synthesizing the biological knowledge of living systems. This goal can be assumed achievable in (at least) two alternative ways, both of which are actually followed by working biosemioticians. Either biosemiotics is considered a philosophical revolution in biological thought that will change biology into an art of interpreting life, i.e. into an appreciation of organisms as semiotic beings whose structure, function, and behaviour is to be understood by participating in the communicative web of life. Or biosemiotics is considered from a model-theoretical point of view as an application of abstract semiotic concepts to a field of empirical research that is foreign to semiotics only for an anthropocentric understanding of signs. This paper sketches both biosemiotic ways of synthesizing biological knowledge and shows how they mirror a fundamental opposition, which has been intensely investigated by Jaakko Hintikka, between two answers to the question whether it is possible that language can be scientifically explained by linguistic means. It is argued that biosemiotics should postulate, as a starting-point of its research programme, a positive answer to this question. A model-theoretical sketch of how to develop a precise definition of organic codes, shall show that this postulate allows to elucidate the inner structure of biosemiotic concepts

Keywords: Methodology of biosemiotics, philosophy of logic and language, hermeneutics, model theory, coding theory

M. Barbieri (ed.), Introduction to Biosemiotics, 209–233.

A biosemiotician rephrases questions on the nature and origin of life in terms of the concepts 'sign', 'information', 'code', and 'communication'. For her, the molecular-biological analysis of the genetic code has taken just a first step on the road to a theory of information processing in living beings. Orientated by general semiotic principles, such a theory must systematically cover topics ranging from inheritance and gene regulation, over intercellular signalling and neural codes, to animal communication and the origin of human language.

> [...A]ll living things are designed on the basis of highly conserved developmental systems that read an (almost) universal language encoded in DNA base pairs. As such, life is arranged hierarchically with a foundation of discrete, unblendable units (codons, and, for the most parts, genes) capable of combining to create increasingly complex and virtually limitless varieties of both species and individual organisms. [...T]he human faculty of language appears to be organized like the genetic code – hierarchical, generative, recursive, and virtually limitless with respect to its scope of expression. (Hauser, Chomsky, and Fitch 2002: 1569)

Structural and functional analogies between sign systems that have evolved in the natural history of life on Earth, stimulate a wide variety of research programmes nowadays – within and beyond the traditional disciplines of biology. Bioinformaticians, for example, announce that they "[...] all have had to become molecular linguists, to learn to respect the unity of biology." (Botstein and Cherry 1997: 5507) The biosemiotic extension of molecular linguistics can enhance, not only the co-operation between the manifold subdisciplines of biology, but also the efficiency of biology as a bridge between the natural sciences and the humanities. The concepts of sign, information, code, and communication seem, however, to transcend the physical world, so that a biologist who understood these concepts in a literal sense would be in constant danger to impute intentionality, or even consciousness, to the objects of his research. But the use of complex languages, such as English or Japanese, by human beings should instead be considered a special case of communication in which the source and the destination of communicated signs normally are intentional and self-conscious agents. The fundamental hypothesis of biosemiotics asserts that the general form of language is an organizational principle which existed, exists, and will exist in the absence of human beings (Eigen 1979).

How can language fulfil this most important function? The philosophical debate on *human* language may be helpful to recognize possible answers to this question. Based on a distinction of Jean van Heijenoort (1967) between logic seen either as language or calculus, the Finnish philosopher Jaakko Hintikka (1997) argues convincingly that two fundamental conceptions of human language, which can be traced back to Leibniz' projects of constructing a *characteristica universalis* and a *calculus ratiocinator*, dominated philosophical thought in the twentieth century (see for a short introduction Kusch 1989: 2ff.). On the one hand, language can be regarded as the universal medium of human existence, its *characteristica universalis*, of which it is not able to break out. Each attempt to do so, e.g. by scientifically objectifying the semantic relation between language and the world, necessarily presupposes the very same relation that is under scrutiny: all possible scientific propositions about semantics are propositions at all only in so far as they are already semantically related by some interpretation to their meanings. If human beings cannot get out of their language, an objective theory of semantics is *a priori* impossible. On the other hand, language can be considered a kind of computational[1] technology of human beings, their *calculus ratiocinator*, which is adaptable to a great variety of specific circumstances. In particular, the syntactic structures of language are semantically re-interpretable in such a way that it is possible to scientifically objectify any semantic system in the framework of some other semantic system (and sometimes even in the very same one).[2]

Why is the contrast between philosophers who regard human language as the universal medium and philosophers who regard human language as a calculus, important for biosemiotics, the general science-to-be of information processing in living systems? The *first main thesis* of this paper is that principal differences between today's approaches to biosemiotics follow from the contrast between those philosophical conceptions of language if the latter are applied to living beings as natural-born semiotic systems. Extending the conception of language as the universal medium to organic information processing in general, the synthesis of biological knowledge through semiotics considers biosemiotics a philosophical revolution in thinking about life that shall change biology into an art of interpreting life, i.e. into an appreciation of organisms as semiotic beings whose structure, function and behaviour can be understood only by participating in the communicative web of life. An important example of this position in current biosemiotics is what can be called 'the Copenhagen interpretation of biosemiotics', since researchers working in the Danish capital are its leading proponents (*section 1*). If language is a kind of computational technology that, on the one hand, pervades all human activities but that, on the other hand, is semantically re-interpretable in various ways, the extension of this conception of language to biosemiotics leads to a model-theoretical perspective on the relation between semiotics and biology. Biosemiotics is, then, regarded as an application of abstract concepts from the general science of signs to a field of empirical research that is foreign to semiotics solely for an anthropocentric understanding of signs (*section 2*).

As a consequence of extending Hintikka's analysis of the contrast between two conceptions of language to biosemiotics, the *second main thesis* of this paper asserts that biosemioticians must affirm, right from the very beginning of their research programme, the idea of language as a kind of computational technology if biosemiotics shall lead to 'the new biological synthesis'. A model-theoretical sketch of how to develop a precise definition of organic codes, shall show that the philosophical postulate according to which language is a calculus, allows to elucidate the inner structure of biosemiotic concepts (*section 3*).[3]

1. INTERPRETING LIFE: LANGUAGE AS THE UNIVERSAL MEDIUM OF ORGANIC EXISTENCE

In the literature on information processing in living systems that has been published under the title 'biosemiotics' so far, the philosophical presupposition that language is the universal medium of organic existence dominates.[4] If, however, all writings that describe phenomena of life semiotically or information-theoretically, were also classified as biosemiotic literature, the philosophical presupposition that language is a calculus would dominate. The reason for this unequal distribution of general conceptions of language in the biological research community is that most of the 'official' biosemioticians see themselves in philosophical revolt against the reductionistic and selectionist explanation of the structure, function, and evolution of living systems that is common in today's biology.

The works of the Copenhagen biologists Jesper Hoffmeyer and Claus Emmeche, which have been of greatest importance for the rise of biosemiotics since the early 1990s, represent a paradigmatic example of the position that language is the universal medium of life.[5] Emmeche, who has intensely reflected on the epistemological status of biosemiotics, distinguishes three different rôles biosemiotics plays in current research on life (Emmeche 1997: 79ff.).

In its *spontaneous* form biosemiotics is nothing but a methodologically unreflected import of semiotic terms into biology. It is not really interested in a systematic investigation of organisms as semiotic systems. Yet the danger of an uncontrolled use of isolated terms, such as 'sign', 'meaning', 'code', and 'information', in biology is that it can be easily critized as relying upon naïve anthropomorphic metaphors. To avoid that such a justified criticism becomes detrimental also to methodical biosemiotics, the reflective biosemiotician has to systematically re-interpret the spontaneous use of semiotic terms in biology, particularly in molecular genetics, as evidence for a kind of "quasi-semiotics" (Emmeche 1999: 275) that underlies biological thinking in general.

In its *revolutionary* form biosemiotics sharply criticizes the predominance of reductionism, mechanicism, and instrumentalism in today's biology. This type of reflective biosemiotics regards the semiotic understanding of life as a scientific alternative to the current biochemical and selectionist explication of living systems. Revolutionary biosemiotics fights, thus, for a change of paradigm in biology, a new kind of biology.

In its *philosophical* form biosemiotics develops a new conception of living beings, or even of nature in general, into which biology is integrated as an epistemologically autonomous science. Its objects, organisms as semiotic systems, are considered ontologically irreducible to purely physical and chemical systems. From this perspective biosemiotics foremost is, not a revolutionary kind of biological science, but a metaphysical re-contextualization of scientific biology in the framework of a new philosophy of life.

How the revolutionary and the philosophical form of biosemiotics are related to each other, is of particular importance for the analysis of Emmeche's and

Hoffmeyer's Copenhagen interpretation of biosemiotics.[6] Reading the first thesis of Hoffmeyer's biosemiotic theory as summarized in Emmeche et al. (2002: 14), "Signs, not molecules, are the basic units in the study of life," one might wonder whether the cited statement belongs to revolutionary or philosophical biosemiotics. Emmeche et al. (2002: 14) explanation of Hoffmeyer's thesis leads to an answer to this question.

By representing an organism merely as a composition of small non-living bodies that interact according to the mechanical forces, or quantum mechanical laws, as established in physics, we may never reach the description of life itself that will correspond to a biologist's intuition about its nature, including concepts like organism, metabolism, ecosystem, reproduction, etc., as these have been understood in a tradition of biological culture. However, if we try to include into a model of an elementary living process all what is required for the process to be a model of life, it appears that the set of features we arrive at will include the features that characterise a *sign*, or a sign process. That is, in order to have a set of physical processes to be characterized as living, these have to be realized, partly or fully, through the mediation of signs; 'signs', of course, in a specific sense, as we are talking about a very general notion of signs, more encompassing than just 'conventional symbols'. And it follows that "if signs (rather than molecules) are taken as fundamental units for the study of life, biology becomes a semiotic discipline" [Hoffmeyer]. This semiotic understanding is also achieved if we include into the features of this model the model-building itself, because models are not the sum of their building blocks but are defined by being about something else; they are complex signs occuring in organisms: "The understanding that biology models the activity of model-building organisms is at the core of biosemiotics, of course" [Hoffmeyer]. Thus, the statement about the basic units not only concerns the method of study, it also concerns ontology. The element of life is the sign, not the molecule.

The following six remarks comment extensively on this explanation of Hoffmeyer's thesis that signs, not molecules, are the basic units in the study of life, in order to show what the cited text implies as to the epistemological status of the Copenhagen interpretation of biosemiotics.

First remark. According to the Copenhagen interpretation, the biosemiotic description of life must capture the "intuition" of biologists by using concepts that proved to be suitable for expressing the biologists' intuition. The adequate understanding of these concepts was developed "in a tradition of biological culture." Since the history of biology – no matter whether one starts with Aristoteles, Lamarck, or anybody else – comprises more than a single tradition of biological thought, the question of which tradition is meant here arises immediately. To put it shortly, it is the tradition of anti-reductionism, anti-mechanicism, and anti-Darwinism (Barbieri 2002). The biosemiotician should, thus, train her understanding of the phenomena of life by participating in this specific tradition of biological discourse. Such a hermeneutical approach to the semantics of biology may lead to a form of cultural relativism in science when it is combined with a picture of science as a repository of traditions.

Second remark. In cultivating her intuitive appreciation of the phenomena of life, the Copenhagen biosemiotician gets used to the concept of "life itself". Broadly speaking, it denotes a synthesis of "physical processes" by semiotic "mediations" between them. Whereas the physical processes occuring in organisms are, of course, describable by the science of physics, their semiotic mediations are seemingly not – so the latter constitute the very essence of life itself. Since "signs" seem to be the "fundamental units" in semiotics they are, not only methodologically the "basic units in the study of life", but also ontologically the true "elements" of semiotical mediations between physical processes in organisms.

Supplement to the second remark. The ontological all-pervasiveness of signs in life is also a necessary consequence of Emmeche's definition of life as "*functional interpretation of signs in self-organized material code-systems making their own umwelts.*" (Emmeche 1998: 11) He comments his definition as follows:

> This definition seems to imply that information (signs, or meaning) is conceptually primary; while organisms, metabolism and evolutionary replication are secondary with respect to the semiotical processes. When we apply the semiotical concepts to natural systems, this is often taken for granted. However, in order to bridge the gap between (physical) nature and (semiotical) culture, we have to develop a theory of the causal nature of sign-interpretation that can account for the generation of the so-called *original meaning* (not just ascribed observer-dependent meaning) as part of the natural activity of physical systems under specific boundary conditions. (Emmeche 1998: 11)

If the difference between physical explanation and semiotic understanding of processes in living systems were overcome in "a theory of the causal nature of sign-interpretation", the ontological supremacy and universality of signs in life would be ultimately established. Copenhagen biosemiotics hopes, thus, to become the fundamental ontology, not just of human existence, but of organic existence in general. To recognize the ontological identity of all living beings, the biosemiotican must understand the concept of life itself in a semantically uniform way: in so far as the semiotic ontology proves to be truly fundamental, it is unreasonable to conceive of a form of life that is different from the one which real organisms experience.[7]

Third remark. Semiotic mediations of physical processes by signs are the essence of life as understood in a certain tradition of biology with which the Copenhagen interpretation affiliates itself. The hermeneutical approach to the intuition of biologists must, therefore, lead to an understanding of the concept of life itself, the content of which cannot be successfully reduced to a combination of other concepts without losing its semantic substance. This irreducibility results in a tautological definition of the concept of life itself. On the most abstract level of her reasoning, the biosemiotician is forced to always repeat that life itself is life itself since the semiotic properties by which she defines life itself are specific to life itself (see Küppers 1990: ch. V.2 for an analogous analysis of holistic definitions of the concept of life). According to the Copenhagen interpretation there exists a singular object, life itself, that shows all of the essential properties which the biosemiotic concept of life denotes. To this singular biosemiotic object the biosemiotician can only point. Therefrom follows that, principally, she cannot vary the semantic relation between her concept of life itself and the instantiations of life itself (see supplement to second remark).

Fourth remark. From the Copenhagen point of view, "model-building" by organisms is an integral part of the biosemiotic model of life itself. Models are to be understood, not as "the sum of their building blocks", but as "complex signs", i.e. semiotic wholes that semantically refer to something. The semiotic mediations of physical processes in an organism lead, in the very same organism, to the construction of models of itself and its environment. As physical processes in organisms are semiotically mediated by signs, all experience of reality by organisms is also semiotically mediated by models as complex signs.[8] For organic existence there does not exist a non-semiotic access to any kind of reality. Semiotics has, thus, a transcendental status in biology; the Copenhagen interpretation sees biosemiotic semantics from a Kantian perspective.

Fifth remark. An analytical approach to models is, according to the Copenhagen interpretation, not adequate since either it proceeds on a purely syntactic level by formally studying the concatenation of signs in the model, or it loses the holistic quality of the semantic dimension of the model as a semiotic whole: the model in its entirety has the quality of being about something. The same is true for biosemiotic models, which describe organisms as semiotic beings: these biosemiotic models are man-made complex meta-signs defined by their semantic reference to complex object-signs made by organic existence. From this follows that a methodological split between more or less irrelevant possibilities of constructing a formal syntax of models and the important analytical inaccessibility of their semantics runs through the object-theory as well as the meta-theory of biosemiotics. Hoffmeyer's invective against the "hegemonial role it [i.e. mathematics] plays in scientific modeling" (Hoffmeyer 1998: 461), is motivated by his reasoning that such a methodological split exists since life cannot be mathematically modelled – at least as mathematical modeling is understood today.

Sixth remark. The model-building of biosemioticians is, according to the Copenhagen interpretation, a special case of the model-building activity of organic existence in general. Biosemioticians interpret life as an interpreting activity. The biosemiotic meta-models of the semiotic processes going on in organisms cannot claim to be something principally different from the object-models they refer to. The tautological definition of the concept of life itself and the singularity of the biosemiotic object (see third remark) are, thus, embedded in a hermeneutical circle in which there does not occur a semiotic cut between, on the one hand, models that are the objects of biosemiotics and, on the other hand, models that are used in biosemiotics. From the perspective of the Copenhagen interpretation it is *a priori* impossible to construct a scientific meta-language by which anybody can speak about the object-language of life from a position 'outside' life, since such an imaginary scientific meta-language would tell the biosemiotician all sorts of things but nothing about her singular object, life itself.

To sum up, the Copenhagen stance on basic problems of biosemiotics is to be characterized by the following six points. *First*, the Copenhagen interpretation takes a hermeneutical approach to biological discourse and may lead to a kind of cultural relativism in science. *Second*, the Copenhagen interpretation uses the concept that denotes its object, the concept of life itself, in a semantically uniform way. *Third*, the Copenhagen interpretation tautologously defines the concept of life itself in the form of a semantically irreducible category and can only point to the singular object, life itself, that is denoted by this category. *Fourth*, the Copenhagen interpretation grants semiotics a transcendental status in biology. *Fifth*, the Copenhagen interpretation splits syntax and semantics apart by acknowledging irrelevant possibilities of constructing a formal syntax of models and by assuming the important impossibility of an analytical semantics of semiotic models. *Sixth*, the Copenhagen interpretation denies the possibility of constructing a scientific meta-language for biology. All these six points amount to the fundamental thesis of the Copenhagen interpretation that signs constitute the universal medium of organic existence. From a philosophical perspective, this thesis expresses a generalized version of the conception of language as the *characteristica universalis* of human beings (see introduction).[9]

Coming back to the question of how biosemiotics should be related to current scientific research on living beings – is it a revolution in biology or rather a philosophical reframing of biology? –, the conclusion from the six remarks above must be that, if a biosemiotician adopts the Copenhagen interpretation, she has necessarily to consider biosemiotics a new philosophical perspective on life that becomes, in the long run, a revolution in the science of biology. Hoffmeyer (1998: 463) interprets this transformation as a process that is dependent on the wealth of biosemiotic knowledge:

> In its present state of underdevelopment, biosemiotics can not yet be seen as an alternative to 'the "hard" science of biochemical biology', it rather takes the role of a meta-biological perspective. But in due time I hope to see biosemiotics fusing with biology to create a semiotically informed science of life.

At present, the 'hard' physico-chemical objectification of living systems can only be complemented from a meta-scientific position by 'soft' biosemiotic descriptions of organisms as subjective agents making sense of their environments (see Emmeche's definition of life above). Eventually, 'soft' biosemiotics might also scientifically prevail against 'hard' biology since the very essence of life that distinguishes biological objects from those of physics and chemistry (see the second remark above), can be studied solely from a biosemiotic perspective.[10]

2. COMPUTING CODES: LANGUAGE AS AN ORGANIZATIONAL PRINCIPLE OF LIVING SYSTEMS

Emmeche's distinction between biosemiotics as a spontaneous and unsystematic activity of biologists, as a scientific paradigm that will revolutionize biology, and as a new philosophical perspective on biology (see section 1), allows to regard a fourth possibility only as the disappointing result of failing to realize the second or third alternative: biosemiotics considered part and parcel of the biochemical and selectionist paradigm prevalent in biology. Most of the molecular geneticists and evolutionary theorists who use semiotic or information-theoretical concepts, can be seen as contributors to biosemiotics understood in this fourth way. Well-known examples for evolutionary theorists who have thought intensely about the rôle of codes in the natural history of life are George C. Williams and his conception of evolution in the codical domain (Williams 1992: 10ff., Williams 1997: 164ff.) as well as John Maynard Smith's and Eörs Szathmáry's theory of major transitions in biological information processing (Maynard Smith and Szathmáry 2000). In the French school of molecular genetics, Jacques Monod's ideas about syntactic randomness, semantic contingency, and pragmatic indefiniteness in the living (Monod 1973, Artmann 2002) as well as François Jacob's sketch of an evolutionary pragmatics of tinkering (Jacob 2000, Artmann 2004) are important contributions both to 'normal' biology and biosemiotics. The Göttingen school of biophysical chemistry thinks about molecular evolution in a semiotic frame of reference (Eigen 1979, Küppers 1990). Even though they might never have heard about a theory called 'biosemiotics', those biologists are talking about the natural existence of semiotic structures that open up new possibilities for the evolution of complex functional organizations. The most famous example for this is, of course, the genetic code that makes possible the information transfer from polynucleotide chains of deoxyribonucleic acid (DNA), more precisely: from the linear sequence of its four nucleotide bases, into polypeptide chains of proteins, more precisely: into the linear sequence of its twenty amino acids (see for a detailed introduction Watson et al. 2004: ch. 15).

The best method through which progress in normal biology can be made by sign-theoretical means is to define semiotic structures as formal as possible and to search for biological phenomena that instantiate these structures. The scientific culture to which such biosemiotic thinking belongs is the tradition of mathematico-mechanistic biology. In theoretical population genetics, this tradition has taken the form of an "algebra of evolution" (Rice 2004: 165) that studies how natural selection and other evolutionary mechanisms work in the gene pools of populations. Ronald Aylmer Fisher (1999: ix), one of the founding fathers of this biological discipline, describes the method of using mathematics in empirical science as follows:

> The ordinary mathematical procedure in dealing with any actual problem is, after abstracting what are believed to be the essential elements of the problem, to consider it as one of a system of possibilities infinitely wider than the actual, the essential relations of which may be apprehended by generalized reasoning, and subsumed in general formulae, which may be applied at will to any particular case considered.

Fisher's methodological approach to mathematics in biology is essentially a model-theoretical one. From the point of view of model theory, mathematical theories consist of sentences whose components are chosen from a set of symbols (a vocabulary) and concatenated according to a set of syntactic rules (a grammar).[11] The set of sentences that can be computed in this way, constitutes a particular formal language. Every theory, i.e. every subset of a language, can be interpreted and re-interpreted semantically by a variety of models, i.e. by entities that fulfil all syntactic conditions of the theory. In the model-theoretical way of speaking about the semantics of theories 'model' means, not a simplified version of complex empirical reality, but an entity whose intrinsic order instantiates the syntactic structure of a theory.[12] The language in which the theory is formulated can be characterized as a syntactic calculus, a code, that is semantically re-interpretable in many different ways. Consequentially, Hintikka (1997) calls the model-theoretical perspective the 'language as calculus-viewpoint' from which language is considered a computational[13] technology of human beings that is, due to its adaptability to a great variety of contexts of use, able to scientifically objectify any semantic system (see introduction).[14] Even in sciences such as mathematics and logic that are under strict syntactic control, there does exist neither a closed universal language in which all particular theories can be discussed, nor a single meta-model of any possible object-model. Instead, the many languages of mathematics, logic, and empirical science are an evolving linguistic population whose members can be distinguished by their abilities to talk about different sets of mathematical, logical, and empirical entities (Hintikka 1998: 204f.).

The linguistic agility of model theory is impressively documented in the approach of the American philosopher Patrick Suppes to the semantic dimension of scientific theories. His approach is to axiomatize a scientific theory by defining a predicate within set theory (Suppes 1999: 249ff., Suppes 2002: 30ff.). A model for the theory thus axiomatized is every entity that satisfies this predicate, so that the syntactic structure of the set-theoretical predicate semantically defines the set of its possible models (Suppes 1999: 253f., Suppes 2002: 33). This does not mean that the relation between syntax and semantics of an empirical theory must be described in a completely formalized way: neither has general set theory to be axiomatized in first- or higher-order logic nor other parts of current mathematics that are used in the definition of a set-theoretical predicate for an empirical theory (Suppes 1999: 249f. and 255, Suppes 2002: 27ff.). Moreover, Suppes does not pretend that set theory is the one and only language in which everything mathematical must be, at least principally, expressed (Suppes 2002: 34, see also Hintikka 1998: 16ff.); instead, he subscribes to "a pluralistic attitude toward the concept of structure." (Suppes 2002: 35)

Suppes' programme of axiomatization by defining a semi-formalized set-theoretical predicate was followed in biology by Magalhães (2000) and *in extenso* by Balzer and Dawe (1997). The latter developed set-theoretical predicates for transmission as well as molecular genetics in order to emphasize the unity of both theories, whose conceptual structures can be generated by refining or specializing a more basic theory. For a biosemiotician, the main problem with Balzer's and Dawe's approach to genetics is that they do not sufficiently take the rôle of the genetic code into consideration. In their axiomatization of molecular genetics, Balzer and Dawe state an axiom, called *AM3*, that introduces a function *ex* (short for 'expression') mapping the set of DNA codons to the set of amino acids (which contains also a dummy element as the value of *ex* for chain-terminating codons) according to the decoding rules of the genetic code, e.g. *ex* (TTT) = *ex* (TTC) = Lysine (Balzer and Dawe 1997: 105f.). Yet the list of all these decoding rules, i.e. axiom *AM3*, does not tell the biologist anything about the genetic code as a code. An answer to the question on what abstract semiotic structure this axiom is based, would be, however, of utmost importance if a biologist were to find other realizations of the very same structure. The biosemiotician has, thus, to axiomatize genetics in a direction that is, so to speak, orthogonal to that of Balzer's and Dawe's.

The model-theoretical approach to semantics means that the relation between a biological theory and the set of its possible objects can be formally explored. This is in contradistinction to the Copenhagen interpretation of biosemiotics, which defines its fundamental concept, 'life itself', tautologously as a semantically irreducible category so that a Copenhagen biosemiotician can only point to the singular object, life itself, that is denoted by this category (see third remark in section 1). Another tenet of the Copenhagen interpretation is also challenged by the model-theoretical perspective on the semantics of biological theories: a Copenhagen biosemiotician uses the concept of life itself in a semantically uniform way (see second remark in section 1). From a model-theoretical viewpoint there does not exist any *a priori* reason why this must be so: if language is a calculus, various semantic relations between a theory and its models are conceivable. To what extent the semantics of a biological theory can vary, is a question that arises naturally in this context. The model-theoretically orientated biosemiotician must, for example, ask what the minimal conditions on semantic instantiations of a concept of living system are. She would be trapped in self-contradiction if she were considering a closed set of standard examples of organisms as the paradigm on which her concept of living systems must be based, or if she regarded her intuitive understanding of organisms as being based on a transcendental category of life itself that *a priori* delimits the set of objects coming under it (see fourth remark in section 1).[15]

The model-theoretical approach just characterized should not be mistaken for another methodological strategy that, at first sight, seems to be closely related but is, in fact, only another version of linguistic universalism: the axiomatization of biological theories as conceived by Joseph H. Woodger. His project of using modern logic, more precisely: the formal system of Russell's and Whitehead's *Principia Mathematica*, as the syntactic frame for a new biological object-language wanted "[...] to provide an exact and perfectly controllable *language* by means of which biological knowledge may be *ordered*." (Woodger 1937: vii) If successful, this would lead to "a scientifically perfect language" (Woodger 1937: viii), which could expel any linguistic (and, thereby, philosophical) confusion from biology. Woodger's project belongs to the tradition of logical empiricism whose logical[16] perspective on scientific language he summarizes as follows:

> A science is an exact science when its language has attained to such a degree of perfection that its syntax is completely known, i.e. when we know and hence can formulate precisely the rules according to which sentences can be constructed in it, and when we know and can formulate precisely the rules according to which a given sentence is regarded as a consequence of others, and, finally, when all sentences have been so arranged that from a small number of initial ones all the remaining ones can be derived by successive applications of the rules. Such a language is called a *calculus* or *axiom-system*. (Woodger 1937: 2)

This definition of an axiomatic system, or calculus, is restricted to syntax since it considers only the deductive power of a set of sentences that are distinguished as axioms from all other possible sentences. Woodger directs his attention solely on the proof-theoretical and syntactic aspect of calculus and leaves its model-theoretical and semantic aspect completely out of consideration. To semiotically foreshorten the concept of calculus in this way, is characteristic of the universalist conception of language, which regards syntax as formalizable but semantics as ineffable (see fifth remark in section 1). The model-theoretical approach to language defines a calculus, or axiomatic system, not only syntactically by its deductive power, but also semantically by its descriptive potential (Hintikka 1998: ch. 1), which can be investigated formally either in another language, or in the same language, or in both ways.

The proponent of the position that language is a calculus argues against the idea that the axiomatization of empirical science is a way to epistemological perfection. Though Woodger's project has not been completely ineffectual,[17] a model-theoretically educated biosemiotician is not lost in day-dreams about a perfect object-language of biology, which could be realized by a thorough formalization of today's muddled talking about life. She rather searches for a meta-language as a formal tool by means of which biological object-languages that seem to be incommensurable, can be compared if they are seen as different terminological 'fillings' of an object-linguistic structure that is used for the scientific explanation of living systems (compare sixth remark in section 1).[18] The biosemiotician, by assigning the function of such a comparative meta-language to model theory, asserts that there may exist different useful and mutually translating object-languages in biology. Model-theoretical comparisons between empirical theories allow, therefore, to overcome the linguistic relativism that is threatening every hermeneutical approach to language (see first remark in section 1). The only alternative to using formal meta-linguistic rules for the semantic interchange between different object-languages is the breakup of biosemiotics into incommensurable ideologies that repeat their old sayings again and again.

Semiotics acts as a scientific language both on the object-linguistic and the meta-linguistic level of biology. Biosemiotics describes biological objects as semiotic systems, and it is described as a biological object-language in the meta-linguistic framework of model-theoretical semantics. Yet the model-theoretical calculi that meta-theoretically compute axiomatizations of object-theoretical concepts, and the semiotic systems (organisms) that are models of the axiomatized biosemiotic concepts, are not supposed to be grounded on a shared ontological experience of what it means to be living (compare sixth remark in section 2). It is, however, possible to consider biological systems as realizations of computing codes and to try to transform them into bio-inspired computational technologies (Sipper 2002), but whether such engineering projects will succeed is a problem that must be empirically decided one by one – and not by sweeping *a priori* arguments.

3. A MODEL-THEORETICAL DEFINITION OF ORGANIC CODES

Before the explanative power of biosemiotics can be discussed seriously, precise definitions of its central concepts have to be developed. In semantics it is particularly important, not only for the Copenhagen interpretation of biosemiotics (see section 1),[19] but also for Marcello Barbieri's (2003) theory of organic codes, to give a clear and accurate definition of the concept of code. Otherwise, biosemioticians would be unable to defend themselves against the criticism that this concept has been, and may still be, useful for explaining how the amino acid sequence of a protein is synthesized but useless for solving any other biological problem, since it does not go beyond the descriptive level of biochemical causality (Blumenberg 1986: 405ff., Godfrey-Smith 2000a).

How does Barbieri define the concept of code, which he uses to describe, not only the genetic code, but also splicing codes and signal transduction codes (Barbieri 2003: 96ff.) – and, in the future, maybe even more organic codes (Barbieri 2003: 218f.)?

A code is defined as *a correspondence between two independent worlds*, and this definition immediately suggests a useful operative criterion. It means that the existence of a real organic code is based on (and can be inferred from) the existence of organic molecules – called *adaptors* – that perform two independent recognition processes. [...] Codes have, in brief, three fundamental characteristics [...]:

(1) They are rules of correspondence between two independent worlds.
(2) They give meanings to informational structures.
(3) They are collective rules which do not depend on the individual features of their structures. (Barbieri 2003: 93f.)

In order to deal with the criticism that this characterization of codes is not clear enough (Levy and Jablonka 2004: 67), Barbieri's idea of organic codes should be stated more precisely. Biosemiotics will not have the slightest chance of convincing biologists to use semiotic concepts, if its analyses of living system do not reach a level of mathematical sophistication that is common in theoretical biology and Artificial Life. To axiomatize Barbieri's theory of organic codes by defining the set-theoretical predicate 'is an organic code' may be a first small step on the route to a more advanced form of biosemiotic reasoning.

From a purely syntactic point of view, a code C of fixed length n (i.e. a block code) over a finite alphabet A of symbols is simply defined as a subset of the Cartesian product of n copies of A, $C \subseteq A^n := A \times A \times \ldots \times A$ (Roman 1996: 34, Walker 2000: 3). The elements of the set C are called 'codewords'. The most simple code alphabet, into which any other alphabet can be transcribed if a change in the length of codewords is allowed, is the binary field $F_2 := \{0, 1\}$ with addition and multiplication modulo 2. The genetic code *GenCo* is, for example, a block code of length 3, whose alphabet consists of the symbols A, C, G, and T (or U). The codewords of *GenCo* are the sixty-four nucleotide triplets (codons), whose binary transcription results in codewords of length 6.

From a semantic point of view, a code is more than the set of codewords C defined in the preceding paragraph: a bijective function$e$, called 'encoding function', from a source alphabet S onto C has to be added (Roman 1996: 34). After some finite set $S = \{s_1, s_2, \ldots, s_n\}$ of symbols is chosen as a source alphabet, the ordered pair (C, e) defines an encoding scheme for S. In semantics, it is appropriate to call these encoding schemes 'codes'.[20]

This semantic definition of code cannot be applied easily to the genetic code *GenCo* since *GenCo* is degenerate: there often exists more than one codon to code for an amino acid (see below), so that the encoding 'function' of *GenCo* is not a function at all. From this fact does not follow that *GenCo* is not uniquely decipherable (see Roman 1996: 37ff. for an introduction to uniquely decipherable codes). There does not exist a *GenCo* codeword that encodes two different amino acids, and, if the starting-point of the decoding process is given, a sequence of nucleotides can be interpreted only in a single way as a sequence of codewords. It is, therefore, better to define *GenCo* as the ordered pair (S, d), i.e. the decoding scheme with the set of amino acids S as source alphabet and the homomorphism $d: C \rightarrow S$ as decoding function from the set C of codons onto the set S of amino acids (including a dummy element, see section 2).[21] Since unique decipherability is one of the most important features an organic code possesses, it is sensible to call, in the context of biosemiotics, these decoding schemes 'codes'. The following paragraphs will, thus, use decoding schemes (S, d) to define codes semantically.

From a model-theoretical point of view, a code (S, d) can be described formally in more detail. First, the source alphabet S and the set of codewords C are to be characterized as parts of structures. Any structure **A** is specified by four components (Hodges 1993: 2f.):

- the domain of **A**, dom**A**, a set whose elements are called 'elements of **A**';
- a subset of dom**A**, whose elements $c^{\mathbf{A}}$ are called 'constant elements of **A**';
- for each integer $n > 0$, a set of subsets of $(\text{dom}\mathbf{A})^n$, whose elements are called 'n-ary relations $R^{\mathbf{A}}$ on dom**A**';
- for each integer $n > 0$, a set of maps from $(\text{dom}\mathbf{A})^n$ to dom**A**, whose elements are called 'n-ary operations $F^{\mathbf{A}}$ on dom**A**'.

A n-tuple from **A**, denoted by a', b', etc., is a finite sequence of length n that consists of elements of **A**. Each constant $c^{\mathbf{A}}$, each relation $R^{\mathbf{A}}$, and each operation $F^{\mathbf{A}}$ may be denoted by one or more constant symbols c, n-ary relation symbols R, and n-ary function symbols F, respectively. The signature $\Sigma_{\mathbf{A}}$ of a structure **A** is specified by, first, the set of constant symbols of **A**, second, separately for each positive integer n, the set of n-ary relations symbols of **A**, and third, separately for each positive integer n, the set of n-ary function symbols of **A**. A structure **A** with signature Σ is called a 'Σ–structure' (Hodges 1993: 4f.).

Where is the place of source alphabets S and sets of codewords C in the signature of structures? Taking as an example again the genetic code *GenCo*, there are two structures **AS** and **AC** whose domains include all possible amino acid sequences and all possible codon sequences, respectively. The elements of dom**AS** are called 'primary structures of proteins', the elements of dom**AC** 'codon chains'. The set of all constant elements of dom**AS** is equal to the set of amino acids, which are denoted by constant symbols coming from the source alphabet S; and the set of all constant elements of dom**AC** is equal to the set of codons, which are denoted by constant symbols coming from the set of codewords C. The binary linkage operations $L^{\mathbf{AS}}$: dom**AS** × dom**AS** → dom**AS**, and $L^{\mathbf{AC}}$: dom**AC** × dom**AC** → dom**AC**, are the only operations on dom**AS** and dom**AC** that are denoted by a symbol. Moreover, they are denoted by the same function symbol L in the signatures of **AS** and **AC**.

The linkage operations L^{AS} and L^{AC} generate an element of dom**AS** and dom**AC**, respectively, by concatenating two elements of the same domain without overlaps or gaps between these elements. Any relation that may be defined on dom**AS** or dom**AC** is not denoted by a relation symbol. The signatures Σ_{AS} and Σ_{AC} of the structures **AS** and **AC** are, therefore, $\Sigma_{AS} = (S, \emptyset, \{L\})$ and $\Sigma_{AC} = (C, \emptyset, \{L\})$.

How does the decoding function connect these structures? To answer this important question, a few more model-theoretical concepts must be introduced. A term t of the signature Σ is (Hodges 1993: 11):

- either a constant symbol of Σ; or
- an expression $F(t_1, \ldots, t_n)$ if n is a positive integer, F is an n-ary function symbol of Σ, and t_i are terms of Σ; or
- a variable, i.e. any symbol v_i, with i a natural number, that is used, e.g., as a temporary label for an element of a structure.

If Σ is a signature and X a set of variables, the term algebra of Σ with basis X is defined as the Σ-structure **T** that has the following properties (Hodges 1993: 14):

- dom**T** is the set of all terms t of Σ whose variables v_i are in X;
- the constants $c^{\mathbf{T}}$ of **T** are equal to c for each constant symbol c of Σ;
- the operation $F^{\mathbf{T}}(t')$ of **T** is equal to $F(t')$ for each n-ary function symbol F of Σ and each n-tuple t' of elements of dom**T**; and
- the relation $R^{\mathbf{T}}$ is empty for each relation symbol R of Σ.

The relations on the signature Σ. are, so to speak, forgotten by its affiliated term algebra; for this reason it is unimportant that there was not denoted any relation on dom**AC** or dom**AS** above.

In a term algebra, the symbols formerly used to denote the components of a structure now denote themselves. A term algebra establishes, therefore, a semiotic level of speaking about a sign system, i.e. about a signature $\Sigma_{\mathbf{A}}$, without considering the objects these signs denote, i.e. the structure **A**. As regards the genetic code, more precisely: its source alphabet and its set of codewords, the difference between the structure containing the domain 'primary structures of proteins' (or the structure containing the domain 'codon chains') and the term algebra defined on the signature of this structure is equal to the difference between the purely biochemical and the purely semiotic level of description of the genetic code. Since biosemiotics is interested in the relation between both levels, it must take structures *and* term algebras into account. If the biochemical and the semiotical levels of description were identified so that a purely semiotic way of describing living systems resulted, a universalist stance on language in biosemiotics, such as adopted by the Copenhagen interpretation (see section 1), would be the consequence. If semiotic term algebras were considered negligible so that only biochemical structures were taken into consideration, the argument of Blumenberg and Godfrey-Smith against the explanative power of the concept of coding (see above) would become true.

A closer look at some of the axioms that describe term algebras will make it clear that the theory of term algebras can describe how strings of source symbols and strings of codewords are generated. Let the structure **T** be a term algebra. Then the following three sentences are true for all term algebras **T** (see for a complete list of first-order axioms for term algebras Hodges 1993: 35). First, there do not exist two constant symbols of **T** that are equal to each other. Code-theoretically this means that neither a source symbol is equal to any other one nor a codeword equal to any other one. Otherwise, it could not be verified whether a given source symbol or codeword really is this particular source symbol or codeword. Second, for every positive n, the term that is generated by applying any n-ary function symbol F of **T** to any n-tuple of terms of **T**, is not a constant symbol of **T**. This axiom says that, as regards the linkage operation L, the concatenation of source symbols (or codewords) does never result in a source symbol (or codeword). Otherwise, unique decipherability could not be guaranteed (see above). Third, for every positive integer n, if the application of any n-ary function symbol F of **T** to any two n-tuples a' and b' results in the same term, then a' and b' are the same tuple. Applied to the linkage operation L, it follows from this axiom again that all strings that are generated by the (repeated) use of L are uniquely decipherable.

It is now shown that the decoding function d of a code (S, d) can be described as a homomorphism from one term algebra to another one. A semantic homomorphism h from a Σ-structure **A** to a Σ-structure **B** is a function from dom**A** to dom**B** that fulfils the following three conditions (compare Hodges 1993: 5):

- for each constant symbol c of Σ, $h(c^{\mathbf{A}})$ is element of the set of constant elements of **B**;
- for each positive integer n and each n-ary relation symbol R of Σ and n-tuple $a'$from A, if $a' \in R^{\mathbf{A}}$ then $h(a') \in R^{\mathbf{B}}$;
- for each positive integer n and each n-ary function symbol F of Σ and n-tuple a', $h(F^{\mathbf{A}}(a')) = F^{\mathbf{B}}(h(a'))$.
- $h(a')$ denotes $(h(a_0), h(a_1), \ldots, h(a_{n-1}))$ for a $n-$tuple $a' = (a_0, a_1, \ldots, a_{n-1})$.

A homomorphism thus defined can exist *per definitionem* only from a structure to another one that share the same signature. To define a decoding function d from the set C of codewords to the set S of source symbols directly as a homomorphism from the term algebra $\mathbf{T}_{\mathbf{AC}}$ of $\Sigma_{\mathbf{AC}} = (C, \emptyset, \{L\})$ to the term algebra $\mathbf{T}_{\mathbf{AS}}$ of $\Sigma_{\mathbf{AS}} = (S, \emptyset, \{L\})$ is, therefore, impossible. Because the set of constant symbols of both signatures $\Sigma_{\mathbf{AC}}$ and $\Sigma_{\mathbf{AS}}$ has to be the same if $\Sigma_{\mathbf{AC}}$ and $\Sigma_{\mathbf{AS}}$ shall be same signature, the constant elements of dom**AC** and dom**AS** must be transcribed into one and the same alphabet, for example the binary field $F_2 := \{0, 1\}$ with addition and multiplication modulo 2 (see above). Then a subset B of F_2 contains the transcribed source symbols as well as the transcribed codewords. This transcription must, of course, preserve the unique decipherability of codewords. For any element n of B that is not a codeword, d acts like an identity function, i.e. $d(n) = n$.

Since, in term algebras, the components of signatures denote themselves, two term algebras of the same signature are one and the same term algebra. The homomorphic decoding function d from the term algebra $\mathbf{T}_{\mathbf{AC}}$ to the term algebra $\mathbf{T}_{\mathbf{AS}}$ becomes, thus, a semantic endomorphism of the term algebra $\mathbf{T}_{\mathbf{AB}}$ of $\Sigma_{\mathbf{AB}} = (B, \emptyset, \{L\})$,

d: $\mathbf{T_{AB}} \rightarrow \mathbf{T_{AB}}$. Notice that this endomorphism d is not constrained by any internal relational order that may exist in **AC** or **AS**.

The model-theoretical approach to organic codes leads to the following definition of the set-theoretical predicate 'is an organic code':

A biological system BS is an *organic code* if and only if there are a term algebra $\mathbf{T_{AB}}$ of signature $\Sigma_{\mathbf{AB}} = (B, \emptyset, \{L\})$ and a semantic endomorphism d of $\mathbf{T_{AB}}$ such that BS is the decoding scheme (B, d) and B the set of codewords and source symbols.

This definition of organic code captures the first fundamental characteristic of organic codes as described by Barbieri (2003: 94, see above). That codes are "rules of correspondence between two independent worlds" means the following: there exist two structures that are, on the one hand, defined by different sets of elements, constants, relations, and functions (they are "two independent worlds"), but that are, on the other hand, connected by an endomorphism ("rules of correspondence") of a term algebra that is constructed on the base of those structures by the procedure described above.

The second characteristic feature of codes according to Barbieri (2003: 94) is that codes "[...] give meanings to informational structures"; as regards the genetic code, the ribotype, i.e. the ribonucleoprotein system of the cell, is the set of meaning-giving molecules in protein synthesis. Barbieri (2003: 94) states that the relation between a codeword and its meaning is contingent, i.e. not motivated by any similarity between both related elements. From this contingency follows that the semantic relation of a codeword to its meaning cannot be predicted if solely the syntactic aspects of a code are known. A molecular biologist who wants to know why a codon codes for a certain amino acid must also reconstruct the contingent physico-chemical boundary conditions in which the code evolved and upon which its evolution reacted. Though the semantics of the primeval genetic code might have been stereochemically determined, its evolution led to semantic contingency: stereochemical affinities between codons and amino acids do not play a causal rôle in decoding any more (Godfrey-Smith 2004b: 204). For a code to be semantically contingent it is sufficient that there could have originated and evolved another code which would be functionally equivalent to the genetic code that has evolved in the natural history of living systems on Earth (Beatty 1995).[22] In the definition of organic code given above, this semantic contingency is captured by the free choice of a decoding function d. What particular function d is chosen, is just restricted by the condition that d must be an endomorphism, i.e. a function that preserves the structure which the linkage operator L defines on the set of binary codewords and source symbols B. This does not imply that the codewords have to follow each other in the same order as the source symbols they encode; it is only implied that the decoding function does not destroy any information about the sequential order of encoding codewords.

Barbieri's third characteristic feature of codes is also already rephrased in the definition of organic code given above. According to Barbieri (2003: 94), codes "[...] are collective rules which do not depend on the individual features of their structures." That a code is a class of "collective rules" means the following: as soon as a decoding scheme (B, d) is established, it is systematically applied on the set B of codewords and source symbols without exception. This feature of organic codes does not contradict the contingency of the decoding function d: the latter maps codewords to strings of source symbols, and this mapping does "[...] not depend on the individual features of their structures." Whereas Barbieri's second feature of organic codes characterizes the decoding function itself, the third feature concerns the use of this function in a decoding scheme.

Though the genetic code is contingent, it seems to have been evolutionarily optimized before fixation, so that its high adaptive value explains its near-universability. This optimization concerns the buffering of mutational errors by letting similar codons code for the same amino acid (so the genetic code is degenerate), or for chemically similar amino acids (Watson et al. 2004: 461ff.). But not only the effects of DNA mutations on the sequence of amino acids in proteins are minimized this way. Moreover, "mistakes in translation don't translate into termination." (Hughes and Ellington 2005) Research into what global effects the artificial change of the genetic code has on the fitness of an organism, suggests "[...] that the genetic code is so supremely optimized that the effects of any folding catastrophe are greatly reduced." (Hughes and Ellington 2005: 1273) 'Folding catastrophe' means that, due to mistakes in transcription or translation, or due to some experimental alterations in the code, proteins are misfolded or even remain unfolded so that they cannot (or just can to a less degree) fulfil their former biological function. Of course, after fixation (i.e., since billions of years) organisms have been highly optimized for the existing genetic code (Hughes and Ellington 2005: 1273), so it is difficult to decide which optimization, the codical or the organizational, explains a specific feature of a living system.

Considering the optimization of the genetic code, Marshall Nirenberg (2005: 204), one of its decoders, comes to the conclusion that "[...] the arrangement of codons and amino acids is not random." Semiotically rephrased, the decoding function of the genetic code is not primarily motivated by any stereochemical similarity between an encoding codon and an encoded amino acid, but secondarily motivated by a structural analogy: a chemical similarity between encoding codons is related to a chemical similarity between encoded amino acids. How can this type of non-randomness be integrated into the definition of organic code? The first step consists in introducing a measure of the distance between two codewords and between two strings of source symbols. As regards the genetic code, a small distance between two encoding codons then implies a small distance between two encoded amino acids, and vice versa. Such a measure of distance is well-known in coding theory. Let x and y be strings of length n over an alphabet A. The Hamming distance between x and y, $H(x, y)$, is the number of places in which x and y differ, i.e. the cardinality of the set $\{i \mid x_i \neq y_i\}$ (Hamming 1950: 154f., Thompson 1983: 8ff., Roman 1996: 104, Walker 2000: 3). Since the Hamming distance H defines a metric on A^n, the closeness of any two strings x and y of A^n can be quantified. After the distances between all strings of A^n have been measured, all pairs (x, y) of strings of A^n can be ordered by a 'less than or equal to'-relation in respect to their Hamming distance.

The non-randomness of the genetic code as stated by Nirenberg means that the decoding function d of the genetic code preserves the ordering of the set of binary codewords B which is based on the Hamming metric on B. The transciption of the constant elements of dom**AC** and dom**AS** into one and the same alphabet, the binary field $F_2 := \{0, 1\}$ with addition and multiplication modulo 2 (see above), must, therefore, get rid of such features of codewords and source symbols that would not allow to reflect the biochemical similarities between the constant elements they are symbols of. As to the genetic code, this is of particular importance for the source alphabet S containing the source symbols (such as *Gly*, *Ala*, *Asp*, *Glu*, etc.) that denote the amino acids: it must be transcribed in such a way that the Hamming distances between the binary source symbols really mirror biochemical similarities between the constant elements of dom**AS** which these symbols stand for. The set of codewords C is more easily transcribed into a binary form since the usual codon notation using the nucleotide symbols A, C, G, and T, combined with the distinction between purines and pyrimidines, adequately reflects the biochemical similarities between the codons, the constant elements of dom**AC**.

An optimized decoding function preserves, not the absolute Hamming distances, but the order of the Hamming distances between the codewords. The definition of the set-theoretical predicate 'is an organic code' that shall axiomatize the theory of organic codes can be completed as follows:

A biological system BS is an *(optimized) organic code* if and only if there are a term algebra $\mathbf{T}_{\mathbf{AB}}$ of signature $\Sigma_{\mathbf{AB}} = (B, \emptyset, \{L\})$ and a semantic endomorphism d of $\mathbf{T}_{\mathbf{AB}}$ such that BS is the decoding scheme (B, d), B the set of codewords and source symbols, and the following axiom is satisfied:
For each two pairs (b_k, b_l) and (b_m, b_n), $b_i \in B$, if $H(b_k, b_l) \leq H(b_m, b_n)$ then $H(d(b_k), d(b_l)) \leq H(d(b_m), d(b_n))$, with H the Hamming distance defined on B.

Apart from this constraint on the choice of the decoding function, there might wait further axioms of organic codes to be discovered. The incompleteness of the definition above is desirable, because the set-theoretical predicate 'is an organic code' shall axiomatize, not an *a priori* intuition of life, but an empirical theory that is still in its infancy.

ACKNOWLEDGEMENT

I thank Tommi Vehkavaraa (University of Tampere) for his helpful comments on a first version of this paper.

NOTES

[1] For a justification of why this technology is called 'computational', see section 2, n. 13.

[2] Though it might appear so, the distinction between both conceptions of human language cannot be mapped one-to-one onto the contrast between the hermeneutical and the analytical tradition in twentieth-century philosophy (for introductory comparisons between these traditions see Habermas 2000

and Engel 2000). Important thinkers who belong to the (pre-)history of analytical philosophy, such as Frege, Wittgenstein, Tarski, and Quine, must be counted to the *characteristica universalis*-party (Hintikka 1997); and at least one important philosopher in the immediate prehistory of hermeneutical philosophy, Husserl, is a member of the *calculus ratiocinator*-party (Kusch 1989).

3 Both theses of this paper are closely connected to the conception of biosemiotics as a structural science, i.e. as a transdisciplinary formalization programme that tries to discover abstract analogies between research problems of different empirical sciences in order to contribute to their solution (Artmann 2005). Well-known examples of structural sciences are cybernetics and semiotics, game theory, information and net theory. These disciplines have proven to be useful beyond the context of their discovery and first application (as regards semiotics: beyond logic and linguistics), so that their possible application is not restricted by frontiers between empirical sciences (here: between linguistics and biology). The present paper explores one important consequence of regarding biosemiotics as a structural science, namely its model-theoretical stance on the semantics of biosemiotic theories and, thus, its methodological stance on ontological problems of biology.

4 This is well-documented in Kalevi Kull's account of the history of biosemiotics from the point of view that language is the universal medium (Kull 2005).

5 Another important author to be mentioned here is Anton Markos, who refers particularly to the hermeneutical movement in philosophy (Markos 2002). The same goes for Andreas Weber (2003).

6 'Copenhagen interpretation of biosemiotics', of course, connotes Niels Bohr's philosophical stance on foundational problems of quantum theory. The question of what relation exists between both Copenhagen interpretations deserves a detailed analysis. Here just a short comment on Bohr's central tenet, the principle of complementarity, is possible. Lecturing on *Light and Life* (Bohr 1933), Bohr proposed that biology must acknowledge a fundamental complementarity in its field of research. Dissecting an organism to analyse its material components and describing an organism as a whole to show its purposiveness are, according to Bohr, mutually exclusive but equally necessary methods in biology: physico-chemical analysis is indispensable for exploring the material parts of living systems but destroys the phenomenon under scrutiny, life itself, so truly biological synthesis is indispensable for holistically comprehending life as an irreducible biological fact. Yet thirty years later, Bohr (1963) had to revise this application of complementarity in view of the success of molecular biology. The principle of complementarity does not any more limit the physico-chemical explanation of life *a priori* but just legitimates a complementary manner of using both reductionistic and holistic concepts in biology as epistemologically sensible for all practical purposes. As regards complementarity, the Copenhagen interpretation of biosemiotics is, at the same time, more and less conservative than Bohr's ideas about biology. Emmeche and Hoffmeyer are, on the one hand, epistemologically less conservative since they hope that some day semiotics will conceptually unify biological knowledge (see below). Emmeche and Hoffmeyer are, on the other hand, ontologically more conservative because they assert that there does exist at least one irreducibly complementary pair of semiotic entities in organisms (see section 3, n. 19 on their thesis of code-duality).

7 This is the main message of Emmeche's 1994 book on Artificial Life research: though Artificial Life considers life "an intrinsically semiotic phenomenon" (Emmeche 1998: 6), it is principally not able to explain the complex semiotic nature of real life.

8 Referring to this, Emmeche draws an analogy to the Copenhagen interpretation of quantum mechanics and its conception of physical measurements: semantic interactions of an organism with its environment (i.e., sensing and controlling) endow the internal syntactic computations of the organism via models with meaning; they are non-computational and contingent means of pragmatic operations directed towards attaining the goals of the organism (Emmeche 1994: 149ff.).

9 It is, therefore, not really surprising that in Hoffmeyer's *Signs of Meaning in the Universe* (1996) important philosophers are cited whose theories strongly support the conception of language as the universal medium. To this group belong, e.g., the late Ludwig Wittgenstein's philosophy of language (Hoffmeyer 1996: 53 and 98) as well as Maurice Merleau-Ponty's philosophy of subjectivity (Hoffmeyer 1996: 51). Of particular importance for the philosophical foundations of the Copenhagen interpretation seems to be the Danish existentialist tradition, namely Søren Aa. Kierkegaard's concept of subjectivity as living self-referentiality (Hoffmeyer 1996: 50f.) as well as Knud E. Løgstrup's theory

of the phenomenological inseparability of human existence and language (Hoffmeyer 1996: 98 and 111), according to which language is "en regenerativ livsytring" (Løgstrup 1997: 198), a regenerative expression of life that mediates hermeneutically between human beings – a Copenhagen biosemiotician would, of course, add: between living beings in general. This biosemiotic generalization is anticipated by Løgstrup himself whose phenomenological philosophy moves from an existentialist to a cosmological orientation (Andersen 1989: 552ff.).

[10] This transformation of the relation between biosemiotics and biology is detectable also in the work of Emmeche. In Emmeche (1997), he still criticizes the revolutionary perspective on biosemiotics because he neither sees today's biology in need of such a paradigm change, nor does he think that the history of science can be simply described as being a Kuhnian process of punctuated equilibria, in which long periods of quiet normal science are interrupted by violent scientific revolutions. In Emmeche (1999: 288, n. 3), he is more cautious about determining the epistemological status of biosemiotics: "Rather than being a competing scientific paradigm, a more likely interpretation of biosemiotics is that it constitutes a part of a whole new theoretical and philosophical footing (not to say foundation) of biology." Emmeche even makes it clear that he wants to test the revolutionary potential of biosemiotics – in the hope of a positive result: "As someone who would like to see the biosemiotic research program flourish, but feels uncertain about its precise epistemological nature with respect to biology, I find it important [...] to see *how far we can go* in positing biosemiotics as an alternative research program that may complement or eventually, in the future, even displace the molecular paradigm. But we should also be prepared to be forced to take more moderate positions, viz. suggesting biosemiotic concepts as a tool box that in certain domains of biology may help to organize our knowledge better, pose more interesting questions, and make alternative testable hypotheses, even though it may not take the role of an alternative paradigm." (Emmeche 1999: 274f.)

[11] This and the following statements are meant to give a first sketch of the model-theoretical perspective on semantics. They are, of course, very informal and do not adequately reflect the inner complexity of the fundamental concepts of model theory (for a detailed introduction see Hodges 1993).

[12] In recent methodological reflections on theoretical population genetics the shift of orientation, from modelling in the classical sense of making simple images of complex reality, to modelling in the modern sense of semantically instantiating syntactic structures, can be clearly diagnosed, mainly in connection with discussions of the Price equation (see Frank 1995: 379, Rice 2004: 166 and 188).

[13] It is necessary to comment on the definition of calculus as a computational technology in the context of the philosophy of language. Hintikka (1997: 5) explains his understanding of 'calculus' as follows: "The analogy between language and a calculus has been used in twentieth-century philosophy to highlight three different things. They are (1) the allegedly purely formal character of language and its laws; (2) the need of doing actual calculus-like manipulations when using language (in the sense of putting it to use, not in the sense of speaking it); (3) the possibility of *re*-interpreting language as freely as interpreting an uninterpreted calculus. [... E]mphasis on (1) is characteristic of believers in the universality of language as of the defenders of the 'language as calculus' conception. It is not what I have in mind here, nor is (2), which is Wittgenstein's reason for comparing language to a calculus. What I am emphasizing in my use of the terminus technicus 'language as calculus' is simply and solely the *re*-interpretability sense (3)." To understand Hintikka's concept of calculus in the sense of a computational technology seems, therefore, to be inadequate. The process of interpreting and re-interpreting is, however, nothing but a semantic form of computation. This paper cannot argue *in extenso* for this hypothesis, but the opposite supposition would lead to a fatal consequence: all questions of whether an entity is a model of a theory, would be undecidable *a priori* by any imaginable formal procedure.

[14] It is, however, not correct to identify the model-theoretical approach to semantics *in toto* with the position that language is a calculus. Alfred Tarski, the father of formal model theory, used it to show that truth is definable for a certain kind of formal language only in a stronger meta-language. From this proof he draw the conclusion that for any such language which has no stronger meta-language, truth is undefinable. Among these languages Tarski counted formalizations of colloquial languages of human beings, such as English or Japanese. Since truth is, in Tarski's model theory, a most important semantic concept, he concluded that the semantics of these languages is not accessible by model theory. Tarski was, thus, a universalist in Hintikka's sense (Hintikka 1997: 12ff.). This is an important example

indicating that the distinction between the conceptions of language as the universal medium and as a calculus is an ideal-typical one (Kusch 1989: 4 and 8f.).

[15] From the model-theoretical perspective, computational research into the evolution of living systems, which began in the 1950s (Fogel 1998), is nothing but the search for the limits of semantic variation in biology. The most radical type of this research belongs to Artificial Life, which "[...] is the study of man-made systems that exhibit behaviors characteristic of natural living systems. It complements the traditional biological sciences concerned with the *analysis* of living organisms by attempting to *synthesize* life-like behaviors within computers and other artificial media. By extending the empirical foundation upon which biology is based *beyond* carbon-chain life that has evolved on Earth, Artificial Life can contribute to theoretical biology by locating *life-as-we-know-it* within the larger picture of *life-as-it-could-be*." (Langton 1989: 1) Langton's modal definition of the object of Artificial Life, life-as-it-could-be, shows the close relation between a model-theoretical view on the semantics of empirical theories and the mathematico-algorithmical exploration of possible worlds – Fisher's above-cited statement is, thus, corroborated (see Artmann 2003).

[16] The empiricistic aspect of Woodger's axiomatization of biology is well documented in his description of how the new biological (more precisely, genetic) object-language should be constructed: "I shall not begin at the top of the existing pyramid of genetical statements and work down; I shall begin with observation records and try to see how the levels of theoretical statements are reached. This process, which I call methodological analysis, thus proceeds in the opposite direction to that followed in axiomatization and is necessary preliminary to it." (Woodger 1952: 95f.)

[17] The most important theory that originated in Woodger's project is the theory of Lindenmayer grammars describing life cycles of organisms (see Lindenmayer 1964).

[18] In cultural semiotics, Umberto Eco has a model-theoretical conception of semantics in mind when he mentions an instrument of comparison between colloquial languages, i.e. a calculus in Hintikka's sense, that is not a colloquial language but can (even if only approximatively) be expressed in any colloquial language so that the latter can act as as a meta-language of all other colloquial languages including itself (see the concluding remarks on translating in Eco 1995).

[19] According to Emmeche et al. (2002: 15), the second thesis of Hoffmeyer's biosemiotic theory is that "[c]odes of living beings are dual." This means that the presence of two types of code – digital on the one hand, analogue on the other hand – is a necessary presupposition of life. The three main characteristics of digital codes (in particular of the genetic code) are resumed by Hoffmeyer as follows (Hoffmeyer 2000, Hoffmeyer 2001). First, the generation of new messages is not constrained by extra-codical requirements (a property that leads, e.g., to recombinations resulting in impossible genetic instructions). Second, digital codes are based on shared rules (see, e.g., the near-universality of the genetic code) and can, therefore, fulfil a memory function (e.g., inheritance). Third, a digital code can refer to itself by constructing meta-messages (that happens, e.g., in regulatory genes). But digital codes are, according to Hoffmeyer, passive: to be biologically relevant, they need analogue codes as semiotic structures of metabolic activity. The latter are based on the biochemical topology of molecular shapes, not on the Boolean algebra of genetic switches. Since, in contrast to digital codes, analogue ones are considered by Hoffmeyer as computationally too demanding to be adequately modelled by mathematics, code-duality is an important case of ontological complementarity in the Copenhagen interpretation (see section 1, n. 6). In contradistinction to the thesis of code-duality, the model-theoretical framework developed in this section presupposes that everything codical is digital, or can be digitalized.

[20] That the concept of code belongs to semantics is crucial to an adequate understanding of information theory and semiotics in biology. Otherwise, the biosemiotician could not differentiate conceptually between, on the one hand, the replication of DNA and the transcription of DNA into messenger ribonucleic acid (mRNA), and, on the other hand, the translation of mRNA into a sequence of amino acids. Whereas the first two processes are directly copying the syntactic information contained in DNA, the latter process is copying the syntactic information contained in mRNA through semantic relations between mRNA codons and amino acids. These relations are mediated by transfer RNAs (tRNAs) acting as adaptors between mRNAs codons and amino acids (Barbieri 2003: 97ff.). In short, replication and transcription is codeless transmission of syntactic information, and translation is transmission of syntactic information via the genetic code. This very important distinction is not drawn by Stegmann's criticism

of the thesis that arbitrariness is a necessary condition for a semantic relation: "[W]hen pairing up during transcription or replication, the nucleotide bases are related in a chemically necessary way to each other. If we accept that semantic information is transferred during replication and transcription, we find that the nucleotide base symbols do not need to be chemically arbitrarily related to transfer this information. One response is to deny that transcription and replication are informational. [...] However, this move would have to explain how mRNAs can contain genetic information when no information has been passed on to them during their transcription. It also would be difficult to see where the genetic information in the offspring's mRNA comes from when no information has been passed on the offspring's DNA during replication." (Stegmann 2004: 218f.). If Stegmann's reasoning were true, the reader of this paper would have to wonder how she can understand what is written here, since printing this book is a non-arbitary process of syntactic replication that leads to a semantic understanding (translation) of arbitrary symbols through the code of the English language.

[21] That Balzer and Dawe choose to axiomatize the decoding function $d\colon C \rightarrow S$ of *GenCo* (see their axiom *AM3* in section 2), is, thus, appropriate.

[22] Contingency is, of course, not a sufficient condition on codes in living systems; the criticism of Stegmann (2004: 219) does not apply to the concept of organic code defined above.

REFERENCES

Andersen, Svend (1989). Sprog og skabelse: Fænomenologisk sprogopfattelse i lyset af analytisk filosofi, med henblik på det religiøse sprog. Copenhagen: Gyldendal

Apel, Karl Otto; et al. (2000). *Un siècle de philosophie 1900–2000*. Paris: Gallimard/Centre Pompidou

Artmann, Stefan (2002). Three types of semiotic indeterminacy in Monod's philosophy of modern biology. In: *Sign Systems Studies* 30(1), 149–161

Artmann, Stefan (2003). Artificial Life as a structural science. In: *Philosophia naturalis* 40(2), 183–205

Artmann, Stefan (2004). Four principles of evolutionary pragmatics in Jacob's philosophy of modern biology. In: *Axiomathes* 14(4), 381–395

Artmann, Stefan (2005). Biosemiotics as a structural science: Between the forms of life and the life of forms. In: *Journal of Biosemiotics* 1(1), 183–209

Balzer, Wolfgang; Dawe, Chris M. (1997). *Models for Genetics*. Frankfurt am Main: Lang

Barbieri, Marcello (2002). Has biosemiotics come of age? In: *Semiotica* 139, 283–295. Reprinted with a new postscript in this volume

Barbieri, Marcello (2003). *The Organic Codes: An Introduction to Semantic Biology*. Cambridge: Cambridge University Press

Beatty, John (1995). The evolutionary contingency thesis. In: Gereon Wolters and James G. Lennox (eds.): *Concepts, Theories, and Rationality in the Biological Sciences*. Pittsburgh: University of Pittsburgh Press, 45–81

Blumenberg, Hans (1986). *Die Lesbarkeit der Welt*. Paperback edition (Original edition 1981). Frankfurt am Main: Suhrkamp

Bohr, Niels (1933). Light and life. In: *Nature* 131: 421–423 and 457–459

Bohr, Niels (1963). Light and life revisited. In: *ICSU Review of World Science* 5: 194–199

Botstein, David; Cherry, J. Michael (1997). Molecular linguistics: Extracting information from gene and protein sequences. In: *Proceedings of the National Academy of Sciences* 94, 5506–5507

Eco, Umberto (1995). *The Quest for a Perfect Language*. Oxford: Blackwell

Eigen, Manfred (1979). Sprache und Lernen auf molekularer Ebene. In: Armin Mohler (ed.): *Der Mensch und seine Sprache*. Berlin: Propyläen, 181–218

Emmeche, Claus (1994). *The Garden in the Machine: The Emergent Science of Artificial Life*. Translated by Steven Sampson. Princeton: Princeton University Press

Emmeche, Claus (1997). Biologi og semiotik: Den biosemiotiske tanke. In: Keld Gall Jørgensen (ed.): *Anvendt semiotik*. Copenhagen: Gyldendal, 62–94

Emmeche, Claus (1998). Defining life as a semiotic phenomenon. In: *Cybernetics and Human Knowing* 5(1), 3–17

Emmeche, Claus (1999). The Sarkar challenge to biosemiotics: Is there any information in a cell? In: *Semiotica* 127(1–4), 273–293

Emmeche, Claus; Kull, Kalevi; Stjernfelt, Frederik (2002). *Reading Hoffmeyer, Rethinking Biology.* Tartu: Tartu University Press (Tartu Semiotics Library Vol. 3)

Engel, Pascal (2000). *L'espace des raisons est-il sans limites?* In: Apel et al. (2000), 231–273

Fisher, Ronald A. (1999). *The Genetical Theory of Natural Selection.* Edited by J.H. Bennett (Original edition 1930). Oxford: Oxford University Press

Fogel, David B. (ed.) (1998). *Evolutionary Computing: The Fossil Record.* New York: IEEE Press

Frank, Steven A. (1995). George Price's contributions to evolutionary genetics. In: *Journal of Theoretical Biology* 175, 373–388

Godfrey-Smith, Peter (2000a). On the theoretical role of 'genetic coding'. In: *Philosophy of Science* 67, 26–44

Godfrey-Smith, Peter (2000b). Information, arbitrariness, and selection: Comments on Maynard Smith. In: *Philosophy of Science* 67, 202–207

Habermas, Jürgen (2000). *Philosophie hérmeneutique et philosophie analytique: Deux variantes complémentaires du tournant linguistique.* In: Apel et al. (2000), 177–230

Hamming, Richard W. (1950). Error detecting and error correcting codes. In: *The Bell System Technical Journal* 26(2), 147–160

Hauser, Marc D.; Chomsky, Noam; Fitch, W. Tecumseh (2002). The faculty of language: What is it, who has it, and how did it evolve? In: *Science* 298, 1569-1579

Hintikka, Jaakko (1997). *Selected Papers.* Vol. 2: *Lingua Universalis vs. Calculus Ratiocinator: An Ultimate Presupposition of Twentieth-Century Philosophy.* Dordrecht/Boston/London: Kluwer

Hintikka, Jaakko (1998). *The Principles of Mathematics Revisited.* Paperback edition (Original edition 1996). Cambridge: Cambridge University Press

Hodges, Wilfrid (1993). *Model Theory.* Cambridge: Cambridge University Press (Encyclopedia of Mathematics and Its Applications Vol. 42)

Hoffmeyer, Jesper (1996). *Signs of Meaning in the Universe.* Translated by Barbara J. Haveland. Bloomington and Indianapolis: Indiana University Press

Hoffmeyer, Jesper (1998). Semiosis and biohistory: A reply. In: *Semiotica* 120(3/4), 455–482

Hoffmeyer, Jesper (2000). Code-duality and the epistemic cut. In: *Annals of the New York Academy of Sciences* 901, 175–186

Hoffmeyer, Jesper (2001). Life and reference. In: *BioSystems* 60, 123–130

Hughes, Randall A.; Ellington, Andrew D. (2005). Mistakes in translation don't translate into termination. In: *Proceedings of the National Academy of Sciences* 102(5), 1273–1274

Jacob, François (2000). *Le jeu des possibles: Essais sur la diversité du vivant.* Paperback edition (Original edition 1981). Paris: Le livre de poche

Küppers, Bernd-Olaf (1990). *Information and the Origin of Life.* Cambrige/MA: MIT Press

Kull, Kalevi (2005). A brief history of biosemiotics. In: *Journal of Biosemiotics* 1(1), 1–25

Kusch, Martin (1989). *Language as Calculus vs. Language as Universal medium: A Study in Husserl, Heidegger and Gadamer.* Dordrecht/Boston/London: Kluwer (Synthese Library Vol. 207)

Langton, Christopher G. (1989). Artificial Life. In: C.G. Langton (ed.): Artificial Life: Proceedings of an Interdisciplinary Workshop on the Synthesis and Simulation of Living Systems Held September 1987 in Los Alamos, New Mexico. Redwood City/CA: Addison-Wesley (Sante Fe Studies in the Sciences of Complexity, Proceedings Vol. 6), 1–47

Levy, Arnon; Jablonka, Eva (2004). Review of Barbieri (2003). In: *Acta Biotheoretica* 52, 65–69

Lindenmayer, Aristid (1964). Life cycles as hierarchical relations. In: John R. Gregg and Francis T. C. Harris (eds.): *Form and Strategy in Science: Studies Dedicated to Joseph Henry Woodger on the Occasion of His Seventieth Birthday.* Dordrecht: Reidel, 416–470

Løgstrup, Knud Ejler (1997). *System og Symbol: Essays.* Second edition. Edited by Viggo Mortensen et al. Copenhagen: Gyldendal

Magalhães, João Carlos Marques; Krause, Décio (2000). Suppes predicate for genetics and natural selection. In: *Journal of Theoretical Biology* 209, 141–153

Markos, Anton (2002). Readers of the Book of Life: Contextualizing Evolutionary and Developmental Biology. Oxford: Oxford University Press

Maynard Smith, John; Szathmáry, Eörs (2000). *The Origins of Life: From the Birth of Life to the Origin of Language*. Paperback edition (Original edition 1999). Oxford: Oxford University Press

Monod, Jacques (1973). *Le hasard et la nécessité: Essai sur la philosophie naturelle de la biologie moderne*. Paperback edition (Original edition 1970). Paris: Seuil

Nirenberg, Marshall (2005). Deciphering the genetic code: A personal account. In: Jan A. Witkowski (ed.): *The Inside Story: DNA to RNA to Protein*. Cold Spring Harbor/NY: CSHL Press, 189–208

Rice, Sean H. (2004). *Evolutionary Theory: Mathematical and Conceptual Foundations*. Sunderland/MA: Sinauer

Roman, Steven (1996). *Introduction to Coding and Information Theory*. New York/Berlin/Heidelberg: Springer

Sipper, Moshe (2002). *Machine Nature: The Coming Age of Bio-Inspired Computing*. New York et al.: McGraw-Hill

Stegmann, Ulrich E. (2004). The arbitrariness of the genetic code. In: *Biology and Philosophy* 19, 205–222

Suppes, Patrick (1999). *Introduction to Logic*. Reprint (Original edition 1957). Mineola/NY: Dover

Suppes, Patrick (2002). *Representation and Invariance of Scientific Structures*. Stanford/CA: CSLI Publications (CSLI Lecture Notes Vol. 130)

Thompson, Thomas M. (1983). *From Error Correcting Codes through Sphere Packings to Simple Groups*. Washington, DC: Mathematical Association of America (Carus Mathematical Monographs Vol. 21)

van Heijenoort, Jean (1967). Logic as calculus and logic as language. In: *Synthese* 17, 324–330. Reprinted in: Hintikka (1997), 233–239

Walker, Judy L. (2000). *Codes and Curves*. Rhode Island: American Mathematical Society (Student Mathematical Library Vol. 7)

Watson, James D.; Baker, Tania A.; Bell, Stephen P.; Gann, Alexander; Levine, Michael; Losick, Richard (2004). *Molecular Biology of the Gene*. Fifth edition. San Francisco and Cold Spring Harbor: Cummings and CSHL Press

Weber, Andreas (2003). *Natur als Bedeutung: Versuch einer semiotischen Theorie des Lebendigen*. Würzburg: Königshausen und Neumann

Williams, George C. (1992). *Natural Selection: Domains, Levels, and Challenges*. New York and Oxford: Oxford University Press

Williams, George C. (1997). *The Pony Fish's Glow and Other Clues to Plan and Purpose in Nature*. New York: Basic Books

Woodger, Joseph H. (1937). *The Axiomatic Method in Biology*. Cambridge: Cambridge University Press

Woodger, Joseph H. (1952). *Biology and Language: An Introduction to the Methodology of the Biological Sciences Including Medicine*. Cambridge: Cambridge University Press

CHAPTER 10

TOWARDS A DARWINIAN BIOSEMIOTICS. LIFE AS MUTUAL UNDERSTANDING

ANTON MARKOŠ, FILIP GRYGAR, KAREL KLEISNER, AND ZDENĚK NEUBAUER

Department of Philosophy and History of Science, Faculty of Sciences, Charles University, Viničná 5, 128 44 Praha 2, Czechia markos@natur.cuni.cz

> *No thing is where the word is lacking, that word which names the given thing. What does "to name" signify? We might answer: to name means something with a name. And what is a name? A designation that provides something with a vocal and written sign, a cipher. And what is a sign? Is it a signal? Or a token? A marker? Or a hint? Or all of these and something else besides? We have become very slovenly and mechanical in our understanding and use of signs.* Heidegger 1982a, 60-1.

Abstract: Contemporary understanding of the evolution of life prefers the existence of mutually isolated lineages which are indifferent to each other, and are interconnected only through their common descent. We argue that a necessary precondition for biosemiotics is communication between all coevals sharing the biosphere. It will embrace mutual custody of existing communication pathways and codes, and of common evolutionary heritage

Keywords: biosphere, semiosphere, coding and reading, Darwinian biosemiotics, life in a scientific perspective, horizontal communication

INTRODUCTION

Understanding, the ability to extract meaning, is a concept inseparable from any version of semiotics or hermeneutics. If we, then, speak of *bio*semiotics, immediately a question arises: *who* is that who understands in this case and how? Of course, the most natural answer is "Living beings, by confronting the context of their lived world with experience and memory gathered from the past (of the individual, of

M. Barbieri (ed.), Introduction to Biosemiotics, 235–255.

the lineage, of the biosphere).” Such an answer would, however, sound very suspicious, even mystical, in the context of contemporary biology which is, these days, all brought and bound in the land of modern science where “laws of physics and chemistry” lie. The basic property of the objects studied by science is *being lifeless*, inert, submitted to rules given from outside. Therefore we should be aware that, by contemplating meaning and understanding, we drift far away from contemporary biology, even if we otherwise anxiously stick to its scientific jargon. Would it be possible to raise a conceptually and lexically different scaffolding, such that would provide us with a more resilient ground for a biosemiotic discourse?

The decisive start to erect such an auxiliary edifice was made by M. Barbieri (2003, 2005) by his layout of semantic biology where matter, information and organic codes unite into a ternary superposition. Acknowledgment of the existence of true codes, i.e. rules given by historical conventions, gives biology a new reliable basis, a platform for putting known things into a new perspective. The novelty of Barbieri’s approach is in acknowledging that information (understood as a nonrandom string of signs) to be utilized, requires also a memory matrix, which is equipped with a set of rules (codes) explaining how to handle the string. Such a matrix (like the ribotype at the level of translation, or the phylotype in ontogeny) resides in bodily structures inherited maternally (i.e. not as pure, virtual information), and its structural transformations are functions of species-specific conventions on how to handle the affairs.

The outfit of his theory allows Barbieri to speak of life as an activity of “artifact making” – a very unusual statement indeed – while still remaining in the safe realm of biology. This is possible because semantics – in contrast to semiotics – is, or can be developed into, an objective, logical science with clearly stated codes, code-translation matrices, and rules (grammar). Semantics can be a matter of the computer world with its hardware and software, or even the world of Jacquard weaving looms with their punched cards. Contraptions of this kind have a meaning imposed from outside, and their doings and outputs will reveal meaning only for the users and the makers (of machines and of programs). Moreover, as coding systems are supposed to be – even should be – complete, a problem arises for ontogeny, how to *reconstruct*, in real (and very short) time, an enormous burden of coding systems implied in, e.g. morphogenesis or learning language (as discussed in general epistemology of G. Bateson, 1980). Here we meet Barbieri’s solutions as a challenge for further elaboration. We argue that above the Barbieri’s platform – that of codes – there exist further levels, e.g. semiotic or hermeneutic ones.

How did the contraption called “living being” come into existence, and who agreed on conventions (sensu Barbieri) allowing its working? The most natural answer “Life itself” would not be welcome within the scope of science. Barbieri explains in his contribution how the living state – up to the level of his platform where it abounds with a plethora of different kinds of codes acting at many levels of description – could have evolved by bottom-up evolution, from simple molecules. Yet we consider the result – agents moving on the platform – to be rather robots than genuine living beings.

We therefore chose an opposite – top-down – approach, sincerely hoping that one day we shall be able to land safely on Barbieri's platform and thus connect both realms of knowledge – of science (i.e. of communication means) and of meaning (of communication). We start with the proposal that meaning, evolution, morphogenesis, imitation, mimicry, pattern recognition, understanding signaling, patterns, symbols from other beings, the ways that lead evolution into new dimensions, creative inventing novelties etc., – are facets and integral part of embodied existence of living beings; beings who *care* about their being, and who maintain uninterrupted corporeal lineages from the very beginnings of life on our planet. They are uniting the extant biosphere into a single, dynamic semiotic space, which is kept together by the mutual interactions and experiences of all its extant inhabitants, fitting and co-fitting (hence 'fitness'), storing memory traces, encoding them into negotiated codes or even in a form that we may regard as digital script. The message of our contribution should read that organic codes are negotiated "from above", from shared language(s). Equivocality, defocusing of all phenomena and all forms of reference to them, allows novelty to precipitate from the field of possibilities. At the same time, the existence of this superposed and commonly shared field allows mutual games of understanding, misunderstanding, cheating and imitation at all levels of the biosphere, e.g. precipitation of the actual version of the fit.

A HISTORICAL EXCURSION

The whole [of god] sees, the whole perceives, the whole hears.
Xenophanes B 24

To set the stage, we shall present here an extremely simplified version of what is called "The Western tradition" of thought, sprouting from ancient Greek civilization. We shall claim that it has always been flanked by two mutually incompatible conceptions of the world. This duality apparently arose in Xenophanes who first undertook an attempt to replace the older mythological worldview with an alternative, which would be rooted firmly in the deity, which is One, Immovable, Rational, and Moral (Kratochvíl 2006). Such a move was able to introduce an order into the world: the deity became the guarantee that the world was governed by eternal and always valid and decipherable laws, based in logic, mathematics, and clear concepts on what was going on. Later, this line of thinking lead to metaphysics, and to science.

In such a layout, the world turned dead, it became a machine blindly "obeying" rules implemented from outside – from the One who is supernatural, i.e. not belonging to the world. History was suspended from such a construction, and all forms of time (mythical, lived, etc.) were reduced to a linear physical dimension enabling the plotting of predictable trajectories of affairs. Only we, humans, are allowed to a certain extent to break ties with this world and also put ourselves "outside", into a god-like position from which we can inspect all minutia of this construed, machine-like, predictable world.

Needless to say, this cozy and transparent worldview was never prevailing in the heterogeneous multicultural and polytheistic Antiquity. It became the reigning view only in the Modern Age, especially when rationalism modified the Xenophantian scheme by expunging Morality and by replacing the deity with impersonal Nature and her "natural laws". Such a framework enabled the founding of natural sciences and the developing of medicine and technologies based on them, thus establishing the metaphysical worldview we live in today.

In spite of many attempts to absolutize this "scientific" picture of the world, it never could fully replace the older thinking in stories, narratives, evolution, fate etc. We (human beings) simply cannot jump out from the legacy of our own history; we are unable to view our condition purely from an outside god-like position; we are always embedded in it. Our past, views, language, interactions are all a matter of never-ending interpretative circles defining the state of our world as it is, constituting its *here and now*, not resting on impartial, universal and supposedly eternal laws. Yet such temporal and singular affairs were regarded as inferior and expelled from "true" genuine sciences, ignored by them. They were exiled into what we call today *humanities*; with a more or less apparent addendum that a day may come when a *consilience* will be reached, such that even these particular, casual and local cultural phenomena and social relations would also find a respectable burial place in the Pantheon of "real", i.e. scientific, knowledge – to find a respectable rest among other objective, alas dead, truths (see, e.g., Markoš 2002).

What was left in an uncertain position was the status of living beings. At the beginning of modern science, the Cartesian concept of *animal-machine* had been successfully developed, raising the hope that biology would establish itself as one of true physical sciences, like physics and chemistry. Such a physicalization of biology culminated in Lamarck and later in the physicalism of the 19th century, and eventually led to contemporary physiology and medicine. Yet the concept was always punctuated by ad hoc assumptions, which had no firm status in science. As Kauffman (2000) shows, the main drawback lies in the fact that, for living beings, no initial and boundary conditions can be stated in advance; consequently, an unequivocal, scientific description of the system is impossible. If no clear beginnings can be defined, you cannot apply the formal analysis required by strong science, and you have to device arbitrary plots – which again will drive to the wide seas of narrativity swarming with fables and free conjectures (which we considered already banished forever – superseded by the rational, scientific approach). Take Lamarck (1994 [1809]) as an example: to do genuine science, he postulated the beginnings of the transformational chains of living beings in *generatio spontanea*. By doing so, he legitimately rejected the curious theory of preformism, then generally accepted. Yet he needed to introduce another kind of germ (*tissue cellulaire* in his terminology), which also had to be already present, to be molded by fluids and physical forces. By doing this, even Lamarck, a behaving follower of deterministic science, fell into the trap of *vitalism*. The trap sprung again and again on many scientists who themselves would vividly rail against the unsavory label of "vitalist" tagged on their names. They too, like in the case of Lamarck, supposed initial *structures and shapes* not required by the general laws or principles of physics (actually, all kinds of programs and codes belong to this category); after doing this, they unconcernedly jumped back to the physics and left their alleged structure to be molded by ordinary, "material" physical forces. Or they introduced some external –often even supernatural – *agency* and left it to affect the ordinary physical world from without. (Note that founders of modern science like Descartes, Galileo, or Newton, took for granted that such a force was emanating from God – later vitalistic attempts can be therefore considered as a kind of surrogate whenever paradoxes arose in the prevailing paradigms.) What all these scientists had in common is that they never raised doubts about the basic presupposition of the Xenophantian-Cartesian scheme. Recall: the world, even living beings as part of it, is "dead", i.e. inert and passive, a kind of a machine blindly "obeying", i.e. executing rules implemented in their very structure, from outside. The world, and life, has no say in the course of events, being but the result of the initial creative force applied to an initial setting of "things", "forces", or "programs". As any modern textbook will tell us, even a crown of creation – the human being – is but a vehicle construed to pass on its genes and memes into the next generation.

Not too many advocates of life can be found in humanities, either. Linguistics, hermeneutics, psychology, history, philosophy, phenomenology and, of course, theology – all became concentrated on our, human, values, silently assuming that the concept of animal-machine for non-human life is fully adequate. Take the now popular movements like biosemiotics or biohermeneutics: a closer examination will usually reveal that what is meant is semiotics (hermeneutics) *of* the living (objects, subjects, humans), instead of semiotics done "*by* the living", i.e. emanating "*from* the living". The vocabulary of many disciplines abounds with *life*, *life world* etc., but they have a negligible impetus on biologists and biology; in reverse, these disciplines are only marginally (or not at all) interested in problems of biology. *Life itself has been expelled unisono by both parties.* We shall return to this problem below, but first let us discuss Darwin and his discovery of evolution.

Recall again: The Xenophantian-Cartesian tradition of science does not allow for singularities brought about by historical time and stories based on the fact that time is historical – based on events, unique chances, irreversible decisions, individual cases and occurrences of "once and never more" type. Whatever *can* exist is already present from the very beginning; often in hidden, but recoverable, calculable, deducible form – in fact in states, aspects, cases of one and the same single category. The whole tradition of this line of thinking does not allow any kind of temporality that differs from monotonous linear succession of ticking; real history in which things do actually happen is excluded. The axioms of such a world are immutable laws and a finite set of entities, which can change *only* in a predictable way, according to decipherable rules, no real novelty can take place! Physics, chemistry, physiology, morphology, and even older theories of evolution[1] were built on the assumption of principles, rules, types, concepts existing behind the incessant milling of phenomena in the sensible world (but available for disclosure).

It was shocking indeed when Darwin cancelled all such axioms by stating the single rule "*Laissez faire, laissez passer, la nature va d'elle même*". Realize that Darwin does not need *any* natural law for his theory! In *Origin of species* he speaks about laws only in the very last paragraph of the book. (Even there it gives an impression of a lip service addressed to some contemporary opponents. It is symptomatic, however, how often this very paragraph is being quoted in textbooks, reviews and theses!).

Not only theologians but almost all sciences rallied against such a heresy represented by the world based on *mere contingency*.[2] Nobody cared to ask *who* was the addressee of the principle: laissez faire, but *whom*? It was life itself that came into play again, more than two millennia after the Xenophantian turn, and it was vindicating its rights in the molding of its world. There are no rules and no goals but those negotiated by critters existing here and now, each bearing the experience of its lineage back to the dawn of life, and laying down the rules for one version of the *adjacent possible* (Kauffman 2000). Living beings do *care* – just allow them to act: laissez les faire!

The novel Darwinian optics thus opened the evolutionary perspective of nature, and the fact could not be denied any more. The decades after Darwin were, therefore, marked by an effort to reconcile the *science* of biology with the accidental nature of life *history*. The solution was ingenious, yet rather disappointing: living beings were once again safely deprived of their autonomy[3] and evolution ceased to be a matter of fact but rather a matter of text – a byproduct of accidents, blunders and misspellings in hereditary prescriptions (genes) for developmental processes affecting survival functions and thus the reproduction rates.[4] We were left with the traditional, unchanged, ideal, objective, normal science with its "laws of physics and chemistry". Biology, nevertheless, still perceives itself as a life *science* with a monopoly and a highest arbiter to all aspects of the living; hence whatever statement about life which is incompatible with the orthodox teaching of biology must be regarded as not true or simply wrong. The idea of acknowledging a genuine autonomy of the critters of this world, of their active participation on making it their home, is definitely not on the agenda of the science prevailing nowadays. It is from the yet marginal biosemiotics – the science of life-signs – whence the "Darwinian reformation of knowledge" may sprout of; having in mind that *The Origin of Species* is *not* a science, is *not* biology in contemporary usage, yet holds a key towards understanding life.

BEING IN THE WORLD

> *Each conscious and surviving individuality — domain has two modes how to become informed: by observation or by participation. It can observe other domains in space, and can also participate in topics, which reach beyond space, and topics from its own past. Because the past (since what is living has never been dead) extends long before the instant of individual birth, it reaches back to the very origins of life and the Universe.*
> Ruyer, 1974, 181.

We would like to introduce living beings as *active* participants in formatting and developing their world; to show that it is them who were, and are, the driving force of evolution. In order to tune the reader into the problem, five parables will be presented, taken from distinguished thinkers who come from very different backgrounds. All layouts will start with a kind of "superposed" states containing the community of extant living beings; *their* version of the world would emerge from such a superposition through the semiotic activity (or games), of its "inhabitants".[5] In order to make such games possible, all players have to understand – in their way – the nature of the others. All must share at least some communication channels as well as some interpretation keys for the messages transmitted. Understanding is possible thanks, first, to the common origin of all lineages (their various degree of kinship and common historical experience), and, second, to an intensive and incessant "horizontal" crosstalk, exchange of messages. Such a mutual understanding – semiotic systems of many sorts – needs continuous attention and tuning; compare with hardwired coding systems, which can be executed even by inert machines indifferent to situation. By "superposition" we mean that no recognizable "wiring", pathway, or script of any kind is decipherable which would reveal the "anatomy" of that state; here solutions are suggested, stored, negotiated, and finally, decisions made, actual states (and stories) precipitate. It is only after habits have been negotiated, rules settled and "artifacts produced", that one can point with the index finger and distinguish "this" and "that", to recognize rules, habits, or even objects. Hence, the whole potential of the living being (or the community) is present in a superposed state, inaccessible from outside; the concrete ways of living being must appear as outcomes for now and here, and only such outcomes will become exposed for the observer(s). This is the moment of safe landing on Barbieri's platform, a place where requirements of science and metaphysics can be met.

A warning is necessary at this point: the systems mentioned below (apart from the last one) were developed as tools to help us to understand *our, human* condition in the world; their presupposition is the presence of language and culture. Our ambition is to corroborate such kind of a world which would contain *all* living beings as its denizens. At this point, however, we are far from daring any re-interpretation of the established reading of those authors; that is why we use the word "parable".

Phenomena

The 20th century phenomenology may be able to give a new impetus to thoughts about life; our first example draws inspiration from M. Heidegger. In his teaching, we human beings (understood as *Dasein*) are always in a state of Being-in-the-world. We are not born as isolated, self-conscious subjects, which afterward clear their way towards the world, things and other human beings. Quite the opposite: we have been thrown into the world in the sense that we *always have been in it*, with things and the Others, we have a fore-understanding of the world and participate, have a concern of its further affairs.[6] The concept of *Dasein* thriving to truth about being, to the supreme event *[Ereignis]* which dwells in the intersection of a *fourfold*, is a beautiful cosmogonic image, which takes us back into the middle of events as *actors and playwrights* of the world affairs. Our being alive, our corporeality, our language will seize us and we seize them.

We shall focus on a special situation of *Dasein* who is always with others; it is in a situation of being-with [Mitsein] and Dasein-with [Mitdasein]: "In clarifying Being-in-the-world we have shown that a bare subject never 'is' proximally, nor is even given. And so in the end an isolated 'I' without Others is just as far from being proximally given." (Heidegger 1962; I., §27, canonic p. 116). Several pages below, we read: "We have shown earlier how the environment which lies closest to us, the public 'environment' already is ready-to-hand and is also a matter of concern [mitbesorgt]. ... This Being-with-one-another dissolves one's own Dasein completely into the kind of Being of 'the Others', in such a way, indeed, that the Others, as distinguishable and explicit, vanish more and more. In this inconspicuousness and unascertainability, the real dictatorship of the 'they' is unfolded. We take pleasure and enjoy ourselves as *they* [man] take pleasure; we read, see, and judge about literature and art as *they* see and judge; likewise we shrink back from the 'great mass' as *they* shrink back; we find 'shocking' what *they* find shocking. The 'they', which is nothing definite, and which all are, though not as the sum, prescribes the kind of Being of everydayness." (ibid, 126–7).

Such, and many similar quotes, should invite anybody who would like to start contemplating about the "superposed" status of living.

There is, however, a great *caveat* in Heidegger's teaching: based on his analysis of being-in-the-world, he claims that a biological attempt for determination of the essence [Wesen] of the Living cannot be successful as, after all, it is us who invented biology. We have no other perspective than in the realm of our own experience, *Dasein*. Understanding is always *ours*, it cannot be based on empirical proofs. Articulation of the ways of Being of the Living is by necessity a mere *human* articulation. It is we who decide what is or is not life; we understand the Living implicitly, before any proposition about it. Before we begin to ask questions about life, we must first understand somehow, what it is to be living; otherwise we cannot ask at all. This pre-understanding is an ontological structure, which anticipates every research.

Despite Heidegger has removed the subject-object dichotomy of the world, unluckily for our argument, by doing so he created a deep gulf between the *Dasein* who is world-forming [weltbildend], and animals (not speaking of other critters) who have a poorer status of "world-impoverished" [weltarm]. We take this starting position as a challenge: might it be possible to develop a similar teaching where what counts for *Dasein* would hold for any living being? We are ready to admit that we *are*, exist differently from, say, an oak tree, a tapeworm or an ox; what we offer is an attempt not to revise Heidegger and put all living beings to a single level, but to uplift the status of life as the entity we share with the others. We shall argue that it is appropriate to broaden the slogan "We the *Dasein*" into "We the living beings", and that it should contribute to a new step in development of sciences: biologization of physics (Markoš & Cvrčková 2002). Example 5 in this section, describing autonomous agents in biospheres, may provide a hint; before that, we continue with another analogy.

Cosmic Reverie

Our second inspiration comes from the *Poetics of reverie* by Gaston Bachelard. Images that arise spontaneously during daydreaming (reverie) do not correspond causally to any hidden archetype or model in the subconscious. They precipitate into the conscious from the "superposed" state, which has been present since before our ability to speak. New images – hypotheses about the world – are born and materialize in the world where they subsequently prove their fitness. A child is born situated in this cosmic reverie inherited and shared by all human beings; only later it is step-by-step lead "out" by its mother and other members of the culture, to construct a meaningful – and unique – image of the world where it can live.

Let us suppose that such an original superposed state of "cosmic dream" may be shared by *all* living beings – due to common origin and billions of years of mutual cohabitation in the world. The state of the biosphere is being continuously decided, or better co-decided, by all actual players of the game. To be a player requires understanding of common rules, common codes, and a lingua franca of a kind. This will bring us again to Heidegger and his parable of the *region* [Gegend] which can be taken as isomorphic to Bachelard's realm of the reverie: "But in thinking, the situation is different from that of scientific representation. In thinking there is neither method nor theme, but rather the region, so called because it gives its realm and free reign [die Gegend gegnet] to what thinking is given to think. Thinking abides in that country, walking the ways of that country. Here the way is part of the country and belongs to it. From the view of the sciences, it is not just difficult but impossible to see this situation. If in what follows we reflect, then, upon the way of thoughtful experience with language, we are not undertaking methodological consideration. We are even not walking in that region, the realm that concerns us." (Heidegger 1982a, 74).

We shall maintain here that similar "languaging" is the state of the whole biosphere-in-evolution. The future is "cleared" in the fitness landscape by *active* effort of all critters living in that landscape. The whole layout is *quite* isomorphic to Kauffman's theory of evolution (1993, 2000): Evolutionary processes as known today, are possible only in a fitness landscape, as it is known today. The question "Whence organisms" should therefore be complemented by "Whence the landscape?" The answer is to be found in co-evolution of both, clearing and cultivating the landscape and at the same time transforming oneself according to the terrain.

Language

Superposition of equally valid and complementary states as known in quantum physics inspired the physicist and theologian Patrick A. Heelan (1998) to propose what he calls "quantum logic". He presupposes: 1. the unformed and inexplicit sphere of speech where, in superposed states, dwell all possible utterances about the world (or its different realms). 2. Any such utterance will bring about a collapse, a projection, of one (or any) explicit and particular statement, a whole set of mutually incompatible, yet true, statements can be uttered in different regimes of language (scientific, poetic, philosophical). Heelan suggests a method where all such statements can be pieced together into a mosaic which, still far from being a reconstruction of the original superposition, will give an idea how they fit into a higher-order picture of reality. He demonstrates his approach on complementary, theory-laden and praxis-laden languages, each representing a "collapse" of superposed manifold of truth about the "object". He takes such collapses as "isomorphic": "By *isomorphism* is meant a one-to-one translatability of any statement in one language into a unique statement in the other language. The two context-dependent languages refer to the same things but from different, often interacting and mutually interfering, perspectives. I have argued that these languages are related among themselves within a lattice structure". (282, ftnt. 25)

This brings us back again to Heidegger and his treatment of language to life. As could be seen above, we take "languaging" as the principal determination of life; we feel free to get inspiration from Heidegger even if he himself would not say such things. The triad "corporeality – life – language" may provide common characteristics of all living beings. But "If we put questions to language, questions about its nature, its being, then clearly language itself must already have been granted to us. Similarly, if we want to inquire into the being of language, then that which is called nature or being must also be granted to us. Inquiry and investigation here and everywhere require the prior grant of whatever it is they approach and pursue with their queries. Every posing of every question takes place within the very grant of what is put in question." (Heidegger 1982a, 71).

If we, living beings, put questions to language, then we should take language as primary, and from these heights allow coding systems to crystallize which are useful for automatic life processes of everyday metabolic functioning. Again, safe and guided (controlled) landing from language to Barbieri's platform of organic codes is a gateway from the realm of living to the realm of contemporary biology. To land smoothly, concepts of semiosphere (Y. Lotman) and biosphere (S. Kauffman) should be of great help.

SEMIOSPHERE

The Russian semiotician Yuri Lotman (1996, 2001) comments on attempts to reduce natural languages (and texts) to a mere code (as in models of information processing provided by molecular biologists), i.e. reducing natural phenomena to natural laws as defined in natural sciences: whatever is beyond the code is ignored. Such an approach, says Lotman, assumes that the user of language is interested only in receiving the relevant messages, by specific selection out of the background noise. All other aspects of the text, its multiple and variable relation to the context, is being ignored. The recipient is "hardwired", and the text plays the role of a mere carrier of transmitted messages; the single goal of a semiotic process is, then, an adequate transmission of the message. It is taken for granted that the meaning of the text remains invariant with regard to the transformations of the text itself. Upon this assumption rests most of the reasoning concerning the relationship between text and meaning.

But, argues Lotman, natural languages are very poorly equipped to fulfill such a role. It would assume that the sender and the receiver of a message have an identical table of codes; such an identity of codes can be achieved only in special cases, to serve very special, narrow purposes, at the price of the language, which is no longer natural. "For a total guarantee of adequacy between the transmitted and received message there has to be an artificial (simplified) language and artificially simplified communicators: these will have a strictly limited memory capacity and all cultural baggage will be removed from the semiotic personality. The mechanisms created in this way will be able to serve only a limited amount of semiotic functions; the universalism inherent to natural language is in principle alien to it." (Lotman 2001, 13)

Thus, artificial languages model not language as such, but only one of its, rather marginal, functions: namely, the ability to transmit a message as formal information. If language happens to be deprived of its additional, and essential, functions, after some time such functions may even become forgotten: language would turn into a sort of algebra and its function would scarcely differ from a mechanical cause–effect relationship. The creative gist of language, however, is the most important factor that would be swept aside by such a process: the text works not only as a transmitter of messages, but also as a generator of new ones; in contrast, neither unambiguous transmissions nor mathematical solutions can be viewed as *new* messages.

The language precedes messages transmitted, as an integral part of them (see also Heidegger referred to above). Although transmitted digital messages can be quantified objectively in the machine language, nothing like this is possible in a natural language. It's because new meaning *can* originate in the process of reading a written text. Hence, when reading a written message, says Lotman, *it is always on the decision of the receiver of the message whether he/she handles the received text as a code, or as a message.* This double function of the text enables even petrified truths of religious, cultural, or scientific communities to escape canonical (i.e. coded) interpretation and allow emergence of novelty. Such truths may breach the narrow hold of previous clichés and start again to circulate in broader contexts. If the text is active in a culture, it will ceaselessly pick up new meanings. This happens again and again in cultural evolution; yet is the biological evolution different in this respect?

Culture is something like a collective personality with a collective memory, mind, and history. Lotman named this entangled web *semiosphere*, a system integrated across all levels of its organization. All participants of communication enter the game with a certain background of experience and memory. Living beings are participating parts, active creators/builders of their own world, they are not merely thoroughly tuned into some given *umwelt*! As we shall see below, Kauffmanian autonomous agents are in a similar situation. Semiosphere is indeed a generator of *new* knowledge (see again below in Kauffman).

The game has an antipode in law. Lotman states that if a goal is given in advance, there is no room for liberty: if the trajectory of a thrown stone can be predicted to the tiniest detail, i.e. if nothing unpredictable can happen during its flight, then there would be no need to throw it. If this holds, then history and time flow would be superfluous, God would not play dice indeed. He would be merely watching readymade videotape – and not even that, since He would be able to see it all at once! A boring spectacle for an annoyed God, indeed! But in a culture, the less expected, the more unusual a phenomenon or an event is, the greater impact it may have. The same obviously holds in all areas of human activity, including science. The Darwinian scenario extended this state of affairs to the history of life and consequently to the whole of nature. This is why concepts borrowed from linguistics like interpretation, translation, evolution etc., nowadays pervade all so-called sciences. This may indicate the end of belief in timeless laws. Moreover, traditional opposites like culture–nature or evolution–history tend to blend.

Biosphere: Expansion into Adjacent Possible

Stuart Kauffman, mathematician and biologist, has an experience with mathematical models, as well as their bodily"incarnations". For him, an ideal mathematical map turns to a mutable and living landscape, when eternal timeless "physical laws" give room to evolution in time, that is to *physis*. In the preamble to his book *Origins of Order* (1993) we read: "Simple and complex systems can exhibit powerful self-organization. [. . .] Yet no body of thought incorporates self-organization into the weave of evolutionary theory. No research program has sought to determine the implications of adaptive processes that mold systems with their own inherent order" (p. vii).

The last sentence could stay as the epigraph for Kauffman's lifelong search for where order comes from in nature and in living beings. He does not take the neodarwinian explanation, rooted in frozen accidents sieved by natural selection and shared in genealogical lines. In such a scenario, organisms "play" a role of passive, ad hoc contraptions; outcomes of historical contingency, and their ontogeny being determined by "blind" genetic programs. Evolution comes out as an opportunistic process, with no room given to spontaneous emergence of order, assisted by active participation of life itself. Kauffman, on the other hand, has an ambition to prove that order is here not *because* of natural selection, but *in spite* of it. The greater the complexity of the system, the less power selection has to change its properties; order emerges not by a random walk but as a result of a system's internal dynamics.

Even more advanced in this respect is another treatise by Kauffman, *Investigations* (2000). Here, he develops further his core idea that the properties of a system cannot be stated in advance, as a kind of finite list. It follows that deterministic laws of physics allowing calculation of the behavior of a system (its configuration space) are not general, but *special* cases: they are autonomous, setting their own rules of behavior – their unique nature. They can be used for describing naturally evolving systems, only if we can state initial and boundary conditions for the system. Newtonian or Einsteinian physics cannot thus be applied in systems with evolution, where this condition cannot be fulfilled. It was demonstrated that general laws for such systems couldn't be stated at all; Kauffman, however, asks whether they couldn't be found at least for a special class of systems – autonomous agents.

The definition of an autonomous agent as a system acting on its own behalf is as follows: "All free-living cells and organisms are autonomous agents. But a bacterium is 'just' a physical system. In its Kantian form, my core question became, what must a physical system be, such that it can act on its own behalf? The stunning fact is that autonomous agents do, every day, reach out and manipulate the universe on their own behalf. Yet that truth is nowhere in contemporary physics, chemistry or even biology. So, what must a physical system be to be an autonomous agent?" (Kauffman 2000, x).

It must embody two features, says Kauffman: autoreproduction and ability to perform a working cycle(s). The last condition is crucial and distinguishes an autonomous agent from the dissipative systems described by Prigogine, such as a flame or a tornado. To perform work in a cycle means to have a contraption – a machine, which is able to return periodically to the initial state. Thus, cyclical processes lie at the heart of the acyclical, historical process of evolution.

To perform work, the autonomous agent must be able to build a machine to lower the degrees of freedom available for the dissipation of energy. Making a machine, however, requires the investment of work. The agents are thus characterized by a cycle (or spiral) of work, and the work extracted may be utilized to reproduce the system or to increase its organization (e.g. by building new machines allowing new kinds of work cycles). It can be used also for mapping the surrounding universe in an active search for resources utilizable to perform work. The author thus leads us towards a kind of hermeneutical circle in nature.

This aspect will become even more accentuated when it comes to communities of autonomous agents – biospheres. By expanding from the actual state into the adjacent possible (defined as a state one time-interval from the actual, the time interval being defined deliberately) the biosphere explores the field of possibilities and accomplishes, or decides on, one of them. The two states may differ in number and/or quality of particles (creating new ones never seen before in the universe) and in new, unpredictable structures. Due to this uncertainty, it is not possible to predict the evolution of a biosphere, even in a single time interval separating the present from the adjacent possible.

Biospheres are thus characterized by a ceaseless flow from the actual into the adjacent possible, *en passant* increasing their organization: "Biospheres, as a secular trend, that is, over the long term, become as diverse as possible, literally expanding the diversity of what can happen next. In other words, biospheres expand their own dimensionality as rapidly, on average, as they can."[7] (Kauffman 2000, xi).

How, then, do biospheres construct themselves? Autonomous agents are ceaselessly measuring selected parameters of the surrounding universe (which is a co-construct of the whole biosphere, where they share their knowledge), detecting the resources utilizable to perform work and canalize it via machines built for this purpose. This means that they acquire useful, relevant knowledge – not just information of *any* kind! To look for the right kind of knowledge, they fish the useful (or at least promising, hopeful) sort of data out of "garbage", this requires interpreting the signs of the surrounding world. We are already amidst semiotic problems: how does Kauffman's "agent" come to know how to build an appropriate machine able to canalize that very type of the energy gradient? The universe offers a continuum of possibilities of qualities that can be distinguished from the background in a certain way and "measured". Only some of them, however, are *relevant* – leading to the recognition of an utilizable energy source that can be coupled to extraction of work or to relevant information. The agent, as well as the community of agents, actively breaks symmetries, looks for and discovers new ways of energy canalization (and puts at stakes its own integrity, its existence), extracts meaning and constructs the adjacent possible. By definition autonomous agents are *endowed with endogenous activity*, they are by no means passive substrates molded by external forces. The co-evolution of autonomous agents then drives them into the adjacent possible along a trajectory, which is non-deterministic, but determining, i.e. selective. By doing so, they create a larger space of possibilities. The definition of the autonomous agent is at the same time the very definition of life, says Kauffman. We – autonomous agents – are co-constructors of our universe.

But how do (can) we perceive our "autonomous activity"? Kauffman provides an answer: "Story is the natural way how we autonomous agents talk about our raw getting on with it, mucking through, making a living. If story is not the stuff of science yet is about how we get on with making our ever-changing living, then science, not story, must change. Our making our ever-changing livings is part of the unfolding of the physical universe." (p. 119).

Storytelling is the most adequate, maybe the only way to store experience. Problems, situations, tasks never repeat themselves exactly the same way. But problems encountered in the past may be of enormous help when one is confronted with a similar situation again. Not because of what is constant, invariant, equivalence, but because of similarities, analogy, correspondence in dealing with novelties – in how to mutually respond (corespond!) to new challenges. One must first be "versed" to be able to converse, with the changing rules of the game. Such experience in versatility cannot be provided (or represented) by static data. It is the 'tune' – the course of the change that makes one tuned to the changing world according to its past trajectory modifications – both gradual and sudden. Thinking in terms of stories seems to be a type of "information processing", which became most effective in evolution. The bounty of life around us represents players of winning strategies in natural games. Darwin is to be read this way!

CORPOREALITY, LIFE, LANGUAGE

> *To speak to one another means: to say something, show something to one another, and to entrust to one another mutually to what is shown. To speak with one another means: to tell something jointly, to show to one another what that which is claimed in the speaking says in the speaking, and what it, of itself, brings to light.* Heidegger 1982b, 122.

Contemporary biology is focused especially on phylogenetical (vertical) aspects of evolution. Modern taxonomy will make sense only in a context of mutually insulated genealogical lineages. Only in the realm of such a paradigm it holds, e.g., that snakes are tetrapods. Such a context also allows the view that handing over the genes to the next generation is the principal "purpose" of living beings. Common ancestry is today the main (because according to contemporary explanatory frames most parsimonious) explanation of biodiversity and taxonomical relationships. It does not mean that pre-Darwinian biologists dwelled in chaos – they only used different criteria for classification. This example illustrates the difficulty connected with choosing appropriate criteria of classification and of explaining phenomena of living nature. In the scope of such an interpretation, the external appearance of living beings is being predominantly explained as a result of advantages emerging from morphological and/or functional adaptation to external conditions, past or present. Here we, however, suggest to consider imitation also to be the result of interspecific information flow. Below, three examples will be outlined and discussed.

The Bacterial Genetic "Internet"

The genetic code, i.e. the correspondence between nucleotide triplets in mRNA, and amino acids entering the polypeptide chain of proteins, is universal throughout the whole biosphere, save minor exceptions. Because of its universality, it is supposed it represent a frozen accident which took place at the beginning of life history, i.e. in the last universal ancestor of contemporary life (for more details, see Barbieri, 2003, 2005). This conjecture is rooted in an a priori view of the present life forms as endpoints of long isolated lineages – branches of the evolutionary tree: were it not for the frozenness and immutability of the code, there would be no obstacle, for each lineage, to develop its own coding tables.

But why not turn this argument upside down? Couldn't it be that the code remained universal just because all living beings were busy to keep it such, to be able to profit from the horizontal exchange of genetic information, and of being "logged" into the planetary network of information exchange? The horizontal transfer of segments of DNA in a prokaryotic biosphere is being now universally accepted, and many cases abound also for the eukaryotes (see, e.g., Thomas 2000, ed., Bushman 2002). Thanks to such a transfer of genetic information, living beings are able to acquire, e.g., capacity to metabolize unusual substrates, resistance against antibiotics, immunity towards bacterial pathogens, pathogenicity towards hosts, etc. (Amábile-Cuevas, ed., 1996). For such messages to be usefully understood and applied, there must exist a strong pressure towards keeping the code unchanged. In close multi-species prokaryotic consortia, mutual understanding of genetic messages is a condition of survival. Even if eukaryotic organisms rely much more on the vertical transfer of information in lineages (species or populations), the universality of the genetic code suggests that even here the necessity of occasional horizontal communication may be of utmost importance.

Modules of Signal Transmission

A communicative game takes place between cells in a body as between individuals in the biosphere. The cells in a body are permanently interconnected by signaling networks like a nervous system, many levels of humoral regulation, morphogens, immune system, etc. It is startling that, again, that signaling modules enabling signal transmission are of a universal nature throughout the biosphere (with a gulf, albeit not insurmountable, between prokaryotes and eukaryotes). Such universality suggests again the presence of intricate interconnectedness between "nodes" like individuals, species, or ecosystems. Symbioses (from pathogenicity up to very intimate connections) exist between organisms, which seemingly have developed independently for hundreds of millions of years (e.g. Carroll et al. 2001). There is no way to explain such an ability to understand a part of something else's network except by ascertaining that, in evolution, often enough possibilities of network interconnections occur, sufficing for all participants of the game to maintain at least a partial inter-penetrability of their internal signaling modules. The same holds for the network of signals like pheromones, odors, or behavioral cues, representing communication between individuals not even of the same kin, but also across phylogenetically distant lineages. Thanks to universal devices, different networks can be interconnected and enter an endless game of informing, misconceptions, broken communication, tapping somebody else's communication lines, cheating, etc.; all this leading to relationships like commensalism, symbiosis, parasitism, etc.

The informational modules provide an expansion of R. Dawkins (1982) concept of extended phenotype, stating that there cannot be anything like a total and final set of gene manifestations. Phenotypic traits are manifold even within a single organism, but they extend even beyond the individual, and practically they can influence anything in the biosphere. The argument holds even if turned upside down: the crosstalk across the biosphere is possible thanks to the existence of a planetary information network which is, to some extent, accessible to all participants of the game: to communicate requires knowing one's partner. Biology is good in reifying means, channels through which this "empathic" current flows, but how to explain the very phenomenon of understanding is the task of semiotics.

Body Plans

In contemporary biology, the external appearance of living beings is being predominantly explained as a result of the advantages emerging from morphological and/or functional adaptation to external conditions present or past. An organ is a functionally specialized part of an organism, and so is the external appearance of whole organisms. Such views, when introduced by Darwinists, rightly upset the older school of idealistic morphology believing in rules of forms with reference to the unity of plan, which become embodied in a particular species and individual. Taking into account the universality of molecular and supramolecular languages ("coding systems"), one may ask whether both idealistic morphology and phylogenetics did not miss another crucial point which we will name *feeling for shape* or, more generally, feeling for being-in-shape, i.e., for being fit (hence fitness – of self and of the others). This will encompass discerning for bodily expressions not merely as signals for pattern-recognition, but as signs for the significance they convey. The ability to communicate or the inborn proneness to establish means for communication (endow signs with meaning) arises quite spontaneously from the common living-together in, and building-together of, a world, a biosphere. Living together turns eventually to a living community which, when grown mature and stable, becomes a culture and we feel justified to study true cultural phenomena. Such naturally formed societies provide models for *natural social studies.*

Bodily shapes seen as social phenomena are not only signs of living community and semantic means for communication but also signs of the common past and means of historical communication with the ancestors – bearers of an evolutionary experience – that is, signs of a past in principle still accessible to understanding and for use. Thus it may not be as far-fetched as it sounds that when predecessors of whales entered the ocean, they did not develop all their necessary adaptations anew: they were able to dig from their phylogenetic memory, age-old experience accumulated already by fish-like ancestors of tetrapods. We are not going to such extremes, but will point out ordinary, yet hardly explainable phenomena from our biosphere.

Animate nature shows variously interconnected networks of sense based on the expression of lineage-specific characters, which provide an opportunity for comparative work resulting, e.g., in taxonomy. On the other hand, such a similarity enables also a recognition of striking phenomena, such as mimicry. The first approach resides in searching for homologies, the second reveals superficial analogies in the organic world. During the last century, homologies were strongly preferred because the quest for tree-like genealogy of life has been regarded as a main goal of biology. Taxonomy, however, cannot explain the evolution of form (particularly within the limits imposed by a pure mutation-selection model). The taxonomists are interested in kinship of living organisms – focusing on phylogenetic relationships only; therefore the material for phylogenetic analyses represent predominantly specially chosen pieces of DNA. Morphological data become increasingly scarce in the effort of inferring phylogenies. The enigma of the frequent preservation of design *in spite of* natural selection on one hand, and the emergence of formal resemblances at within unallied lineages, e.g. mimetism on the other; in other words, the phenomena of homoplasy,[8] apparently evokes negligible interest. For taxonomist, homoplasy is a nuisance – a negative informational noise blurring the desired, clear-cut phylogenetic analysis. For those who are engaged in the research of form, it represents an opportunity to examine the persistence of morphogenetic systems and ontogenetic pathways in the course of evolution. Such latent morphogenetic modules (or latent homologies) remain incorporated in bodies of organisms waiting to be expressed, to emerge unexpectedly in unrelated lineages. From the biosemiotic perspective, homoplasy is not newly derived: rather it is a result of the persistence of a morphogenetic system, which became re-awakened or re-invented in an unusual context. Hence, if organisms need to evolve a new adaptive structure, they may activate remote morphogenetic systems and under the actual circumstances an unexpected shape will appear. From the biosemiotical point of view such "realization" refers both to understanding (becoming aware of a "knowledge" – of the existence of an engram), recognition of its significance (acknowledgement of a mute sign) and its bodily interpretation (the know-how of the developmental process).

A classical example of hesitation when confronted with semblances in unrelated lineages is mimicry. Prevailing interpretations are often counterintuitive and implausible, except, perhaps, for aposematism or crypsis. But there exist many other examples of mimicry, which cannot be explained that easily (for details and references, see Komárek 2003). For example *Batesian mimicry* denotes the imitation of shapes belonging to individuals of some well-protected species, by individuals of some species unprotected (e.g. a butterfly looking like a wasp). The unprotected species (the mimic) thus performs a kind of "semetic parasitism" by taking on a semblance of the protected species (the model). This way, the semes of the model contribute to the spread of memes if in return, the mime will pay for the service by spreading the semes of the model to its progeny; we call such a relationship between two species "a genetic-semetic reciprocal ring". In semetic jargon: "If you want to take advantage of my protection, please accept and dissipate my semes." In *Müllerian mimicry* two or more protected species mutually imitate each other forming a concatenated ring of semetic relationships. Finally, in *Wasmannian mimicry*, we are confronted with the semblance of termitophylic or myrmecophylic inquilines to their termite or ant hosts. Our explanation of the phenomenon is semetic again: whereas some non-semetic inquilines pay for their stay in the hill by serving the hosts some food or tidbit, some of them enter the semetic ring and pay by distributing the semes of their hosts.

Many counterparts of animal semetic relationships can, of course, be found in human societies: for example, the termite – inquiline[9] relation is similar to that between humans and their domestic animals or even plants. Our semes tend to realize mental images by projecting them to the nonhuman partners of man; imposing human shape to other beings with different bodily layout is but a special case; the mimiking activity may originate from the domesticant.

Similar semetic transfers occur also in the parent-child relationship. The upbringing of a child takes a lot of energy from the side of the parents; hence the child is in a position of a "domesticant" of a kind, and is invited or even "forced" to pay back to fulfill visions of its parents.

We consider imitation a result of interspecific "informing" based on a very similar feeling for shape in critters sharing the world since time immemorial. It could be compared to fashion, a phenomenon of collective imitation (or refusal) of practically anything (for more details, see Kleisner & Markoš, 2005). Fashion, as we know, is a luxury phenomenon applied mostly to external appearance (clothes, jewelry, cars, housing). It has no rule (except perhaps that of consequent rejecting the fashion of the immediate past) and it cannot be imposed by authority. Fashion is rooted in imitation, and its prerequisite is the ability to understand others, of having an internal feeling of how one can be seen by others, and a freedom to decide to look – to a certain extent of course – like them. In non-human beings, psychobiology, ethology and memetics brought evidence of how imitation works in young, in primates, carnivores, or domestic animals – hence mainly birds and mammals. We would like to be more general here, by taking interconnectedness of life in the biosphere for granted.

Our thesis is that imitation is a common phenomenon, interconnecting living beings "horizontally". As in the case of fashion, imitation is a matter of free decision – it is imposed from inside, it is a matter of selected superficial traits and of unique casuistic. For example, in the environment of the anthill some of the symbionts will understand the message and will take on *some* superficial traits typical for ants. Such mimicry supposes – like in the case of fashion – a reflected dictate.

We suggest a notion of the *semetic ring* (see also Kleisner & Markoš 2005), to name a shared, but non-kinship semblance (e.g., mimicry), which we shall take as a case of horizontal communication between organisms. With Portmann (1960) we shall suppose that self-expression (*Selbstdarstellung*) of living beings belongs to their shared experience of being alive and that it presupposes the existence of a recipient of the message who will perceive it in a competent manner. Hence, their external appearance or even distinctive characteristics (*eigentliche Erscheinungen* in Portmann) represents a value, a raison d'être in itself. Organisms may even strive to dissipate their own semblance (*image*) throughout the world (literally like media stars do). In addition to reproduction and gene amplification, also semes (as imaginary units of semblance) influence the evolutionary process. Instead of self-reproduction for gene transmission, semetic self-realization through self-dissipation (self-transcendence) may be regarded as the true/real primemover/propeller of evolution. Accepting such a proposition would explain a broad range of similarities in living nature, because semblance is not genetically bound to the bodies which invented the original shape. Again, we move in a biosphere network of mutual understanding, as in the case of gene and signal promiscuity discussed above.

CONCLUSION

Darwinian evolution of the fittest can be understood as a semetic process depending on mutual understanding in the biosphere, by setting future layouts, by interpreting the actual situation and by the scrutiny of historical experience of all beings present at the given time. Such a semetic process presumes mutual understanding by the participants, and a continuity of communication across many levels of organization. Only part of such a crosstalk is not semetic, i.e. has been assigned to automatisms like metabolic pathways, feedback circuits, or coded symbolical interactions. The semetic processes, however, rule over the whole biosphere. They represent a real analogy of a natural language, and it is *this* level of communication, which is decisive for the proceedings of natural history.

ACKNOWLEDGEMENTS

This research was supported by the Research Program CTS MSM 0021620845, and the Program Global Perspectives in Science and Spirituality 10012 (coordinated by Université Interdisciplinaire de Paris and The University of Elon, with funding from The John Templeton Foundation).

NOTES

[1] Evolution *sensu* Lamarck, or in a sense of deterministic development like evolution of a disease,"where each stage is known, or can be known, in advance.

[2] The myth of "science against religion" is of a later date, when science found methods how to tame Darwin's message and enchain it in this or that form of Darwinism. For a plastic narrative from the position "50 years after", see Rádl 1905, 1913 (the English, abridged version 1930).

[3] With a single, and, as everybody believed, temporary concession in case of sexual selection.

[4] It resembles the fate of the second of the two historical events that were ever allowed into science: the cosmology of Big Bang. Yes, it *is* a history, but all the rules were negotiated (settled) in the first 10 to minus zillionth of second – after that, the rest of universe's time passed happily in the good old realm of normal science with its deistic background.

[5] Note that we shall leave out seemingly similar concepts of various kinds of biospheres and noospheres where living beings are not agents of the game but mere passive pawns manipulated by forces or goals external to them.

[6] It follows, that rationalization and objectification of the world is secondary and requires a special effort.

[7] This quotation holds as a definition of the "4th thermodynamic law".

[8] Homoplasy indicates derived similarities which cannot be explained from shared common ancestry.

[9] Inquilines are inhabitants of ant or termite hills.

REFERENCES

Amábile-Cuevas, C.F. (ed.): *Antibiotic Resistance: From Molecular Basics to Therapeutic Options.* Chapman & Hall, London 1996.

Bachelard, G.: *The poetics of reverie. Childhood, language, and the cosmos.* Boston: Beacon Press 1971.

Barbieri, M.: *The organic codes. An introduction to semantic biology.* Cambridge: Cambridge University Press 2003.

Barbieri, M.: Life is "artifact-making". *J. Biosemiotics* 1, 81–101, 2005.

Bateson, G.: *Mind and Nature-A Necessary Unity.* Toronto: Bantam Books 1980.

Bushman, F.: *Lateral DNA transfer. Mechanisms and consequences.* Cold Spring Harbor Laboratory Press 2002.

Carroll, S. B., Grenier, J.K., & Weatherbee, S. D.: *From DNA to diversity: Molecular genetics and evolution of animal design.* Malden, Mass: Blackwell 2001.

Dawkis, R.: *The extended phenotype.* Freeman 1982.

Heelan, P.A.: The Scope of Hermeneutics in Natural Science. *Stud. Hist. Phil. Sci.* 29, 273–298, 1998.

Heidegger, M.: *Being and time.* London: SCM Press 1962.

Heidegger, M: The nature of language. In: *On the way to language.* San Francisco: Harper 1982a, 57–110.
Heidegger, M: The way to language. In: *On the way to language.* San Francisco: Harper 1982b, 111–138.
Heisenberg, W.: *Physics and philosophy. The revolution in modern science.* St. Leonards, Australia: Allen & Unwin 1959.
Kauffman, S.A.: *Investigations.* Oxford: Oxford UP 2000.
Kauffman, S.A.: *Origins of order.* Oxford: Oxford UP 1993.
Kleisner, K., Markoš, A.: Semetic rings: towards a new concept of mimetic resemblances. *Theory Biosciences* 123, 209–222, 2005.
Komárek, S.: *Mimicry, aposemantism and related phenomena. Mimetism in nature and history of its study.* München: LINCOM 2003.
Kratochvíl, Z.: *The philosophy of living nature.* In preparation 2006.
Lamarck, J. B.: *Philosophie zoologique,* Gallimard 1994.
Lotman, Y.M.:*Universe of the mind. A semiotic theory of culture.* London: Tauris. 2001 [1990].
Markoš, A., Cvrčková, F.: Back to the science of life. *Sign System studies* 30, 127–149, 2002.
Markoš, A.: *Readers of the book of life.* Oxford: Oxford UP 2002.
Portmann, A.: Animal forms and patterns, New York: Schocken Books 1967.
Ruyer, R.: *La gnose de Princeton* [The Princeton gnosis], Paris: Fayard 1974).
Thomas, C. M., ed: *The horizontal gene pool. Bacterial plasmids and gene spread.* Harwood Academic Publishers 2000.

CHAPTER 11

FROM THE LOGIC OF SCIENCE TO THE LOGIC OF THE LIVING

The relevance of Charles Peirce to biosemiotics

TOMMI VEHKAVAARA

Department of Mathematics, Statistics, and Philosophy, FIN-33014 University of Tampere, Finland

Abstract: Biosemiotics belongs to a class of approaches that provide *mental models of life* since it applies some semiotic concepts in the explanation of natural phenomena. Such approaches are typically open to anthropomorphic errors. Usually, the main source of such errors is the excessive vagueness of the semiotic concepts used. If the goal of biosemiotics is to be accepted as a science and not as *a priori* metaphysics, it needs both an appropriate source of the semiotic concepts and a reliable method of adjusting them for biosemiotic use. Charles S. Peirce's philosophy offers a plausible candidate for both these needs. Biosemioticians have adopted not only Peirce's semiotic concepts but also a number of metaphysical ones. It is shown that the application of Peirce's basic semiotic conceptions of sign and sign-process (*semiosis*) at the substantial level of biosemiotics requires the acceptance of certain metaphysical conceptions, i.e. *Tychism* and *Synechism*. Peirce's method of pragmaticism is of great relevance to biosemiotics: 1. Independently of whether Peirce's concepts are used or even applicable at the substantial level of biosemiotics, Peirce's method remains valuable in *making* biosemiotics and especially in adjusting its basic concepts. 2. If Peircean semeiotic or metaphysics is applied at the substantial level of biosemiotics, pragmaticism is valuable in clarifying the meaning and reference of the applied Peircean concepts. As a consequence, some restrictions for the application of Peirce in biosemiotics are considered and the distinction of Peirce's philosophy from the 19th century idealistic *Naturphilosophie* is emphasized

Keywords: Biosemiotics, method, concepts, Peirce, semeiotic, metaphysics, pragmaticism

1. BIOSEMIOTICS AND PEIRCE

1.1. Biosemiotics as a Mental Model of Life

The word 'biosemiotics', being a compound of 'bio' and 'semiotics', refers literally to the union of the studies of (biological) life and signs. Because semiotics is

M. Barbieri (ed.), Introduction to Biosemiotics, 257–282.

understood as a science, study, or doctrine of signs, biosemiotics is often characterized as a 'science of signs in living systems' (e.g. Kull 1999: 386). Semiotic concepts are commonly used, depending on the semiotic tradition, to refer to epistemological, linguistic, psychological, social, or cultural phenomena, i.e. usually to some specifically *human* phenomena. In the tradition of biosemiotics, these concepts (or their modifications) are nevertheless used in reference to non-human or not specifically human living phenomena too. Biological life is seen therefore to be analogous to mental life or to human sociality, notwithstanding the fact that the human mind and sociality are essential parts of the biological life of the human species.

The recognition of an analogy between mind and living nature has produced two kinds of approaches or research strategies, both risky in their own peculiar way. The *naturalized models of mind* focuses on mind and tries to naturalize it. This includes evolutionary psychology, sociobiology, 'neurophilosophy' (Churchland 1989), and a form of evolutionary epistemology which studies scientific progress (EET).[1] They tend to commit *naturalistic fallacies* by using too economical or restrictive explanatory principles resulting in a too simple and distorted picture of the complexity of mental phenomena. The primary problem is not to do with the simplistic character of the models in themselves but with its origins: that this simplicity follows from the insufficient methods behind its construction —or from some *a priori* decided physicalistic principles (cf. Barbieri: Editorial, this volume)— and *not* from the studied reality itself.

While naturalistic models of mind pursue often a somewhat reductionistic strategy, the other kinds of approaches, the *mental models of life* —to which biosemiotics belongs—pursue typically a holistic strategy. They focus on natural phenomena and try to model them on concepts that originally referred only to the human mental or social sphere. Consequently, they fall easily into *anthropomorphic fallacies* by predicating properties or qualities exclusive to humans to non-human natural phenomena. The outcomes of such fallacies are either simply *false* descriptions or, (more commonly) so utterly vague sketchings that it is extremely difficult or impossible to judge their validity and other than their moral, religious, ideological, or emotional significance.

Anthropomorphisms *per se* are not avoidable — not even in the extreme naturalistic or physicalistic studies. *All* our concepts, even the ones of mainstream physics, can be argued to have their origin in anthropomorphic metaphors or analogies, many of them ultimately rooted in the aspects of our bodily self-experience (cf. Lakoff & Johnson 1980). Metaphorical *origins* of scientific concepts are not problematic, but when these concepts are *abstracted* and *redefined* for scientific purposes, anthropomorphic errors may arise. The chief problem is how to identify and recognize these anthropomorphic errors, i.e. the illegitimate uses of such redefined concepts that are insufficiently, incompletely, or erroneously *abstracted.*

Besides biosemiotics, also an evolutionary epistemology which studies cognitive mechanisms (EEM) and the *Naturphilosophie* of the 19th century German idealism provide mental models of life. In all of these fields, some kind of continuity (even if only in the form of gradual steps) between non-human biological life and human mental (i.e. logical, psychical, social, or 'spiritual') life is assumed. The forms of mind and sign processes that one can find in biological life are often assumed to be somewhat more primitive, simple, or general than that found in our own minds. However, these three approaches differ from each other in other respects. For example, while Schelling's *Naturphilosophie* (Schelling 1984, orig. 1804) appeals to transcendental arguments *a priori* in its reasoning, the evolutionary epistemologies of Donald T. Campbell (1974, 1997) and Konrad Lorenz (1973) aim to *naturalize* the concepts of knowledge and knowing when generalizing and abstracting them and extending their domain of reference into the animal world and even further. The basic explanatory scheme in evolutionary epistemology is the (Neo)-Darwinian conception of natural selection.[2]

The longing for an all-inclusive metaphysical vision that would experientially unite the nature of man with the nature of his/her environment, and a desire for more narrowly scientific and naturalistic biosemiotic theories have both been present in biosemiotic literature, and are presumably visible also in this volume. Quite often, the tension between these somewhat divergent forces can be found under the surface of biosemiotic discourse and practices — the actual degree of biosemioticians' self-awareness about the motives and purposes of their making biosemiotics evidently varies. I have argued elsewhere (Vehkavaara 2002, 2003) that if biosemiotics is made as a science, it has to be practiced through certain kinds of naturalistic methods, not necessarily (or hopefully) of a physicalistic, reductionistic, or computational kind. The adopted semiotic concepts have to be abstracted, extended, and adjusted appropriately for biosemiotic use so that the used semiotic or mentalist concepts are first naturalized, operationalized, or formalized before they are applied in biology. However, such naturalistic biosemiotics faces the double risk, i.e. committing both naturalistic and anthropomorphic fallacies at the same time.[3] In any case, biosemiotics have to find appropriate and legitimate *methods* of redefining the semiotic or mental concepts it uses in describing living phenomena (cf. Barbieri: Editorial, this volume). Still, there is a great disagreement among the biosemioticians over what the correct standards of such legitimation are.

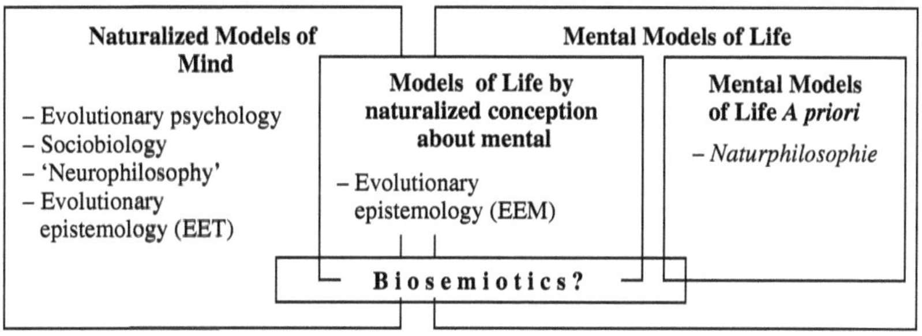

Figure 1. Scientific approaches based on the analogy between mind and life

1.2. Dynamical vs. Structural Approaches

The semiotic or mental concepts applied in biosemiotics have been appropriated from various sources, not only from the traditions of semiotics, but also from hermeneutics, semantics, linguistics, psychology, and from ordinary common sense, i.e. from 'folk psychology' or 'folk biology'. That tradition of biosemiotics which has first recognized and named itself as biosemiotics, as put forward by Thomas Sebeok (1963) and 'microbiologisized' later by Jesper Hoffmeyer and Claus Emmeche (Hoffmeyer & Emmeche 1991; Emmeche & Hoffmeyer 1991; Hoffmeyer 1993) has followed the semiotic tradition originated by Charles S. Peirce (1839–1914).[4] Why has Peirce been chosen in this 'Copenhagen-Tartu school of biosemiotics' as a point of departure rather than the other major founding father of semiotics, Ferdinand de Saussure (1857–1913)? The reasons (or causes) are probably at least partly accidental, i.e. partly due to the intellectual developments and milieu of the thinking of Sebeok, Hoffmeyer, and other dominant figures. Nevertheless, some substantial reasons can be found too.

Perhaps the most striking difference between Saussure's *semiology* and Peirce's *semeiotic*[5] is that Saussure emphasized the role of the static *synchronic* system of signs (*langue*) and defined his signs as having the *dual* character of *signifiant* and *signifié* (i.e. signifier and signified). Saussure centered on social linguistic communication, i.e. how *individual* psychical meanings become *socially* shared and communicated through speech. Saussure's prototype for the concept of sign was *speech*, the uttered (and heard) phoneme, word, sentence, message, etc. (Saussure 1919).

Peirce's starting point, in turn, was human cognition or cognition in general (ability to learn and investigate), how and when the increase in knowledge is possible. For him, the prototype of sign was *thought*, a thought as a *representation*. Peirce concentrated on dynamic sign *processes* (*semiosis*) and defined his concept of sign as an irreducibly triadic composition of a *representamen*, its *object*, and its *interpretant*.[6] The *irreducibility* of this triadic composition means that its three components have no identity *as* an object, representamen, and interpretant independently of the whole sign they are part of. To put it simply, when a (*first*) thing or event is cognized as a *representamen* of some sign, it is recognized as referring to *another* (*second*) thing or event, the *object* of that sign.[7] This act of recognition is

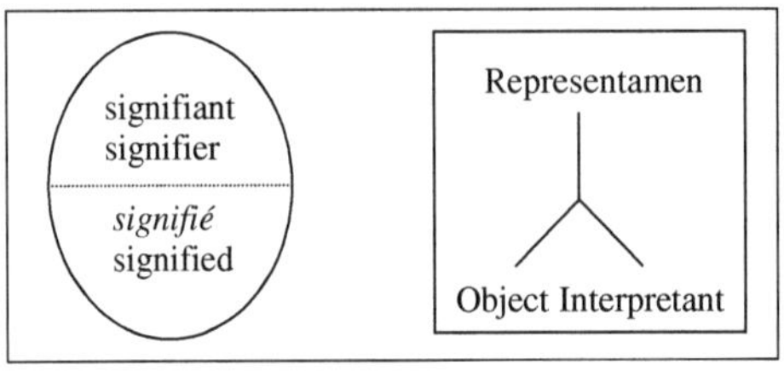

Figure 2. Basic forms of Saussurean and Peircean conceptions of sign

manifested by the *production* of a *third* thing or event in the mind of a recognizer, the *interpretant* of the sign.

In its creation as an interpretation of the representamen, the interpretant gains a *potentiality* to function as another sign of the same object. And if this potentiality is actualized, i.e. the interpretant is actually thought as the thought-*sign*, it will obtain another interpretant that will in turn function as a new thought-sign and so forth potentially *ad infinitum.* A whole chain of signs follows from a singular recognition of a sign in temporal order (CP 2.228, c.1897). This process is called *semiosis.*

The unique character of *semiosis* is that it tends to be a *progressive* process. New signs in the chain expose piecemeal the whole information content about the object that the original representamen contained more or less hidden. In principle, there is an obvious limit to this increase, a limit that may but need not be actually reached. The ultimate end of the series, its *final logical interpretant,* is the *full* embodied conception about the object, the conception that exhibits the whole cognitive content mediated by the sign. There is nothing more to add to this final interpretant, it does not receive a new interpretant anymore and therefore it does not have the nature of sign. Instead, it appears as an undeniable or self-evident *belief,* as a *habit of action,*[8] the habit that is *informed* about the object via the chain of signs. Thus, the Peircean conception of *semiosis* provides a theory or an analysis of how new habits can be adopted, or the old habits can be modified. It is a theory or a description of a rational learning process or gathering of information, a process of *self-controlled habit-formation.* (EP 2:418, 1907.)

It is said that the central task of biosemiotics is to introduce some concept of *meaning* into biology (e.g., Barbieri, 2002, and Editorial, this volume). In Saussure's semiology, meanings become determined merely as differences within the synchronic system of signs, i.e. they are identified only as differences between the meanings expressible in the system. The structuralist approaches, having kinship with (and the origin in) Saussure's semiology, fit best to such biosemiotic applications where biological meanings are considered as the stable ready-made possibilities of material objects or structures. The best biosemiotic example is the case of

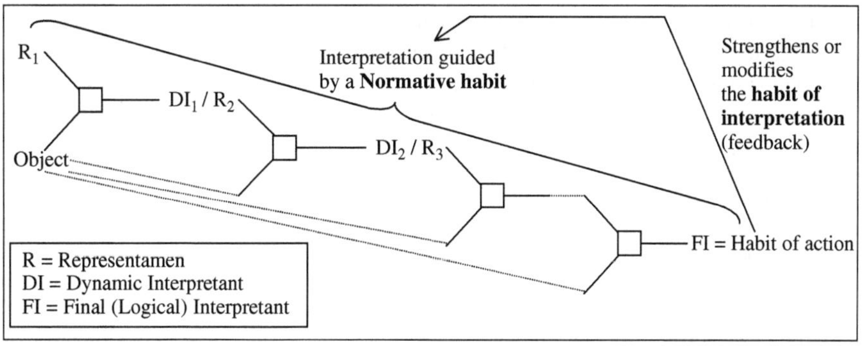

Figure 3. The chain of signs in Peircean conception of *semiosis*

genetic code where basic amino acids are defined as the meanings of corresponding codons.[9] The question of a biosemiotic structuralist would be: What is the semiotic structure to which the organism or other biosystem conforms?

However, if the focus shifts to the *processes* of life, a different question must be asked, namely: How do organisms *change* their structures (or their environments)? And the biological world is full of many kinds of processes: phylogenesis, ontogenesis, conditioning and other forms of learning, protein synthesis, photosynthesis, gaining the resistance to diseases, nest building, etc. Some of them are unique or even potentially endless (e.g. phylogenesis) and others common and infinitely repeated (e.g. protein synthesis). Moreover, it can be argued that living beings, both organisms and their organizations (like ecosystems), do not merely participate in various processes of life but that they are *ontologically* processes rather than 'things' or static structures. Living systems are dissipative systems, thermodynamically far from equilibrium and therefore they have to maintain themselves continuously by their own action if they are going to preserve their stability and identity (cf. Vehkavaara 2003, Bickhard 1998). If the attention in biosemiotics is paid to the regularities of processes rather than to the ones of the structures, a Peircean dynamic approach may appear a more promising starting point than structural ones.

1.3. Peirce and Semeiotic as Logic

Charles Sanders Peirce was born in 1839 as the son of Benjamin Peirce, the leading mathematician in the USA and a professor of mathematics in Harvard. From his early childhood, he became acquainted with scientific community and under his father's guidance and support, he got the best available education in mathematics, philosophy, and sciences (especially in chemistry, astronomy, and biology). Peirce was trained to become a mathematical natural scientist and he earned his living for over 30 years as an experimental physicist.[10] Although he studied and published in the various fields of mathematics, chemistry, geodesy, metrology, astronomy (stellar spectroscopy), cartography, psychology, and history of science, Peirce's significance is evidently the greatest when it comes to his *theoretical philosophy*, i.e. to his logic and metaphysics. Peirce himself thought logic as the science where his greatest expertise is and his most durable achievements stand (Fisch 1982, xxiii and Brent 1998, 38–39). Peirce describes his intellectual development and character with the following modesty:

> From the moment when I could think at all, until now, about forty years, I have been diligently and incessantly occupied with the study of methods [of] inquiry, both those which have been and are pursued and those which ought to be pursued. For ten years before this study began, I had been in training in the chemical laboratory. I was thoroughly grounded not only in all that was then known of physics and chemistry, but also in the way in which those who were successfully advancing knowledge proceeded. I have paid the most attention to the methods of the most exact sciences, have intimately communed with some of the greatest minds of our times in physical science, and have myself made positive contributions — none of them of any very great importance, perhaps — in mathematics, gravitation, optics,

chemistry, astronomy, etc. I am saturated, through and through, with the spirit of the physical sciences. I have been a great student of logic, having read everything of any importance on the subject, devoting a great deal of time to medieval thought, without neglecting the works of the Greeks, the English, the Germans, the French, etc., and have produced systems of my own both in deductive and in inductive logic. In metaphysics, my training has been less systematic; yet I have read and deeply pondered upon all the main systems, never being satisfied until I was able to think about them as their own advocates thought. (CP 1.3, c.1897.)

Peirce is perhaps best known as one of the originators of modern formal logic, of semiotics, and of the first originally American philosophical school, *pragmatism.* Though much of his logical studies falls under current mathematical logic, for Peirce, logic was principally a philosophical science.[11] Peirce included both his *semeiotic* and his version of pragmatism in the science of logic. Traditionally, logic has been vaguely defined as an *art of reasoning* (cf. EP 2:11, 1895) but Peirce wanted to develop logic as a *science* of reasoning that provides *theories* about (the art of) reasoning (e.g. EP 2:30, 1898). More specifically, Peirce defined logic as the science of *deliberate* or *self-controlled thought.*[12] The special character of logic —that distinguishes it from metaphysics and cognitive psychology as well as from mathematics— is its normativity. Logic was defined as a *normative* science of thought, a science that provides criteria for the goodness or badness of thought, i.e. similar to ethics that functions as a normative science of *action.* The *semiosis* is described as a *self-normative* process, where the continuous comparison of the changing representamens with the object directs the sign-process internally — no external authority, normativity, or criterion is needed.

Peirce's conception that the emerging modern logic should be expressed in terms of general semiotic, is based on his argument that *all thought is mediated by signs.*[13] He had a number of reasons to think that thought and signs are intertwined. Firstly, the peculiar character of signs was *defined* to be exactly their ability to mediate thought or meaning. Secondly, Peirce insisted that only *embodied* thoughts can be considered and that the embodiment of thought is a sign (EP 2:256, 1903). Thirdly, from the very beginning of his philosophy, Peirce opposed all forms of foundational intuitionism. He forcefully argued that no intuition, no more sensuous than intellectual, could guarantee an unconditionally or absolutely certain foundation for knowledge. If all 'intuition', i.e. direct or non-mediated reference to the object of thought, is impossible, as Peirce argued, all thought have to be mediated by signs (cf. CP 5.213–215, 251–253, 1868).

2. HOW PEIRCE'S SEMEIOTIC CAN BE APPLIED IN METAPHYSICS AND BIOSEMIOTICS

2.1. From Logic to Metaphysics

Presumably, Peirce was originally studied by biosemioticians because of his *semeiotic*, but it seems that his metaphysical insights (and the metaphysical reading of his *semeiotic*) have inspired more influentially the biosemioticians of the Copenhagen-Tartu school.[14] The Peircean concept of sign and scheme of *semiosis* are nevertheless primarily *logical* conceptions, the prototype of *semiosis* is clearly an *inquiry* or the *scientific* process of investigation.[15] How could such scheme of *semiosis* be applicable also in *natural processes* that seem to be —at least on the surface— of a quite different nature? Would not such an attempt lead to some apparent anthropomorphic error? As a normative science of thought, logic *per se* cannot take into account where its concepts are applied but they need to get firstly a metaphysical interpretation that can be further applied in different special processes of life. Can such application escape being in its heart just one more *a priori* system of rationalistic metaphysics, comparable with the ones of Descartes, Leibniz, Spinoza, Wolff, Hegel, and Schelling?

Part of the fascination of Peirce's metaphysics is that it includes elements that can be applied —or abused— by both those who are attracted to the transcendental argumentation of *a priori* philosophy and those who are more naturalistically minded. It seems to fulfil both underlying intellectual needs of biosemioticians: a longing for an *experientially* understood metaphysical union of man and nature as well as a need for an *experimentally* relevant (and justified) biosemiotic theory. Two questions need to be answered:

1. How can Peirce's theory of rational inquiry be thought to be applicable in modelling the natural processes of life?
2. Can such application be something more relevant than just one more *a priori* system of metaphysics?

2.2. Chance and Continuity

In the second half of the 19th century, the main rivals in metaphysics were the naïve mechanistic materialism of classical positivism and the teleological determinism of absolute or religious idealism. Peirce rejected both the mechanistic and teleological doctrines of inevitable predestination, instead, he proposed a hypothesis that *pure chance* is a real force in nature. This notion that absolute chance is a factor of the universe Peirce called *Tychism* (CP 6.201, 1898). That nature has the element of unpredictable spontaneity, does not nevertheless mean that there would not be real regularities or *laws* of nature too.[16] The regularity of phenomena is not denied but the assumption of the exactitude and absoluteness of natural laws is seen unjustified. At least, there can not be any empirical evidence for that:

> Those observations which are generally adduced in favor of mechanical causation simply prove that there is an element of regularity in nature, and have no bearing whatever upon the question of whether such regularity is exact and universal or not. Nay, in regard to this *exactitude*, all observation is directly *opposed* to it; and the most that can be said is that a good deal of this observation can be explained away. Try to verify any law of nature, and you will find that the more precise your observations, the more certain they will be to show irregular departures from the law. We are accustomed to ascribe these, and I do not say wrongly, to errors of observation; yet we cannot usually account for such errors in any antecedently probable way. Trace their causes back far enough and you will be forced to admit they are always due to arbitrary determination, or chance. (CP 6.46, 1892.)

Peirce's rejection of mechanical determinism is thus based partly on his experience as an experimental scientist and specifically on his awareness about the limitations of the *methods* of experimentation and statistical generalizing. The deterministic demand on the exactitude of natural laws is judged as a mere *a priori* assumption, but if it is not assumed *a priori*, the observed inexactitude of laws can be accepted to be partly due to their very nature as well. This is the doctrine of *Synechism*, which is the most characteristic feature of Peirce's evolutionary metaphysics and of which Tychism is only a corollary (CP 8.252, 1897). It accepts "that being is a matter of more or less", that there is *real vagueness* in nature, i.e. vagueness that is not due to our unclear conceptions and imperfect knowledge. As a regulative principle of logic, it refers to "the tendency to regard everything as continuous" (CP 7.565, 1892)[17] and the reasons to accept it also as a metaphysical doctrine are logical as well. According to Peirce, the atomistic assumption that the nature is a composition of in principle inexplicable ultimate parts leads to the pernicious expectance that the perfect and complete knowledge is in principle achievable. Synechism, in turn, "amounts to the principle that inexplicabilities are not to be considered as possible explanations." (CP 6.173, 1902.) The synechistic hypothesis makes it possible to conjoin law and chance, scientific realism and Tychism. It is a matter of science to study which part of all observed inexactitudes and irregularities are due to our cognitive insufficiency and which part to *real vagueness*.

Although some of Peirce's arguments for Tychism, Synechism, and his other metaphysical hypotheses can no more be judged so forcing as before, the general world view that they draw has become more easily acceptable in the light of the contemporary theories of chaos, dynamic systems, self-organization, and catastrophe than before these mathematical theories were known. The origin of Tychism and Synechism are, however, in those achievements in the science of 1850's —most notably the statistical mechanics (of gases) and Darwinian evolution— that exploited or included the assumption about influential chance. Especially in Darwinian evolution, all the *novelties* come from spontaneous, 'random' variation.[18] But the possibility of the reality of such spontaneity was denied by the Newtonian deterministic world view. As hypotheses, Tychism and Synechism were created to fill that gap. They were *not* intended to legitimate the appealing to miracles as an explanatory principle,[19] but to participate in the general explanation of observed regularities, moreover, of the apparent *novelties* and the *increase* in complexity and diversity in nature:

> But my hypothesis of spontaneity does explain irregularity, in a certain sense; that is, it explains the general fact of irregularity, though not, of course, what each lawless event is to be. At the same time, by thus loosening the bond of necessity, it gives room for the influence of another kind of causation, such as seems to be operative in the mind in the formation of associations, and enables us to understand how the uniformity of nature could have been brought about. (CP 6.60, 1892)

The acceptance of Tychism (or some other equivalent rejection of determinism) is vital for biosemiotics, and especially for such biosemiotics that strives to apply Peirce's semeiotic. In the deterministic world where no genuine choices are possible, whatever is called meaning or meaningful would not have any significance *as* meaning — meanings would be reduced to mere epiphenomena. Only in the world equipped with genuine choices, there is any sense in talking about the success or failure of interpretation — if there were no possibility to err (or to avoid errors), the sign processes would not have their distinctive character. Some amount of freedom or indeterminacy is a prerequisite for any genuine normativity required by *semiosis*. Thus, the acceptance of Tychism opens up the *possibility* of applying semeiotic in natural processes without dictating to what extent it is possible.[20]

Even if it may look at first glance that natural processes and mental processes are of totally different kinds, this glance may prove to be an illusion due to our too concrete and biased level of consideration. If the concept of mind or thought is abstracted far enough[21] so that only the joint features of natural and mental processes are left in its redefinition, then the logical concepts of sign and *semiosis* may be applicable both in the psychical processes of rational thought and in *some* natural processes. The task of biosemiotics is firstly to make such abstraction and redefinition of its basic concepts and then to study which natural processes are semiotic in nature and which ones are not (or to what extent they are semiotic). Still, the self-critical task of biosemiotics is not to be forgotten, the task to detect anthropomorphic errors in its own argumentation and concepts.

2.3. Making Biosemiotics and Peirce

Biosemioticians have adopted some concepts, ideas, and slogans from Peirce's semiotic and metaphysical writings and proposed that they are applicable in the theories about cognition and mutual communication of animals, prokaryotes, plants, and even intracellular communication. However, it is still an open question whether this kind of application will eventually prove insufficient or distorted — a hidden fatal anthropomorphic error cannot be excluded. Although such failure at the substantial level of biosemiotics would be realized, Peirce's semeiotic might still remain valuable at the *methodological* level, i.e. if it is applied in the *making* of biosemiotics, and especially, in the formation of its basic theoretical concepts. This aspect of Peirce's philosophy has so far been mostly neglected by biosemioticians. Whatever the best applicable source of basic semiotic concepts is — be it Peirce, Saussure, Bateson, Lotman or others— Peirce's philosophy offers us a method of adjusting them properly. Namely, one of the main purposes of his whole semiotic was to develop methods of *how to make our ideas clear*.[22] Within his Synechism, Peirce accepted that the world in itself contains (or may contain) some real vagueness. This, however, does not mean that we should be satisfied with the usual vagueness of our *conceptions*, but only that there is no inherent exact meanings hidden in our vague ideas — we have no 'clear intuitions' to appeal to. The general purpose of all scientific inquiry is to provide us the definite and well defined scientific concepts that are transparent in both their reference and meaning. How they can be developed from the vague ideas of our mind, how the necessary vagueness of our concepts could be diminished is a task of logical studies, i.e.

of semeiotic. Every new scientific endeavour consists necessarily of mere vague ideas at the beginning and biosemiotics is still at its beginning. The basic semiotic concepts used in biosemiotics are usually far from definite or clear and desperately need some grounding in concrete observations and experiments.

Moreover, besides being relevant (1) in the *making* of biosemiotics, when basic biosemiotic concepts are formed and defined, the understanding of Peirce's methodological principles would be relevant (2) in *understanding Peirce himself*, when the proposed substantial theory of biosemiotics applies concepts with the Peircean origin. Because Peirce is a far from an easy thinker to make sense, it should be more than clarifying to acknowledge what Peirce's own attitude toward his concepts and arguments was. This can be approached by considering how Peirce himself applied his own methodical principles when he composed his concepts and theory.

3. HOW TO MAKE OUR IDEAS CLEAR — PRAGMATICISM

3.1. Pragmatic Maxim as a Definition of Meaning

Peirce called his general methodology for science *pragmatism,* or more specifically, *pragmaticism.*[23] For him, pragmaticism is not a system of philosophy but only a method of thinking (CP 8.206, c.1905), "a method of ascertaining the meanings of hard words and of abstract concepts" (CP 5.464, 1907). The core of pragmaticism is thus merely *a definition of meaning.*

> In order to ascertain the meaning of an intellectual conception one should consider what practical consequences might conceivably result by necessity from the truth of that conception; and the sum of these consequences will constitute the entire meaning of the conception. (CP 5.9, 1907)

This cryptic definition, the *pragmatic maxim,* requires some explications. Firstly, the pragmatic maxim was designed to define *only* the meanings of *intellectual concepts,* i.e. conceptions that are in principle open for somewhat deliberate adoption or rejection (CP 5.467, 1907).

Secondly, Peirce's pragmaticism did not declare that practical utility would be the ultimate value or that the meaning of a conception would be its *realized* practical consequences. Pragmaticism should not be confused with forms of utilitarianism or instrumentalism. The full meaning of a conception is not reducible to any *actual* consequential events, instead, it contains also those possible consequences that *will not* but *would be* actualized if the circumstances were differently.

> Intellectual concepts [...] essentially carry some implication concerning the general behaviour either of some conscious being or of some inanimate object, and so convey [...] the "would-acts," "would-dos" of habitual behaviour; and no agglomeration of actual happenings can ever completely fill up the meaning of a "would-be." (CP 5.467, 1907).

Thirdly, since the meaning of a concept is not any individual event or thing but a group of certain kinds of 'would-bes', it must be another *conception*, an *anticipative conception* that anticipates or refers to the *possible future effects* of the concept. Moreover, this anticipation is about some 'habitual behaviour', either of *our* action, or of the action of something else, the habitual behaviour of which *we* can adapt *our* action self-controlledly.[24] This is the meaning of 'practical' in the definition.

Fourthly, all this does not, however, mean that the meaning of a proposition, say, about the big bang would be emptied in its direct practical applications in our life. Quite the opposite, the pragmatic maxim was designed to provide the conception of meaning especially for those intellectual concepts towards which our interest is purely theoretical. Our activity 'to find out', to *make experiments*, is also included in such human conduct, to which a proposition that exposes the meaning is applicable:

> [...] that form of the proposition which is to be taken as its meaning [...] must be simply the general description of all the experimental phenomena which the assertion of the proposition virtually predicts. For an experimental phenomenon is the fact asserted by the proposition that action of a certain description will have a certain kind of experimental result; and experimental results are the only results that can affect human conduct. (CP 5.427, 1905)

3.2. Pragmaticist Biosemiotics

If biosemiotics would meet this criterion for meaningfulness, i.e. if the biosemiotic theory could provide some *experimental results* that no other kind of theory could, that would legitimate the biosemiotic approach in an instant. It may be that this criterion is too demanding for contemporary biosemiotics thus far, but as a guiding goal in developing biosemiotic concepts, it is worth attempting. At least, a simple 'armchair test' of the meaningfulness of the biosemiotic concepts should be made by comparing them with their non-biosemiotic alternatives. Does a biosemiotic explanation or point of view bring anything really differing at the level of experimental testing or of practical applications?

> For the maxim of pragmatism is that a conception can have no logical effect or import differing from that of a second conception except so far as, taken in connection with other conceptions and intentions, it might conceivably modify our practical conduct differently from that second conception. (EP 2:234, 1903)

The main obstacle to making such comparisons is the abstract, vague, and metaphorical character of the mostly used semiotic concepts in biosemiotics. The possible experimental or practical bearings of the concepts of that kind are impossible to be 'conceived' with accuracy. The meanings that they are intended to carry are mere blurry feelings and as such as they are difficult to identify and control. The temptation to speak vaguely is understandable, since it leaves the backdoor open for excuses and corrective additions that would specify —or even construct— the vague or partly unconscious *ad hoc* meaning.

> It is easy to speak with precision upon a general theme. Only, one must commonly surrender all ambition to be certain. It is equally easy to be certain. One has only to be sufficiently vague. (CP 4.237, 1902.)

What can be done in order to make the biosemiotic ideas clearer so that they might be put in an experimental test? In the original formulation of the pragmatic maxim,[25] the intellectual meaning of a concept consists of its 'conceivable practical bearings', but those 'practical bearings' are the ones of the *object* —i.e. the referent— of the conception. Thus, in order to determine the pragmatic meaning of a scientific concept, to get the better control over its possible *future* products, i.e. over its potential *interpretants*, we need a control over the intended (or assumed) *objects* of the concept too. This hidden demand is underlined in another formulation of the maxim of pragmaticist:

> The elements of every concept enter into logical thought at the gate of perception and make their exit at the gate of purposive action; and whatever cannot show its passports at both those two gates is to be arrested as unauthorized by reason. (EP 2:241, CP 5.212, 1903)

Thus, the control over the *formation* of our concepts constitutes an essential part in the anticipation of its whole pragmatic meaning. Concepts are derived from some kind of *perception* and the circumstances of the observation of that perception may become structured in the concept. The observation plays a double role — scientific concepts are originated by observation and their meaning is dependent on the would-be observation of the would-be results of their would-be experimental testing. For instance, the majority of cognitive (or communicative) concepts are originally based on the observation of some common internal experience of sensing, knowing, understanding, intending, etc. They are nevertheless later abstracted or formalized and, especially in biosemiotics, extended to refer also to such non-human phenomena (like animal cognition) about which we cannot have internal experience. But if the concepts are abstracted without clear awareness about their derivations, some hidden presumptions may remain in the structure of these concepts, the presumptions that make them not extendable beyond a human sphere. Therefore, scientific concepts cannot be accepted merely as (culturally or intuitively) 'given' — their 'derivations' remain more or less hidden with the consequence that also the meanings of such culturally given everyday concepts remain too vague for scientific use.

The concepts we use even in science are originally vague, but they can be made 'clear and distinct' and one method of achieving this is to analyze the path of the formation of concepts and the observations (or experiences) that are their points of departure. This analysis does not determine the referents of the concepts under scrutiny, they are after all abstracted, but it may *suggest* the possible referents of the concepts, and most of all, exhibit the errors that stay easily hidden.

3.3. Pragmaticism Applied Back to Itself

However, there is one remaining problem, whether Peirce's 'tychastic Synechism' is more than a metaphysical system *a priori.* The differences between Peirce's approach and *a priori* philosophy like Schelling's *Naturphilosophie* or Kant's transcendental idealism can be best illustrated if we apply the pragmatic maxim to Peirce's concepts themselves. We should therefore scrutinise carefully what kind of invisible structural presuppositions are built into his concepts prior to their application to biological theory. This cannot be done properly here (see some details in Vehkavaara 2006), but some guidelines can nevertheless be given. We can study 1. what kind of *practical bearings* he conceived his logical and metaphysical concepts as having and 2. what kind of *perception* Peirce's derivation of his concepts starts from.

If we consider specifically the pragmatic maxim i.e. the conceivable practical consequences of the pragmatic maxim, Peirce himself characterized pragmatism (in a dictionary article) as being the "opinion that metaphysics is to be largely cleared up by the application of the [pragmatic] maxim for attaining clearness of apprehension" (CP 5.2, 1902). So, the intended practical bearing of the pragmatic maxim itself was to free us from floppy *a priori* metaphysics. It was hoped to wipe them away by showing that endless disputes without any conceivable practical —i.e. *experimental*— differences are senseless.[26] The pragmatic maxim provides quite hard criteria of intellectual meaning especially for metaphysical and logical concepts. The fact that the maxim was originally designed to settle the stubborn nonsensical quarrels of metaphysical and religious doctrines does not, however, limit its potential practical bearings, which Peirce clearly intended to cover all scientific or rational thought. The open question of biosemiotics is, can the pragmatic maxim be applied, extended, or further abstracted so that it could work as a base for *biological meaning* too?

Next we have to ask, in the light of the pragmatic maxim, what kind of perception Peirce's logical and metaphysical concepts are based on and what consequences we can draw from that.

4. CONSEQUENCES OF PRAGMATICISM IN UNDERSTANDING PEIRCE

4.1. Observation in Sciences — Metaphysics is not the 'First Philosophy'

Peirce expressed quite explicitly what kind of perception or experience the elements of philosophical concepts are derived from. He recognized three kinds of observation that separate the three classes of Theoretical science.[27]

1. *Pure mathematics* is based on the observation of *imagined* objects without any guarantee of their application in the actual world. It can describe only the *possible forms* that things (including thought) *may* take in our universe. It is a pure science of *hypotheses* providing no *positive* information about the *actual* reality of our universe. As such it is the *negative science*. (CP 2.782, 1901, CP 1.247, 2.77, 1902.)
2. *Theoretical philosophy* (*Philosophia prima*) draws its conclusions from the observation of *universal phenomena* that "come within the range of every man's normal experience, and for the most part in every waking hour of his life" (CP 1.241, 1902). The findings of philosophy should thus be derivable from *familiar experience* common to everyone.

3. *Special sciences* are based on the *special experience* aided with instruments and other special arrangements and on the analysis of its minute details. Special sciences discover *new* phenomena by expanding the ordinary limits of human experience.[28]

These three classes form a nested hierarchy according to the *abstractness* of the *objects* of study specific to each science (CP 1.180, 1903). All sciences may use the same experiential content as the 'data' for their inquiries, but they *observe* different facts from that 'data', the facts that lead to generalizations at the different levels of abstraction (CP 8.297, 1904). Each special science observes from this 'data' the *special* information peculiar to it — astronomy pays attention to astronomically relevant data, etc. Philosophy observes general information that could in principle have been achieved from any other 'data' too. Because the observed universal experience is present in *any* experience, also in those special experiences which special sciences observe, the familiar every day experience suffices for the observational basis of philosophy. Mathematics, in turn, extracts a mere possible form from the'data', the form the properties of which are in that sense independent on any actually perceived 'data' that merely imagined 'data' would suffice for the source of mathematical inquiries.[29]

The general principle of this hierarchy is that lower sciences rest for their *principles* upon (some of) the higher ones that, in turn, draw their *data* in part from the lower ones and furnish them with applications (EP 2:35, 1898, EP 2:458, 1911, cf. also Kent 1987: 18). The subclasses of each class of Theoretical science inherit this principle. With regard to the two major subclasses of philosophy, Peirce kept logic a more abstract science than metaphysics (EP 2:35–36, 1898). Since metaphysics, the *philosophical science* of the *most general facts of the reality*, is based on the observation of universal experience, it can be asked how such knowledge is possible or whether it is possible at all. On what grounds the correct metaphysics could be argued for, since such knowledge —because of being the most general kind— should be independent on any particular observation or experience and compatible with all possible experience. Peirce followed Immanuel Kant's solution of this problem by rejecting the traditional idea about metaphysics as the 'first philosophy', instead, the basic metaphysical concepts should be applied logical ones (EP 2:30–31, 1898) — i.e. logic is prior to metaphysics. The biosemiotic practice of applying Peirce's logical concepts with a metaphysical tone is thus in principle compatible with Peirce's own application. Nevertheless, such applications are always vulnerable to anthropomorphic errors and excess vagueness.

The independence of logic from metaphysics means that the reality of thought is not a logical question — the science of logic cannot decide whether there is any thought in animals, for instance. The logical concept of sign should be independent of embraced metaphysical principles as well as of the findings of natural sciences, For example, the acceptance of Tychism or Synechism is not required for the acceptance of the Peircean conception of sign. However, the *application* of logical concepts in metaphysics and in biology *is* dependent on embraced metaphysical and biological conceptions. Many (though not all) forms of ontological physicalism, for instance, are incompatible with all possible applications of Peircean conception of sign. At least some amount of *real* indeterminacy (which does not necessarily mean free will) is required in order to make real sign processes genuinely normative.

The independence of special sciences on logic and metaphysics gives a quite demanding criterion for the generality of philosophical propositions: philosophical generalizations should be in accordance with *all experiential data of all kinds.* Although special sciences cannot provide any *principles* for philosophy, they may provide critique (even if indirect) for philosophical conceptions. The new findings in special sciences may demonstrate that the philosophical concepts derived from familiar experience have not been abstracted enough but that they are after all formulated in unnecessarily concrete or intricate terms (cf. CP 2.75, 1902) — i.e. that they include naturalistic or anthropomorphic errors.

4.2. Transcendental and Objective Perspectives

In the tradition of transcendental philosophy put forward by Immanuel Kant, the concepts of metaphysics are grounded on *transcendental logic*. In Kantian scheme, all our knowledge is admitted to begin with experience, but not so that all of it would arise out of experience. Instead, Kant assumed that

our empirical knowledge is a compound of that which we receive through impressions, and that which the faculty of cognition supplies from itself (Kant 1781/1787: 1 [B1]).

That part of our knowledge which is in our faculty of cognition itself, must in this Kantian scheme be independent on the empirical or *a posteriori* part, i.e. it is *a priori*, prior to senses. The term *transcendental* refers in the Kantian tradition to concepts focusing on such *a priori* forms of all possible knowledge:

I apply the term *transcendental* to all knowledge which is not so much occupied with objects as with the mode of our cognition of these objects, so far as this mode of cognition is possible *à priori*. A system of such conceptions would be called *Transcendental philosophy*. (Kant 1781/1787: 15 [B24].)

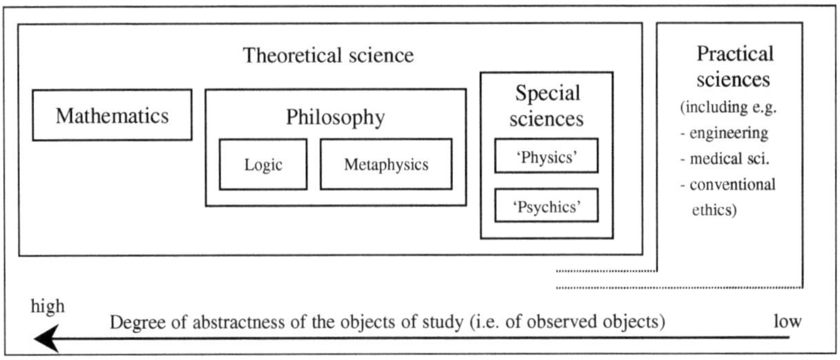

Figure 4. Overview to Peirce's conception about the relations of sciences (before c. 1902)[30]

Peirce's conception about the philosophical observation on which theoretical philosophy is based is an apparent descendant of Kantian conception of transcendental philosophy. The philosophical observation of universal experience *included in any experience* is obvious counterpart for the Kantian idea of transcendental *a priori* form of cognition. Because logic should be derivable from any experience (plus mathematics), i.e. from familiar every day experience, it becomes intimately bound with 'our' perspective and ordinary life. The 'positive facts' that logic can tell us concern the form of our internal epistemic relation with the world we live. This may bring some restrictions on the biosemiotic applicability of Peircean logical concepts.

For instance, the Peirce's original derivation of his concept of sign (CP 1.545–559, 1867) was the construction of the concepts of representamen, object, and interpretant and their irreducible triadic coalition as a sign. The derivation was executed by directing the investigating thought into itself in order to find out *how* it can refer to its object and state something about it. A present thought is directed to observe itself, i.e. directed toward its origin, toward its *object,* to find truth about it, and at the same time, it becomes transformed into another *more self-aware* thought-sign about itself, into its *interpretant.* The interpretant was produced as a means of grasping the true knowledge about the object of thought-sign. Because the interpretant is *constructed* by looking for *truth* about the object, the *aim toward truth* —the logical normativity— is already built in the construction of the triadic structure of sign, no matter how it will be considered or what will be considered in it. This analysis suggests —if it is correct— that Peircean concept of sign may not be as general concept that is often assumed. Consequently, its applicability in biosemiotics is restricted, if it is accepted that the ultimate criterion of goodness for living systems is *survival* or *sufficient fitness* and not so much the truthfulness of their representations. (See more in Vehkavaara 2006.)

However, besides the above described *transcendental perspective*, Peirce employed also a perspective that could be called the *objective perspective*, because within it, a sign is no more considered merely from the perspective of its own, but the whole *chain* of signs, the whole semiosic *process* or succession of signs, is taken as the object of study. The investigating mind is methodically split into the 'observer-mind' and 'observed-mind' so that a present investigating thought in observer's head is no more considered as a part of the object of study. The objective perspective is inevitable for any biosemiotic application of semiotic, because it can be applied to study other minds — it frees us to study and think about non-human minds and non-conscious sign processes. The description of *semiosis* in Chapter 1.2 (see Figure 3) was a description from the objective perspective. The distinction between these two perspectives is essential when Peirce's statements concerning signs are considered. (See more in Vehkavaara 2006.)

Peirce's metaphysics inherits the similar 'transcendental' character from logic. As metaphysics draws its positive content only from the universal features of ordinary experience, the most general facts that it describes must concern their *accessibility* to us, i.e. the form that they 'necessarily' take in *our mind* (independently on their more concrete content). For instance, in his paper "Evolutionary Love" (EP 1:352–371, 1893), Peirce demonstrates there being two other kinds of evolution besides a tychastic one by fortuitous variation. The other two were *anancastic* evolution by mechanical necessity and *agapastic* evolution by 'evolutionary love' i.e. by force of a self-organizing habit. They were all considered as real powers in the world but they are, however, only three *possible forms* that real processes *may* take, three real possibilities that should not be excluded *a priori* when some specific real process is investigated. Whether or not an individual process (be it chemical, geological, celestial, phylogenetic, epigenetic, metabolic, psycho-dynamic, communicative, etc.) is dominated by 'creative love', for instance, is not properly a metaphysical quarrel. It is dependent on the observation of the appropriate *special* phenomenon and therefore belongs under the corresponding special science.

4.3. Fallibilism — Transcendental but not *a Priori*

Although Peirce's conception about philosophical observation gives to his philosophy a kind of 'transcendental flavour' that even the adoption of objective perspective cannot completely strike out, it is nevertheless not transcendental *a priori*. Whereas for Kant the *a priori* conditions of cognition have to be undeniable "with the full guarantee for validity" (Kant 1781/1787: 16), for Peirce, the philosophical observation is by no means infallible although its pitfalls are different than what special sciences face.[31] Because *all* sciences and *all rational thought* is based on the observation of some kind of perception, there are strictly speaking no *a priori* concepts at all, no concepts 'prior to senses'. For Peirce, there is no pure knowledge *a priori*, i.e. no purely *a priori* certain knowledge in a Kantian sense,

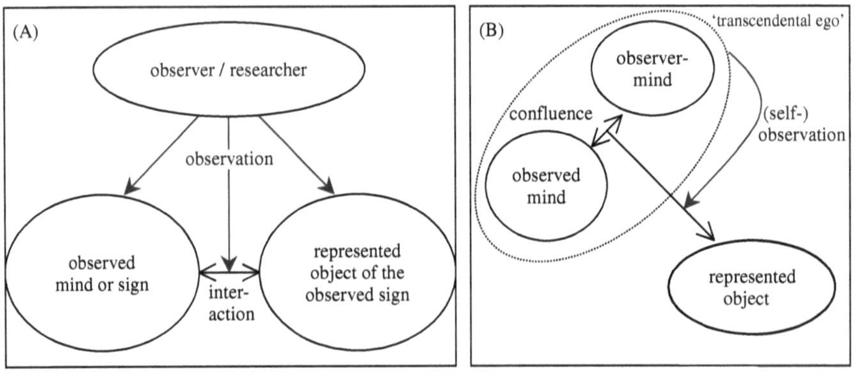

Figure 5. Observation from the objective perspective (A), and from the transcendental perspective (B)

because there is no such knowledge with which "no empirical element is mixed up" (Kant 1781/1787: 2).[32]

With this assumption about the observational origin of all the elements of knowledge, Peirce breaks with the mainstream of modern philosophy. Namely, Peirce argues that no absolute infallibility or certainty is accessible in *any* human endeavour.[33] The fallibility of human sensations is widely accepted (and experientially supported), but it is quite rare in the philosophy of western tradition that this fallibilism is extended even into mathematics.[34] If even mathematics is accepted as fallible without any commitment to nihilism or extreme scepticism, then philosophy, i.e. logic and metaphysics, certainly follows. Peirce did not expect that internal senses, on which mathematics, philosophy and, parts of the psychical special sciences lean, would provide any more epistemically privileged information than external senses. Inconceivability, unimaginability, or indubitability, which are often appealed to when rationalistic *a priori* doctrines are tried to justify, are historically proven to be far from infallible.

But that which has been inconceivable today has often turned out indisputable on the morrow. Inability to conceive is only a stage through which every man must pass in regard to a number of beliefs [...]. His understanding is enslaved to some blind compulsion which a vigorous mind is pretty sure soon to cast off. (CP 6.49, 1892)

The task of philosophy is not to dwell on infallible truths or any other *a priori* certain foundation for scientific knowledge. It would be a grave error to read Peirce as if he would aim to provide a foundational basis of science, a new foundation upon which the house of the new science would be built. Throughout his scientific career, Peirce opposed such a Cartesian dream about the absolutely certain foundation of knowledge that has been governing modern philosophy at least since the days of Descartes (e.g. CP 5.264–265, 1868). Instead, he called his attitude toward philosophy a 'laboratory-philosophy' contrasting it with the philosophies of 'theological seminaries' (CP 1.129, 1905) referring especially to Hegel and Schelling, the major representatives of idealistic *Naturphilosophie*.

[M]y attitude was always that of a dweller in a laboratory, eager only to learn what I did not yet know, and not that of philosophers bred in theological seminaries, whose ruling impulse is to teach what they hold to be infallibly true. (CP 1.4, 1897)

Peirce describes his attitude of 'laboratory-philosophy' as follows:

Thus, in brief, my philosophy may be described as the attempt of a physicist to make such conjecture as to the constitution of the universe as the methods of science may permit, with the aid of all that has been done by previous philosophers. I shall support my propositions by such arguments as I can. Demonstrative proof is not to be thought of. The demonstrations of the metaphysicians are all moonshine. The best that can be done is to supply a hypothesis, not devoid of all likelihood, in the general line of growth of scientific ideas, and capable of being verified or refuted by future observers. (CP 1.7, 1897)

Despite all his rejection of *a priori* knowledge, encouragement to 'laboratory-philosophy', and mocking of the philosophies of 'theological seminaries', Peirce nevertheless appreciated some of the *results* of the great German Idealists: Kant, Hegel, and Schelling (CP 1.21, 1903). However, even if their *a priori* method of reasoning would produce some valid conclusions, such success would more or less be due to an accident — Peirce could not see much sense in the argumentation by which they reasoned to obtain these results. Such *a priori* method does not provide tenable means to distinguish the apparent errors that they include from the valid conclusions. Moreover, their aim or longing to provide some *a priori* certain foundation for science leads to an understandable but harmful habit of giving an infallible status to such *a priori* results, i.e. removing them outside of the target area of inquiry.

For Peirce, the *results* do not make a science but the *way they are produced*, and this applies to logic and metaphysics too. Perhaps the most important lesson that biosemioticians should learn from Peirce is his attitude toward science, science that includes also metaphysics and semiotic. It is the attitude of potential impermanence of all the scientific results that we ever can derive, the attitude that emphasizes the making our scientific concepts and ideas clearer and clearer, i.e. that their pragmatic meaning and conceptual structure would be better and better exposed. It is the attitude that whatever our embraced logical and metaphysical principles are, they are not *believed* and taken as the premises of our inquiry but they are considered as *hypotheses* that have to be argued for instead. For Peirce, beliefs should have no role in scientific argumentation, except as a *source* of *hypotheses* but as such, they are strictly speaking no more fully believed. All convictions and beliefs are judged by Peirce as harmful in science:

> I hold that what is properly and usually called *belief* [...] has no place in science at all. We *believe* the proposition we are ready to act upon. [...] But pure science has nothing at all to do with *action*. The propositions it accepts, it merely writes in the list of premisses it proposes to use [...] and the whole list is provisional. The scientific man is not in the least wedded to his conclusions. He risks nothing upon them. He stands ready to abandon one or all as soon as experience opposes them. Some of them, I grant, he is in the habit of calling *established truths*; but that merely means propositions to which no competent man today demurs. [...] Still, it may be refuted tomorrow; and if so, the scientific man will be glad to have got rid of an error. There is thus no proposition at all in science which answers to the conception of belief. (CP 1.635, 1898.)[35]

The pure scientific attitude that Peirce forcefully expounded is that we should not allow some general world view based on our more or less uncontrolled impressions[36] and 'seemings' dictate what kind of explanations, descriptions, and concepts are in principle acceptable in science. That attitude, if applied to the *making* of biosemiotics, would eventually make it a science.

ACKNOWLEDGEMENTS

I wish to thank Stephen Pain for reading the manuscript and for helping with the copy-editing.

NOTES

[1] Michael Bradie has distinguished two interrelated but distinct programs which both go by the name 'evolutionary epistemology'. 'EET' is an abbreviation of the 'evolutionary epistemology of theories', which "attempts to account for the evolution of ideas, scientific theories and culture in general by using models and metaphors drawn from evolutionary biology." 'EEM', the 'evolutionary epistemology of mechanisms' in turn, attempts "to account for the characteristics of cognitive mechanisms in animals and humans by a straight-forward extension of the biological theory of evolution to those aspects or traits of animals which are the biological substrates of cognitive activity, e.g., their brains, sensory systems, motor systems, etc." (Bradie 1986: 403.)

[2] More properly, the explanatory principle in evolutionary epistemology is the general *selection theory* that is abstracted from the principle of natural selection (Campbell 1997: 7).

[3] In Vehkavaara (2002), I suggested a method of *semiotic naturalism* that would minimize the risk to fall on these fallacies.

[4] Peirce is not the only source of the semiotic ideas of this school of biosemiotics. Jakob von Uexküll and Gregory Bateson have obviously been just as influential as Peirce and some concepts have been borrowed from the cultural semiotics of Juri Lotman (like 'semiosphere').

[5] The term 'semiotic' here refers to the overall field of discourse or discipline that concerns signs. 'Semeiotic' and 'Semiology' are used to refer to Peirce's and Saussure's particular semiotic theories.

[6] The emphasis on sign-action dominated Peirce's later and more mature views on sign and semiosis. However, especially in his early papers (most notably in "On a New List of Categories", CP 1.545-559, 1867), the concept of sign was viewed and derived as a kind of transcendental concept which can hardly be interpreted as a dynamic one. (cf. Vehkavaara 2006). About the distinction between *sign-object* and *sign-action* see e.g. Deledalle (2000: 38–39).

[7] The object of a sign does not have to be any concrete particle or other material thing — it can be anything (a material thing, perception, idea, lawful behavior of nature, dream, etc.) which excites the mind to search some better or fuller representation of it. However, this requires that the object is somehow beforehand, by some *collateral observation*, acquainted — the sign cannot provide the sole access to the object (cf. EP 2:408-9,429, 1907). Still, the sign does not necessarily only draw the interpreter's attention to the object, but that it may also provide some *new* information about the object.

[8] The conception of a *belief* as a deliberate or partly conscious habit of action is one of the core conceptions in pragmatism (cf. CP 5.12, 1907). "A belief in a proposition is a controlled and contented habit of acting in ways that will be productive of desired results only if the proposition is true." (EP 2:312, 1904)

[9] See especially Artmann (in this volume) and also Barbieri (2002) on Morse code and his organic meaning.

[10] This intimate participation in the science of his time influenced greatly both his conception of science and the content of his philosophy. Peirce's metaphysics took its inspiration —besides from modern logic and mathematics— from the latest achievements of the natural sciences of the 19th century. The most important of these were Kirchoff's Spectroscopy in 1859, Mendeleev's Periodic Table in 1852, Rankine's, Clausius', and Kelvin's thermodynamics from 1850 onward, Pasteur's findings in microbiology from 1848 onward, and the most of all, Darwin's natural selection in 1859.

[11] "There is a mathematical logic, just as there is a mathematical optics and a mathematical economics. Mathematical logic is formal logic. Formal logic, however developed, is mathematics. Formal logic, however, is by no means the whole of logic, or even its principal part. It is hardly to be reckoned as a part of logic proper." (CP 4.240, 1902.) In 'logic proper', Peirce included both epistemology (*Speculative grammar*, "the general theory of the nature and meanings of signs") and general methodology (*Methodeutic* or *Speculative rhetoric*, "which studies the methods that ought to be pursued in the investigation") along with *Logical Critic*, logic in narrow sense, "which classifies arguments and determines the validity and degree of force of each kind" (EP 2:260, 1903, CP 2.206–207, 1901).

[12] Peirce used the term 'logic' as the *name* of the logical science and *not* to refer to its *object of study* which is another common use of the term 'logic' (e.g. in phrases 'women's logic' or 'logic of the

universe'). Logic in this latter meaning is, especially if considered as a *description* of a *real* phenomenon, rather a question of metaphysics or psychical sciences than of logical science.

[13] This is Peirce's solution to the problem that thought is, in itself, quite an abstract and vague concept, which is hard to grasp because of its internal, immaterial, temporary, and flexible characters. The 20th century western philosophy, almost every branch of it, has tried to solve this problem in another way, by making a 'linguistic turn', by considering only linguistically expressible thoughts and language as *the* medium of thought. Structuralism based on Saussure's semiological vision is a one form of it.

[14] See Hoffmeyer (1993: 25–27) and Brier (2003: 74). These insights have not necessarily been adopted straight from Peirce's writings but some of them might have been already adopted from other sources (e.g., Gregory Bateson) and Peirce's writings are just found to appear as compatible with them.

[15] It can be argued that logic of science was not only Peirce's starting point (or *motive*) but also one of his main *purposes* of his theory (cf. Vehkavaara 2006).

[16] One of the central characteristic of Peirce's metaphysics is his strong 'three category realism' or "extreme scholastic realism" (CP 8.208, c.1905). This consists of the acceptance that not only singular existent events are real, but that also possibilities and *some* general objects (like laws and habits) are real. This does not mean that *all* generals were real since, according to Peirce, nobody ever thought that "but the scholastics used to assume that generals were real when they had hardly any, or quite no, experiential evidence to support their assumption; and their fault lay just there, and not in holding that generals could be real" (EP 2:342, 1905). The counterpart to 'real' is not 'ideal' but 'figment' or 'illusion': "A figment is a product of somebody's imagination; it has such characters as his thought impresses upon it. That those characters are independent of how you or I think is an external reality. [...] Thus we may define the real as that whose characters are independent of what anybody may think them to be." (EP1:136, 1878) This, in turn, does not mean that what is relative to thought cannot be real. "*Red* is relative to sight, but the fact that this or that is in that relation to vision that we call being red is not *itself* relative to sight; it is a real fact" (EP 2:343, 1905).

[17] According to Synechism, we must not say

- "that the sum of the angles of a triangle exactly equals two right angles, but only that it equals that quantity plus or minus some quantity which is excessively small for all the triangles we can measure"
- "that phenomena are perfectly regular, but only that the degree of their regularity is very high indeed"
- "being is, and not-being is nothing." like Parmenides but "that being is a matter of more or less"
- "I am altogether myself, and not at all you," i.e. "synechism recognizes that the carnal consciousness is but a small part of the man. There is, in the second place, the social consciousness, by which a man's spirit is embodied in others, and which continues to live and breathe and have its being very much longer than superficial observers think." (CP 7.568–575, 1892.)

[18] Peirce viewed Darwinian evolution as an example of *tychastic* evolution. The other possible types of evolution considered by Peirce were *anancastic* evolution by mechanical necessity and *agapastic* evolution by 'evolutionary love' i.e. by force of a self-organizing habit. Lamarckian evolution was Peirce's example of agapastic evolution. (See more later, and in CP 6.300–302, 1893.)

[19] "[A]n explanation should tell *how* a thing is done, and to assert a perpetual miracle seems to be an abandonment of all hope of doing that, without sufficient justification" (CP 2.690, 1878).

[20] "[B]y supposing the rigid exactitude of causation to yield, I care not how little — be it but by a strictly infinitesimal amount — we gain room to insert mind into our scheme, and to put it into the place where it is needed" (CP 6.61, 1892).

[21] Peirce himself defines the most abstract sense of 'mind' as following: "Mind has its universal mode of action, namely, by final causation. The microscopist looks to see whether the motions of a little creature show any purpose. If so, there is mind there." (CP 1.269, 1902.) The general abstracted concept of mind or thought do not contain any assumption of its self-consciousness or of free will though Peirce by no means rejects the real possibility of self-conscious mind equipped with free will: "Thought is often supposed to be something in consciousness; but on the contrary, it is impossible ever actually to be directly conscious of thought. It is something to which consciousness will conform, as a writing may conform it. Thought is rather of the nature of a habit, which determines the suchness of that which may come into existence, when it does come into existence." (EP 2:269, 1903.)

[22] "How To Make Our Ideas Clear" is the title of his perhaps best known paper (CP 5.388–410, 1878). In it, the principle of pragmaticism occurs the first time as presented (though the words 'pragmatism or pragmaticism do not occur).

[23] The idea of pragmatism was developed in early 1870's, in the conversations of the 'Metaphysical Club', a small group of young Cambridge philosophers (and lawyers) lead by Peirce, William James, and Chauncey Wright. It was not until 1898 when James, Peirce's life long friend and both philosophical and financial supporter in the last years of his life, first brought the term 'pragmatism' before the public, which led to the tremendous popularisation of pragmatism around the turn of the century. Although Peirce was probably the originator of the basic idea of pragmatism, Peirce and his pragmaticism (renamed in 1904, in order to be safe from 'kidnappers', cf. CP 8.194, 5.414) was hardly known at all. The leading pragmatists, James and John Dewey, nevertheless gave the honour to Peirce as the originator of it, though even Peirce hesitated whether it was him or James who first used the term 'pragmatism' (CP 8.253, 1900).

[24] "The meaning of a proposition is itself a proposition. Indeed, [...] it is a translation of it. But of the myriads of forms into which a proposition may be translated, what is that one which is to be called its very meaning? It is, according to the pragmaticist, that form in which the proposition becomes applicable to human conduct, not in these or those special circumstances, [...] but that form which is most directly applicable to self-control under every situation, and to every purpose. This is why he locates the meaning in future time; for future conduct is the only conduct that is subject to self-control." (CP 5.427, 1905)

[25] The first written description about the pragmatic maxim was published in 1878: "Consider what effects, that might conceivably have practical bearings, we conceive the object of our conception to have. Then, our conception of these effects is the whole of our conception of the object." (CP 5.402, 1878.)

[26] "Pragmaticism, then, is a theory of logical analysis, or true definition; and its merits are greatest in its application to the highest metaphysical conceptions." (CP 6.490, 1910)

[27] Peirce's main scientific interest and attention concentrated on what he called Theoretical science. Theoretical science differs from Practical sciences according to the most general end of inquiry, the end that functions as an ultimate criterion for the successfulness of the inquiry. Theoretical science is ultimately guided and valued by the *intrinsic* end of inquiry, truth, while practical sciences are guided by various practical ends, *extrinsic* for the inquiry. Theoretical science has two subbranches *Heuretic Sciences* or *Sciences of Discovery* and *Science of Review* or *Retrospective Science*. Mathematics, Philosophy, and Special sciences are the three classes of Heuretic Science. Wheras Heuretic Science is studying 'directly' the phenomena, the Retrospective Science is studying phenomena mediately, collecting and uniting the results of different Heuretic Sciences. Thus, *Synthetic Philosophy* represented by Alexander Humboldt's *Cosmos*, Auguste Comte's *Philosophie positive*, and Herbert Spencer's *Synthetic Philosophy* are classified under the Science of Review. In addition, all considerations that relate different sciences in general, e.g. histories and classifications of sciences belong under the branch of Science of Review. Peirce named synthetic philosophy as *Philosophia ultima* in order to make difference with *Philosophia prima*, the theoretical philosophy of Heuretic Sciences (EP 2:372–373, 1906).

[28] The special sciences consist of two subclasses, physical and psychical. The difference between physical and psychical special sciences is that 'physics' sets forth the workings of efficient causation and 'psychics' of final causation (CP 1.242, 1902). Physical and psychical phenomena are not independent on each other since Peirce did not see final and efficient causation as alternatives, but some chain of efficient causes is always involved in any event guided by a final cause (cf. CP 1.212, 1902).

[29] Thus, every science has its mathematical part, but that part alone studies such forms from which all concrete elements and references to the reality of the actual world are abstracted away (CP 1.133, 1894).

[30] Before the first years of 20th century, Peirce recognized only two subdisciplines of theoretical philosophy, logic and metaphysics, but in 1901-1902, Peirce's conception about the philosophical science deepened. He found out that theoretical philosophy actually contains a couple of other sciences that he previously had not recognized it containing. According to this new conception, theoretical philosophy divides into three subdisciplines, to *phenomenology* (later also *phaneroscopy*), *normative sciences,* and *metaphysics*. Normative sciences divide further into three: to *esthetics*, *ethics* (renamed later as *practics*), and *logic* (or *formal semeiotic*).

[31] "[...] the observational part of philosophy is a simple business, compared, for example, with that of anatomy or biography, or any other special science.

To assume, however, that the observational part of philosophy, because it is not particularly laborious, is therefore easy, is a dreadful mistake, into which the student is very apt to fall, and which gives the death-blow to any possibility of his success in this study. It is, on the contrary, extremely difficult to bring our attention to elements of experience which are continually present. For we have nothing in experience with which to contrast them; and without contrast, they cannot excite our attention." (CP 1.133–134, 1894.)

[32] This rejection (or doubt) of the reasonableness of the whole conception about *a priori* can be seen in Peirce's critique of Kant's starting point:

"Immanuel Kant asked the question, "How are synthetical judgments *a priori* possible?" [...] By *a priori* judgments he meant such as that all outward objects are in space, every event has a cause, etc., propositions which according to him can never be inferred from experience. [...] But before asking *that* question he ought to have asked the more general one, "How are any synthetical judgments at all possible?" How is it that a man can observe one fact and straightway pronounce judgment concerning another different fact not involved in the first?" (CP 2.690, 1877.)

This latter question was Peirce's starting point in the original derivation of the concept of sign (CP 1.545–559, 1867).

[33] "Though infallibility in scientific matters seems to me irresistibly comical, I should be in a sad way if I could not retain a high respect for those who lay claim to it, for they comprise the greater part of the people who have any conversation at all. When I say they lay claim to it, I mean they assume the functions of it quite naturally and unconsciously. The full meaning of the adage *Humanum est errare*, they have never waked up to. In those sciences of measurement which are the least subject to error — metrology, geodesy, and metrical astronomy — no man of self-respect ever now states his result, without affixing to it its *probable error*; and if this practice is not followed in other sciences it is because in those the probable errors are too vast to be estimated." (CP 1.9, c.1897)

[34] "Theoretically, I grant you, there is no possibility of error in necessary reasoning. But to speak thus "theoretically," is to use language in a Pickwickian sense. In practice, and in fact, mathematics is not exempt from that liability to error that affects everything that man does. [...] The certainty of mathematical reasoning, however, lies in this, that once an error is suspected, the whole world is speedily in accord about it." (CP 5.577, 1898)

[35] On the other hand, beliefs are far from forbidden for a scientist, quite contrary, they are *indispensable* in his/her practical life. Even scientists have to cope with the life world of his/her own and in practical decisions everyone should rely more on his/her instincts and beliefs rather than reason. Especially in matters of vital importance, it would be *unwise* to rely chiefly on reason — reason is too slow and fallible in practice if compared with instincts (no matter whether being culturally or biologically fixed) that are tested in practice by past generations. (CP 1.633–639, 1898.)

"Here we are in this workaday world, little creatures, mere cells in a social organism itself a poor and little thing enough, and we must look to see what little and definite task our circumstances have set before our little strength to do. The performance of that task will require us to draw upon all our powers, reason included. And in the doing of it we should chiefly depend not upon that department of the soul which is most superficial and fallible — I mean our reason — but upon that department that is deep and sure — which is instinct." (CP 1.647, 1898)

[36] The high standards of validity that Peirce gave to the philosophical science is underlined by the modesty with which Peirce judged his own vocation to logic: Peirce claimed that he would not have achieved much scientific results about signs, but that most of his propositions were based only on "a strong impression due to a life-long study of signs" (EP 2:413, 1907). Three years before his death, he still denied having tenable grounds for his "sundry universal propositions concerning signs" (EP 2:462, 1911). It is the insufficient amount of *rational self-control* that makes impressions, even if based on life-long study and even if correct, not enough for true science. Impressions are derived directly from intuitive feelings and the estimation of their validity is beyond rational self-control. Impressions of a scientist are good only for hypotheses, but any claim or *belief* about their validity *because* they are 'due to life-long study' do not belong to science.

BIBLIOGRAPHY

Artmann, Stefan (2006). Two Alternative Ways of Synthesizing Biological Knowledge through Semiotics. This volume.

Barbieri, Marcello 2006. Editorial. This volume.

Barbieri, Marcello 2002. Has Biosemiotics come of age? *Semiotica*, 139(1): 283–295. Reprinted in this volume with a Postscript.

Bickhard, Mark H. 1998. A process model of the emergence of representation. In: Farré, George L. & Oksala, Tarkko (eds.), *Emergence, Complexity, Hierarchy, Organization (Selected and edited papers from ECHO III)*. (*Acta Polytechnica Scandinavica* 91.) Espoo: Finnish Academy of Technology, 263–270.

Bradie, Michael 1986. Assessing Evolutionary Epistemology, *Biology and Philosophy* 1(4): 401–459.

Brent, Joseph 1998 (1st ed. 1993). *Charles Sanders Peirce. A Life*. (2nd ed.) Bloomington: Indiana University Press.

Brier, Søren 2003. The Cybersemiotic Model of Communication: An Evolutionary View on the Threshold between Semiosis and Informational Exchange. *Triple C* 1(1): 71–94. Electronic journal: http://triplec.uti.at/files/tripleC1(1)_Brier.pdf.

Campbell, Donald T. 1974. Evolutionary epistemology. In: Campbell, *Methodology and Epistemology for Social Science. Selected Papers*. (Ed. E. Samuel Overmann.) Chicago: University of Chicago Press 1988.

Campbell, Donald T. 1997. From Evolutionary Epistemology Via Selection Theory to a Sociology of Scientific Validity. (Ed. Celia Hayes & Barbara Frankel.) *Evolution and Cognition* 3(1): 5–38.

Deledalle, Gérard 2000. *Charles S. Peirce's Philosophy of Signs*. Bloomington: Indiana University Press.

Emmeche, Claus & Hoffmeyer, Jesper 1991. From language to nature: the semiotic metaphor in biology. *Semiotica* 84(1/2): 1–42.

Fisch, Max H.1982. Introduction. In: Kloesel *et al.* (eds.), *Writings of Charles S. Peirce*. Vol. 1. Bloomington: Indiana University Press, xv-xxxv.

Fisch, Max H. 1986. *Peirce, Semeiotic, and Pragmatism*. (Ed. Kenneth L. Ketner and Christian Kloesel), Bloomington (Ind.): Indiana University Press.

Hoffmeyer, Jesper 1996. *Signs of Meaning in the Universe*. (*En Snegl På Vejen: Betydningens naturhistorie*, 1993; transl. Barbara Haveland.) Bloomington: Indiana University Press.

Hoffmeyer, Jesper & Emmeche, Claus 1991. Code-duality and the semiotics of nature. In: Anderson, Myrdene & Merrell, Floyd (eds.), *On Semiotic Modeling*. Berlin: Mouton de Gruyter, 117–166. Reprinted with annotations in *Journal of Biosemiotics* 1 (2005): 35–85.

Kant, Immanuel 1781/1787. *Critique of Pure Reason*, (*Kritik der reinen Vernunft*, transl. J.M.D. Meiklejohn), Prometheus Books; Amherst (NY) 1990.

Kent, Beverley 1987. *Charles S. Peirce. Logic and the Classification of the Sciences*. Kingston & Montreal: McGill-Queen's University Press.

Kull K. (1999). Biosemiotics in the twentieth century: a view from biology. *Semiotica* 127(1/4), 385–414.

Lakoff, G. & Johnson, M. 1980. *Metaphors We Live By*. Chicago: University of Chicago Press.

Lorenz, Konrad 1973. *Die Rückseite des Spiegels: versuch einer Naturgeschichte menschlichen Erkennens*. München: Piper.

Peirce, Charles S. 1931–1935, 1958. *Collected papers of C. S. Peirce*. Vols. 1–6 (eds. Charles Hartshorne & Paul Weiss). Vols. 7–8 (ed. Arthur W. Burks). Cambridge: Harvard University Press. [Cited as CP.]

Peirce, Charles S. 1992, 1998. *Essential Peirce. Selected Philosophical Writings*. Vols. 1–2. (Ed. Nathan Hauser *et al.*) Bloomington: Indiana University Press. [Cited as EP.]

Peirce, Charles S. 1982–1986. *Writings of Charles S. Peirce*. Vols. 1–3 (Ed. Christian Kloesel *et al.*) Bloomington: Indiana University Press. [Cited as W.]

Saussure, Ferdinand de 1916. *Course in General Linguistics*. (Transl. Roy Harris, 1983). Chicago & La Salle (Ill.): Open Court 1997.

Schelling, Friedrich Wilhelm Joseph 1985. *Ausgewählte Schriften*. Vols. 1–6 (ed. M. Frank). Frankfurt: Suhrkamp.

Sebeok, T.A. 1963. Communication in Animals and Men. *Language* 39: 448–466.

Sebeok, T.A. 1976 *Contributions to the Doctrine of Signs*. Bloomington: Indiana University Press.

Vehkavaara, Tommi 2002. Why and how to naturalize semiotic concepts for biosemiotics. *Sign Systems Studies* 30(1): 293–313.

Vehkavaara, Tommi 2003. Natural self-interest, interactive representation, and the emergence of objects and *Umwelt*: an outline of basic semiotic concepts for biosemiotics. *Sign Systems Studies* 31(2): 547–587.

Vehkavaara, Tommi 2006. Limitations on applying Peircean semeiotic. Biosemiotics as applied objective ethics and esthetics rather than semeiotic. *Journal of Biosemiotics* 1(2) (in print).

CHAPTER 12

TOWARDS A STANDARD TERMINOLOGY FOR (BIO)SEMIOTICS

MARCEL DANESI
Department of Anthropology, Victoria College University of Toronto, Toronto, Ontario M5S 1K7 Canada

Abstract: Semiotic theory has often been burdened by terminological inconsistencies and especially by the use of concepts and definitions in idiosyncratic ways. This paper aims to provide a framework for eliminating such inconsistencies and idiosyncracies by putting forward a simple system of terms based on Modeling Systems Theory, as developed by the late Thomas A. Sebeok, that takes into account semiotic behavior across species. The four basic forms proposed here (singularized, composite, cohesive, and connective) are defined, illustrated, and defended as solutions for standardizing semiotic terminology and for bringing general semiotic theory more in line with the biosemiotic movement

Keywords: Semiotic terminology, biosemiotics, semiotic theory, modeling systems theory

INTRODUCTION

When all is said and done, the fundamental goal of theoretical semiotics is to figure out how signs are constituted and what kinds of functions they encompass. The terminological frameworks developed by Ferdinand de Saussure (1857–1913) and Charles Sanders Peirce (1839–1914) stand, to this day, as the standard ones for pursuing this objective. But over the last five decades, a perusal of the major writings in semiotic theory and practice reveals that terminological inconsistencies, partisan factions (Saussurean vs. Peircean), and a host of *sui generis* neologisms have sprung up that currently tend to tarnish the image of semiotics as a true scientific enterprise. Even the fact that there is still no real agreement as to what the science itself should be called—*semiology* or *semiotics*—is a symptom of the general vagueness and terminological inconsistency that has beset the field throughout its recent history (Deely 2003). Aware of this state of affairs and of the historical factors

M. Barbieri (ed.), Introduction to Biosemiotics, 283–298.

that brought it about, the late Thomas A. Sebeok (1920–2001) initiated a project of trying to standardize the *modus operandi* of the field starting in the 1960s (e.g. Sebeok 1963, 1972). The project was itself inspired by the biosemiotic movement, which Sebeok promoted with a great deal of enthusiasm. Called the *global semiotic* project, it is still an ongoing concern (Sebeok 2001). The monumental four-volume handbook of semiotic theories and practices, which Sebeok instigated and helped to bring to fruition (Posner, Robering, and Sebeok, 1997–2004), is one of the first concrete indications that global semiotics has started to shape trends in theory and practice. The handbook is, fundamentally, an attempt to shift the focus of semioticians away from the human world of signs to the relation that inheres among the human (anthroposemiotic), animal (zoosemiotic), and plant (phytosemiotic) domains of semiosis. The premise held by Sebeok is that many (if not most) of the disparate terms and concepts used within semiotics can be easily integrated into a simple framework called *Modeling Systems Theory* (MST).

The point-of-departure of MST is a renaming of the basic components of the sign, which is generally defined as the relation **[A stands for B]**, or **[A = B]** for short, in an unambiguous fashion. The **[A]** part is called a *form*, and the **[B]** part the *referent*. The linkage of the two dimensions produces a *model*, the **[A = B]** relation itself. Models exist across species; signs (in the traditional semiotic sense) do not. The crux of MST is that the **[A = B]** structure varies not only according to the modeling process utilized (simulation, indication, etc.), but also according to function. The purpose of this brief chapter is to present the main features of MST in schematic form, since I believe that they cohere into an approach that can truly provide a terminological framework for uniting future work in biosemiotics and semiotics generally.

FORM AND REFERENT

Before discussing MST, it is necessary to justify the use of the terms *form, referent,* and *model*, since these lie at the core of the MST approach. These terms are preferred to terms such as *signifier, signified,* and *signification* (among many others) because they are not only devoid of any partisan view of semiosis, but they overlap considerably with terminological practices in philosophy, psychology, mathematics, and linguistics, thus allowing semiotics to engage in a true meaningful interdisciplinary dialogue with these cognate fields. In Gestalt psychology, for instance, the notion of *form* is central to the understanding of cognition. Gestalt psychologists believe that form is the most important part of experience, giving coherence to each individual element of experience. A referent is simply what a form represents in virtually all fields. Finally, as in mathematics, a *model* is defined as any structure (formula, diagram, etc.) used to explain or describe relationships.

A salient characteristic of organic life is the fact that it has the capacity to produce forms to communicate needs, urges, etc. and (in most cases) to make reference to events and objects in the immediate environment. This capacity goes under the rubric of *semiosis*, and its distinguishing trait is that it allows organisms to model to the world they inhabit in species-specific ways. Human forms are differentiated from all other kinds of forms in that they are imbued with what is generally called "meaning." Indeed, the word *meaning* comes up constantly in semiotics, which is typically defined as the "science of meaning." But, then, what is *meaning*? As Ogden and Richards showed in their pivotal 1923 work, titled appropriately *The Meaning of Meaning*, there are at least 23 meanings of the word *meaning* in English.

To avoid such vagueness and ambivalence, the terms *reference, sense*, and *definition* are often used by cultural semioticians. Reference is the process identifying and naming something real or imaginary; sense is what that something elicits psychologically, historically, and socially; and definition is a statement about what that something refers to. Words may have the same (or similar) referents, but they also have different senses. For example, the "long-eared, short-tailed, burrowing mammal of the family Leporidae" can be called *rabbit* or *hare* in English. Both words refer essentially to the same kind of mammal. But there is a difference of sense—*hare* is the more appropriate term for describing the mammal if it is larger, has longer ears and legs, and does not burrow. Another difference is that a *rabbit* can be perceived as a "pet," while a *hare* is unlikely to be recognized as such.

The German philosopher Gottlob Frege's (1879) emphasis on the role of sense in theories of meaning became shortly thereafter a major area of discussion within both philosophy and semiotics (and continues to be so). Frege argued that sense was as central to meaning as was reference. His famous example was that of the "fourth smallest planet and the second planet from the Sun" as being named both *Venus* and the *Morning Star*. The two terms referred to the same thing, he observed, but they had different senses—*Venus* refers to the planet in a straightforward referential way (nevertheless with implicit allusions to the goddess of sexual love and physical beauty of Roman mythology), while *Morning Star* brings out the fact that the planet is visible in the east just before sunrise. Knowledge of forms (words, figures, etc.), clearly, includes awareness of the senses that they bear in social and historical context—a fact emphasized further by philosopher Willard O. Quine (1953). In his now-classic example, Quine portrayed a linguist who overhears the form *Gavagai* from the mouth of a native informant when a rabbit is sighted scurrying through the bushes. But the linguist, Quine goes on to remark, cannot determine if the word means "rabbit," "undetached rabbit parts," or "rabbit stage," all of which are senses of that word. The sense of the form, therefore, will remain indeterminate unless it can be inferred from the context in which *Gavagai* occurs.

Definition, as mentioned, is a statement about what something means by using words and other signs (for example, pictures). As useful as it is, the act of defining leads inevitably to circularity. Take the dictionary definition of *cat* as "a small carnivorous mammal domesticated since early times as a catcher of rats and mice and as a pet and existing in several distinctive breeds and varieties." One of the problems that emerges from this definition is the use of *mammal* to define *cat*. In effect, one term has been replaced by another. So, what is the meaning of *mammal*? A *mammal*, it states, is "any of various warm-blooded vertebrate animals of the class Mammalia." But this definition is hardly a viable solution. What is an *animal*? The dictionary defines *animal* as an *organism*, which it defines, in turn, as an individual form of *life*, which it then defines as the property that distinguishes living *organisms*. Alas, at that point the dictionary has gone into a referential loop, since it has employed an already-used concept, *organism*, to define *life*. This looping pattern surfaces in all definitions. It suggests that signs can never be understood in the absolute, only in relation to other signs.

In contemporary semiotics, the terms *denotation* and *connotation* are preferred to reference and sense. Consider, again, the word *cat*. The word elicits an image of a "creature with four legs, whiskers, retractile claws," etc. This is its *denotative* meaning, which is really a mental picture of *cat* in terms of specific features that are perceived to define cats in general—"retractile claws," "long tail," etc. The denotative meaning allows users of signs to determine if something real or imaginary under consideration is an exemplar of a "cat." Similarly, the word *square* elicits a mental image characterized by the distinctive features "four equal straight lines" and "meeting at right angles." It is irrelevant if the lines are thick, dotted, 2 meters long, 80 feet long, or whatever. If the figure has "four equal straight lines meeting at right angles," it is denotatively a square. The word *denotation*, incidentally, is derived from the compound Latin verb *de-noto* "to mark out, point out, specify, indicate." The noun *nota* ("mark, sign, note") itself derives from the verb *nosco* ("to come to know," "to become acquainted with" and "to recognize").

All other senses associated with the words *cat* and *square* are connotative meanings—that is, they are meanings that are derivational or extensional and thus culture-specific. Some connotative senses of *square* can be seen in expressions such as the following:

She's so *square*.	=	"old fashioned"
He has a *square* disposition.	=	"forthright," "honorable"
Put it *squarely* on the table.	=	"evenly," "precisely"

Notice that an old-fashioned person, an honorable individual, and the action of laying something down evenly nevertheless imply the denotative meaning of "square." The concept of *square* is an ancient one and, thus, probably known by everyone (hence "old-fashioned"); it is also a figure with every part equal (hence "forthright"); and it certainly is an even-sided figure (hence "evenly"). Connotation encompasses all kinds of senses, including emotional ones. Consider the word *yes*. In addition to being a sign of affirmation, it can have various emotional senses, depending on the tone of voice with which it is uttered. If one says it with a raised tone, as in a question, "Yes?" then it would convey doubt or incredulity. If articulated emphatically, "Yes!" then it would connote triumph, achievement, or victory.

Connotation is the operative sense-making and sense-extracting mode in the production and decipherment of creative texts such as poems, novels, musical compositions, art works—in effect, of most of the non-technical texts that people create. But this does not imply that meaning in technical (information-based) domains unfolds only denotatively. On the contrary, many (if not all) scientific theories and models are constructed connotatively, even though they end up being interpreted denotatively over time. Above all else, it should be emphasized that connotation is not an option, as some traditional philosophical and linguistic theories of meaning continue to sustain to this day; it is something we are inclined to extract from a form. Even something as apparently denotative as a common digit invariably induces connotative meanings. The numbers 7 and 13 in many cultures reverberate with meanings such as "fortune," "destiny," "bad luck," and so on. These are hardly fanciful or dismissible. They tend to have real-world consequences, despite their apparent superstitious senses. This can be seen, for instance, in the fact that many high rise buildings in North America do not label the "thirteenth floor" as such, but rather as the "fourteenth," in order to avoid the possibility of inviting the bad fortune associated connotatively with the number13 to the building and its residents.

Abstract concepts, such as *motherhood, masculinity, friendship,* and *justice*, are particularly high in connotative content. In 1957, the psychologists Osgood, Suci, and Tannenbaum showed this empirically by using a technique that they called the *semantic differential.* The technique allows investigators to flesh out the connotative (culture-specific) meanings that abstract concepts elicit. It consists in posing a series of questions to subjects about a particular concept—*Is X good or bad? Should Y be weak or strong?* etc. The subjects are then asked to rate the concept on seven-point scales. The ratings are subsequently collected and analyzed statistically in order to sift out any general pattern they might bear.

Suppose that subjects are asked to rate the concept "ideal American President": for example, *Should the President be young or old? Should the President be practical or idealistic? Should the President be modern or traditional?* and so on:

Young							*old*
	1	2	3	4	5	6	7
practical							*idealistic*
	1	2	3	4	5	6	7
modern							*traditional*
	1	2	3	4	5	6	7
Attractive							*bland*
	1	2	3	4	5	6	7
friendly							*stern*
	1	2	3	4	5	6	7

A subject who feels that the President should be more "youngish" than "oldish" would place a mark towards the *young* end of the top scale; one who feels that a *President* should be "bland," would place a mark towards the *bland* end of the *attractive-bland* scale; and so on. If we were to ask a large number of subjects to rate the President in this way, we would get a "connotative profile" of the American presidency in terms of the statistically significant variations in sense that it evokes. Interestingly, research utilizing the semantic differential has shown that the range of variations is not a matter of pure subjectivity, but reveals, rather, a socially-based pattern. In other words, the connotations of many (if not all) abstract concepts are constrained by culture: for example, the word *noise* turns out to be a highly emotional concept for the Japanese, who rate it consistently at the ends of the scales presented to them; whereas it is a fairly neutral concept for Americans, who tend to rate it on average in the mid-ranges of the same scales. Connotation is not, therefore, open-ended; it is constrained by a series of factors, including conventional agreements as to what signs mean in certain situations. Without such constraints, our forms would be virtually unusable. All semiosis (whether it is denotative or connotative) is a relational and associative process—that is, signs acquire their meanings not in isolation, but in relation to other signs and to the contexts in which they occur.

The distinction between denotation and connotation is analogous to Frege's distinction between reference and sense. And indeed these terms are used interchangeably in the relevant semiotic literature, as are Rudolf Carnap's (1947) terms intension (= denotation) and extension (= connotation). While there are subtle differences among these terms, it is beyond the present purpose to compare them. Suffice it to say that in current semiotic practice they are virtually synonymous:

reference	=	denotation	=	intension
sense	=	connotation	=	extension

The use of the denotation vs. connotation dichotomy is often credited to philosopher John Stuart Mill (1806–1873) but, in actual fact, it can be traced back to the medieval Scholastics, and in particular to William of Ockham (c. 1284-c. 1347). In both Ockham and Mill, however, connotation is used is a specialized way—it designates the specific referents to which a term can be applied. The distinction between denotation and connotation as we understand it today, on the other hand, was used for the first time by the American linguist Leonard Bloomfield in his seminal 1933 book called *Language*, a distinction elaborated by the Danish linguist Louis Hjelmslev (1970) a little later. Although Hjelmslev's treatment is a highly abstruse and largely confusing one, it nevertheless had the effect of putting this basic distinction on the semiotic agenda once and for all. Especially relevant is Hjelmslev's characterization of connotation as a "secondary semiotic system" for expressing subjective meanings. The French semioticians Roland Barthes (1977) and Algirdas J. Greimas (1987) subsequently made it obvious that connotation was an inbuilt feature of sign.

In MST the terms denotation and connotation continue to be used. However, the referents that they imply, called *denotata* and *connotata* respectively, are not unless they are required. Connotation is a uniquely anthroposemiotic phenomenon. If the case arises then a referent can be further specified as denotative or connotative. A referent is, thus, anything that is given form; it can be a mating request signaled by a coo, a territoriality sign (such as urination), and so on and so forth. Most referents manifest themselves as signals in animals; only in the human world do they crystallize as true signs (symbols, words, etc.) with connotative values. The relation itself, **[A = B]**, is a model, a form that is connected to a referent in some discernible way. In MST, therefore, it can stand for a signal, a sign, a text, etc.—anything that stands for something other than itself in some specific way and according to the biology of the species that emits and is capable of receiving it.

For example, most male birds sing from a series of perches that outlines their territory (the area claimed and defended). A typical song has two main purposes. (1) It warns other males of the same species to stay out of the territory; (2) it attracts a mate. To human ears, the songs of all the birds of a particular species may sound alike. However, each bird's voice sounds different to the other members of the species. Even in a crowded colony, parent birds can single out the voices of their chicks, and chicks recognize those of their parents. In MST terms, this is so because their forms are recognizable by each species as are their referents.

In 1974 Marcel Florkin suggested that the concepts of *signifier* and *signified* were equivalent to *genotype* and *phenotype* respectively, proposing that biosemiotics assumes the basic Saussurean conception of semiosis. Barbieri (1985, 2003) has, however, insinuated later that this proposal is not completely satisfactory since a cell has a triarchic structure consisting of genotype, phenotype and ribotype dimensions (the ribotype is the ribonucleoprotein system). So, rather than use terms that refer to human semiosis (signifier and signified, for instance), and which have a particular tradition within semiotics proper, the basic terminological proposal by MST is that simple terms such as *form* and *referent* are preferable because they allow for a larger inclusivity of phenomena. In effect, form is the genotype, referent the phenotype, and model the ribotype minus the specific biological connotations that such terms imply.

Model-making is especially prolific and creative in anthroposemiosis. Before building a house, a constructor will make a miniature model of it and/or sketch out its structural features with the technique of blueprinting. Explorers will draft a map of the territory they anticipate investigating. A scientist will draw a diagram of atoms and subatomic particles in order to get a "mental look" at their physical behavior. Miniature models, blueprints, maps, diagrams, and the like are so common that one hardly ever takes notice of their importance to human life; and even more rarely does one ever consider their *raison d'être* in the human species. Human model-making constitutes a truly astonishing evolutionary attainment, without which it would be virtually impossible for people to carry out their daily life routines. All this suggests the presence of a *modeling instinct* that is to human mental and social life what the physical instincts are to human biological life. In effect, the main tenet of MST is that the forms made by humans to understand the world result from this instinct. The modeling process can be simulative, whereby the form is designed to simulate its referent in some way (through resemblance, imitation, etc.). It can also be relational. The pointing finger is a relational form that is designed to show the spatial relation of some referent to the pointer or to some other referent. And, of course, it can be based on cultural conceptualizations. The form made with the index and middle finger in the shape of a "V" to stand for "peace" is the outcome of specific cultural events (which need not concern us here).

But modeling instincts occur in other species as well. As various contemporary biosemioticians have cogently argued (e.g. Hoffmeyer 1996, Sebeok 2001), such "instincts" are really strategies that allow members of the same species to convey information. For instance, the mating behavior of a fish called the three-spined stickleback includes many examples of instinctive modeling behavior. The male stickleback chooses a mating area and drives other fish from it. He then collects plants and shapes them into a small mound. He wriggles through the completed mound, creating a tunnel, which is slightly shorter than the fish. The mound is a mating nest. Meanwhile, his normally dull-colored body undergoes a change in color—his belly becomes bright red and his back bluish-white. The male then starts to court females. Whenever a female enters his mating area, he swims towards her and performs a zigzagging "dance." He continues dancing until a female follows him to his nest, where she lays her eggs. The female then swims away, and the male fertilizes the eggs, staying near the nest to protect the eggs and, later, the neonates. Mound, tunnel, and dance are all examples of models in the biosemiotic sense being discussed here. Each male stickleback is born with the capacity to make such models built into his nervous system.

At this point, it is clearly obvious that it is necessary to distinguish between forms and models as they occur in Nature and as they are developed in cultural settings. The products of instinctive modeling tendencies can be called, simply, *natural forms* and those shaped by culture *artifactual forms*, i.e. forms made intentionally to stand for something. In animal species the range of artifactual modeling will, clearly, be much more limited than it is in the human species.

There are four general types of forms: *singularized, composite, cohesive*, and *connective* (Sebeok and Danesi 2000). In traditional biological theory *singularized forms* are called *signals* across species (in anthropo-, zoo-, and phytosemiotics). The term *sign* is used instead to set human semiosis apart from animal semiosis. In MST, on the other hand, a sign is classified in the same category as a signal, since it is defined as a singularized form that is specific to the human domain that models a singular (unitary) referent or referential domain. Singularized forms can be verbal or nonverbal. The English word *cat*, or the Spanish word *gato*, for example, are verbal singularized forms standing for the referent "carnivorous mammal with a tail, whiskers, and retractile claws;" a drawing of a house cat is its nonverbal (visual) equivalent. Now, a description of the same referent as "a popular household pet that is useful for killing mice and rats" constitutes, clearly, a different kind of form. This is known traditionally as a descriptive *text*. In MST, a text can be defined, more exactly, as a *composite form*; i.e. as a form that has been made to represent various aspects of a referent or referential domain—"household pet," "killing mice," etc.—in a composite (combinatory) manner. Although texts do not exist in other species, composite forms do. The well-known example of the "bee dance" is a case-in-point. Classifying a *cat* in the same category as a *tiger, lion, jaguar, leopard, cheetah*, etc. exemplifies a third type of modeling strategy—namely, the tendency to perceive certain forms as *cohering* into a general model of something. This is known traditionally as a *code*. In MST, a code can be defined as a system of forms that allows for the representation or communication of referents perceived to share common traits—e.g. *cat, tiger, lion, jaguar*, etc. = the feline code. Codes can also be natural or artifactual. The best-known example of a natural code (literally Nature's Code in this case) is the genetic code. The genetic code lies in the order of the bases in the DNA molecule. This order of bases is passed on from one generation of cells to the next, and from one generation of an organism to the next. It makes a dog give birth to a dog, not a cat. It is this order that determines the color of eyes, the shape of ears, and thousands of other traits. Finally, the use of the word *cat* in an expression such as "He's a cool *cat*" is the result of a fourth type of modeling strategy, known traditionally as *figural*. In MST, the term *connective form* is preferred instead, because a figural sign (a metaphor, a metonym, etc.) is more precisely a model connecting one type of referent (or referential domain) to another. In the above case, a human referent is connected to a feline referent. Connective forms are unique to human semiosis.

These four types of modeling strategies are not mutually exclusive. Indeed, they are highly interdependent—singularized forms go into the make-up of composite forms which, in turn, are dependent upon the forms that cohesive systems make available. MST thus provides a simple framework for comparing human semiosis with animal and plant semiosis.

MODELING SYSTEMS THEORY

Although MST has roots in the work of various twentieth-century semioticians, it has never really blossomed forth as a comprehensive theoretical and methodological framework for general use in semiotics until Sebeok's pivotal work, which itself is really a particular interpretation of ideas found in the writings of the biologist Jacob von Uexküll (1909) and of various members of the so-called Tartu School of Semiotics, of which Yuri Lotman (1991) is probably the best known. Four general principles underlie the MST perspective:

1. Species-specific understanding of the world is indistinguishable from the forms used to model it (the *modeling principle*).
2. Modeling unfolds in various ways, from simulation to indication and symbolism (the *variability principle*).
3. Models and their referential domains are interconnected to each other (the *interconnectedness principle*).
4. All forms display the same pattern of structural properties (the *structuralist principle*).

Needless to say, it is not possible to go here into the many interesting philosophical problems related to what is knowledge. The *modeling principle* implies simply that in order for something to be known and remembered, it must be assigned some form. The *variability principle* implies that modeling varies according to the referent and to the function of the modeling system. The *interconnectedness principle* asserts that a specific form is interconnected to other forms (words to gestures, diagrams to metaphors, etc.). The *structuralist principle* claims that certain elemental structural properties characterize all forms (selection, combination, etc.). These are well known and need not be discussed here.

The first true scientific-philosophical study of signs was the one carried out by St. Augustine (354–430 AD). This philosopher and religious thinker was among the first to distinguish clearly between *natural* and *conventional* (artifactual) signs, and to espouse the view that there was an inbuilt *interpretive* component to the whole process of representation (Deely 2001, 2003). It was, as well known, the British philosopher John Locke (1632–1704) who introduced the formal study of signs into philosophy in his *Essay Concerning Human Understanding* (1690), anticipating that it would allow philosophers to understand the interconnection between representation and knowledge. But the task he laid out remained virtually unnoticed until Saussure and Peirce took it upon themselves to provide a scientific terminology that made it possible to envision even more than what Locke had hoped for—namely, an autonomous field of inquiry centered on the sign. Their terminology, as argued here, is particularly useful in understanding human semiosis in and of itself. Peircean terminology has currently arisen to be the dominant form, probably because it clearly expands the semiotic paradigm (at leas implicitly) to include basic semiosic propensities such as simulation and indication. The key concept in both Saussurean and Peircean views of the sign is that no single form can bear meaning unless it enters into systematic connections with other forms. A primary goal of MST, and of biosemiotics generally, is to show how these connections crystallize.

Incidentally, the origin of biosemiotics as a distinct contemporary mode of inquiry can be traced to the work of Jacob von Uexküll (1909), since it was von Uexküll who was the first to describe communication processes in an interspecies fashion (at least to the best of my knowledge). His basic proposal was that the Innenwelt (inner world) of an organism is well adapted to interpret the Umwelt (the outer world it inhabits) in a specific way and thus to generate species-specific models of it. His use of *model* and *form* as generic terms is at the basis of MST.

Let us now look more closely at the four types of forms. The function of a singularized form, as mentioned, is to make reference to single objects, unitary events, individual feelings, etc. in some way (Thom 1975, Sebeok 1994). Throughout the history of semiotics, there have been several attempts to identify and classify them. Among these, Peirce's typology with 66 varieties is surely the most comprehensive, far-reaching, and sophisticated of all such attempts. In the verbal domain, one can also mention Roman Jakobson's (1970) classificatory system, which has shed considerable light on the minutiae of singularized verbal modeling. A singularized form that results from an attempt at simulation is known in Peircean theory as an *icon*. In this case the form **[A]** is a simulative model of **[B]**, its referent: i.e., it results from a modeling strategy that attempts to capture some sensory property of the referent through resemblance, imitation, etc. In human semiosis, Roman numerals such as I, II, and III are iconic forms because they are designed to resemble their referents in a visual way (one stroke = one unit, two strokes = two units, three strokes = three units); onomatopoeic words (*boom, zap, whack*, etc.) are also products of iconic modeling because they constitute attempts to portray referents in an acoustic way; commercially-produced perfumes that are suggestive of certain natural scents are likewise iconic because they attempt to model scents in an artificial way; and so on.

There are endless manifestations of iconicity in zoosemiosis, involving all types of sensory channels—chemical, auditory, visual, etc. Unlike human iconicity, however, they are (in all likelihood) tied to biological functions. An elegant (if sometimes disputed) example of iconic animal behavior is graphically described by Kloft (1959). Kloft suggested that the hind end of an aphid's abdomen, and the kicking of its hind legs, constituted, for an ant worker, iconic models, standing for the head of another ant together with its antennae movement. The ant can purportedly identify the likeness (the near end of the aphid) with its meaning (the front end of an ant), and act on this information, i.e. treat the aphid in the manner of an effigy. Camouflage too is a natural iconic modeling system. Many animals are difficult for enemies to see because they resemble their surroundings. For instance, a dark moth lying against the brown or black bark of a tree is hard to see. However, that same moth would be clearly visible if it sat on a green leaf. A number of animals can change their colors and thus remain camouflaged even when moving among backgrounds that have different colors. The chameleon, a type of lizard, is green when surrounded by leaves but turns brown when moving slowly on bark or on the ground. The ptarmigan, an Arctic bird, is brown in summer but becomes white in winter, when snow covers the ground.

A singularized form that results from an attempt to indicate some relation is known as an *index*. Indexical forms do not resemble their referents, like iconic ones do. They indicate or show where they are in relational, spatial, or temporal terms. In this case the modeling process consists in putting the form user in relation to a referent, or several referents in relation to each other. Words such as *here*, *there*, *up*, *down*, etc. have this quality. They indicate the location of the form-user in relation to the referent (as near, up, etc.). Arrows used as semaphores and the index finger used to point out things are other examples of indexical forms.

Natural indexicality is also manifest in various animal species. In fact, a vast map of indexical marks is printed overnight by animals of all sorts (Ennion and Tinbergen 1967). Tracks, scents, and other forms that identify the location or existence of a certain animal in a species-specific way are all indexical forms. For example, a wolf marks out its territory by urinating on bushes, rocks, and other objects. The scent of urine warns intruders of the wolf's presence and the risk of an encounter. The more aggressive forms of defense generally are used when the intruder is especially persistent.

A singularized form that results from some culture-specific convention is known, of course, as a symbol: e.g. a cross figure standing for "Christianity;" a V-sign made with the index and middle fingers standing for "peace;" the color white standing for "cleanliness," "purity," "innocence," and so on. Several societies may use the same symbols, but they will tend to stand for different referents. In many societies, for example, the color red symbolizes war and violence. But in China red represents marriage. Among American Aboriginal peoples, it stands for the East. Red symbolizes life in the Shinto religion of Japan, but law schools in France. In effect, a symbol has only the meaning that people have given it. In their mythologies, people have used symbols to help explain the world. The Greeks symbolized the sun as the god Helios driving a flaming chariot across the sky. The Egyptians represented the sun as a boat. Animals, human beings, and plants have all stood for ideas and events. Some groups adopted the serpent as a symbol of health because they believed that by shedding its skin, the serpent becomes young and well again. The Greeks portrayed Asclepius, the god of healing, holding a staff with a serpent coiled around it. The staff is often confused with the caduceus of the god Mercury, which has two snakes coiled around it. Today, both are used as symbols of the medical profession.

Symbols are rare in the zoosemiotic domain, but not completely absent from it. Examples such as the stickleback one mentioned above seem to verge on the symbolic. Indeed, rudimentary manifestations of natural symbolism can be found in Nature if one looks for them (Pitts and McCulloch 1947, Haldane 1955, Sebeok 1973, Jacob 1974). A rhesus monkey, for instance, shows fear by carrying its tail stiffly out behind; baboons convey fear by carrying a vertical tail. However, the converse is not necessarily true: "a mother of a young infant [baboon] may hold her tail vertical not in fear but to help her infant balance on her back; and the tail may also be held vertical while its owner is being groomed in the tail region" (Rowell 1972: 87). This is, arguably, symbolic behavior given that tail orientation stands in an indirect fashion for an emotion. Consider, further, the behavior of the insects of the carnivorous family Empididae. In a species of dipterans of this family, the male offers the female an empty balloon prior to copulation (Huxley 1966). The evolutionary origin of this seemingly bizarre gesture has been unraveled by biologists. But the fact remains that the gift of an empty balloon is a wholly symbolic act, designed simply to reduce the probability that the male himself will fall prey to his female partner.

Composite modeling, as mentioned, is the activity of representing complex (non-unitary) referents by combining various forms in some specifiable way. Drawings, narratives, theories, conversations, etc. are all examples of composite forms in the human domain. These are constructed with distinct singularized forms that fit together structurally, but which are, as a whole, different from any one taken individually. In analogy to atomic theory, a singularized form can be compared to an atom and a composite form to a molecule made up of individual atoms, but constituting a distinct physical form in its own right. Salt is made up of sodium and chlorine, but is not a simple combination of the two.

Composite modeling occurs in all facets of human life, allowing people to envision distinct bits of information and real-world referents as integrated wholes. Such modeling is also found in animals. A striking example can be seen in the behavior of a small family of cerophagous picarian birds (*Indicator indicator*). This species has developed a remarkable symbiotic relationship with certain mammals—ratels, baboons, and humans—by employing a strategy that guides other birds to the vicinity of wild bees' nests. A would-be guiding bird will come to, say, a person, and chatter until followed, but keep out of sight of the pursuer most of the time (Friedmann 1965).

As mentioned, a cohesive model is known in traditional semiotic theory as a code. The Roman numeral system is a simple example of a cohesive system fashioned in part iconically. This system consists of seven symbols for representing all numbers from 1 to 1,000,000: I for 1, V for 5, X for 10, L for 50, C for 100, D for 500, and M for 1000. An example of an indexical code is the system of street signs used typically to regulate and guide traffic. These provide such information as the distance of certain places from specific locations, the direction one is traveling in, etc. Essentially codes are organizational grids of referents and referential domains.

Natural cohesive modeling is found in the animal kingdom, albeit not in the same way that it is found in human semiosis. A remarkable example of a cohesive model can be observed in the mound constructions that are engineered by common termites. These social insects have the ability to construct extremely hard walls from bits of soil cemented with saliva and baked by the sun. Inside the walls numerous chambers and galleries are constructed by these ingenious engineers, interconnected by a complex network of passageways. Ventilation and drainage are provided, and heat required for hatching the eggs is obtained from the fermentation of organic matter, which is stored in the chambers serving as nurseries. Of more than 55 species common in the United States, the majority builds its nests underground. The subterranean termites are extremely destructive, because they tunnel their way to wooden structures, into which they burrow to obtain food. Now, upon close examination, the mound structure simulates the constituents of the termite's social evolution, even after the colony itself has become extinct—i.e. the mound visually mirrors the social organization of these architect insects. This is a dramatic example of unwitting cohesive iconicity manifesting itself in Nature as a property of a species' social behavior (Sebeok 1991).

Finally, connective forms are the result of associative reasoning—a cognitive capacity that is unique to humans. The ever-burgeoning literature on what has come to be known as *conceptual metaphor theory* (e.g. Lakoff and Johnson 1980, 1999, Lakoff 1987, Johnson 1987, Gibbs 1994, Goatley 1997, Fauconnier and Turner 2002) has made it obvious that such forms permeate human communicative and representational behavior. A connective form results when abstract referents are linked to concrete ones. The formula *thinking is seeing*, for example, is a connective form because it delivers the abstract concept of "thinking" **[A]** in terms of the physiological processes associated with seeing **[B]** This underlies utterances such as:

1. I do not *see* what possible use your *ideas* might have.
2. I can't quite *visualize* what that new *idea* is all about.
3. Just *look at* her new *theory*; it is really something!
4. I *view* that *idea* differently from you.

A specific metaphorical statement uttered in a discourse situation is now construable as a particular externalization of a connective form. So, when we hear people using such statements as those cited above, it is obvious that they are not manifestations of isolated, self-contained metaphorical creations, but rather, specific instantiations of a form connecting thinking with seeing.

A connective form may also be the product of metonymic modeling. Metonymy entails the use of an entity to refer to another that is related to it. A metonymic model results when part of a domain starts being used to represent the whole domain (Lakoff and Johnson 1980: 35–40):

5. She likes to read Dostoyevski (= the writings of Dostoyevski).
6. He's in dance (= the dancing profession).
7. My mom frowns on blue jeans (= the wearing of blue jeans).
8. Only new wheels will satisfy him (= car).

Each one of these constitutes an externalization of a metonymically-derived model: (5) is an instantiation of *the author is his or her work*, (6) of an *activity of a profession is the profession*, (7) of a *clothing item represents a lifestyle*, and (8) of a *part of an object represents the entire object*.

CONCLUDING REMARKS

The disciplinary status of semiotics as a "science" has always been a topic of debate. Indeed, many semioticians have refused (and continue to refuse) to call their field a science, preferring to define it with terms like "activity," "tool," "doctrine," "theory," "movement," "approach" (Nöth 1990: 4). However, just as many perceive semiotics to have the necessary characteristics to qualify it as a scientific enterprise. Umberto Eco (1978), for one, argues that semiotics meets five basic criteria that characterize any science:

1. it is an autonomous discipline;
2. it has a set of standardized methodological tools;
3. it has the capability of producing hypotheses;
4. it affords the possibility of making predictions;
5. its findings may lead to a modification of the actual state of the objective world.

Lacking from this list, however, is the presence within semiotics of a set of terms for classifying semiotic observations and then discussing and communicating them in unambiguous ways. MST forces all semioticians to reflect upon the foundational notions of their science and, more precisely, on how these can be communicated and discussed in true scientific fashion. Sebeok left us an exciting new agenda for doing exactly this.

REFERENCES

Barbieri, M. (1985). *The Semantic Theory of Evolution*. London: Harwood Academic Publishers.

Barbieri, M. (2003). *The Organic Codes. An Introduction to Semantic Biology*. Cambridge: Cambridge University Press.

Barthes, R. (1977). *Elements of Semiology*. New York: Atlantic Books.

Bloomfield, L. (1933). *Language*. New York: Holt, Rinehart, and Winston.

Carnap, R. (1956 [1947]). *Meaning and Necessity: A Study in Semantics and Modal Logic*. Chicago: University of Chicago Press.

Danesi, M. (2003). *Poetic Logic: The Role of Metaphor in Thought, Language, and Culture*. Madison, WI: Atwood Publishing.

Deely, J. (2001). *Four Ages of Understanding: The First Postmodern Survey of Philosophy from Ancient Times to the Turn of the Twentieth Century.* Toronto: University of Toronto Press.
Deely, J. (2003). *Why Semiotics?* Ottawa: Legas Press.
Eco, U. (1978). Semiotics: A Discipline or an Interdisciplinary Method. In: A. Sebeok (ed.), *Sight, Sound, and Sense*, 73–88. Bloomington: Indiana University Press.
Ennion, E. R. and Tinbergen, N. (1967). *Tracks.* Oxford: Oxford University Press.
Fauconnier, G. and Turner, M. (2002). *The Way We Think: Conceptual Blending and the Mind's Hidden Complexities.* New York: Basic.
Florkin, M. (1974). Concepts of Molecular Biosemiotics and Molecular Evolution. In M. Florkin and E H. Stotz (eds.), *Comprehensive Biochemistry*, Molecular *Evolution*, 1–124. Amsterdam: Elsevier.
Frege, G. (1879). *Begiffsschrift eine der Aritmetischen nachgebildete Formelsprache des reinen Denkens.* Halle: Nebert.
Friedmann, H. (1955). The Honey-Guides. *U.S. National Museum Bulletin* 208. Washington, D.C.: Smithsonian.
Gibbs, R. W. (1994). *The Poetics of Mind: Figurative Thought, Language, and Understanding.* Cambridge: Cambridge University Press.
Goatley, A. (1997). *The Language of Metaphors.* London: Routledge.
Greimas, A. J. (1987). *On Meaning: Selected Writings on Semiotic Theory.* P. Perron and F. H. Collins (trans.). Minneapolis: University of Minnesota Press.
Haldane, J. B. S. (1955). Animal Communication and the Origin of Human Language. *Science Progress* 43, 385–401.
Hoffmeyer, J. (1996). *Signs of Meaning in the Universe.* Bloomington: Indiana University Press.
Huxley, J. (1966). A Discussion of Ritualization of Behaviour in Animals and Men. *Philosophical Transactions of the Royal Society of London* 251, 247–526.
Jacob, F. (1974). *The Logic of Living Systems: A History of Heredity.* London: Allen Lane.
Jakobson, R. (1970). Language in Relation to Other Communication Systems. In C. Olivetti (ed.), *Linguaggi nella società e nella tecnica*, 3–16. Milano: Edizioni di Comunità.
Johnson, M. (1987). *The Body in the Mind: The Bodily Basis of Meaning, Imagination and Reason.* Chicago: University of Chicago Press.
Kloft, W. (1959). Versuch einer Analyse der Trophobiotischen Beziehungen von Ameisen zu Aphiden, *Biologische Zentralblatt* 78, 863–870.
Lakoff, G. (1987). *Women, Fire and Dangerous Things: What Categories Reveal about the Mind.* Chicago: University of Chicago Press.
Lakoff, G. and Johnson, L. (1980). *Metaphors We Live By.* Chicago: Chicago University Press.
Lakoff, G. and Johnson, M. (1999). *Philosophy in the Flesh: The Embodied Mind and Its Challenge to Western Thought.* New York: Basic.
Lakoff, George and Núñez, R. E. (2000). *Where Mathematics Comes From: How the Embodied Mind Brings Mathematics into Being.* New York: Basic Books.
Locke, J. (1690). *An Essay Concerning Human Understanding.* London: Collins.
Lotman, Y. (1991). *Universe of the Mind: A Semiotic Theory of Culture.* Bloomington: Indiana University Press.
Nöth, W. (1990). *Handbook of Semiotics.* Bloomington: Indiana University Press.
Ogden, C. K. and Richards, I. A. (1923). *The Meaning of Meaning.* London: Routledge and Kegan Paul.
Osgood, C. E., Suci, G. J., and Tannenbaum, P. H. (1957). *The Measurement of Meaning.* Urbana: University of Illinois Press.
Peirce, C. S. (1931–1958). *Collected Papers.* Cambridge, Mass.: Harvard University Press.
Pitts, W. and McCulloch, W. S. (1947). How We Know Universals: The Perception of Auditory and Visual Forms. *Bulletin of Mathematical Biophysics* 9, 127–149.
Posner, R. Robering, K., and Sebeok, T. A. (eds.) (1997–2004). *Semiotik/Semiotics. Ein Handbuch zu den zeichentheoretischen Grundlagen von Natur und Kultur/A Handbook on the Sign-Theoretic Foundations of Nature and Culture,* 4 vols. Berlin–New York: Mouton de Gruyter.
Quine, W. V. O. (1953). *From a Logical Point of View.* Cambridge: Harvard University Press.
Rowell, T. (1972). *The Social Behaviour of Monkeys.* Harmondsworth: Penguin.

Saussure, F. de. (1916). *Cours de linguistique générale.* Paris: Payot.
Sebeok, T. A. (1963). Communication among Social Bees; Porpoises and Sonar; Man and Dolphin. *Language* 39, 448–466.
Sebeok, T. A. (1972). *Perspectives in Zoosemiotics.* The Hague: Mouton.
Sebeok, T. A. (1994). *Signs: An Introduction to Semiotics.* Toronto: University of Toronto Press.
Sebeok, T. A. (2001). *Global Semiotics.* Bloomington: Indiana University Press.
Sebeok, T. A. and Danesi, M. (2000). *The Forms of Meaning: Modeling Systems Theory and Semiotics.* Berlin: Mouton de Gruyter.
Thom, R. (1975). *Structural Stability and Morphogenesis: An Outline of a General Theory of Models.* Reading: W.A. Benjamin.
Uexküll, J. von (1909). *Umwelt und Innenwelt der Tierre.* Berlin: Springer.

CHAPTER 13

INFORMATION THEORY AND ERROR-CORRECTING CODES IN GENETICS AND BIOLOGICAL EVOLUTION

GÉRARD BATTAIL

E.N.S.T., Paris, France (retired), E-mail: gbattail@club-internet.fr

Abstract: As semiotics itself, biosemiotics is concerned with semantics. On the other hand, the scientific study of communication engineering led to the development of information theory, which ignores semantics. For this reason, many biologists thought that it would be useless in their disciplines. It turns out however that problems of communication engineering are met in biology and thus can only properly be dealt with using information theory. As an important example, the faithful transmission of genetic information through the ages is a difficult problem which has been overlooked by biologists. Cumulated errors in the DNA molecule due to radiations and even to its own indeterminism as a quantum object actually perturb its communication through time. A simple information-theoretic computation shows that, contrary to the current belief, the genomic memory is ephemeral at the time scale of geology. The conventional template-replication paradigm is thus not tenable. According to a fundamental theorem of information theory, error-correcting codes can perform almost errorless communication provided certain conditions are met. Faithful conservation of genomes can thus be ensured only if they involve error-correcting codes. Then the genomes can be recovered with an arbitrarily small probability of error, provided the interval between successive generations is as short (at the time scale of geology) as to almost always avoid that the number of cumulated errors exceeds the correcting ability of the code

This paper presents an intuitive outline of information theory and error-correcting codes, and briefly reviews the consequences of their application to the problem of genome conservation. It discusses the possible architecture of genomic error-correcting codes, proposing a layered structure referred to as 'nested codes' which unequally protects information: the older and more fundamental it is, the better it is protected. As regards the component codes of this system, we notice that the error-correcting ability of codes relies on the existence of constraints which tie together the successive symbols of a sequence. It is convenient in engineering to use mathematical constraints implemented by physical means for performing error correction. Nature is assumed to use to this end 'soft codes' with physico-chemical constraints, in addition to linguistic constraints that the genomes need for directing the construction and maintenance of phenotypes. The hypotheses that genomic error-correction means exist and take the form of nested codes then suffice to deduce many features of the living world and of its evolution. Some of these features

M. Barbieri (ed.), Introduction to Biosemiotics, 299–345.

are recognized biological facts, and others answer debated questions. Most of them have no satisfactory explanation in current biology. The theoretical impossibility of genome conservation without error-correcting means makes these consequences as necessary as the hypotheses themselves. The direct identification of natural error-correcting means is still lacking, but one cannot expect it to be performed without the active involvement of practising geneticists. The paper also briefly questions the epistemological status of the engineering concept of information and its possible relation to semantics. Roughly stated, information appears as a necessary container for semantics, providing a bridge between the concrete and the abstract

Keywords: Biological evolution, error-correcting codes, genome conservation, genomic channel capacity, information theory, nested codes, soft codes

1. INTRODUCTION

It has been recognized during the last decades that recording, communication and processing of information play a paramount rôle in the living world, at any spatial and temporal scale, through a wide range of physical and chemical means. Moreover, it has become more and more apparent that the true divide between the non-living and the living things is precisely that the latter make extensive use of information, while the former do not. These statements legitimize the concept of biosemiotics and provide motivation for it. The researchers in the field naturally relied on the already established semiotics, especially following the pioneering works of Ferdinand de Saussure, Roman Jakobson, Charles Peirce, and others. Understandably, the semantic problems are at the heart of biosemiotics, as they are central to 'classical semiotics'.

Quite independently, the scientific study of communication engineering led more than half a century ago to the development of *information theory* (Shannon, 1948). It was intended to solve the technical problems associated with the communication of a message from some sender to some addressee, without any care of its semantic content: a messenger has indeed not to know about the meaning of the message he carries. Only outer characteristics of the physical support of this message (e.g., its spatial dimensions and its weight if it consists of a solid object) are relevant to him. Completely ignoring semantics, information theory introduced a quantitative measure of information perfectly fitted to communication engineering and very successfully used in this field.

One may wonder why information theory has been yet so sparsely used in biology. The early attempts made by biologists to use concepts of the 'classical' information theory (i.e., as introduced and developed by Shannon (Shannon, 1948)) almost invariably came to a sudden end with the remark that the entity referred to in the theory as 'information' is very restrictive with respect to the ordinary meaning of the word, especially insofar as it ignores semantics. They thought that a better fitted 'organic information' should be used instead. However, they were unable to appropriately define it and they preferred to wait until somebody could do so.

This point of view would be tenable only if no problems of communication engineering were met by living beings. Communication is so familiar to humans that the necessity of physical supports and devices in order to perform it is very often overlooked. Engineers, who have to design systems and devices intended to communicate, are on the contrary fully conscious of this necessity. As an engineer, I observe that the engineering problems of communication are far from being foreign to the living world, so information theory has actually much more to offer to biologists than most of them believe. That information theory impoverishes the concept of information with respect to the common meaning of the word is undeniable. But is it a reason for rejecting it? As regards the definition of fundamental entities in sciences which use a mathematical formalism, it often occurs that *less is more*. It turns out that the admittedly restrictive concept used in information theory probably captures the most important features of information, at least as far as engineering functions like recording and communication are concerned. Moreover, the simplicity of its definition enabled extremely wide and successful developments. By 'successful' I mean not only that it enabled the onset of information theory as a consistent new science but that it has been experimentally confirmed in a striking manner through the countless engineering applications of information theory. At its beginning, however, information theory appeared as weakly connected with engineering practice for lack of an available implementation technology. The solutions that information theory offered to engineering problems looked by far too complicated to be reliably implemented at reasonable costs. But a few decades later the semi-conductor technology had made such progresses that this implementation became possible and fruitful. Although almost all the basic concepts of information theory were already contained in Shannon's work (Shannon, 1948), a very valuable collective experience was gained in its applications, which unfortunately is only shared within the community of communication engineers. I believe that the *a priori* rejection of the classical information theory deprived the biological community of a wealth of potentially useful concepts and results of fundamental importance. Stated in more adamant words, it amounted to throw out the baby with the bathwater. Among these ignored results, those related to the protection against errors have been generally overlooked by biologists. The firmly established, but paradoxical, theoretical possibility of *errorless communication in the presence of errors* is ill-known, as well as the main properties of the technical means which implement it, namely, the *error-correcting codes*. Laymen as well as scientists of other disciplines than communication engineering often ignore their existence, let alone how they work, although they make a daily use of them, e.g., as a necessary ingredient of mobile telephony.

I show indeed, and this is a major topic dealt with in this paper, that the template-replication paradigm is unable to account for the genome conservation. Although phenotypic membranes shield the genome from mechanical and chemical aggressions, errors in the DNA sequence of nucleotides inevitably occur due to radiations (of cosmic, solar or terrestrial origin), and even because molecules are basically indeterministic quantum objects. Information theory tells the limits of what is possible as regards communication. The limit which is imposed on any communication by the presence of errors is called 'channel capacity'. It defines a horizon beyond which no communication is possible. Merely computing the channel capacity associated with the genomic memory shows that the template-copying paradigm of today's genetics has a far too short horizon in time, hence is unable to account for the faithful transmission of genetic information through the ages. The genomic memory actually appears as ephemeral at the geological time scale. In order to faithfully communicate the genetic information, the genome must act as an error-correcting code and be regenerated from time to time, after an interval as short as to ensure that the number of occurring errors is very unlikely to exceed its error-correcting ability. Besides being a trivially known fact, that nature proceeds by successive generations appears as an absolute necessity in the light of information theory. 'Generation' assumes here the strong meaning of genome *regeneration*. Notice that the parameters which determine the error-correcting ability of the code on the one hand, and those which control the time interval between regenerations on the other hand, are presumably unrelated. Depending on these parameters, a variety of situations as regards the permanency or mutability of species results. Their proper matching results from natural selection, making the time interval as short as to ensure the conservation of the species which we observe in the living world, but long enough to leave room for the variability needed to fit environmental changes.

Conservation of the genome then no longer appears to be the rule and replication errors, the exception. On the contrary, the genome conservation can only result from a dynamic process where error-correcting codes necessarily play a major rôle. This reversal of point of view has deep consequences on our understanding of the living world and the means by which it came to existence, i.e., biological evolution. Many features of the living world and of its evolution can actually be deduced from the main hypothesis that the genomes actually involve error-correcting means, together with the subsidiary one that they involve 'nested codes', i.e., combine several codes into a layered architecture. Some of these features are well-known biological facts, others suggest answers to debated questions. Most of them are not satisfactorily explained by current biology. In a sense, one may rightfully think of such results as speculative as mere consequences of hypotheses. However, the information-theoretical impossibility of genome conservation without error-correcting codes makes these hypotheses necessary, and so are the consequences derived from them.

As regards how nature implements error-correcting codes, I first recall that the ability of a code to correct errors results from the dependence between the symbols of its words which has been induced by the encoding operation. Engineers use to this end constraints of mathematical character because they are well defined and easily implemented by physical devices. Constraints of other kind can be used for the purpose of error correction, however, although mathematical constraints only are actually implemented by engineers due to their convenience. I thus assume that, at variance with engineers, nature uses physico-chemical constraints which make the successive nucleotides of DNA mutually dependent hence endowed with error-correction ability. I also contemplate constraints of linguistic character: in order to direct the phenotype construction, the genome must use some kind of language which necessarily implies constraints. Notice that the possibility of error correction then appears as a by-product of other biological constraints, according to an approach which is typical of nature and can be referred to as 'tinkering'[1]. No explicit encoding operation is needed. In fact the direct experimental identification of natural error-correcting means is still lacking, but one cannot expect it to be performed without the active involvement of practising geneticists.

2. AN INTUITIVE OUTLINE OF INFORMATION THEORY AND ERROR-CORRECTING CODES

2.1 Shannon's Paradigm, Variants and Interpretations

Before we can deal with the possible use of error-correcting codes for genomic communication, we need to introduce some information-theoretic concepts originating in communication engineering. We shall also propose variants and interpretations of these concepts which hopefully will make them useful in the context of genetics. These concepts and interpretations were to a large extent left implicit in our previous works about the application of error-correcting codes to genetic communication (Battail, 1997–2006), so the present section may be thought of as a key to help their understanding. We shall then provide an overview on error-correcting codes.

2.1.1 Shannon's paradigm

The basic scheme of a communication in the engineering literature (sometimes explicit but more often implicit) is Shannon's paradigm, a variant of which is represented in Fig. 1. A *source* generates an information message (an arbitrary sequence of symbols, a symbol being an element of some given finite set named alphabet[2]) intended to a *destination*. The only link between the source and the destination is a *noisy channel* which propagates the signals representing the symbols of its input message in an imperfect fashion, so a received symbol can differ from the transmitted one (an event referred to as *error*) with nonzero probability. The device labelled *channel encoder* transforms the message that the source generates into the one which enters the noisy channel. The *channel decoder* operates the inverse transformation and delivers to the *destination* a message which is hopefully identical to the one the source generated.

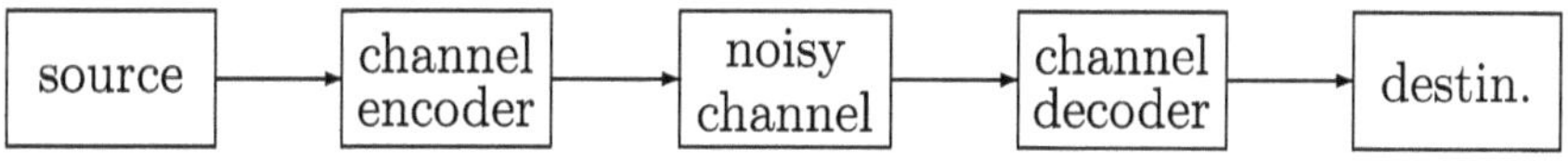

Figure 1. Shannon's paradigm, restricted to the functions of channel encoding and decoding

The encoded message must contain a larger number of symbols than the source output, a property referred to as *redundancy* which is essential for protecting the information message against the channel noise, i.e., to perform *error correction.* According to the *fundamental theorem of channel coding*, if the redundancy is high enough (depending on the channel noise and being the larger, the worse is the channel), the joint operation of the encoder and the decoder can ideally overcome the channel errors. Then the message that the destination receives is identical to the one originally delivered by the source, and error-free communication is achieved. The error probability after decoding cannot exactly equal zero for a message of finite length, but increasing the message length and properly designing the encoder and its associated decoder can make the probability of a decoding error vanish when the message length approaches infinity, regardless of the channel error probability. More on this topic will be found below (Sec. 2.3.2).

Shannon's paradigm is so important as a background in communication engineering that we now give some comments about it. We believe it conveys a spirit which biologists should share with engineers in order to actually realize and exploit the potential of information theory in their own discipline, as well as to be aware of its limits.

2.1.2 Some preliminary comments on Shannon's paradigm

Notice first that the success of Shannon's paradigm in engineering is due to a large extent to its *flexibility*. The borders of the blocks in Fig. 1 should not be considered as rigidly fixed, but on the contrary as possibly redefined at will so as to best fit any problem at hand. We shall try below to use this latitude in order to make it useful in the context of genetics. The only requirement is that everything that perturbs the communication between the source and the destination, to be referred to as *noise*, should be located in the channel. In order to prevent misunderstandings, let us stress that the concepts of information and noise are by no means absolute. On the contrary, the message from the source is considered as bearing information only insofar as it can be used by the *destination*. The error events are interpreted as noise with respect to this particular choice of the information message. The usefulness to the destination is the *only* criterion which differentiates the information message from the noise, meaning that their distinction is *arbitrary*. They may well exchange their rôles. For example, the sun is a source of noise for a receiver intended to the signal of a communication satellite. However, the signal from the satellite perturbs the observation of a radioastronomer who studies solar radiation, hence it is a noise source for this particular destination. When we refer to the usefulness or harm that a sequence of symbols has for the destination, we necessarily consider its *purpose*, hence we cannot avoid some kind of teleology or subjectivity.

The entities at the extremities of the communication link in Shannon's paradigm, namely the source and the destination, are *devices* or *living beings*. In an engineering context, the devices are man-made artifacts so the living beings ultimately involved are human, either directly or indirectly through their engineering products. Insofar as we consider nature as an engineer, we may think of the source and destination as extended to other living beings or nature-made devices. On the other hand, the channel is a physical device or system.

As it has been described and represented in Fig. 1 Shannon's paradigm is unidirectional, in the sense that it only considers the source sending messages to the destination, but not the destination sending messages to the source. A conversation between humans, for instance, is better represented as a bidirectional scheme. Such a scheme is also used in many cases of interest to communication engineers. Then each of two locations contains a source and a destination merged into a single entity, and two channels are provided between the source of one of the entities and the destination of the other one.

2.1.3 Communication through space or through time

In communication engineering, the source and the destination of Fig. 1 are located at a distance and the message has thus to be communicated through space. It consists of a succession of symbols in time. However, the same scheme can depict the situation where the source and the destination are separated in *time*. Then the message is a sequence of symbols written on some support extending itself in *space* and read later, so the 'channel' of the figure refers in this case to the support on which the message is written. The channel errors result from the degradation that this message possibly suffers during the time interval which separates the instant of writing the message and that of its reading. In genetics, we are concerned with such a communication through time and the support of information is a DNA (or RNA) unidimensional polymer. In such a case, no communication channel can exist from the future to the past (such a channel would violate causality), so only the unidirectional version of Shannon's paradigm, as depicted in Fig. 1, is valid to represent genetic communication. Similarly, if the source is a gene and the destination the molecular machinery which generates a polypeptidic chain, the central dogma of molecular biology asserts that only the unidirectional version of Shannon's paradigm is valid, although no violation of causality would result from the availability of a channel in the reverse direction. For these reasons, we shall in the following exclude the bidirectional scheme and consider only the unidirectional version of Shannon's paradigm as fitted to genetic applications.

2.1.4 Variants of Shannon's paradigm intended to genetic applications

In Fig. 1 which describes a usual engineering situation, the source is clearly identified as the origin of the message. Sending an information message is intentional, which is compatible with the above remark that both the source and the destination are ultimately human beings. We now propose a variant of Shannon's paradigm which better fits the absence of purpose which is typical of the biological context (where, at least, we do not know if there is an intention).

This variant of the basic scheme of Fig. 1 is depicted in Fig. 2, where we just merged the first two blocks of Fig. 1 into a single block (inside a dashed box) labelled 'redundant source', and its last two blocks into a single one (again inside a dashed box) labelled 'redefined destination'. Then the message delivered by the

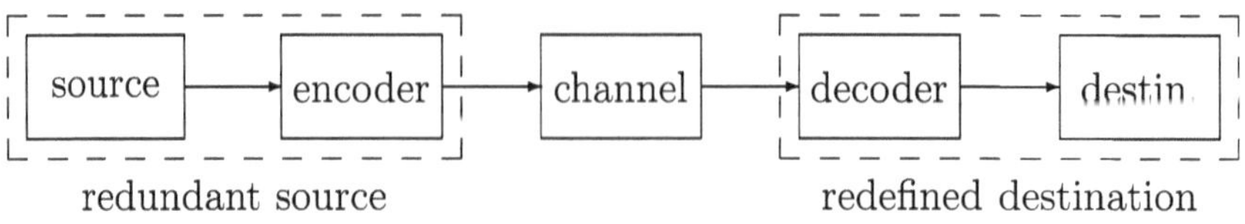

Figure 2. A variant of Shannon's paradigm

redundant source has error-correcting capabilities which can be exploited by the redefined destination.

Let us now consider the case of a genetic message incurring successive replications. Figure 3 depicts the case where two channels are successively used (i.e., where two replications successively occur). The error-correcting properties of the encoded message enable merging the decoder which follows the first channel with the encoder which precedes the second one into a single entity named 'regenerator'. The concept of *regeneration* thus appears as better fitted to the context of genetics than the engineering concept of decoding which refers to an explicit 'information message' to be communicated, and its very possibility relies on the redundancy of the initial encoding.

We may now describe a chain of successive replications incurred by a registered message (e.g., the genetic message of DNA) as in Fig. 4. An original redundant source at left delivers a redundantly encoded message which is written on a first support (labelled 'channel 1'), regenerated in the device labelled 'regen. 1', written again on 'channel 2', etc. The last step considered is the i-th regeneration, where i is a finite but possibly large number.

The redundant source has been depicted in Fig. 2 and regenerators in Fig. 3. If the number of replication steps i is very large, the initial encoded message from

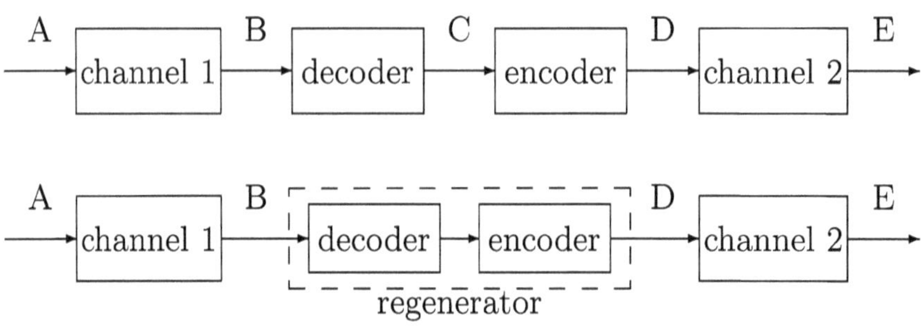

Figure 3. The regeneration function. Channels 1 and 2 have to be used successively. The upper picture is relevant to a conventional error-correcting code. The sequences found at the points designated by letters are: in A, an encoded sequence; in B and E, received sequences; in C, an information message; and in D, the sequence which results from the encoding of the first decoded information message, hence the restored initial sequence if no decoding error occurs. In the lower picture, the decoder and encoder have been merged into a single entity labelled 'regenerator' where the information message no longer appears

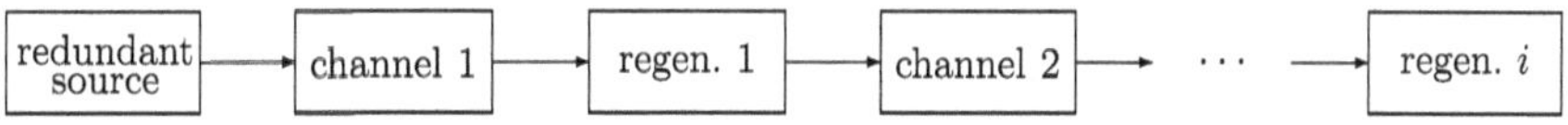

Figure 4. A chain of i successive regenerations

the redundant source is likely to have been modified and even maybe 'forgotten', insofar as regeneration errors likely occurred. As i increases, the current encoded message thus depends less and less on the original redundant source, and more and more on the regeneration errors which occurred, i.e., on a succession of contingent events. In the context of genetics, of course, natural selection operates on the corresponding phenotypes, letting only the fittest ones survive. This illustrates well the ambivalence of regeneration errors. On the one hand, they should occur very unfrequently for the sake of genome conservation, which is necessary for maintening a species and multiplying the number of the individuals which belong to it. On the other hand, they give rise to new phenotypes when they occur, thus exploring at random the field of what is possible. These new phenotypes are the rough material which makes evolution possible, as targets of natural selection.

Remember that, as we noticed above, the rôles of information and noise are relative to each other, entirely depending on their usefulness or harm for the destination in Shannon's paradigm. In the context of genetics, the destination consists of the cell machinery which processes the genetic message borne by a DNA molecule. Therefore, the regeneration errors *create new genetic messages* in a very objective sense. There is no paradox in this statement since there is no difference of nature between information and noise, as we already noticed. Moreover, because we liken the error-correcting codes to the multiple constraints which make DNA strings admissible as genetic messages (see Sec. 5 below), such strings having suffered regeneration errors are just as plausible to the cell machinery as the error-free ones.

2.2 An Outline of Information Theory

We shall now use Shannon's paradigm in order to introduce the main quantities defined by information theory and discuss their relationship. We first introduce the basic measures of information (proper and mutual) and the quantities which are associated with the blocks of Shannon's paradigm: the entropy of the source and the capacity of the channel. We introduce also the basic functions of source coding (intended to replace an initial message by a shorter one bearing the same information) and channel coding (aimed at protecting a message against transmission errors), and show that the entropy and the capacity acquire an operational significance as defining the limits of what is possible in source- and channel-coding, respectively. We also mention an alternative definition of information seen as a measure of complexity which gave rise to the algorithmic information theory (Kolmogorov, Solomonov, Chaitin). It provides another understanding of information but does not really question the validity of the basic classical results. We believe that Shannon's information is better fitted to the actual application of information theory to sciences like biology but our discussion below of the relationship of information and semantics will also be inspired by the algorithmic information theory.

2.2.1 *Basic information-theoretic measures*

This section is intended to provide an insight into the information-theoretic quantities used in the paper, especially the channel capacity. Of course, much more on these topics can be found in Shannon's seminal work (Shannon, 1948) and in textbooks like (Gallager, 1968; Cover and Thomas, 1991).

The *information* brought by the occurrence of an event is measured by its *unexpectedness*. If an event x occurs with probability p, then its unexpectedness can be measured by $1/p$ or by some positive increasing function of it, $f(1/p)$. If two events x_1 and x_2 occur independently of each other with probabilities p_1 and p_2, respectively, it is reasonable to measure the information associated with the occurrence of both by the sum of the information measures separately associated with each of them. The probability of the outcome of both events is $p_1 p_2$ since they were assumed to be independent, so the function f is chosen such that

$$f(1/p_1 p_2) = f(1/p_1) + f(1/p_2).$$

The logarithmic function satisfies this equality and has many desirable mathematical properties, especially continuity and derivability, and moreover can be shown to uniquely satisfy a set of axioms which are plausible for an information measure. It has thus been chosen, so the above equality becomes

$$\log(1/p_1 p_2) = \log(1/p_1) + \log(1/p_2) = -\log(p_1 p_2) = -\log(p_1) - \log(p_2).$$

The information associated with the occurrence of the event x of probability p is thus measured by

$$i(x) = \log(1/p) = -\log p.$$

Choosing the base of the logarithms defines the unit of information quantity. Following Shannon, the most usual choice of this base is 2, and the unit is then referred to as the *bit*, an acronym for *bi*nary digi*t*. However, a digit and an information unit are distinct entities. A binary digit does not necessarily bear information; if it does, it bears at most a binary information unit. To avoid any confusion with the more common use of the acronym 'bit' in order to mean 'binary digit', we prefer to rename 'shannon' the binary unit of information, instead of 'bit' as Shannon originally did.

Let us now consider a repetitive random event X having its q possible outcomes denoted by $x_1, x_2, \ldots, x_q$, each occurring with probability $p(x_1) = p_1, p(x_2) = p_2, \ldots, p(x_q) = p_q$, respectively. That one of the q outcomes necessarily occurs results in

$$p_1 + p_2 + \cdots + p_q = \sum_{i=1}^{q} p_i = 1.$$

The quantity of information brought in the average by the occurrence of X is thus

$$(1) \quad H(X) = p_1 \log(1/p_1) + p_2 \log(1/p_2) + \cdots + p_q \log(1/p_q) = -\sum_{i=1}^{q} p_i \log(p_i),$$

a positive quantity if none of the probabilities $p_1, p_2, \ldots, p_q$ is 0 or 1, i.e., if X is actually random. In information theory the quantity $H(X)$ is referred to as the entropy of X, but since the same word is used in physics with a different but related meaning we prefer to name it *prior uncertainty*. It measures the uncertainty which precedes the event and is resolved by its occurrence. The event X which consists of choosing a symbol among an alphabet of size q with probabilities $p_1, p_2, \ldots, p_q$, to be referred to as the *source event*, thus provides an average amount of information equal to $H(X)$. The maximum of $H(X)$ is achieved when $p_1 = p_2 = \ldots = p_q = 1/q$. This maximum equals $\log q$.

When the successive outcomes of X are not independent, the prior uncertainty (entropy) per symbol of a stationary source[3] is defined as the limit:

$$(2) \qquad H = \lim_{n\to\infty} \frac{1}{n} H_n,$$

where

$$H_n = -\sum_{\underline{s}} p(\underline{s}) \log p(\underline{s}),$$

where $\underline{s}$ is any sequence of length n output by the source and $p(\underline{s})$ is its probability. The summation is made over all possible sequences $\underline{s}$. The assumption that the source is stationary suffices to ensure that the limit in (2) exists.

An important property of the entropy of a redundant source is that the set of its outputs can be divided into two disjoint categories: the *typical* and atypical sequences. The number of typical sequences is about q^{nH}, where q is the source alphabet, n is the sequence length and H the entropy of the source expressed using logarithms to the base q. For long enough sequences, the probability that the source output is atypical vanishes, which means that the probability that the source generates a typical sequence approaches 1. Remember that the maximum entropy of a q-ary source is $\log q$ and that $\log_q q = 1$. For a redundant source, i.e., such that its prior uncertainty or entropy differs from the maximum, its value H when using logarithms to the base q is less than 1. The q^{nH} typical sequences are thus a minority among all possible strings of n q-ary symbols, whose number is q^n, but the actual source output sequences almost surely belong to this minority.

Now, a *channel* can be considered as a means for observing the source event X, referred to as its input, through the outcomes $y_1, y_2, \ldots, y_r$ of another event Y, its output, which is probabilistically related to X. By 'probabilistically related', we mean that when the outcome of X is x_i, $1 \le i \le q$, then the probability of a particular outcome y_j of Y, $1 \le j \le r$, assumes a value $p(y_j|x_i)$ which depends on x_i. Such a probability is referred to as the conditional probability of y_j given x_i.

It will be convenient for us, although it is not necessary, to assume that $r \geq q$. We cannot observe directly the source event which would provide to us the quantity of information $H(X)$, but only the channel output Y which depends on the source event only through the set of conditional probabilities $p(y_j|x_i)$, $1 \leq i \leq q$, $1 \leq j \leq r$. The channel output Y then provides about the source event X a quantity of information equal to the uncertainty which is *resolved* by the outcome of X, $H(X)$, *minus* the uncertainty about X which *remains* when the outcome of Y is known, denoted by $H(X|Y)$. This quantity of information is thus expressed as

$$(3) \qquad I(X;Y) = H(X) - H(X|Y),$$

referred to as the *mutual information* of X and Y, where $H(X|Y)$ is given by

$$H(X|Y) = \sum_{i=1}^{q}\sum_{j=1}^{r} p(x_i, y_j) \log p(x_i|y_j),$$

where $p(x_i, y_j) = p(y_j|x_i)p(x_i)$ is the probability of occurrence of both x_i and y_j, referred to as the joint probability of x_i and y_j, and

$$p(x_i|y_j) = \frac{p(x_i, y_j)}{p(y_j)} = \frac{p(x_i, y_j)}{\sum_{i=1}^{q} p(x_i, y_j)},$$

where $p(x_i)$ and $p(y_j)$ are the probabilities of x_i and y_j, respectively. $H(X|Y)$ is referred to as the conditional uncertainty (or conditional entropy in the usual information-theoretic vocabulary) of X given Y. The word 'mutual' expresses the easily proved symmetry of $I(X;Y)$, namely that

$$I(X;Y) = H(X) - H(X|Y) = H(Y) - H(Y|X) = I(Y;X).$$

The mutual information $I(X;Y)$ is a nonnegative quantity, as expected for a measure of information.

The extension to sequences where the successive outcomes of X are not independent, similar to the one which led to the expression (2) of $H(X)$, is used to define $H(X|Y)$ in this case, hence $I(X;Y)$ according to (3).

Another important extension of the mutual information concerns the case of continuous variables. Let now X denote a real random variable which assumes a value belonging to the infinitesimal interval $(x, x+\mathrm{d}x)$ with probability $p(x)\mathrm{d}x$. The function $p(x)$ is referred to as the probability density function of X and is such that $\int p(x)\mathrm{d}x = 1$, where the integral is taken on the whole real axis. At variance with the case where X assumes discrete values, it is not possible to define its entropy by (1). However, if we formally replace in (1) the summation by an integration, the probability p_i being replaced by the probability density function $p(x)$ of X, we obtain the quantity

$$(4) \qquad h(X) = -\int p(x) \log[p(x)]\mathrm{d}x$$

which is referred to as the *differential entropy* of X. This quantity exhibits some of the properties of the entropy of the discrete case, but not all. For instance, it may be negative and depends on the unit which measures X. However, it turns out that the mutual information of two continuous variables can still be defined according to the formula homologous to (3), namely

$$(5) \qquad I(X;Y) = h(X) - h(X|Y),$$

where the proper entropies have just been replaced by the differential entropies defined by (4), with $h(X|Y)$ defined as

$$h(X|Y) = \int_x \int_y p(x,y) \log p(x|y) \mathrm{d}x\mathrm{d}y,$$

where $p(x,y) = p(y|x)p(x)$ is the joint probability density function associated with the pair of variables $\{x,y\}$, and

$$p(x|y) = \frac{p(x,y)}{p(y)} = \frac{p(x,y)}{\int p(x,y)\mathrm{d}x},$$

where $p(x)$ and $p(y)$ are the probability density functions of x and y, respectively.

The *channel capacity* $C(X;Y)$ is defined as the maximum, for all probability distributions $\{p_1, p_2, \ldots, p_q\}$ on X, of the mutual information $I(X;Y)$. Its computation is very simple in the discrete case for symmetric channels (like those given as examples) since then the maximum of the mutual information is obtained for equally probable outcomes of X, i.e., $p_i = 1/q$ for any i such that $1 \le i \le q$. Then

$$(6) \qquad C(X;Y) = H(Y) - H(Y|X)$$

for equally probable outcomes of X. The conditional probabilities involved in the computation of $H(Y|X)$ are the transition probabilities of the channel and $H(Y)$ is easily computed in terms of the transition probabilities since $p(y_j) = \sum_{i=1}^{q} p(y_j|x_i)p(x_i) = (1/q)\sum_{i=1}^{q} p(y_j|x_i)$.

Discrete channels can be represented by diagrams where arrows indicate the transitions from the input alphabet symbols to those of the output alphabet. A transition probability is associated with each arrow. Three such diagrams are drawn in Fig. 5. The first one, **a**, is drawn for $q = 4$, the second one, **b**, for $q = 2$, and the third one pertains to the case where the alphabet symbol considered is a pair of complementary binary symbols (to be interpreted below as nucleotides), their non-complementarity being interpreted as an erasure denoted by ε. The capacity of the channels represented by Figs 5 **b** and **c** will be computed in Sec. 3.3, Eq. (11) and (12), respectively.

Thanks to the extended definition (5) of the mutual information, the capacity of a channel with discrete or continuous input and continuous output can be defined, *mutatis mutandis*. The case of discrete input and continuous output is especially important as describing the typical channel in digital communication, where a finite number of waveforms represent the alphabet symbols and where the channel perturbations are continuous, e.g., thermal noise modelled as Gaussianly distributed.

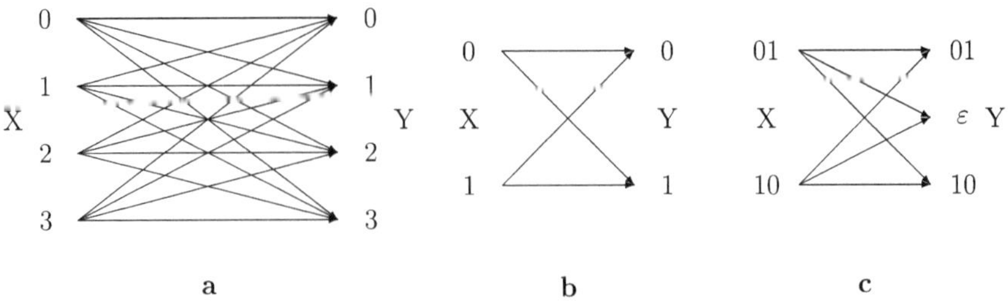

Figure 5. Example of channels. **a**: Transitions for input and output alphabets of size q. The transition probability associated with the horizontal branches equals $1-p$, where p is the errror probability, that associated with any oblique branch is $p/(q-1)$. The figure has been drawn for $q=4$. **b**: Same diagram when $q=2$. The horizontal branches are labelled $1-p$, the oblique ones p. **c**: Complementary binary symbols. The case where the two symbols of the received pair are not complementary (i.e., $Y=00$ or $Y=11$) is interpreted as a third symbol denoted by ε. If the probability of error of a single symbol is p, the probability of a transition from 01 or 10 to ε is $2p(1-p)$, the probability of transition from 01 to 10, or 10 to 01, is p^2, and that associated with the horizontal branches is $(1-p)^2$

2.2.2 *Basic coding functions, source coding*

The output of a source can be transformed into another fully equivalent string of symbols so as to either reduce its length (compress it), or make it resilient to errors. These encoding operations are referred to as *source coding* and *channel coding*, respectively. The two fundamental theorems of information theory state the limits of what is possible for each of these codings. The shortest possible length after source coding of a string of n q-ary symbols is nH, where H is the source entropy as defined by (1) or (2), expressed using logarithms to the base q, the alphabet size of the source, so we have $H \leq 1$. This means that it is possible to encode the output of a source of entropy H so as to reduce its length by a factor of H and still be able to recover its initial output, while the same output can not be recovered from any string of length reduced by a factor less than H. We shall not develop here the topic of source coding. Let us just notice that the entropy of a source can then be interpreted as the ratio of the length of the shortest possible compressed string which can be attained by source coding to the length of the original source output, the compressed string remaining equivalent to it in the sense that it enables an exact recovery.

As regards channel coding, errorless communication is possible if the source entropy H is less than the channel capacity C but not if it is larger than it. It turns out moreover that source coding increases the vulnerability of the message to errors and that channel coding increases the message length as introducing redundancy, so the main two coding operations appear in a duality relationship. Expressing the limits of what is possible in terms of the information measures defined above, namely, the entropy of the source and the capacity of the channel gives these information measures an operational significance. We shall in Sections 2.3.1 and 2.3.2 discuss at greater length the topic of channel coding, which involves error-correcting codes. We shall especially insist on the paradoxical but experimentally confirmed possibility of errorless communication in the presence of channel errors.

2.2.3 Algorithmic information theory

Another way of defining information measures is provided by the *algorithmic information theory*. Given a *universal computer* like the Turing machine, historically the first theoretically conceived computer, or any modern general-purpose computer, the basic idea of the algorithmic information theory consists of defining the information measure associated with a string of symbols as the length of the shortest programme which lets the computer generate an output identical to this string (the computer input and output alphabets and that of the string are assumed to be the same, usually binary). This measure is generally referred to as the *algorithmic complexity* of the string. (The length of the programme depends to some extent on the computer, but mainly on the string itself and thus is a valid measure of its actual complexity, at least for sufficiently long strings.) The algorithmic information theory uses the algorithmic complexity as the basic measure of information. A recent book by Chaitin provides an excellent and highly readable account of this topic (Chaitin, 2005).

At first sight, the probabilistic information theory developed by Shannon and the more recent algorithmic information theory have little in common. The former deals with probabilistic ensembles of symbol strings, while the latter seems to work on individual strings. This point of view is however too simplistic. The only objective way to determine the parameters of the ensembles of symbol strings considered by the probabilistic information theory consists of deriving them from frequency measurements, which necessarily involve a number of realizations which is very small as compared with the total number of strings they contain (remember that the total number N of possible binary strings of length n is 2^n; for instance for $n = 1,000$ we have $N = 2^{1,000} \approx 10^{301}$, a number which exceeds by far the estimated number of atoms in the visible universe, namely, 10^{80}). As regards the algorithmic information theory we assumed that, given some string, a programme shorter than it can be found which lets the computer generate an output identical to this very string. Such a programme must involve its description in machine language. Any language is a combinatoric object which implicitly refers to an ensemble of realizations. Although there seems to be little relationship between the information measure associated with a source in the conventional information theory, namely its entropy (1) or (2), and the measure of complexity provided by the algorithmic information theory, it turns out that both are asymptotically related in the sense that, for very long strings, the complexity approaches the entropy. Despite their obvious difference, the conventional information theory and the algorithmic information theory interestingly appear as describing two facets of a same fundamental entity.

The main practical weakness of the complexity as defined by algorithmic information theory is its uncomputability. Given a source output, knowing its complexity would imply the availability of a means to *optimally* compress it (in the sense of source coding, see Sec. 2.2.2). However, when an algorithm is known to compress a source (i.e., a computer programme which results in the same output as the given source output and is shorter than it), it is impossible to know if this specific algorithm is the shortest possible one. Only an upper limit on the algorithmic complexity is thus known. On the other hand, the availability of accurate probability measures is obviously needed in order to compute the quantities defined by the conventional information theory. Frequency measurements are the ordinary means for obtaining plausible probability distributions but they necessarily rely on comparatively few samples, hence have a limited precision.

2.3 On Error-correcting Codes

2.3.1 An introduction to error-correcting codes

Error-correcting codes have a long and complicated history. The most successful codes yet known, the *turbo codes*, can be considered as practically implementing 'error-free communication', the paradoxical promise of Shannon's information theory made no less than 45 years before the invention of turbo codes. Turbo codes can indeed be described in a very simple way which gives intuitive insight into the reason of their success and ignores the twists and turns which preceded their invention. We shall give below an intuitive description of turbo codes as an introduction to the needed properties of good error-correcting codes but we need first introduce the information-theoretic result that error-free communication is possible as a necessary background.

To begin with, we define as above an alphabet as a given collection of symbols in finite number, say q, referred to as the alphabet size. These symbols can be arbitrary signs or objects provided they can be unambiguously distinguished, like letters, digits, electric voltages, signal forms, or molecules... Due to the necessity that its symbols be distinguishable, the smallest size of an alphabet is $q = 2$. The main properties of codes can be understood if we assume, as we shall do most often in this section, that the alphabet is binary, i.e., that its size equals this minimum, an assumption which entails little loss of generality. The symbols of the binary alphabet will be denoted by 0 and 1.

Let us now define a *word* of length n as a sequence of n symbols from a given alphabet of size q. Since each symbol of a word can assume q distinguishable values, the total number of possible different words is q^n, say 2^n in the binary case. It will be very convenient to represent an n-symbol word as a point in an n-dimensional space, each coordinate of this point being one of the n symbols of the word. For instance, if $q = 2$ and $n = 3$, there are $2^3 = 8$ different possible words, each of which being represented as a vertex of a cube. The useful values of n are much larger, but there is no difficulty in extending this definition to an n-dimensional space with $n > 3$. We may define the *Hamming distance* d between two words as the number of coordinates where their symbols differ. For instance, if $n = 7$, the distance between 1101000 and 0110100 is $d = 4$. We refer to the space endowed with this distance measure as the n-dimensional Hamming space. An error-correcting code is a subset of all possible n-symbol words such that the minimum distance between any two of its words is larger than 1. The minimum distance between any two different n-symbol words is only 1 since they may differ in a single coordinate, so an error-correcting code is a strict subset of all n-symbol words. The property that not any n-symbol word belongs to the error-correcting code is referred to as *redundancy*. In the case where $n = 3$, we may define a code as having even weight, the weight of a word being defined as the number of symbols '1' it contains. Here is the list of its codewords: 000, 011, 110 and 101. Its minimum distance is $d = 2$. A code with the largest possible minimum distance for $n = 3$, i.e., $d = 3$, only contains two words, for instance 000 and 111. To communicate a message of length k, with $k < n$ to ensure the code redundancy, we must establish a one-to-one correspondence, or *encoding rule*, between the 2^k messages of length k and the 2^k n-symbol words which belong to the code. Little loss of generality results if the message explicitly appears, e.g., in the first k positions in the word (or in any k determined positions). Then the encoding rule is said to be *systematic*, the symbols at the selected k positions are said *information* symbols, and the remaining ones, which are completely determined by the information symbols, are said *check* or redundancy symbols. For instance, with $n = 3$ and $k = 2$, if the 2 first symbols represent the message, we have a single check symbol which results from adding modulo 2 the information symbols (addition modulo 2 is the same as ordinary addition except that 1 plus 1 = 0 modulo 2).

In a communication system using an error-correcting code, only words belonging to this code may be transmitted. As an example, let us assume that a binary code is used and that an error in the channel consists of changing a 1 into a 0 or vice-versa. Then the channel errors result in a received word which possibly differs from the transmitted one. Moreover, the received word is at a Hamming distance from the transmitted one which equals the number of errors which occurred, say e, to be referred to as the *weight* of the error pattern. For a binary symmetric channel, i.e., if we may characterize it as making an error with a constant probability $p < 1/2$, independently, on each symbol of the word, then the probability of a particular error pattern of weight e is simply $P_e = p^e(1-p)^{n-e}$. For $p < 1/2$, P_e is a decreasing function of e, so a given error pattern is the more probable, the smaller its weight. There is no loss of generality in assuming $p < 1/2$ since, being arbitrary, the labelling of the received symbols by '0' or '1' can always be chosen so that this inequality holds provided $p \neq 1/2$. The case $p = 1/2$ is not relevant since it is equivalent to the absence of any channel. We may thus use as the best possible rule for recovering the transmitted word: *choose the word of the code the closest to the received word.* Its use determines the word of the code which has the most probably been transmitted. This *regeneration* rule is expressed in very simple geometrical terms in the n-dimensional Hamming space thanks to the distance defined between its words.

The mere statement of this rule enables us to understand the most important properties that an error-correcting code must possess in order to be efficient. The words of a code must be far from each other, so they should be very few as compared with all possible n-symbol words, i.e., the redundancy should be large.

But they should also be as evenly distributed in the n-dimensional space as possible, since any concentration of codewords would reduce their mutual distances with respect to the case of a more even distribution. For a given amount of redundancy, endowing a code with this property is by far the most difficult task in the design of an error-correcting code, although its necessity is quite intuitive and its statement is easy. We shall see below that the best known method to provide evenly distributed points in the Hamming space actually consists of choosing them *at random*, as strange as it may look.

2.3.2 *Error-free communication over a noisy channel is theoretically possible*

It was convenient in the above examples to consider small values of the word length n. Let us now go to the other extreme and assume that n is very large. Then, the *law of large numbers* tells that the weight of an error pattern is very probably close to its average, namely np (in other words, the frequency of errors measured in a large sample is with high probability close to the error probability). In geometrical parlance, this means that the received point is with high probability close to the 'surface' (an $(n-1)$-dimensional volume) of the n-dimensional sphere of radius np centred on the transmitted word. If the radius np is smaller than half the minimum distance d between any two words of the code (simply referred to as its *minimum distance*), then clearly the received word is with high probability closer to the truly transmitted word than to any other, so the above regeneration rule succeeds with high probability. Morever, the probability of a regeneration error vanishes as n approaches infinity. On the contrary, if $np > d/2$ a wrong codeword may be closer to the received word, in which case the regeneration rule above fails with very high probability. Notice the paradox: for a given probability p of channel error, increasing the word length n also increases the average number of erroneous symbols in the received word. Nevertheless, increasing n decreases the probability of a regeneration error provided $p < d/2n$. If this inequality holds, *errorless* communication of a message through an *unreliable* channel is possible. This result itself is paradoxical, and nobody imagined it could be reached anyway before its possibility was proved by information theory. It started the researches on error-correcting codes and remained later a very strong incentive to them.

The problem of designing an optimal error-correction code having M words of length n using a q-symbol alphabet for a given channel has no known general solution. However, choosing $M = q^k$ words at random within the n-dimensional space, with $k < n$ to provide redundancy, results in a code close to the optimum. This method, referred to as *random coding*, was used by Shannon in the proof of the fundamental theorem of channel coding (Shannon, 1948). This theorem asserts that 'errorless' communication is possible if, and only if, the information rate $R = k/n$ is less than a limit which depends on the channel error probability p (decreasing as p increases), referred to as the *channel capacity* C (see Sec. 2.2.1). 'Errorless' means that, provided $R < C$, a vanishing probability of error can result from using an adequate (but not explicitly specified) code, as n approaches infinity.

The main virtue of random coding is to statistically ensure that the codewords are as evenly distributed in the Hamming space as possible. Further elaboration of this fundamental theorem led to stronger results which, loosely speaking, tell that an arbitrarily chosen code is good with high probability. In a more adamant style: *all codes are good.* The problem of almost optimum error-correction coding *seems* thus to be solved, and moreover in an unexpectedly simple way.

However, a formidable problem remains. Remember that implementing the regeneration rule above implies to find the codeword the closest to the received word. In the absence of any structure, a code is an arbitrary set of M n-symbol words. There is no other way for implementing this regeneration rule than to compare *each* of the M codewords with any *single* received (possibly erroneous) word to be regenerated. The trouble is that for useful values of the codeword length, i.e., n as large as to make the probability of a regeneration error small enough, M is a huge number. For example, in a binary code with $n = 1,000$ and $R = 1/2$, we have $M = 2^{500} \approx 10^{150}$. (Remember that the number of atoms in the visible universe is estimated to about 10^{80}.) Implementing regeneration when an arbitrary code is used thus bumps against a complexity barrier. This problem cannot actually be solved unless the code is given some structure intended to alleviate the complexity of regenerating its codewords. A large number of codes and code families having a strong mathematical structure were invented, but their results were invariably far from the promise of the fundamental theorem of channel coding, namely error-free communication at a rate close to the channel capacity. Most experts believed that finding good codes having a tractable structure was hopeless due to an intrinsic incompatibility of goodness and structure. This widely shared opinion was summarized in the folk theorem: *all codes are good, except those we can think of.*

This opinion was by far too pessimistic. For instance, I noticed in 1989 that the sole criterion used in order to design a good code was to endow it with a minimum distance *as large as possible.* I criticized this seeming dogma, and suggested that a better criterion could be to look for *random-like* codes, i.e., codes such that the distribution of distances between their words is close in some sense to that of random codes (regardless of their actual minimum distance) but constructed according to a deterministic process (Battail, 1989, 1996). (Analogously, easily generated pseudo-random sequences which mimic truly random sequences are known and widely used in simulation.) Codes designed according to this criterion should have performance close to the optimum.

2.3.3 Error-free communication can be practically approached: turbo codes

In 1993, soon after the random-like criterion was stated, the pessimistic opinion above was definitively ruined with the advent of the *turbo codes* (Berrou et al., 1993; Berrou and Glavieux, 1996). Turbo codes actually meet the random-like criterion, although they were not explicitly designed in order to fulfil it (Battail, Berrou and Glavieux, 1993). Their implementation is comparatively simple and well within the possibilities of current technology. Besides being the best codes presently available, turbo codes have a performance close enough to the theoretical limit (the channel capacity) to be considered as almost optimal, at least from a practical point of view.

In the brief description of turbo codes to be presented now, we shall restrict ourselves to the binary alphabet, with its symbols denoted by 0 and 1, endowed with the structure of binary field, i.e., where two operations are defined: multiplication (the same as in ordinary arithmetic: $0 \times x = 0$, $1 \times x = x$, for $x = 0$ or 1) and modulo 2 addition to be denoted by $\oplus$ (the same as ordinary addition except that $1 \oplus 1 = 0$). The structure of a turbo encoder is depicted in Fig. 6. (It is astonishing that such a simple device can provide a good approximation of random coding, the implementation of which is of prohibitive complexity.)

We assumed that the necessary redundancy is obtained by generating three symbols every time an information symbol enters the encoder. In other words, the code rate R defined as the number of symbols of the information message divided by the number of actually transmitted symbols is $R = 1/3$. The choice of this particular rate will make turbo codes easy to understand, but several technical means enable generating turbo codes having different rates. For the ease of its description, the encoder is depicted here as having a single input and three outputs, but these outputs are easily transformed into a single one with a symbol rate three times larger, an operation called 'parallel-to-serial conversion'. A sequence U of N binary information symbols enters the encoder. One of the three output sequences is identical to the input sequence U. The other two output sequences, denoted by V_1 and $V_2(\Pi)$, are generated by two, possibly identical, rate-1 encoders. By 'rate-1 encoder' we mean a device which computes a binary output symbol as the sum modulo 2 of the binary symbol which enters it and of certain of the m preceding ones at well defined positions. The 'memory' m is a small number (typically in the range of 3 to 5) in order to avoid an excessive complexity of the decoding devices. Moreover, each of the rate-1 encoders is assumed to be *recursive*, i.e., its output is added modulo 2 to the entering binary symbol. The first rate-1 encoder directly receives the input information sequence U. The second one receives an *interleaved* version $\Pi(U)$ of U where the symbols of U are reordered by a device named interleaver. For example,

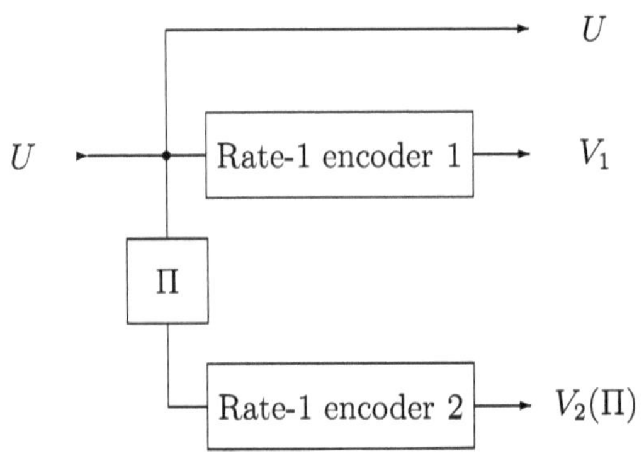

Figure 6. Rate-1/3 turbo encoder

if $N = 7$, the interleaver Π may reorder the symbols initially numbered 1234567 according to the order 3517426. Then $U = 1101000$ results in $\Pi(U) = 0010110$ and $U = 0110111$ in $\Pi(U) = 1101011$. The actually useful values of N are much larger, say 1000 or more, so the number $N! = N \times (N-1) \times (N-2) \cdots \times 1$ of possible different interleavers is huge.

The code generated by this encoder is referred to as 'linear', which means that it only implements the operations of the binary field. A linear code always contains the all-0 word and one easily checks that the set of Hamming distances between all the distinct codewords reduces to the set of its distances to the all-0 word, i.e., to the weights of its words. Due to the assumed recursivity of the rate-1 encoders, an input consisting of a single 1 would result in an output of infinite weight should no truncation occur, but it is actually limited by the size of the blocks considered, namely N symbols. We shall call 'large weight sequences' the sequences which would be of infinite weight should no truncation of the block size occur, and the other ones as 'small weight sequences'. The response of the rate-1 encoders to information messages of weight larger than 1 is not always of large weight, but for a properly designed encoder the proportion of sequences of small weight is no greater than 2^{-m}, m being the encoder memory. For a randomly chosen interleaver Π, the probability that the encoded sequences V_1 and $V_2(\Pi)$ are both of small weight is only 2^{-2m}. The total weight of a codeword so generated is the sum of the weights of U, V_1 and $V_2(\Pi)$. Since the small weight sequences generated by the rate-1 encoders are few, and since moreover the weight of most of them is not very small, we obtain that the 'weight spectrum' of the code, i.e., the set of weights which is obtained for all the 2^N possible input information sequences, closely resembles the set of weights of random sequences of length $3N$, so very few of them have little weight. The turbo encoder has thus generated a 'pseudo-random' code which satisfies the random-like criterion alluded to above.

Interestingly, the turbo encoder of Fig. 6 exhibits the three main properties which can be expected from a good encoding device. Splitting the input into several branches (three in the figure) amounts to repeat each of the entering symbols, which is the most elementary form of *redundancy*. The rate-1 encoders introduce *dependence* between the successive symbols they receive which, jointly with redundancy, is used in the decoding process. And the permutation that the interleaver operates introduces some form of *randomness* which, contrary to Shannon's random coding, is easily undone since the inverse permutation is immediately available. Not surprisingly, a cleverly designed decoding device can use all these features to closely approach the theoretical limit, and we may think of the turbo code scheme as a kind of paradigm.

2.3.4 Decoding turbo codes

Besides its good weight (distance) properties, a very interesting feature of the turbo code scheme is its comparatively easy decoding. More precisely, a reasonably good approximation of its optimal decoding has low complexity. A key concept here is that of *extrinsic information* on a symbol. Let us denote by C_1 the constraints which tie together the symbols of U and V_1, as created by the first rate-1 encoder. Similarly, we denote by C_2 the constraints due to the second rate-1 encoder, which tie together the symbols of U and $V_2(\Pi)$. Now consider some binary symbol u which belongs to U. Due to C_1, symbols of V_1 and of U (u excepted) contain information about u, besides u itself, which is referred to as its *extrinsic information*. Since it belongs to U, the *same* information symbol u is somewhere in the sequence $\Pi(U)$ which enters the rate-1 encoder generating $V_2(\Pi)$, and we know its location since the interleaver Π is known as a part of the whole encoder. Therefore, due to C_2, symbols of $V_2(\Pi)$ and of U (u excepted) also bear extrinsic information about u.

Let us now consider the corresponding received symbols and sequences. Let us denote by u' the received symbol which corresponds to u and by U', V_1' and $V_2'(\Pi)$ the received sequences which correspond to U, V_1 and $V_2(\Pi)$, respectively, 'received' meaning that the channel noise makes the symbols erroneous with a certain probability. Due to the channel noise, the receiver does not know u with certainty but only its *a priori* probability. It is intended to evaluate the probabilities $\Pr(u=0)$ and $\Pr(u=1)=1-\Pr(u=0)$, and to take as decoding decision about u the binary value which corresponds to the largest of these two probabilities, in terms of the sequences U', V_1' and $V_2'(\Pi)$. Besides the known *a priori* probability that $u=0$, the receiver can reassess the probability $\Pr(u=0)$ in terms of the extrinsic information due to C_1, using algorithms which exploit the code constraints. These algorithms are easy to implement if the memory m of the rate-1 encoders is not too large. The reassessed probability of error is less than the initial one. The receiver can also use the extrinsic information associated with C_2. The interleaving performed by Π makes the extrinsic information associated with C_1 independent from that associated with C_2, so the probability $\Pr(u=0)$ as reassessed in terms of C_1 can be used as the *a priori* probability for reassessing the same probability in terms of C_2. Moreover, and this is the most powerful tool for decoding, this process can be *iterated*, with the newly reassessed probability in terms of C_2 being used as the *a priori* probability for a reassessment in terms of C_1, and so on. This process is repeated as many times as needed (there are criteria for stopping this iteration).

If the channel is not too bad (in more formal words, if the code rate is smaller than some threshold which is itself smaller than the channel capacity), this iterated decoding process converges to a 'hard decision' (i.e., all the reassessed symbol probabilities approach 0 or 1) which is very likely the best possible decoding decision. The precise analysis of this process is not easy, and the design of the interleaver so as to optimize the overall performance remains an open problem. However, the decoding mechanism as a whole is well understood and the performance of turbo codes so decoded is much closer to the theoretical limit, i.e., the channel capacity defined in Sec. 2.2.1, than previously obtained by the use of other codes. The capacity thus can be considered as actually reached for most practical purposes. The reader is referred to (Guizzo, 2004) for an excellent description of the turbo codes in non-technical terms and the history of their invention. The word 'turbo code' which was coined by Berrou and Glavieux to designate these codes was actually inspired by the iterated decoding process where the result of a first decoding is used again in the next step, in a way reminiscent of a car turbo charger which uses its own exhaust to force air into the engine and boost combustion. This iteration process can be interpreted as a kind of feedback.

3. CONSERVING THE GENOME NEEDS ERROR CORRECTION

After these lengthy but necessary preliminaries, we are now able to apply concepts from information theory and error-correcting codes to genetics. To begin with, we compute the probability of error of a nucleotide as a function of time, and then the corresponding capacity as the main parameter which measures the genome ability to communicate information through time. This computation shows that the genomic capacity decreases exponentially fast down to zero due to the accumulated errors, hence that the genome is ephemeral at the time scale of geology. The faithful communication of genetic information can thus only rely on error-correcting codes.

3.1 The Genome as a Sequence of Symbols

Applying information-theoretic results to genomes implies as a first step the identification of their alphabet. The quaternary alphabet {A, T, G, C} having as symbols the DNA nucleotides may seem obviously relevant, but experimental data show that genomes, or regions of genomes, are more or less '(G+C) rich'. The (G+C) density is even assumed to have an important genetic rôle (Forsdyke, 1996), and how such an offset with respect to an equal frequency of nucleotides is conserved through time needs moreover to be explained. Using instead the binary alphabet {R, Y} which only keeps the chemical structure of the nucleotides (purine R, double-cycle molecule, i.e., A or G, or pyrimidine Y, single-cycle molecule, i.e., T or C) presumably better fits reality since the genomes are experimentally known to have the same average number of purines and pyrimidines. We shall in the sequel make all calculations with an alphabet size equal to some number denoted by q, but for the purpose of illustration we shall assume that the binary alphabet {R, Y} is considered, i.e., $q = 2$.

The integrity of a genome is mainly threatened by chemical reactants and radiations. Cellular and nucleic membranes can provide an adequate shielding against chemical agents, but not against radiations of solar and cosmic origin, or due to natural radioactivity. Moreover, the DNA molecule belongs to the quantum world according to two of its dimensions but, as a long string of nucleotides, it extends itself in the third dimension at the macroscopic scale. It can support a definite information only provided its intrinsic indeterminism as a quantum object is corrected by genomic codes.

To take into account the possibility of genomic symbol errors, let us now consider a situation where a symbol from an alphabet of size q has been chosen to bear some information but may, or not, be replaced by (or changed into, or received as) another symbol of the same alphabet[4], an event to be referred to in general as a *transition*, or to an *error* when it results in a symbol different from the initial one. Assuming that a given symbol is randomly subjected to error with a constant probability per unit of time, ν, we shall compute in Sec. 3.2 its probability of error as a function of time. We shall then use this result to show that the corresponding channel capacity decreases exponentially fast down to zero as time tends to infinity.

The computation presented in Sec. 3.2 only concerns a single type of errors, where a symbol different from the correct one is substituted for it. Errors consisting of an erasure (a symbol is not recognized as belonging to the alphabet), a deletion (a symbol has been removed from the message) or an insertion (an extraneous symbol has been appended) may occur. Our restriction to errors by substitution is mainly motivated by the fact that this case has been much more studied by information and coding theoretists than the other ones, and that the design and implementation of error-correcting codes for deletions and insertions is significantly more difficult. Taking account of other types of error than substitutions would complicate the discussion although presumably not entailing very different conclusions. Moreover, taking account of other types of errors can but worsen the situation, except as regards the erasures which are milder than substitutions. Even if all errors consisted of erasures, an utterly optimistic assumption, the capacity would still exponentially decrease down to zero as a function of time, as shown in Sec. 3.3.3.

3.2 Symbol Error Probability as a Function of Time

Remember that, as discussed in Sec. 2.3.1, the typical information-bearing event in any communication system is the choice of a particular symbol among a given alphabet. To assess the communication performance of a genome, let us consider its nucleotides as symbols of an alphabet of size q. Let us assume that such a symbol incurs an error during the infinitesimal time interval $\mathrm{d}t$ with probability $\nu\mathrm{d}t$, where ν is a frequency provisionally assumed to be constant. We assume that an error affecting a symbol consists of replacing it by another one, chosen with probability $1/(q-1)$ among the other $q-1$ symbols of the alphabet (this is the worst probability distribution in case of an error by substitution, according to information theory).

Let $P(t)$ denote the probability that a given symbol differs from the initial (correct) one at some time $t \geq 0$. The given symbol is identical to the initial one with probability $1-P(t)$, and in this case the probability of error increases during the interval $(t, t+\mathrm{d}t)$ by an amount of $\nu\mathrm{d}t$. But if the given symbol is already in error, an event of probability $P(t)$, the probability of error *decreases* by an amount of $\nu\mathrm{d}t/(q-1)$ since the newly occurring error can recover by chance the initial symbol. We can thus express the probability $P(t+\mathrm{d}t)$ as

$$P(t+\mathrm{d}t) = P(t) + \nu\mathrm{d}t[1-P(t)] - \nu\mathrm{d}t\frac{P(t)}{q-1} = P(t) + \nu\mathrm{d}t\left[1 - \frac{q}{q-1}P(t)\right].$$

This equality is equivalent to the differential equation

$$(7) \qquad P'(t) = \nu\left[1 - \frac{q}{q-1}P(t)\right],$$

where $P'(t)$ denotes the derivative of $P(t)$ with respect to time. Its solution satisfying the initial condition $P(0) = 0$ is

$$(8) \qquad P(t) = \frac{q-1}{q}\left[1 - \exp\left(-\frac{q}{q-1}\nu t\right)\right].$$

Figure 7 represents this error probability when $q = 2$.

The slope of the graph of $P(t)$ at the origin, $P'(0)$, equals ν, and $P(t)$ tends to the horizontal asymptote $P(\infty) = (q-1)/q$. This asymptotic behaviour for t approaching infinity means that after a long enough time the given symbol no longer depends on the initial one and becomes random with uniform probability over the alphabet.

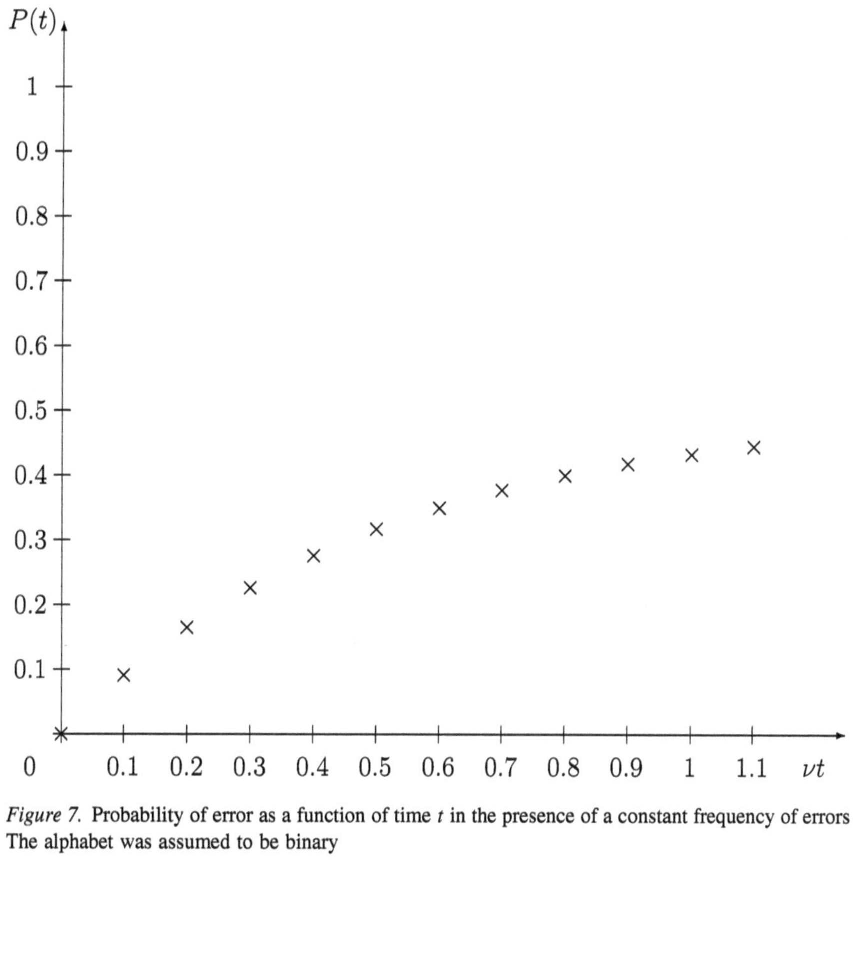

Figure 7. Probability of error as a function of time t in the presence of a constant frequency of errors ν. The alphabet was assumed to be binary

If we now consider a genome consisting of a sequence of n symbols, each of them being independently affected by errors with probability $P(t)$, the average number of erroneous symbols at time t in this genome is then $N_e(t) = nP(t)$. Replacing $P(t)$ by its expression (8) results in

$$N_e(t) = nP(t) = \frac{(q-1)n}{q}\left[1-\exp\left(-\frac{q}{q-1}\nu t\right)\right].$$

If the sequence considered is a word of an error-correcting code of length n and minimum distance d, remember that it will be corrected with a probability the larger, the larger is n, which moreover approaches 1 as n tends to infinity, provided $N_e(t) < d/2$.

If the symbol error frequency varies as a function of time, say $\nu(t)$, one should just substitute $\nu(t)$ for ν in (7). No simple solution like (8) can then in general be obtained, but $P(t)$ remains an increasing function of time since (7) shows that its derivative $P'(t)$ is positive, and it tends to the same asymptotic value $P(\infty) = (q-1)/q$. If $\nu(t)$ remains larger than some known value ν_0, then (8) written for $\nu = \nu_0$ provides a lower bound $P_0(t)$ to the actual probability of error, and the capacity computed in terms of $P_0(t)$ is larger than the actual capacity.

3.3 Capacity Associated with Nucleotides

3.3.1 Capacity of a nucleotide in a single-strand genome

As stated in Sec. 2.2.1, information theory defines the *channel capacity* as the largest amount of information which can be communicated in the average by a symbol. It equals $\log q$ in the absence of errors (with the base of the logarithms defining the information unit), but it also accounts for the information loss due to a nonzero probability of error. Its importance results from the fundamental theorem which states that errorless communication using an (n, k) error-correcting code (each word of which is n-symbol long and bears k information symbols) can be achieved asymptotically for n approaching infinity if, and only if, the information rate k/n of the code is less than the channel capacity expressed using q-ary units.

In the case of q-ary symbols affected with probability p by errors of the type specified above, this capacity, computed using (6), reads:

$$\text{(9)} \qquad \begin{aligned} C_q &= \log_2 q + p\log_2 p + (1-p)\log_2(1-p) - p\log_2(q-1) \\ &= \log_2 q + p\log_2\left(\frac{p}{q-1}\right) + (1-p)\log_2(1-p) \end{aligned}$$

shannons per symbol (remember that we name 'shannon' the binary unit of information; see Sec. 2.2.1). The subscript q in C_q is intended to remind the alphabet size. Letting in the above expression $p = P(t)$ as given by (8), thus assuming again that ν is constant, results in

$$(10)\qquad C_q(t) = \frac{q-1}{q}\left[1-\exp\left(-\frac{q}{q-1}\nu t\right)\right]\log_2\left[1-\exp\left(-\frac{q}{q-1}\nu t\right)\right] + \frac{1}{q}\left[1+(q-1)\exp\left(-\frac{q}{q-1}\nu t\right)\right] \log_2\left[1+(q-1)\exp\left(-\frac{q}{q-1}\nu t\right)\right]$$

which expresses the capacity of the genomic channel as a function of time in the presence of a constant error frequency ν. Notice that the error probability $P(t)$ and hence the capacity $C(t)$ depend on time in (8) and (10) through the product $\nu t = \tau$, a dimensionless quantity which can be interpreted as a measure of time using $1/\nu$ as unit. The formula (10) accounts for the degradation of the channel capacity due to the accumulation of errors. It decreases from $\log_2 q$ for $\tau = 0$, with a slope equal to $-\infty$, down to 0, *exponentially* for τ approaching infinity.

Let us assume again that the relevant alphabet is binary (say, {R, Y}, the purine/pyrimidine chemical structure of a nucleotide). The capacity given by (10) for $q = 2$, namely

$$(11)\qquad C_2(t) = \frac{1}{2}\{[1-\exp(-2\nu t)]\log_2[1-\exp(-2\nu t)] + [1+\exp(-2\nu t)]\log_2[1+\exp(-2\nu t)]\},$$

has been plotted in terms of $\tau = \nu t$ in Fig. 8 where it is represented by crosses. Similar shapes would be obtained with alphabets of other size. For a single-strand genome, a binary error-correcting code can be used in order to ensure errorless communication (asymptotically for large n) provided $C_2(t)$ remains larger than its rate k/n, hence if t is small enough.

3.3.2 *Capacity of a pair of nucleotides in a double-strand genome*

A pair of complementary nucleotides in double-strand DNA has however a larger capacity. If it happens that the two available nucleotides are not complementary, it is clear that one of them is in error although which is wrong is not known, so no decision about the value of one of them can be taken. This case is referred to in information theory as an 'erasure'. It is less harmful than a wrong decision since it warns that the considered symbol is unreliable. Taking account of such erasures results in an increased capacity as exploiting the informational equivalence of the complementary strands. Let us assume that the errors independently affect the nucleotides of a complementary pair and let p denote the error probability of a single nucleotide. An error occurs only when both nucleotides are simultaneously in error, an event of probability p^2, and an erasure when one of them is in error but

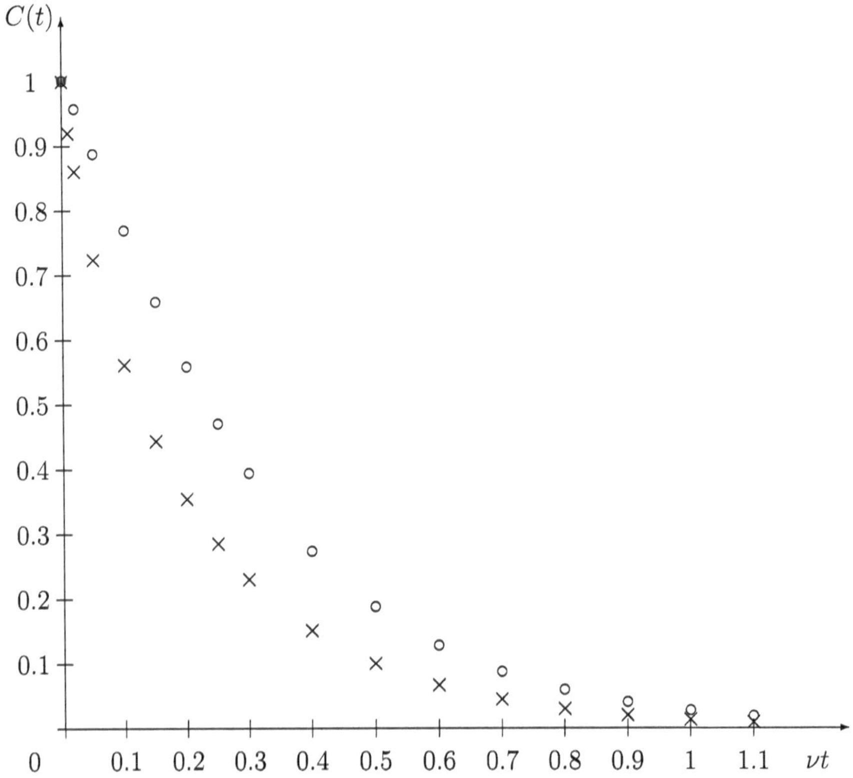

Figure 8. Genomic capacity as a function of time t in the presence of a constant frequency of errors ν, in shannons (binary units) per symbol. The alphabet is assumed to be binary. Points represented by crosses refer to a single DNA strand, while those represented by circles are plots of the capacity taking into account the availability of the two complementary DNA strands

the other one is not, which occurs with probability $2p(1-p)$. Let $C_{q,\mathrm{ds}}$ denote the capacity of this channel, where the subscript 'ds' stands for 'double strand'.

Assuming again for the purpose of illustration that the alphabet is binary, i.e., {R, Y}, the capacity $C_{2,\mathrm{ds}}$, computed using again (6), is expressed in shannons as:

$$C_{2,\mathrm{ds}} = (1-2p+2p^2)[1-\log_2(1-2p+2p^2)]+2[p^2\log_2 p+(1-p)^2\log_2(1-p)],$$

or, after replacing p with the probability of error at time t, $P(t)$, expressed in terms of νt according to (8):

$$(12) \qquad C_{2,\mathrm{ds}}(t) = \frac{1}{2}\{[1-\exp(-2\nu t)]^2\log_2[1-\exp(-2\nu t)] + [1+\exp(-2\nu t)]^2\log_2[1+\exp(-2\nu t)] - [1+\exp(-4\nu t)]\log_2[1+\exp(-4\nu t)]\}.$$

This capacity is also plotted in terms of $\tau = \nu t$ in Fig. 8 (circles). As expected, it is larger than the single-strand capacity $C_2(t)$ given by (11). The slope near the origin is -2 instead of $-\infty$. When τ approaches infinity, $C_{2,\mathrm{ds}}$ approaches $2C_2$. Taking into account the availability of two complementary DNA strands thus results in a moderate improvement of the capacity (by a factor less than, and asymptotically equal to, 2). Remember that the capacity $C_{2,\mathrm{ds}}(t)$ given by (12) measures the largest possible information rate per symbol of the genetic channel for the binary alphabet {R, Y}, and that an (n, k) error-correcting code (each word of which is n-symbol long and bears k information symbols) can provide errorless communication (asymptotically for n approaching infinity) only if $k/n < C_{2,\mathrm{ds}}(t)$.

3.3.3 *Capacity in the presence of erasures only*

The above capacities were computed assuming that an error affecting the genome consists of substituting a wrong symbol for the correct one. A milder kind of error would consist of simply *erasing* it, i.e., identifying it as not belonging to the alphabet. A very optimistic assumption would be that only erasures may occur. Let v denote the probability of an erasure. One easily shows that the corresponding capacity is $(1-v)\log_2 q$ shannons. The same reasoning as in Sec. 3.2, but assuming that once a symbol has been erased it remains so, shows that if the probability that an erasure occurs within the infinitesimal time interval $\mathrm{d}t$ is $\nu \mathrm{d}t$ where the frequency ν is assumed to be constant, then the probability of erasure as a function of time is simply $v(t) = 1 - \exp(-\nu t)$. The capacity as a function of time is thus, in the single-strand case:

$$C_{q,\mathrm{er}} = \exp(-\nu t)\log_2 q,$$

which is again an exponentially decreasing function. Using the double strand structure would reduce the probability of erasure to $v^2(t) = [1 - \exp(-\nu t)]^2$, i.e., that of simultaneous erasures on both strands, finally resulting in the capacity:

$$C_{q,\mathrm{er,ds}} = \exp(-\nu t)[2 - \exp(-\nu t)]\log_2 q$$

which asymptotically equals twice the single-strand capacity so it still exponentially decreases down to 0 when t approaches infinity.

3.4 How can Genomes be Conserved?

3.4.1 *General consequences of the capacity calculations*

The curves of Fig. 8 (or those which can be drawn when other assumptions about the alphabet size are made) clearly show that the capacity becomes negligible after a time interval close to $1/\nu$, meaning that the genomic channel is completely inefficient at the time scale of geology for plausible values[5] of ν. Means for regenerating the genome must be available and necessarily take the form of error-correcting codes endowing the genome with the necessary property of *resilience to errors*. The genome regeneration must moreover be performed after a time interval as small as to avoid the genomic channel to degrade beyond the code correction ability (see Sec. 2.3.2).

The results plotted in Fig. 8, based on the capacity computations of Sec. 3.3, thus show that the genomes quickly (at the geological time scale, of course) bear less and less information about the original message in the absence of a regeneration process. Conservation of the genome is not the rule and error is not the exception. This implies a reversal of the onus of proof: it is the conservation of distinct genomic features which needs to be explained. We shall develop this remark below (Sec. 4.1) but we may already stress that it plainly contradicts a basic assumption of today's genetics, underlying almost all its arguments but left implicit as probably believed obvious.

3.4.2 Main and subsidiary hypotheses

That genomic error-correcting codes exist will be referred to in the sequel as our *main hypothesis*, although information theory shows it is necessary, not merely speculative. A subsidiary hypothesis must furthermore be introduced in order to fit properties of the living world as well as nature's approach. It consists of assuming that a genomic code combines several codes according to a layered architecture referred to as *nested codes*.

The assumption that genomes are words of error-correcting codes is tenable only if they are redundant, as such codes should be. In information theory, 'redundancy' does not merely mean that several copies of something are available but the far more general property that the number of symbols which are used in order to represent the information exceeds that which would be strictly necessary. Genomes are in fact extremely redundant since the number of distinct genomes of length $n = 133$, 4^{133}, approximately equals the estimated number of atoms in the visible universe, namely, 10^{80}. In the absence of redundancy, the number of possible genomes of length n would be 10^n, an inconceivably large number for n of a few millions as in bacteria, let alone for n of a few billions as in many plants and animals. Even the shortest genomes, that of viruses, are a few thousands of nucleotides long and thus still appear as very redundant.

3.4.3 Subsidiary hypothesis: nature uses 'nested codes'

Our subsidiary hypothesis is that nature uses nested codes. By 'nested codes', we mean a system which combines several codes as follows (it is convenient here to assume that the encoding rule is systematic, as defined in Sec. 2.3.1). A first information message is encoded according to some code. Then, a second information message is appended to the codeword which resulted from the first encoding, and the resulting message is encoded again using another code.

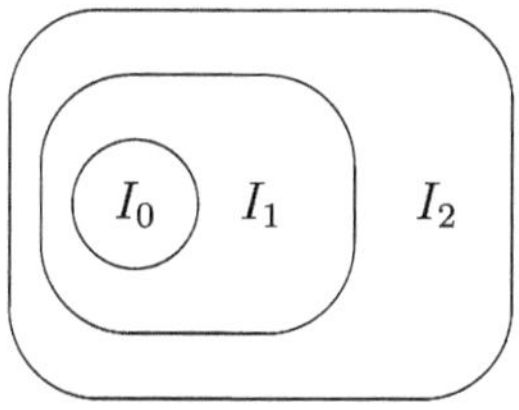

Figure 9. The fortress metaphor. A code is represented as a closed wall which protects what is inside it. I_0, I_1, I_2 and I_3 are successive information messages. I_0 is protected by 3 codes, I_1 by 2 codes, I_2 by a single code and I_3 is left uncoded

This process is repeated several times, the last information message being left uncoded. The resulting nested system is depicted in Fig. 9 using the fortress metaphor where each code is represented as a wall which encloses the information message it contains, assuming a three-layer scheme. Clearly, an information message is the better protected, the closer to the centre it is in this picture. Of course, the walls represent here purely abstract obstacles to errors seen as attackers. Notice that a very efficient protection of the most central information does not demand very efficient individual codes: the multiplicity of enclosing walls provides a much higher degree of safety than each of them separately.

Notice that we assume that the different coding layers appeared successively in time, meaning that the nested structure of ancestral forms of life has a number of layers less than that of more recent ones. The appearance of the successive coding layers may well, at least for certain of them, coincide with the 'major transitions in evolution' (Maynard Smith and Szathmáry, 1995; Barbieri, 2003). The onset of the nested structure can be understood since DNA can be replicated. If its copy is appended to the initial genome instead of being separated from it, then an increase in the genome length results and the longer genome can evolve so as to increase its error-correcting capability. Using old means to create new ones, as assumed here, is a typical feature of nature's approach often referred to as *tinkering*.

The hypothesized nested structure is plausible if we notice that certain parts of the genome like the *HOX* genes are conserved with astonishing faithfulness in many animal species, with the consequence that the organization plans of the corresponding phenotypes are very faithfully conserved. At variance with the extreme permanency of the *HOX* genes, however, it turns out that some genomic variability has advantages as witnessed by the evolutive success of sex as a means for creating new combinations of alleles. It is thus likely that genomic information is unequally protected against errors, and the nested structure is the simplest way to do so. Moreover, since we assumed that the codes appeared successively in time, the genomic information is the better protected, the older it is, so that the variability mainly concerns the most peripheral layers of the nested structure. We now undertake to draw consequences from the main and subsidiary hypotheses we made.

4. HOW GENOMIC ERROR-CORRECTING MEANS FIT IN WITH THE LIVING WORLD AND ITS EVOLUTION

We claim that our main and subsidiary hypotheses explain many features of the living world and of its evolution. We do not feel exaggerated to say that they explain very basic properties of life, including some left unexplained by today's biology for lack of relevant answers, or even for not realizing that they need an explanation. Other consequences of our hypotheses provide arguments about controversial points, especially those discussed in Sec. 4.3. We first examine the consequences of the computations presented in Sections 3.2 and 3.3, showing that the genome conservation needs frequent enough regenerations. Then we consider the consequences of our hypotheses as regards the living world and, since its present state cannot be understood without considering the way it came into existence, as regards biological evolution.

4.1 Genome Conservation and Time Between Successive Regenerations

We have seen in Sections 3.2 and 3.3 that the probability that a nucleotide is in error, hence the corresponding capacity, are functions of time: the probability of error increases, and the capacity exponentially decreases down to zero as time increases. An obvious consequence of this fact is that the genomes must act as words of an error-correcting code (our main hypothesis) and, moreover, that they must be regenerated (in the sense of Sec. 2.3.1) before the accumulated errors result in uncorrectable error patterns. As a consequence, the time interval between two regenerations must be short enough.

The minimum distance between the codewords in the system of nested codes, assuming the subsidiary hypothesis to hold, is the minimum distance d of the code which corresponds to the outmost wall of Fig. 9, hence to the species level. The time interval t_r between successive regenerations should be such that the average number of errors $nP(t_r)$ in a codeword (where n is the codeword length and $P(t)$ denotes the symbol error probability as a function of time as in Sec. 3.2) remains significantly less than half the minimum distance of the code, $d/2$, so that regeneration is almost surely successful. Similarly, but now for the best possible code of a given rate k/n, the capacity $C(t_r)$ should remain larger than this rate.

The genome conservation depends on an external parameter: the nucleotide error frequency ν, and two internal parameters. A parameter of the genome itself: its correcting ability as a code, measured by its minimum distance d, on the one hand; and a parameter of the living being which contains it: the time interval between regenerations, t_r, on the other hand. Both internal parameters may be assumed to have evolved independently, their mutual matching in the extant living world resulting from natural selection. We may assume that evolution eventually resulted in a proper adjustment of these parameters for most species, but let us consider what happens in case of a mismatch. Let us first assume that t_r is too large. Then regeneration errors are frequent so living beings in this situation belong to unstable species, with many new forms appearing and disappearing in comparatively short time intervals. The Cambrian explosion may have resulted from a mismatch of this type. If on the contrary the time interval t_r is too small, then we have very conservative species which can remain identical to themselves during long periods but lack flexibility to evolve in the presence of a changing environment, hence risk to get extinct for lack of fitness. Of course, the actual picture is much more complicated if we take account of our subsidiary hypothesis that the error-correcting codes assume the form of nested codes. Roughly speaking, we may think that the more numerous are the code layers, i.e., the more recent is the considered genome, the higher is the global error-correcting ability of the code, meaning that more variation may be accepted and even favoured in the most peripheral layers. Therefore we may expect that more recent, better protected genomes can accept a time interval between regeneration much longer than that of earlier and less protected genomes. If we assume that the lifetime of individuals is equal to the time interval t_r (in the case of bacteria) or proportional to it (e.g., for animals), we may thus explain why it is much shorter for ancestral forms of life than for highly complex more recent beings.

We do not actually know when the regeneration process takes place. In animals with sexual reproduction one may plausibly assume that it occurs during meiosis. Then, regeneration coincides with generation and, besides being a trivial biological fact, that nature proceeds by successive generations appears as an absolute necessity in the light of information theory. But other mechanisms may be contemplated. For instance, the recent finding in *Arabidopsis thaliana* of 'non-Mendelian inheritance' (Lolle et al., 2005) could be explained by assuming that, in this species and probably in other plants, the regeneration process is not systematically performed each time the genome of a gamete is replicated, but is sporadically triggered from the outside by some kind of stress.

4.2 Discreteness and Hierarchical Taxonomy of the Living World

We now assume that evolution resulted in a proper adjustment of the two parameters d and t_r which control the genome conservation and the mutation rate. The hypothesis that genomic error-correcting codes exist immediately implies that the genomes are far from each other in terms of the Hamming distance, separated by at least the minimum distance d of the code. If we except the small individual differences due to the information left uncoded, genomes are at least at a distance d from their closest neighbours, which implies the *existence of distinct species*. In the absence of genomic error-correcting properties, the living world would appear as a collection of chimeras.

The picture becomes more complicated but more realistic when we take into account the subsidiary hypothesis of a system of nested codes. Let d_0 denote the minimum distance of the most central (ancestral) code. Geometrically, this means that the messages which pertain to the initial information I_0 can be represented by points which are initially at least d_0 apart in the Hamming space. The further encoding by a code with minimum distances d_1 replaces each of these points by a constellation of points centred on it but the distance of the points representing the messages of I_0 becomes at least $d_0 + d_1$. After a third encoding, the points corresponding to I_0 become $d_0 + d_1 + d_2$ apart, etc. The points which correspond to I_1 are only $d_1 + d_2$ apart, those corresponding to I_2 only d_2 apart. Every time a new encoding is performed, the minimum distance between the points representing the previously encoded words is enhanced by the minimum distance of the new code.

We just described the succession of events which resulted in the construction of the nested code system. Simultaneously to the construction of this system, regeneration errors occur at random and are the more frequent, the distance between the points in the Hamming space is the lesser. But the points are the more distant in this space, they represent words which belong to the more central layers of Fig. 9. A large distance implies that the corresponding regeneration error pattern has larger weight, thus presumably gives rise to a phenotype more different from the original than an error pattern of smaller weight[6]. Clearly, besides the discreteness of species which results from the main hypothesis, the layers of Fig. 9 delineate a *hierarchical taxonomy* among them which results from the subsidiary hypothesis.

But why should the multiple layers of the nested codes appear successively in time? Appending a new error-correcting code to those already in use results in a diminished probability of error, hence in an increased permanency, so it provides an *immediate* evolutive benefit. Indeed, doing so increases both the length and the redundancy of the overall code and the increase of these parameters reduces the probability of regeneration (decoding) error. At the same time, increasing the genome length gives more room for specifying the phenotypes, which may thus be more complex, hence potentially better fitted as regards natural selection in its conventional form. Appending a new code thus both immediately improves the genome permanency and indirectly enables enhancing the phenotype fitness in subsequent evolution. The next section develops these remarks in more general terms.

4.3 Consequences of the Hypotheses as Regards Evolution

Besides the necessity of using error-correcting codes so as to ensure the faithful conservation of genomes, we see that consequences of our hypotheses, which assume that error-correcting codes are actually implemented in the process of transmitting the genomic information and moreover take the form of a system of nested codes, closely match known features of the living world. They also hint at features of the biological evolution.

4.3.1 Trend of evolution towards complexity

The subsidiary hypothesis of a nested structure is not even necessary to explain the trend of evolution towards complexity, a puzzling feature for present biological theories. Our main hypothesis alone implies the trend of evolution towards complexity as a mere consequence of the rather paradoxical information-theoretic fact that the longer the code, the smaller can be made the error probability after decoding, even if the code rate remains constant. Hence increasing the genome size can result in increasing its permanency.

We saw in Sec. 2.3.2 above that the error-correcting codes are means for performing *reliable* communication over *unreliable* channels. Here, 'reliable' is intended to mean that the error probability can be made as small as desired, regardless of the initial error rate, by increasing the length of the codewords, subject to the necessary condition that the codes are *redundant* enough. This key property is not merely a paradoxical theoretical result, but it is fully supported by experiment as countless engineering applications of error-correcting codes were made possible by the tremendous progress of semi-conductor technology. As a striking example, mobile telephony would simply not exist without the availability of sophisticated long codes. If nature uses efficient enough codes (and we may safely assume that the Darwinian mechanisms resulted in almost optimal codes, as products of evolution having a prominent rôle in the genome conservation), then we may think that increasing the genome length results in diminishing the error rate of the genome replication, hence increasing its permanency. However, increasing the genome length while keeping the redundancy rate constant increases the quantity of information which is borne by the genome, thus giving room for specifying more complex (and, thanks to natural selection, better fitted) phenotypes. Indeed, although information theory ignores semantics, information can be thought of as a *container for semantics* (see Sec. 7 below). The availability of more information thus enables to specify more phenotypic features, so basic results of information theory explain the yet poorly understood trend of evolution towards an increased complexity.

4.3.2 Evolution proceeds by jumps

The hypothesis that the genomes behave as words of error-correcting codes, hence are distinctly far apart in the Hamming space, entails that, as resulting from regeneration errors, mutations change genomes to distinctly different ones, which implies that evolution proceeds by jumps. The view of evolution which we propose is thus clearly saltationist, giving an unequivocal answer to this rather controversial point.

4.3.3 Genetic information has a random origin

The accumulation of errors tends to make the genomic message less and less dependent on the original one. The information-theoretic quantity which measures this dependence, the *channel capacity*, has been computed as a function of time in Sec. 3.3 and plotted in Fig. 8. As time increases, it exponentially decreases down to zero. If an error-correcting code is present, the genomic message is exactly regenerated provided the correcting ability of the code is not exceeded, which occurs with high probability if the genome regeneration (decoding) is performed at short enough time intervals. The genomic message only varies when a regeneration error occurs. Such an event is very unfrequent, but it results in a burst of at least d erroneous symbols when it occurs (d denotes as above the minimum distance of the genomic code), the new genome thus becoming significantly different from the initial one. The genomic code then ensures the conservation of this 'wrong' genome exactly as it does for the initial 'correct' one. Instead of a genome gradually becoming less and less dependent on the original genome due to the occurring errors, we obtain that it remains a long time faithfully conserved but suddenly becomes markedly different from the original when a regeneration error occurs. Next regeneration errors increase the difference in discrete steps. Continuing this process during a long enough time has thus the same ultimate consequence on the genome as if no error-correcting code is used: the original genomic message is progressively forgotten, but according to a much slower pace depending on the time interval between regenerations. Another difference is that, when an error-correcting code is employed, the genomes resulting from replication errors are conserved as efficiently as the original one was. Then each genome, whether original or affected by errors, remains identical to itself during an average time interval the average of which depends only on the probability of a decoding error. Each decoding error may be thought of as originating a separate species (excluding errors occurring in the most peripheral, uncoded layer of the nested codes scheme, which only account for differences between individuals of a same species). Another important consequence of our hypotheses is that the extant genomic information originated from replication errors since the original DNA molecule is presumably forgotten for long but, of course, these products of *chance* were strongly filtered by the *necessity* of natural selection acting on the corresponding phenotypes. Only information at the most central position in the nested codes system, hence very old and fundamental, is a possible remnant of the common origin of the extant living beings.

5. GENOMIC ERROR-CORRECTING CODES AS 'SOFT CODES'

5.1 Defining Soft Codes

It would be naïve to expect that the error-correcting codes that nature uses closely resemble those that engineers design. The latter are defined as a set of words which obey constraints expressed by deterministic mathematical equalities. Looking for error-correcting codes of natural origin, we were led to the concept of 'soft code', where the constraints may be expressed as inequalities or forbidding rules as well as mathematical equalities, and may be probabilistic as well as deterministic. Having thus extended the concept of error-correcting codes, we may think of the many mechanical, steric and chemical constraints obeyed by the DNA molecule, or the protein for which it 'codes', as inducing soft codes. Even linguistic constraints may be considered since in a sense the genome describes the construction of a phenotype, which needs some kind of language.

5.2 Potential Genomic Soft Codes

We gave elsewhere a rather comprehensive list of the potential genomic soft codes which result from the several constraints which the genome obeys (Battail, 2005).

For the sake of completeness we list here more briefly these soft codes and then give more emphasis on comments.

A first kind of potential soft codes are associated with structural constraints of DNA. As a sequence of nucleotides, a DNA molecule is clearly subjected to mechanical and chemical constraints due to its spatial structure, its bonding with proteins like histones and, in eukaryotes, its packing in nucleosomes and higher-order structures. Researches in this direction even suggested more precisely that the DNA molecule as packed in chromatin can be interpreted as a kind of 'soft turbo code', in both prokaryotes and eukaryotes (Carlach, 2005). Genomes (especially the human one) often contain very short sequences (typically 3-base long) which are repeated thousands or even millions of times. Such sequences bear almost no information. Such 'junk' DNA may however play a rôle in an error-correction system as separating along the DNA strand more informative sequences which, due to the 3-dimentional structure of the DNA molecule, may be spatially close to each other and share mechanical or chemical constraints (a function which loosely resembles that of interleaving used in engineering). Interestingly, interleaving has an important function in turbo codes, as described in Sec. 2.3.3. That this 'separator' conserves its structure of a short motif repeated many times hints at a function which needs to be maintained, in contradiction with the word 'junk' used to qualify it. Similarly, the conservation of the (G+C) density at a value different from the average 1/2 which would be expected from pure randomness, must be explained as resulting from some kind of error-correcting means.

In the portions of the genome which specify proteins, i.e., in genes in a restricted sense, the sequence of codons (triplets of nucleotides) is furthermore constrained as are the proteins themselves: the structural constraints of proteins induce soft codes on the sequence of codons which correspond to the amino-acids according to the 'genetic code'[7]. Physiologically active proteins are not fully characterized by the sequence of amino-acids (the polypeptidic chain) that the sequence of gene codons specifies. They are made of a number of 3-dimensional substructures (α helices, β sheets, which are themselves included into higher order structures named 'domains') which impose strong constraints of steric and chemical character on proteins. Moreover, proteins owe their functional properties to their folding according to a unique pattern, which implies many chemical bonds (especially disulphur bridges) between amino-acids which are separated along the polypeptidic chain but close together in the 3-dimensional space when the protein is properly folded. The sequence of amino-acids is thus subjected to many constraints, which in turn affect the codons through the inverse 'genetic code'. Due to the universal rôle of DNA for specifying proteins, such constraints must be present in any living being.

At a high level of generality, we mentioned above that soft codes may be induced by linguistic constraints, too. We already noticed that the message which is needed for unambiguously identifying a biological species and even an individual inside it is very much shorter than the actual genomes, even the shortest ones like those of viruses (see Sec. 3.4.2). Genomes are thus highly redundant, a necessary condition for them to possess significant error-correcting properties. From another point of view, this redundancy has rather obvious reasons: the genome does not merely identify a living being. Modern biology interprets it as a *blueprint* for its construction. The genome of any living being needs actually contain the *recipe* for its development and its maintenance. Besides those parts of the genome which direct the synthesis of proteins, i.e., the genes in a restricted sense, and the associated regulatory sequences which switch on or off their expression (i.e., make the gene direct or not the synthesis of the protein it specifies), the genome must somehow *describe* the succession of operations which results in the development and the maintenance of a living thing. This demands some kind of *language*, and a language involves many morphological and syntactic constraints which may be interpreted as generating redundant soft codes having error-correcting capabilities. Moreover, the linguistic constraints appear at several different levels, so a written text assumes the structure of nested soft codes which we were led to hypothesize for the genetic message. In order to illustrate the error-correcting ability of languages, notice that we can correctly perceive the spoken language in extremely noisy acoustic surroundings like vehicules or crowded places. By 'correctly perceive', we do not mean to grasp the meaning, which concerns the semantic level, but simply recover without error the uttered speach as a sequence of phonemes. It turns out that the individual phonemes are identified with a large error probability, but the linguistic constraints together with the high processing power of the human brain eventually result in errorless communication in the presence of noise. We can say that the daily experience of a conversation experimentally proves the ability of the human language, as a highly redundant soft code, to behave like good error-correcting codes designed by engineers.

The number and variety of constraints indeed suggest that many potential genomic error-correcting mechanisms actually exist, which moreover are organised as nested soft codes. The resulting system of nested soft codes closely resembles Barbieri's organic codes (Barbieri, 2003), although it is merely intended to cope with the necessity of protecting the DNA molecule against radiations and quantum indeterminism which no phenotypic shielding can ensure. Barbieri's concept of organic codes, on the other hand, does not refer to the necessity of error correction but results from a deep reflection on biological facts. He considers as an organic code the correspondence which exists between unidimensional[8] strings of completely different molecules (as a famous example, think of the relationship between triplets of nucleotides and the 20 amino-acids which make up proteins, referred to as the 'genetic code'). This correspondence does not result from any physical or chemical law, but can be considered as a pure convention or artifact, just like conventional rules are found in linguistic or engineering. Such rules are maintained thanks to 'semantic feedback loops' (Battail, 2005). According to our point of view, the specific constraints of each of the strings which are tied together by a correspondence rule act as soft codes with error-correcting ability. Barbieri's organic codes actually assume the structure of nested codes. Especially significant in this respect is Fig. 8.2 in (Barbieri, 2003), p. 233, to be compared with Fig. 9 above which uses the fortress metaphor to illustrate the concept of nested codes. This rather unexpected convergence provides a mutual confirmation of both points of view, which appear as complementary. This may also be thought of as an illustration of 'tinkering' as a typical method of nature, where some biological objects are used to perform functions completely different from those they initially performed. However, since many biological problems take the chicken-and-egg form, a word like 'multivalence' could be more appropriate than 'tinkering' (although less picturesque) in the absence of a clear knowledge of the chronology.

5.3 Some Further Comments about Genomic Soft Codes

Soft codes do not exactly fit the properties of error-correcting codes which were described in Sec. 2.3.1. Since their definition involves probabilistic constraints and constraints expressed as inequalities, the mutual distances between their words become random, and especially the minimum distance d which accounts to a large extent for the performance of a code. On the other hand, when discussing in Sec. 3.4.2 the consequences of our hypotheses we assumed that the properties of genomic error-correcting codes were those of conventional codes. This may be thought of as a simplifying assumption. One may moreover argue that, if the soft codes combined into the nested scheme are numerous enough, and if moreover their words are long enough, then the law of large number results in a small-variance minimum distance which can rightfully be approximated by a deterministic quantity.

Let us also notice that the soft code concept implies that the biological constraints are also those which enable error correction, at variance with the uncoded case but also with that of hypothetic codes obeying purely mathematical constraints. This may mean that the genomes which are admissible as words of a genomic error-correcting code also specify viable phenotypes. If this is true, decoding (regeneration) errors produce viable, possibly hopeful, monsters. This remark makes rather plausible the explanation of the Cambrian explosion which we suggested in Sec. 4.1.

6. IDENTIFICATION OF GENOMIC ERROR CORRECTION MEANS

6.1 Indirect Evidence of Genomic Error Correction Means

6.1.1 Spectral and correlation properties of genomes

The experimental analysis of DNA sequences has shown they exhibit long-range dependence. First of all, their power spectral density has been found to behave as $1/f^{\beta}$, asymptotically for small f, where f denotes the frequency and β is a constant which depends on the species: roughly speaking, β is the smaller, the higher the species is in the scale of evolution; it is very close to 1 for bacteria and significantly less for animals and plants (Voss, 1992).

Another study of DNA sequences first restricted the quaternary alphabet of nucleic bases {A, T, G, C} to the binary one {R,Y} by retaining only their chemical structure, purine or pyrimidine (as we did above, too). An appropriate wavelet transform was used to cancel the trend and its first derivative. Then the autocorrelation function of the binary string thus obtained has been shown to decrease according to a power law (Audit et al., 2002). This implies long-range dependence at variance with, e.g., Markovian processes which exhibit an exponential decrease. Moreover, in eukaryotic DNA the long-range dependence thus demonstrated has been shown to depend on structural constraints (Audit et al., 2002). The double-strand DNA is actually wrapped around histone molecules acting as a spool (making up together a 'nucleosome'), which implies bending constraints along the two turns or so of the DNA sequence in each nucleosome.

The $1/f^{\beta}$ behaviour of the spectrum and the long-range dependence of the DNA sequence restricted to the purine/pyrimidine alphabet are of course compatible with each other. Moreover, they both denote (at least if further conditions are fulfilled) the existence of a fractal structure, meaning that the DNA sequence is in some sense self-similar. In other words, a basic motif is more or less faithfully repeated at any observation scale. We may therefore think of the message borne by the DNA strand as resulting from 'multiple unfaithful repetition' which could in principle enable the use of many low-reliability replicas of the basic motif symbols in order to get reliable decisions for the purpose of regeneration. This implies a very large redundancy, indeed an obvious property of the DNA message which we already noticed. The existence of such a regeneration process, possibly approximated by majority voting, is as yet a conjecture. It remains to determine whether, and how, nature implements regeneration based on long-range dependence at some stage of the DNA replication process (Battail, 2003). Moreover, the long-range dependence is compatible with the turbo code structure which has been conjectured to exist in genomes (Carlach, 2005).

6.1.2 *Distance properties of eukaryotic genes under evolutive pressure*

Forsdyke formulated in 1981 (Forsdyke, 1981) the interesting idea that in eukaryotic genes the introns are made of check symbols associated with the information message borne by the exons so as to make up words of a code in systematic form (as defined in Sec. 2.3.1). The literature generally states that introns are more variable than exons. A counterexample was however provided in 1995, again by Forsdyke, who experimentally found that the exons are more variable than the introns in genes which 'code' for snake venoms (Forsdyke, 1995).

It turns out that both the generally observed greater variability of introns and Forsdyke's counterexample can be explained by the assumption that the system of exons and introns actually acts as an error-correcting code in systematic form where the exons constitute the information message (which directs the synthesis of a protein) and the introns are made of the associated check symbols. Interpreted as a regeneration error, a mutation occurs in favour of a codeword at a distance from the original word equal to the minimum distance of the code or slightly larger. If the exons 'code' for a protein of physiological importance, which is by far the most usual case, the evolutive pressure tends to the conservation of this protein so the regeneration errors are mostly located in introns. If however the evolutive pressure tends to make the protein highly variable, as in the arms race of snakes and rodents, then the regeneration errors will be mostly located in exons and the introns will be conserved (Battail, 2004). Strictly speaking, this does not prove that exons and introns together constitute a codeword in systematic form. At least, we can say that the experimental evidence does not disprove this statement.

6.2 Lack of Direct Identification of Genomic Codes

Error-correction means are necessary for counteracting the lack of permanency of the genome pointed out in Sec. 3.3. We showed moreover in Sec. 3.4.2 that assuming their existence enables to derive a number of properties which the living world actually possesses, some of them being so familiar and general that biologists did not even try to explain them. We just mentioned above indirect experimental evidence of this existence. The direct identification of genomic error-correcting codes would be highly desirable as an experimental proof of their existence, but it is still lacking. A necessary condition for identifying these codes is of course that geneticists look for them, which implies their active involvement. Moreover, succeeding in this task needs more than superficial knowledge and understanding of error-correcting codes and information theory (Battail, 2006).

6.3 Identifying the Regeneration Means: an Open Problem

The problem of genomic regeneration (decoding) is left for future researches. The principle of the regeneration can be stated: the genome replication process aims at creating a new genome, hence subjected to all the constraints that a genome should obey. On the other hand, it should replicate the old genome which presumably suffered errors. These conflicting requirements must be solved in favour of the *constraints*. Since we used constraints of biological origin to define the genomic codes, obeying these constraints amounts to correct errors. We may thus think of the replication process as necessarily performing regeneration by *providing the approximate copy of the old genome which best fits the genomic constraints.* Replacing 'old genome' by 'received codeword' in the above statement just describes the engineering function of regeneration, as defined in Sec. 2.1.4. An intriguing feature of regeneration as implementing this rule is that its operation demands that the regenerator (decoder) possesses a full description of the constraints at any level, whether they are of linguistic character or originate in physico-chemical properties of molecular strings. This is again a chicken-and-egg problem, and it is impossible (and maybe meaningless) to know what came first: the description of constraints or the onset of molecular strings with their physico-chemical properties.

As regards the implementation of regeneration, it must be stressed that the full knowledge of a code does not *ipso facto* entail that adequate means for its decoding are known. Moreover, there exist several decoding processes for a given code which more or less approximately implement the optimum rule stated in Sec. 2.1.4

which are generally the more complex, the closer to optimality. Still more than the identification of genomic error-correcting codes, that of the means actually implemented by nature for their decoding (i.e., genome regeneration) is thus difficult and challenging. Remember that we used above the human language as an example to illustrate the error-correcting properties of certain soft codes: the means implemented in the brain for this task are presumably very complex and still unknown. Also, it is possible that existing mechanisms believed to perform 'proof-reading' actually implement some kind of genome regeneration. (Incidentally, proof-reading can only check that the copy is faithful to the original, hence correct errors intrinsic to the replication process. It is of no use if the original itself suffered errors.)

A rather fruitful and general interpretation of decoding consists of noticing that each symbol in an encoded message is represented in two ways. Obviously, on the one hand, by itself; but on the other hand, less obviously, by other symbols due to the constraints which tie together the encoded symbols and provide 'extrinsic information' about the considered symbol as introduced in Sec. 2.3.4. In good codes, the constraints enable to compute each symbol in terms of many other ones, according to many different combinations which provide *replicas* of it (Battail et al., 1979; Battail, 2003). In the presence of errors, an erroneous symbol can be corrected by majority voting or possibly by an improved decision rule where the replicas are weighted in terms of their *a priori* probability of error, and moreover taking into account the possible presence of a same symbol in several replicas. An erroneous symbol is then corrected with high probability if the error rate is as small as to let few replicas be wrong. However, a single erroneous symbol combined with others to compute a replica suffices to make wrong this replica, so the correction process becomes inefficient if the error rate increases too much.

7. ON THE EPISTEMOLOGICAL STATUS OF INFORMATION AND ITS RELATION TO SEMANTICS

7.1 The Word and the Concept of Information

The purpose of Shannon was to develop *a theory of communication*, as stated in the title of his seminal papers (Shannon, 1948). Soon after their publication, this theory has however been referred to as *information theory*, and some regret that the word 'information' has been used to designate it. I think that it was necessary to name what is communicated or, more precisely, what can be quantitatively measured in what is communicated. Despite its vagueness and polysemy, the word 'information' is quite acceptable to this end provided one reminds that, once a word belongs to the vocabulary of science, its meaning becomes precise at the expense of more or less severe semantic restrictions (Servien, 1931). It turns out that, in its restricted scientific meaning, information is an entity of its own.

Information theory uses easily understood and well defined information measures, but does not define the very concept of information. I try now to outline such a definition, consistent of course with information theory. I even attempt to clarify the relationship of information and semantics. This subject is still debated. While the review of information theory given in Sec. 2 expresses a consensus among information theoretists, I give here my rather naïve personal point of view as a mere contribution to the debate about the epistemological status of information and its relation to semantics. These personal views have been moreover strengthened and made more precise with the help of the algorithmic information theory, especially as discussed in (Chaitin, 2005). This book presents in nontechnical terms powerful arguments shedding light on the very meaning and limits of scientific explanations, which necessarily rely on a finite quantity of information. (A very short outline of the algorithmic information theory has been given in Sec. 2.2.3 above.)

7.2 An Information as an Equivalence Class Needing a Physical Inscription

Let us consider first an arbitrary string of symbols from some alphabet. We may transform it by channel coding into a longer string which will resist errors within certain limits, or by source coding so as to shorten it. In both cases, the initial string can be exactly recovered since the encoding and decoding operations are assumed to be stricly reversible. Since the size of the output alphabet of an encoder can differ from the input one, the encoding operations can also perform alphabet conversions. Moreover, the physical support of the encoded messages is itself arbitrary. What is needed is only that a number of states of some physical system equal to the alphabet size can be unambiguously distinguished, regardless of the phenomena which are involved. We may thus think of *an* information as the equivalence class of all the messages which can be transformed into one another by any source and channel coding operation, using any alphabet and borne by any physical support.

Defining an information as an equivalence class is rather abstract and I disagree with the statement made by Landauer in (Landauer, 1991) and many subsequent papers that 'information is physical'. I consider however, without contradiction, that recording or communicating an information necessarily involves some physical support bearing one of the members of the equivalence class to which it belongs, as a string of symbols of some alphabet. I thus believe that information has no objective existence without a physical support, at variance with a more or less implicit idealistic point of view. I believe that even the information recorded and dealt with in the human mind has a physical support in neuronal devices and processes inside the brain, although they remain still largely unknown. The embodiment of information into some support, which is necessary for its recording, communication and processing, will be referred to below as its *physical inscription*.

7.3 Information as a Container for Semantics

Defining an information as an equivalence class, we have been able to avoid any reference to semantics, in accordance with the total separation between information and semantics that information theory assumed since its beginning. That information theory has been fully developed without having recourse to any notion of semantics *a posteriori* legitimates the initial separation, and any reflection about the relationship of information and semantics must account for it.

Having identified *an* information with an equivalence class, semantics results from associating some meaning with some information according to some convention. As an example, assume that I want to identify a plant. To know the species to which it belongs, I shall use a plant guide which contains a series of questions about the shape of the leaves, the number of petals of the flowers, and other distinctive features. These questions are dichotomic, i.e., can be answered by yes or no. To each binary information associated with an answer corresponds some semantic content which results in differentiating plant species. Clearly, the more numerous the questions, the larger the set of different species which can be distinguished. If the questions are independent, answering n questions enables distinguishing 2^n different species. A set of n dichotomic questions bears an information measured by at most n binary units or shannons (see Sec. 2.2.1) which means in a sense that more information can contain 'more' semantics. Of course a quantitative measure of semantics is meaningless in general, but this example shows that semantics can be associated with information by conventions and moreover that more information enables to make finer semantic distinctions. In a computer context as in (Chaitin, 2005), we may similarly consider that each symbol of a binary programme asks the computer to make one operation among two possible ones, which is equivalent to answer a question by yes or no. The shortest possible programme for a given computer output (the length of which defines its algorithmic information measure, see Sec. 2.2.3) then answers independent dichotomic questions. Information appears here as a *container for semantics*, providing a bridge between the physical world through its necessary physical inscription, and the world of abstraction through the conventions which associate meaning and information.

That information acts as a container for semantics is maybe best illustrated by a counterexample. It dates back to 1670 and its author is Molière. In his play *Le Bourgeois gentilhomme*, Monsieur Jourdain is a bourgeois infatuated with nobility. His young neighbour Cléonte is in love with his daughter, but cannot expect to marry her unless Monsieur Jourdain believes he is noble. He thus disguises himself as the son of the Turkish sultan. Not only he wears Turkish attire, but he uses a forged, allegedly Turkish, language that his servant Covielle, similarly disguised, is assumed to translate into French. Here is a very short excerpt of the play:

Cléonte (in Turkish attire, playing the sultan's son and speaking to Monsieur Jourdain)
— *Bel-men.*
Covielle (Cléonte's servant, playing a Turkish interpreter)
— *Il dit que vous alliez vite avec lui vous préparer pour la cérémonie, afin de voir ensuite votre fille, et de conclure le mariage.* (He says that you should go fast with him to prepare yourself for the ceremony, in order to later see your daughter, and to celebrate the marriage.)

The audience intuitively perceives that something is wrong and laughs: indeed, so many semantic instances cannot be contained in a message bearing so little information.

8. CONCLUSION

Probably the most important contribution of this paper is the statement that the paradigm of genome replication by copying a template is wrong and should be replaced by that of genome regeneration based on its intrinsic error-correcting properties. For cultural reasons, biologists and communication engineers have little interaction, although both communities abundantly use the word 'information' (but rarely agree on the *concept* of information). Wrongly believing that conservation of the genome is the rule and error, the exception, biologists consider natural selection as the sole factor which shapes the living world. For instance, the book *Mendel's demon* by Mark Ridley is entirely devoted to the problem of the genome conservation and evolution but does not contemplate other mechanisms than template-replication and natural selection (Ridley Mark, 2000). The possibility of any intrinsic error-correcting ability of genomes is not alluded to, and the book contains no reference at all to the literature on error-correcting codes. Similarly, Lolle et al. simply look for RNA templates in order to explain their observations (Lolle et al., 2005), although RNA is notoriously less stable than DNA.

Biologists actually have much to gain in learning information theory (Yockey, 2005; Battail, 2006). The deep change of point of view it would provide can generate new means for understanding the living world and prompt a vast amount of researches in unexpected directions. That its basic paradigm is refuted by information theory shows that a complete renewal of genetics is needed, a wide and demanding task. Although we established above that the existence of genomic error-correcting codes is an unavoidable consequence of the faithful conservation of genomes, almost nothing is known as yet as regards the error-correcting means that nature actually implements. Even though some plausible assumptions can be made about the constraints which provide the genomic error-correction ability, how they are actually used to the benefit of the genome conservation remains entirely to be discovered. The multiplicity of codes combined into the hypothesized nested structure hints at a variety of encoding constraints and regeneration processes. It is thus a very wide field of research that information theory opens to genetics and evolution theory. Besides biology itself, communication engineering should in turn learn much from the eventual understanding of these mechanisms, just like aeronautic engineering learned much from knowing how birds and insects fly. A collaboration between communication engineers and biologists should thus be highly beneficial for both communities. Let us wish that the present paper can help them in setting up such a fruitful interaction.

NOTES

[1] One may even think of the various biological constraints as by products of the necessary error-correcting means.
[2] We shall in the following denote by q the number of symbols of the alphabet, and they will be referred to as q-ary. Most examples will use the binary alphabet, i.e., $q = 2$.
[3] 'Stationary' means that the probabilities involved in the source operation do not vary with time.
[4] Or an alphabet of larger size which contains all the symbols of the initial alphabet plus some others. An example of this case is provided in Sec. 2.2.1 where an erasure is represented as a third symbol appended to the binary alphabet.
[5] The lack of a reliable estimate of the error frequency ν unfortunately forbids to quantitatively exploit these results.
[6] We assume here that the more different are genomes, the more different are the corresponding phenotypes. A kind of isomorphism between the genomes and the phenotypes is thus assumed although it can only be approximative. The same assumption legitimates the use of the Hamming distance for reconstructing phyletic trees, a current biological practice.
[7] We use quotes, here and in the sequel, in order to express that it is not truly a code in the information-theoretic sense, but rather a mapping in the mathematical vocabulary.
[8] Unidimensionality is a common feature of messages in engineering and in biology. It appears as necessary for unambiguous semantic communication.

REFERENCES

Audit B, Vaillant C, Arneodo A, d'Aubenton-Carafa Y, Thermes C (2002) Long-range correlation between DNA bending sites: relation to the structure and dynamics of nucleosomes. J. Mol. Biol. 316: 903–918

Barbieri M (2003) The organic codes. Cambridge University Press Cambridge, UK

Battail G, Decouvelaere M, Godlewski P (1979) Replication decoding. IEEE Trans. Inf. Th. IT-25(3): 332–345, May 1979

Battail G (1989) Construction explicite de bons codes longs. Annales Télécommunic. 44(7–8): 392–404, July-August 1989

Battail G, Berrou C, Glavieux A (1993) Pseudo-random recursive convolutional coding for near-capacity performance. Proc. GLOBECOM'93, Communication Theory Mini-Conference, Houston, USA, Nov. 29-Dec. 2, 4: 23–27

Battail G (1996) On random-like codes. Information Theory and Applications II J.-Y. In: Chouinard P. Fortier Gulliver TA (eds) Lecture Notes in Computer Science No. 1133, pp 76–94, Springer

Battail G (1997) Does information theory explain biological evolution? Europhysics Letters 40(3) Nov. 1st: 343–348

Battail G (2001) Is biological evolution relevant to information theory and coding? Proc. ISCTA'01, Ambleside, UK, 343–351 July 2001

Battail G (2003) Replication decoding revisited. Proc. Information Theory Worshop 03, Paris, France, Mar.–Apr. 2003, 1–5

Battail G (2004) An engineer's view on genetic information and biological evolution. Biosystems 76: 279–290

Battail G (2004) Can we explain the faithful communication of genetic information? DIMACS working group on theoretical advances in information recording: 22–24 March 2004

Battail G (2005) Genetics as a communication process involving error-correcting codes. Journal of Biosemiotics 1(1): 103–144

Battail G (2006) Should genetics get an information-theoretic education? IEEE Engineering in Medicine and Biology Magazine 25(1): 34–45, Jan.-Feb. 2006

Berrou C, Glavieux A, Thitimajshima P (1993) Near Shannon limit error-correcting coding and decoding : turbo-codes. Proc. ICC'93, Geneva, Switzerland, May 1993, 1064–1070

Berrou C, Glavieux A (1996) Near optimum error correcting coding and decoding: turbo codes. IEEE Trans. on Communications 44: 1261–1271, Oct. 1996

Carlach J-C (2005) Comparison of structures and error-detecting/correcting properties between DNA molecules and turbo codes, private communication

Chaitin G (2005) Metamath! Pantheon Books, New York

Cover TM, Thomas JA (1991) Elements of information theory. Wiley, New York

Forsdyke DR (1996) Different biological species 'brodcast' their DNA at different (G+C)% "wavelengths". J. Theor. Biol. 178: 405–417.

Forsdyke DR (1981) Are introns in-series error-detecting sequences? J. Theor. Biol. 93: 861–866

Forsdyke DR (1995) Conservation of stem-loop potential in introns of snake venom phospholipase A_2 genes. An application of FORS-D analysis. Mol. Biol. and Evol. 12: 1157–1165

Gallager RG (1968) Information theory and reliable communication. Wiley, New York

Guizzo Erico (2004) Closing in on the perfect code. IEEE Spectrum INT-41(3): 28–34, March 2004

Landauer R (1991) Information is physical. Physics Today. May 1991, 23–29

Lolle SJ, Victor JL, Young JM, Pruitt RE (2005) Genome-wide non-mendelian inheritance of extra-genomic information in *Arabidopsis*. Nature 434(7032): 505–509, March 24, 2005

Maynard Smith J, Szathmáry E (1995) The major transitions in evolution. Oxford University Press, Oxford, UK

Ridley Mark (2000) Mendel's demon: [gene justice and the complexity of life]. Weidenfeld & Nicholson, London

Servien Pius (1931) Le langage des sciences. Payot, Paris

Shannon CE (1948) A mathematical theory of communication. BSTJ 27: 379–457, 623–656, July and October 1948. These papers have been reprinted, with comments by Weaver W as a book entitled The mathematical theory of communication. University of Illinois Press, Chicago 1949

Voss RF (1992) Evolution of long-range fractal correlation and $1/f$ noise in DNA base sequences. Phys. Rev. Lett. 68: 3805–3808, June 1992

Yockey HP (2005) Information theory, evolution, and the origin of life. Cambridge University Press, New York

CHAPTER 14

RNA AS CODE MAKERS: A BIOSEMIOTIC VIEW OF RNAi AND CELL IMMUNITY

MARCELLA FARIA[1,2]

[1] *Departamento de Bioquímica, Instituto de Química, Universidade de São Paulo, Caixa Postal 26077, São Paulo 05513-970, São Paulo, Brazil*

[2] *Laboratório Espeial de História da Ciência, Instituto Butantan, São Paulo, Brazil*

Abstract: The development of the adaptive immune system as it is known in vertebrates relies on the highly coordinated program of cell differentiation achieved by such multicell organisms during their embryonic development, as well as during their functional physiology. This paper discusses the acquisition of an immune response by means of cell function specialization (recognizers, presenters, killers) in the light of biosemiosis. In particular, it will be argued that self/nonself differentiation rises in multicell organisms by a switch of organic codes and operating logic. In fact, double-stranded RNA molecules that induce a highly specific and selective mRNA degradation in non-vertebrates bring about an ubiquitary silencing of transcription and translation in differentiated vertebrate cells. This last response requires elements which are common to cell immunity, the so called interferon response machinery which is responsible by preserving cell genomes from mobile DNA fragments often generated during viral infection. This particular phenomenon will be extensively discussed to show the general point of how a major evolutionary change - invertebrates to vertebrates, in this particular case – requires the development and fixation of new organic codes. The pattern of embryonic and functional cell differentiation achieved by vertebrates' immune system will only be possible whenever, in evolution, cells are able to discriminate, recognize and integrate signs. We propose that the performance of these increasingly complex skills by cells is the hallmark of different levels of stabilization for living systems, the levels of CELL/SELF/SENSE. The way double-stranded RNA is dealt with by each of the levels proposed will be analyzed as a case study of a broader phenomenon: the contextual meaning of molecular signs as fixed by the combination of natural convention and natural selection as component mechanisms of the evolutionary process

Keywords: Cell immunity, RNAi, Organic codes, Natural conventions, tinkering

M. Barbieri (ed.), Introduction to Biosemiotics, 347–364.

La simplification, n'est pas dans le but dans l'art. On y arrive malgré soi en voulant faire des choses réelles qui ne soient pas la carcasse que nous voyons, mais ce qu'elle nous cache.

Constantin Brancusi

INTRODUCTION

The attempt to provide biological knowledge with a more explanatory conceptual framework is possibly Biosemiotics main motivation nowadays. Even though such enterprise, as mentioned in the editorial of this book, is not a homogeneous one. In such context, a dangerous trap for anyone trying to develop a research program in Biosemiotics becomes the difficulty of defining not only "how" this particular structural science (Artmann, 2005) can help Biology, but also, "which" particular brunch of Biosemiotics we are using as a structural science. Therefore, we should take the time to address the "which" question briefly, as a matter of methodological choice, and focus on the analytical development of the "How" question, where we do hope to make some concrete contributions.

There are many concepts in biology, which the widespread use seems to legitimate and vulgarize, but are still very problematic, lacking a precise definition. In the present work, we shall mention three of them: meaning, complexity and contingency. The discussion will be summary, only to establish formal links between different interpretations of these concepts and the various schools of structural sciences they refer to. By doing so, we should clarify to which sources in Biosemiotics and sciences of complexity we are related in the search for more precise and operational definitions of meaning, complexity and contingency in biological systems.

The difficulty dealing with the notion of meaning in biology is as remote as it has been overlooked. Here, we shall adopt the theory of organic codes (Barbieri, 2003) to address the question. According to this view, living systems are semiotic unities in the sense that they have the triadic structure of "sign, code, and meaning". All biological systems share conventional rules of correspondence between two different worlds (codes) that build up dimensional information (meaning) from linear information (signs). The cell and its triadic organization (genotype, ribotype, phenotype) should, in this scope, be understood as the simplest semiotic unity, maybe the first to be originated in the evolutionary process, but not the only one. From that perspective, the systematic search for collective rules that are not determined by individual features in their structures (organic codes) and the identification of functional unities of increasing complexity which convert signs into meanings by codification becomes a feasible research agenda. As pointed out in a previous chapter of this book (Artmaan, Computing codes versus Interpreting life), the key feature of this school of biosemiotics is its model-theoretical perspective on languages that are axiomatically described as computing codes. The emphasis, therefore, when investigating biological meaning is in the identification of organic codes, formally and systematically, and not – as opposed to other views – in the quest for hermeneutic formulas that would allow us to interpret life itself in a rather transcendent way.

Complexity, as defined from a strictly informational standpoint, is the ability of some opened systems to use energy in order to increase its own order, creating a chain of information transfer (Shannon, 1948). If this is the concept of complexity one shall accept, it becomes almost natural for a biologist to understand the central dogma of molecular biology as a chain of causality leading from information in DNA segments to structure and function in protein polypeptide chains. In such perspective it would be acceptable to try the reduction of complexity to underlying causes, in fact the power of reductionist practices in providing scientific basis to our knowledge of natural phenomena is undeniable. Nevertheless, in agreement with Cohen's analysis of the subject (Cohen, 2004), we can assume that there are limits to the use of reductionism in the investigation of complexity in biological systems. These limits can be formulated in various terms and were indeed discussed by many authors (Brent and Bruck, 2006; Salthe, 2004; Westernhoff and Palson, 2004; Aderem and Hod, 2001; etc.). For our purposes, we should only stress the fact that an informational account of complexity does not take into consideration the role-played by codification in the building and maintenance of multiscale and self-organized biological systems. Coding is crucial for our understanding of meaning in biology and that is the reason we shall try to integrate it to any concept of complexity adopted.

In biology, it seems, boundary conditions are ever changing and are ever restricted to contingent resources inside their history (the narration of their uses). In biology, as opposed to Physics, time is not just a parameter but also the determinant one. In terms of evolution, time would be the ground for compromise between contingency and coherence, a compromise that has various ways such as replication, recombination, mutation, synchronization and hierarchization, yet "biological consistency" can only occur and is determined by the temporal scale of organic cycles, by this particular "cyclic-story-telling" temporal pace. **Figure 1** illustrates the ideas of "time" assumed in some of the Physics and Biology attempts to model the nature in movement. As for this article is concerned, we should just stick to the notion that biological time acts by diversification of agents, as a differentiation process generating specificity (or discriminatory competencies) and, in consequence, generating hierarchic levels. Arrows in a metabolic pathway, a philogenetic tree, or a signal transduction map do not establish equivalence between the points they connect, arrows in biology stand for realization of potentialities or, at the opposite direction, for the indetermination of potentials. In both cases a precise sense of time is at work, integrating it in a greater picture is an inevitable and inviting task for contemporary thinkers. It is our working hypothesis that the operational link between agents, states, structures and/or functions in biology accounts for the generation of meaningful information, based on the codification process that connects instances with no necessary-mechanic association (or material cause). In the study of complex systems the process we are referring to as codification is normally treated as emergence, a term very charged with philosophical enquires on

Figure 1. Time in Physics and Biology

how causality operates to build multiscale systems. It can be argued that codification is indeed a special instance of emergent phenomena but, since codification is much simpler to define and seems to be sufficient to the scope of biological complexity we are interested in, we shall adopt this concept. Nevertheless, there are some terms we can borrow from the sciences of complexity that are helpful for the understanding of scalar hierarchies in biology, the most useful is the notion of physical attractor: long-term stable states towards which complex systems tend (Huang et al., 2005 and Cohen, 2004). This definition should be addressed in more detail further.

Philosophical approaches to the question of stabilization levels in living systems also provide valuable contributions to the understanding of biological complexity. The theory of levels of reality and its various formulations has been reviewed in a recent paper (Polli, 2001). Initially developed by contemporary authors (Spencer, Morgan and Alexander) the categorization of reality into levels attempted to give the theory of evolution a metaphysical framework. Levels such as "Matter, Life and Mind" or ontological regions such as "Nature, Consciousness and Society" (Husserl) will follow this same rationale. These original levels were put forward by many thinkers, from Hartmann's "phylogenetic" layers where levels would be defined by their constitutive unites (atoms, molecules, cells, etc.) and corresponding structures, to Polli's "systemic" levels defined by groups of suitable categories and their underlying dynamics. To this last definition of levels and the theoretical framework it seeds we shall from now on refer to as "Dynamic Ontology". Such view has an enormous heuristic power. The possibility of building up reality levels according to sensible classification of the dynamic categories, sets the scientist free from the fixed boundaries of material causes, but it also demands a new type of imagination, new ways of measuring, modeling and manipulating reality.

Edward Wilson, an eminent contemporary biologist, also recognized by his contributions into the fields of philosophy of science and methodology, has defined "complexity theory as the search of algorithms used in nature that display common features across many levels of organization" (Wilson, 1998). Assuming the terminology we have been using in the present work, this attempt would be equivalent to the search for organic codes in every scale in which living systems shall adopt long-term stability states (the previously mentioned notion of physical attractor).

The idea of contingency underlies both of the concepts that are essential in Darwinian theory of evolution: Natural selection and adaptation. It is also present in the new – Darwinian notion of exaptation (Gould) and, although not explicitly defined in any case, contingency intuitively accounts for the role played by chance during evolution. Once again the problem of such definition in the framework of our analysis is that it does not take into consideration coding, or natural convention, as one of the mechanisms of evolution. Therefore, we will try to define biological contingency as related to the previous definitions adopted for biological meaning and complexity in the framework of the organic codes theory.

François Jacob has proposed the notion of "evolution as tinkering" in the mid 70's (Jacob, 1976). He claimed that the way living things are shaped by evolution is not a balance of teleonomic coherence, replicative invariance and chance variation (as stated in Monod's chance and necessity and broadly accepted), but rather by "the constant reuse of the old to make new" (Jacob, 1986). Tinkering, as opposed to engineering, has to deal with the contingency of resources and their history; therefore, it does not and cannot aim a predetermined output. Stefan Artmann and other structuralist semioticians tried to develop the tinkering concept from a semiotic perspective (Artmann, 2004). There, the materials to be recycled by tinkering become signs and their syntax, semantics and pragmatics. The theory of tinkering assumes the concept of process consistency as the relation between contingency and coherence, as the formal determinant of evolution. This would be equivalent, using Barbieri's terminology of the organic codes theory, to admit that evolution proceeds by natural convention and by natural selection. The pragmatics of any evolution (of living beings, living institutions, living theories etc.) can be analyzed by means of its consistency. Biopragmatics, as a research agenda, should be the search for the set of coherent transformation of contingent boundaries given limited resources. Such investigation done by semiotic means becomes the search for context-dependent transformation of all processes that diversify potentials in the precise sense of originating organic codes.

Our understanding of biological meaning, complexity and contingency is intimately linked to the theory of organic codes in Biosemiotics by one hand, and to some accounts of multiscale emergence formulated by sciences of complexity and philosophy of sciences by the other hand. These notions will frame our discussion of the acquisition of cell immunity and the role played by repetitive RNA sequences in different levels of organization displayed by living systems. By doing so, we will come up with a new attempt to classify functional unities of life into categories of increasing complexity. Kinetic constants, structural limits, and ontological drifts no longer define the frontiers amongst levels. The frontiers become rather a matter of dynamics as the ground for the origin and evolution of semiotic systems as stable states.

The notion of categories that bear a formal correspondence with the stabilization levels adopted by biological systems is seminal in our reasoning and will be developed in detail on the next sections. Briefly, we propose that there are three major levels of stabilization for the living:

1) the CELL, whereas by the discriminatory competences of a semiotic unity a functional autonomy towards the environment is first achieved, The CELL level is able to provide environmental change with biological meaning;
2) the SELF, whereas recognition tasks are added to the previously acquired discriminatory competencies and more complex semiotic unities rise, being able to couple environmental change and Cell fate either by the triggering of life cycles or differentiation programs;
3) The SENSE, whereas cognition skills are added to previously acquired discrimination and recognition ones, giving rise to more complex semiotic unities that display metabolic, developmental and somatic autonomy towards the environment, being able to make dynamic use of information to remodel their own function and structure.

Typically, in each level of organization there is no unique solution for the dynamics of a system compatible with the production of long-term stability states. In articulating our analysis, a dichotomized repertoire of solutions will be discussed for the three categories proposed (Cell/Self/Sense): a Fancy one and a Frugal one. The fancy/frugal distinction refers to alternative pathways taken by living unities under the pressure of natural selection and the synthetic power of natural convention. In both cases, the dynamics of constitutive elements bring about structures and behaviors at higher levels. Nevertheless, the way higher level dynamics constrain lower level structures and behaviors seem to be different in each case. Downward determination is streamlining in frugal solutions, conversely, when fancy pathways are adopted higher levels tend to be more permissive in the determination they perform, pragmatically that will allow to more flexibility concerning structures and behaviors to be naturally selected and conventionned. Therefore, we shall propose this instrumental dichotomy between "Frugal/Fancy" to proceed through the analysis of different levels of life organization. **Figure 2** produces a formal representation of alternative configurations adopted by the living at the levels of: Cell, Self and Sense as categories in an analytical framework (Körner, 1999).

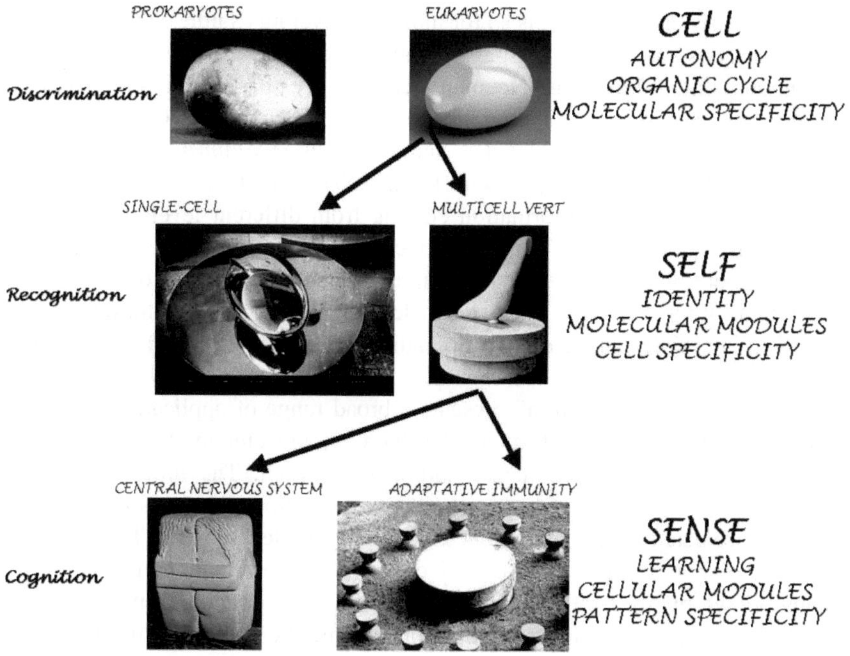

Figure 2. The Cell/Self/Sense categories, levels of stabilization adopted by living systems

UNITY OF LIFE – CELL MAKING

Let us take the cell as the first (in time and space) dynamic configuration that behaves as a living unity. The cell is the minimal thing conserved through evolution capable of:

- Multiplication, variation and heredity (life definition by Herman Muller 1966);
- Assembling functioning units in a structural hierarchy that has acquired trough evolution the ability to store and process the information necessary to its own reproduction (Lila Gatlin, 1972);
- Transforming an external energy and matter flow into an internal flux of self-maintenance and self-reproduction (Varela and Maturana, 1974);
- Assembling and perpetuating artificial structures from natural ones (Barbieri, 2002).

Regardless of the definition of life we shall adopt, just to illustrate a few possibilities, any cell (prokaryotic or eukaryotic, autonomous or living in an organism, differentiated or not) will fulfill the criteria. A cell is a unity of life, a whole which dynamic configuration displays stability relative to its elements.

The organic cycles based on ATP recycling were developed and fixed using cells as photographic paper. Energy gradients tend to dissipate by organized and/or periodic means in nature, so that relative to the energy flow a cell is also a plausible level of synthesis, as much as it is so for molecular cycling (Salthe, 2005). Different levels of analysis and of synthesis, at the same time.

Assuming that, following Neuman terminology based on the work of Bateson, cells are recursive-hierarchical systems (Neuman, 2004) that enable organized/periodic use of molecules and electrons, we must admit causality to proceed bottom-up and top-down (Ellis, 2005a and b and El-Hani, 2005). The compromise between information coming from different levels in hierarchic systems exists everywhere (thanks to feedback loops, patterns are recursive in nature), but only living systems, can use this compromise to dynamically change their own behavior in various levels (thanks to "evolution" understood as the possible output of two operating mechanisms: "natural convention" and "natural selection").

The term "natural convention" presents a broad range of applications that goes beyond the central dogma of molecular biology, presumably the first organic code fixed by natural selection and natural convention. The intersection in an imaginary Venn diagram displaying these two evolutionary processes would be the actual (selected and conventioned) evolving unites of life. The synthetic integration of bottom-up causation and downward determination in hierarchical systems allows by its own nature – synchronic determination from the bottom and diachronic constraining from the top – multiple solutions in the higher levels. At the dynamic level of cells, which is under analysis here, solutions as diverse as non nucleated Eubacteria and Archaebacteria, or the nucleated cells arranged as single cell and as multicell organisms are equally compatible with supporting life.

Cells' autonomy relies on their creating compartments to make cyclic use of energy. The prokaryotic solution is frugal in that its streamlining nature constrains further changes in form, despite their remarkable adaptability to changing environmental conditions. The eukaryotic solution can be referred as fancy, in that its extra-compartmentalization opens up windows of opportunity for alternative controls. In eukaryotic cells form is not as constrained from within, the structure seems permissive to adaptation to the same extent as it is the case in prokaryotes, but it is also permissive to complexification into new logical typing (Bateson, 2002), into creation of new forms and patterns.

At the Cell stabilization level, double-stranded RNAs (dsRNAs) are constitutive structures in all RNA species of the intracellular environment: messenger RNA, transporter RNA, ribosomal RNA and small nuclear RNA. Base-pairing between complementary regions of different RNA molecules, or even intramollecular links, seem to be essential for many control-steps of RNA metabolism, namely: translation initiation and termination; messengers stability; messengers editing (only in nucleated cells); and transcription termination (Lewin, 2000). It is textbook common sense that local RNA-RNA interactions at the Cell level are RNA metabolism signs. In Prokaryotes, these controls are restricted to the steps of protein synthesis, in nucleated cells they also account for RNA processing events.

UNITY OF LIFE – SELF-MAKING

The following level of integration from the perspective of nucleated cells deals with the setting of increasingly abstract compartments. Autonomy towards the indiscriminate external world is necessary and sufficient for creatures to live, but eukaryotic cells could and did discriminate further: between cell types and between cell types in time. We shall refer to that kind of discriminatory property as "recognition" (which literally means "an awareness triggered by contact"), a property essential for what will be called "self-making". Two very different general strategies seemed to be selected to cope with the cell-to-cell discrimination/recognition problem. Unicellular eukaryotes, as yeast and parasites, have taken the frugal way and multicell organisms have taken the fancy one.

The making of self in single cells seems to require differentiation: alternative cell stages attuned in adaptative life cycles. Environmental conditions become integrated into signals that control growth, but also, functions of a diverse logical typing as differentiation, migration, latency, mating, invasion, which are not clonal. Such processes encompass the positioning of single cells in their own life cycle. By the comparison between alternative stages of the same cell and among different cells and their pattern of contact: the notion of identity unfolds in each and every cell.

The fancy path leads to bigger wholes; many cells are assembled in organisms. Here the making of self also requires differentiation, but in organisms there seem to be synchronic life cycles for different cell types. Populations of cells as they dynamically associate in tissues, organs, systems, follow rather diverse programs of differentiation, latency, senescence, programmed cell death. The notion of identity unfolds in a cell-to-cell basis but emerges for the whole organism as well. Once again, the fitness of frugal and fancy strategies is equivalent, but the fancy of multicellularity broadens the spectra for future change. The Cambrian explosion, for example, illustrates the diversity of forms triggered in multicell organisms whenever the fancy path was the substrate for further change.

The role of RNAs as code-makers has been previously stated by Barbieri (Barbieri, 2003) in the scope of the central dogma, bridging the gap between DNA and protein, essential in the making of the Cell level, as mentioned in the previous section, local dsRNA structures, in particular, act as RNA metabolism signals at the Cell level. We shall analyze some mechanisms that cells have developed to deal with double-stranded RNA in different context to elucidate the role of RNAs as code-makers also in the making of Self. RNA interference (RNAi) is a physiological phenomenon widely conserved through evolution by which double-stranded RNA (dsRNA) triggers the silencing of cognate genes (reviewed in Faria et al., 2004). The process was first observed in *Caenorhabditis elegans* after the realization that the injection of dsRNA into this worm brought about the specific degradation of homologous endogenous mRNAs. The evidence of other dsRNA induced homology-dependent gene-silencing mechanisms as chromatin remodeling, chromosome rearrangements, genome *de novo* methylation and translation inhibition emerged later, making it compulsory to enlarge the scope of the investigation (for extensive review see Agami, 2002). According to the currently accepted model, dsRNA can trigger RNAi following their conversion into small, 21–25 nucleotide (nt), interfering RNAs (siRNAs) by members of two families of enzymes: the *rde-1* (for RNAi defective)/*ago-1* (for Argonaute) family and the Dicer multidomain RNAse-III family. The siRNAs will then guide another enzyme complex, the RNA-induced silencing protein complex (RISC) to homologous mRNAs and induce their cleavage and degradation. It is worth mentioning that dsRNAs are physiological intermediates of processes as diverse as viral infection, the expression of transgenes, and the transcription of repetitive sequence gene arrays (endogenous or exogenous, single or multi copy). We will develop the idea that the way different cells deal with such a "polisemic" signal will ultimately reflect their tolerance against genome instability. In the case of differentiated vertebrate cells, dsRNAs induce the interferon response, which activates protein kinase R (PKR) and 2′5′(A)$_n$-synthetase and triggers, as final consequences, the ubiquous inhibition of translation and the induction of mRNA degradation, respectively (Leaman et al., 1998 and Clemens et al., 1997). The toxic effects of dsRNAs in somatic vertebrate cells can be overcome by the use of siRNAs (the shorter versions of dsRNAs) as the input signal to trigger specific-gene silencing. Interestingly, bypassing the interferon response shows that RNAi, thought not visibly triggered by long dsRNAs sequences (the interferon response is just prevalent), is still perfectly functional after cell differentiation (Elbashir et al., 2001).

Two enzymes seem critical for the logical shift that takes place during vertebrates somatic cell differentiation. In single cell eukaryotes and invertebrates PKR homologues do not exist. As for 2′5′ (A)$_n$-synthetase, the enzyme is highly conserved amongst vertebrate, but only poorly homologous putative sequences are found restricted to two species of sponges, among the invertebrates. In embryonic and stem cells, the response to dsRNA is restricted to the silencing of homologous endogenous genes because these two classes of enzymes are inactive or not expressed. Let us dissect the functional structure of PKR, which is, at present, better characterized than 2′5′(A)$_n$-synthetase. PKR is a kinase dependent on dsRNA binding for its activation, the catalytic kinase activity lies in a C-terminal domain and the dsRNA binding is mediated by a N-terminal domain (Lemaire et al., 2005). Upon dsRNA binding PKR undergoes auto-phosphorylation and dimerization, once activated it phosphorylates the eukaryotic initiation factor eIF2α and inhibits translation initiation, in addition PKR induces proinflammatory genes (such as type I interferon) by activating the NF-κB pathway (this issue will discussed in more detail in the "sense making" section of the article). Interestingly, the catalytic domains of other kinases that phosphorylate eIF2α such as HRI, GCN2 and PEK, are highly conserved, but their regulatory domains are different (Rothenburg et al., 2005). It seems that the association of a dsRNA binding activity with a catalytic kinase domain in the same enzyme enabled differentiated vertebrate cells to connect the presence ds-RNA necessarily to translation inhibition, PKR links two otherwise separated sets of information, this happens by natural convention. Many proteins are composed of modular functional units which combined through evolution achieve the conformational flexibility required for regulation without sacrificing the specificity essential for catalysis. In that sense, neither dsRNA recognition, nor translation initiation factors phosphorylation are major evolutionary novelties, but their assembly into the same protein that is alternatively expressed depending on the identity of the cell is new, it is exclusive to vertebrates and it stands for a new organic code.

Yeast, single cell parasites, invertebrates and non-differentiated vertebrate cells are Eukaryotic cells, but based on "single cell" logic. Their response to double-stranded RNA is selective to its sequence; RNAi operates by inhibiting the expression of cognate messengers without killing the triggered cell. The notion of self unfolds allowing to some plasticity of the genome in the behalf of keeping cell stability. At the multicell level of mature vertebrates, the selective response to ds-RNA is no longer enough; these systems would rather spare the affected cell than risking genome stability. The notion of self unfolds privileging the stability of the bigger whole, the organism.

The similarities between invertebrates and vertebrates are very striking for many dimensions of self development that were not mentioned in the present work and should be discussed in depth in the future, namely: body plans, embryology and the pattern specificity of most organs and systems. The dichotomizing exception is the development of more or less complex cell-mediated adaptative immunity and of central nervous systems, exclusive to vertebrates. The differentiated response of vertebrates to dsRNAs segregates along with their acquisition of adaptative cell systems able to produce somatic change, memory and learning. Some of the consequences of such achievements we shall examine in the following section.

UNITY OF LIFE – SENSE MAKING

The following level of integration from the perspective of multicell organisms deals with the setting of compartments increasingly abstract. Identity provided by discrimination between cell types and synchronic differentiation programs are essential features in self-making, but organisms could and did discriminate further, by building classes of differences and dealing with hierarchic levels of classes by integrating simplified versions of those (coded information, memory and decision-making). We shall refer to discriminatory properties of that kind as "cognition", they are essential to what we will call "sense making".

Two very different general strategies seemed to be selected to cope with the hierarchic multicell integration/cognition problem in vertebrates. The complexification of a central nervous system is the frugal solution. Adaptative cell immunity is the fancy way.

Nervous and Immune system development, both require the differentiation of very specialized cells to mediate somatic adaptation to integrated signals and learning. The two systems care for protecting the whole organism against foreigners and for its body maintenance, only they use very different topological strategies. In Cohen's formulation "The nervous system houses spatially fixed non-renewable neurons, with a hard-wired network geometry. The immune system is composed by constantly renewing, physically flowing population of cells" (Cohen, 2004).

In the nervous system the prototypical cells are neurons, despite differences due to the nature of the specific sensorial structures they connect, these cells share minimal features concerning their structure and function. In every case, dendritic region, cell body, axons and synapses will be respectively responsible for the reception, integration, conduction and propagation of the nervous impulse. These cells of ectodermic origin undergo three main irreversible transitions during embryogenesis until becoming functional neurons. First, there is the determination to a neuronal pathway, then the migration and, eventually the synaptogenesis. The three events are controlled by cell context sensed as neurotrophic factors and cell adhesion molecules relative presence, at each step there is a decrease in the potential destinies the cells can follow. The frugality of vertebrate's nervous system has nothing to do with their complex functionality. They respond as a robust network of information processing and integration which plasticity is only comparable with that of their own immune system. The economical nature of nervous system has to do with the relatively constrained form of their cellular unities, and their being unable to regenerate after differentiation.

As the immune system ontogeny evolves, the adaptative system accumulates a population of mesodermic origin lymphocytes equipped with unique surface receptors able to recognize nonself epitopes in cognate interactions. Recognition will trigger proliferation and further differentiation (clonal selection) and after stimulation by cytokines or other by-products of innate immunity the lymphocytes progeny acquire effector's functions. The antigen-specific receptors, Immunoglobulins and T-cell receptors are generated in a somatic process of gene rearrangement that constructs the variable part of the molecule bearing specificity towards the epitope. It is worth mentioning that a complex selection check, by means of the major histocompatibility complex molecules presented, avoids self-recognition. Therefore, when in action, adaptive immunity accounts for specificity and memory. The fancy of the system is not much in these properties, shared by the nervous system, but in its functional organization. The unfolding of responses to antigens in vertebrates is a clear example of somatic evolution at the scale of ontogeny and at the scale of physiology; cells are under the pressure of the same laws of mutation and selection as individuals in a species (Du Pasquier and Flajnik, 1999). The immune system of vertebrates also exploits innate immunity and the nature of some of the mediators that are common to innate and adaptative pathways. This cross talk will be of particular interest to our understanding of RNAs role in the building of sense. Innate response to double-stranded RNA (integrated by the system as a sign of viral infection) includes inhibition of viral replication (by PKR and $2'5'(A)_n$-synthetase activation) and a canonical inflammatory response, shared by invertebrates. The novelty is that Interferon and interleukins secreted as part of the innate response by infected cells will trigger signal transduction pathways guiding alternative differentiation of lymphocytes B and T, NK and TAP, and ultimately recruiting the adaptative cell immunity into the scene. In this sense, dsRNA is also dealt with by triggering the rearranging machinery to generate antibodies against the cognate viral antigens. A link that has become necessary by natural convention/selection based on the modular association of receptors in immune cells and the modular association of cell types into discrete functions.

Even in invertebrates, RNAi seems to play a role in sense making, as it starts emerging in evolution. In *C. elegans* a remarkable aspect of the RNAi process is its ability to spread throughout the target gene beyond the sequence homology region harbored by the dsRNA trigger molecule, a phenomenon called transitive RNAi (Sijen et al., 2001). Besides, in the worm, RNAi also spreads throughout the organism, suggesting a mechanism to forward the signal from cell-to-cell. The sid-1 gene product is a Trans membrane protein that could act as a channel for such systemic silencing (Winston et al., 2002). Following this same rationale of amplification by spreading of the RNAi silencing to homologous sequences in the genome, is their targeted methylation or the methylation of associated histones. Two recent studies have shown that in the fission yeast the integrity of RNAi machinery is required for epigenetic silencing at centromers, and for initiation of heterochromatin formation at the mating locus, being also important for proper regulation of chromosome dynamics during cell division by meiosis and mitosis (Volpe et al., 2002 and Hall et al., 2002). We shall propose that by promoting intercellular communication, all attempts to spread the RNAi phenomenon could be regarded as incipient sense-making strategies. In "single-cell oriented" organisms the operating strategy for dealing with dsRNA is only the specific silencing of homologous sequences. "Sense making" appears by as the spreading of this strategy by multiple mechanisms to as many cells as possible. By the other hand, the possibility of displaying alternative reactions to dsRNA will be a privilege of "multicell oriented organisms" only to be fully realized along with the development of vertebrates immune and nervous adaptative systems. The Venn diagram in **Figure 3** summarizes the molecular partners of dsRNA associated with the different cell responses that can be triggered depending on cell context, the comparison stands for the differences between differentiated vertebrate cells and non-differentiated vertebrate cells or invertebrate cells.

As a concluding remark on the contextual nature of dsRNA signs in sense-making, we must discuss some evidence on the mechanisms that control dendritic protein synthesis in neurons. In 2002 it has been proposed that translational control could be achieved by means of ribosomal/mRNA interactions (Mauro and Edelman, 2002). In what the authors called "The ribosome filter hypothesis", the sub cellular localization of particular mRNAs would be a result of the complementarity between their non-coding regions and sequences on rRNAs associated with 40S ribosomal subunit, resulting in local dsRNA structures. More recently, the presence of cytoplasmic

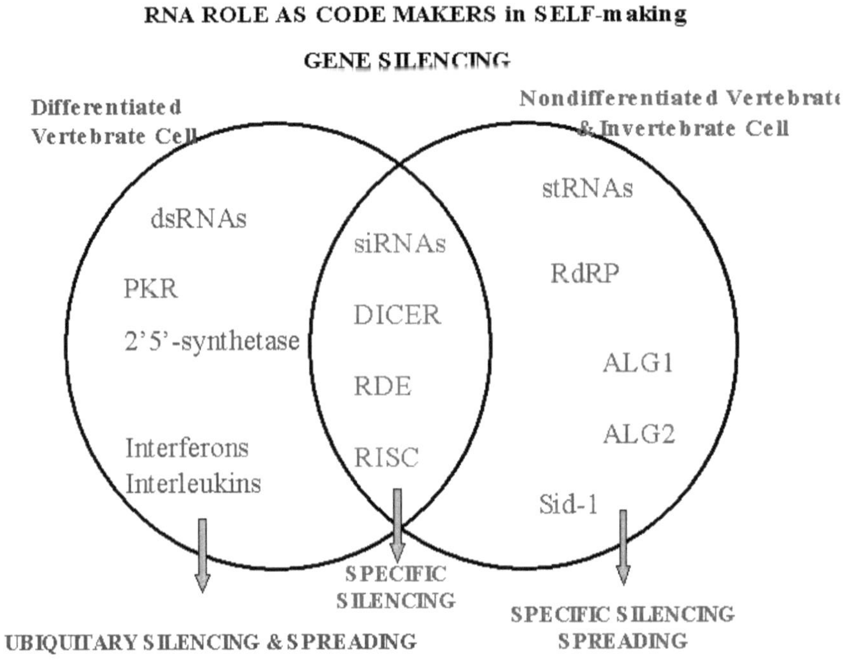

Figure 3. Molecular machineries associated to cell response to dsRNA, there are common and exclusive codes amongst different classes of cells

RNA granules has been associated with mRNA/rRNA interactions (Anderson and Kedersha, 2006) and their putative role in translational control is reinforced by the fact that such structures are restricted to certain cell types and cell regions where the selective translation of recruited messengers is carried out. Of particular interest is the fact that, in neurons, structures of that kind (neuronal granules) have emerged as important players in the targeting of specific protein synthesis to dendritic regions. The local translation performed in neurons seems to be dependent on microtubules integrity, mRNA/rRNA local double-stranded formation and RISC pathway integrity (Cristofanilli et al., 2006; Ashraf et al., 2006 and Pinkstaff et al., 2001). Moreover, such pattern of gene expression control is associated with long-lasting forms of memory, at least in Drosophila (Ashraf et al., 2006). The data is far from being conclusive, but the evidence suggests that dsRNA might have a precise role in nervous system sense-making, by targeting protein synthesis to synaptic regions and by favoring specific paths of cell cognition. In **Figure 4** we can see a Venn diagram illustrating that panic response is a common feature of differentiated vertebrate cells, and that neurons and immune cells have developed different pathways for dealing with dsRNA. In Neurons, these molecules can trigger dendritic protein synthesis, while in immune system they will trigger somatic cell differentiation.

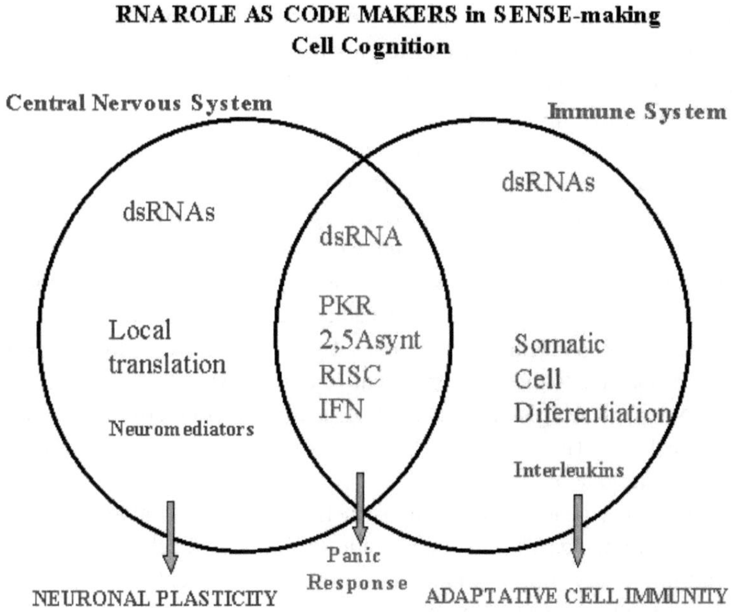

Figure 4. Molecular machineries associated to cell response to dsRNAs, there are common and exclusive codes amongst vertebrates Nervous and Immune Systems

CELL, SELF, SENSE – CONCLUSION AND RESEARCH PERSPECTIVES

Levels of reality based on denoting categories imply that the structuring of such levels "does not respect *a universal principle of linearity,* then one is forced to *restrict* the multidynamic frames *to their linear fragments*" (Polli, 2001, emphasis mine). Because of this assumption we have the fact that properties of higher hierarchical levels bear a causal dependency towards lower hierarchic ones, but are categorically independent from those. Another way to phrase the same statement, only adopting other terminology (El-Hani and Queiroz, 2005), is to say that properties from higher hierarchical levels in biological systems are not reducible to lower level ones from a synthetical standpoint, but are reducible from an analytical standpoint. The only research agenda that seems fruitful assuming these dynamics of multiple causalities in hierarchic systems comes from a balance of analytical and synthetical procedures, of descriptive and categorial classifications. Methodologically one shall proceed through analytical reductionism in order to identify lapses of the living system that can be linearly explored, but then, the integration of such "horizontal cut" into the greater picture to build up complexity, restore context and probe deductibility, will be no less than necessary.

The CELL/SELF/SENSE account of the unity of life is an attempt to pull forward the type of research agenda mentioned on the previous period.

The analysis of the molecular partners RNAs are able to recruit as:

a) mediators of genetic coding into proteins
b) mediators of sequence-specific gene silencing by RNAi;
c) mediators of global cell response to integrative signs;

is clearly an analytical reductionist approach. The integration of each RNA-molecular machinery-"partnership" into categorial frameworks (CELL/SELF/SENSE), the classification of properties and dynamics accordingly, in respect to the categories they are embodied in, is clearly a deductive categorial approach. The biological meaning of repetitive RNA sequences evolves by means of the physiological processes that are associated to their presence at different levels:

a) dsRNA are RNA processing signs at the CELL level, able to recruit either only translational machinery (in Prokaryotes) or translational machinery and splicing machinery (in Eukaryotes);
b) dsRNAs are selective gene silencing signs at the SELF level, able to recruit either only selective nucleases (in single or non-differentiated cells) or selective nucleases and ubiquitary transcriptional and translational machinery (in differentiated cells of multi-cell organisms);
c) ds RNAs are cell cognition signs at the SENSE level, able to trigger localized protein synthesis modulation (in multi-cell organisms neurons) or the recruitment of adaptative cell immunity for targeted cell destruction (in all other differentiated systems of multi-cell vertebrates)

Research in theoretical biology aims testing the explanatory, predictive and heuristic power of scientific theories. In this scope, the following steps in our research program would be testing the proposed categorical framework by means of:

- the analysis of other case-studies that could validate the "organic code/ level transition hypothesis";
- the formalization of the attributes that segregate into each category in a less natural language;
- the application of the Cell/Self/Sense categories to other disciplines in search for overcoding.

In fact, these three approaches are currently under investigation. Meanwhile, let us summarize some of the principles that are conclusive in the study of contextual meaning of dsRNAs and shall be seminal for future projects.

The assembly of dynamic configurations into stabilization levels applies to complex systems in general. This tendency to build tangled hierarchies as means to accommodate energy flows, could be the missing link (and the common material ground) between Physics and Biology. Therefore, it seems, if one wishes to attack the emergence of biological tinkering and of biological timing (as canonical indexes of biological contingency and complexity), it might be worth analyzing how natural hierarchization (the assembly of levels) rise. In particular, it would be helpful to investigate the specificities of coded hierarchies. Following Barbieri's formulation, this track will lead us to the first organic codes, the first semiotic unity and its minimal conformation: a cell.

Copying first and coding later (replication, mutation, differentiation, and others being just instantiations of these two relational patterns) are new functions, restricted to the realm of living things. Once again, this two relational patterns are reducible in analysis to their physical grounds, though not strictly deducible from them. Copying and coding are new, emergent properties, coherent once contingent, and once coherent and contingent, necessarily consistent. Natural selection, Natural convention, adaptation, evolution and even life itself would be corollaries of those relational patterns originated some 4.5 billion years ago with the first triadic cells (composed by genotype, ribotype, phenotype).

REFERENCES

Aderem A and Hood L, 2001, Immunology in the Post-Genomic era. Nature Immunology, 2, 5, 373–5.

Agami R., 2002, RNAi and Related Mechanisms and Their Potential Use For Therapy. Curr Opin Chem Biol. 2002 Dec, 6 (6): 829–834.

Anderson, P and Kedersha, N. RNA Granules. The Journal of Cell Biology 2006, vol 172, 6: 803–808.

Artmann, S. 2004, Four Principles of Evolutionsry Pragmatics in Jacob's Philosophy of Modern Biology. Axiomathes 14:381–395. Kluwer Academic Publishers.

Artmann, Stefan (2005). Biosemiotics as a structural science: Between the forms of life and the life of forms. Journal of Biosemiotics 1 (1): 183–209.

Ashraf SI, McLoon AL, Sclarsic SM and Kunes S. Cell 124, 191–205, January 13, 2006.

Barbieri, 2003, The Organic Codes. An Introduction to Semantic Biology. Cambridge University Press, Cambridge, UK.

Bateson G, 2002, Mind and Nature. Necessary Unity. Hampton Press, Inc., USA.

Brent R and Bruck J, 2006, Can computers help to understand biology? NATURE|Vol 440|23 March 2006.

Clemens MJ., 1997, PKR-A Protein Kinase Regulated by Double-Stranded RNA. Int J Biochem Cell Biol. 1997 Jul, 29 (7): 945–949. Review.

Cohen, IR. Tending Adam's Garden (Elsevier Academic Press, 2004).

Cristofanilli, M, Iacoangeli, A, Muslimov, IA, Tiedge, H. Neuronal BC1 RNA: microtubule-dependent dendritic delivery. J Mol Biol. 2006 Mar 10; 356 (5): 1118–23.

Curr Med Chem – Central Nervous System Agents, 2004, 4: 243–253.

Du Pasquier L and Flajnik M, 1999, Origin and Evolution of the Vertebrate Immune System. Fundamental Immunology, Fourth Edition. Lippincott-Raven Publishers, Philadelphia.

Elbashir SM, Harborth J, Lendeckel W, Yalcin A, Weber K, Tschul T, 2001, Duplexes of 21-nucleotide RNAs Mediate RNA interference in Cultured Mammalian Cells. Nature. 2001 May 24, 411 (6836): 494–498.

El-Hani CN and Queiroz, J, 2005, Modos de Irredutibilidade das Propriedades Emergentes. Scientiae Studia, vol 3, 1, 2005.

Ellis, G.F.R., 2005a, Physics, Complexity and Causality. Nature, vol 435/9 June 2005.

Ellis, G.F.R., 2005b, Physics and the Real World. Science and Religion: Global Perspectives, June 4–8, 2005, in Philadelphia. www.metanexus.net

Faria M, Giovannangeli C, Ulrich AH, 2004, DNA and RNA Code-reading Molecules as Potential Gene Silencers in Neurobiology – What Are They and What Are They God For?

Gatlin LL, 1972, Information Theory and the Living Systems, Columbia University Press, New York, USA.

Gould, SJ. Ever Since Darwin. Reflections in Natural History (New York, W.W. Norton, 1977).

Hall IM, Noma KI, Grewal, SIS, 2002, RNA Interference Machinery Regulates Chromosome Dynamics During Mitosis and Meiosis. PNAS, January 7, 2003, vol 100, 1:193–198.

Hall, VP and Edelman, GM, The ribosome filter hypothesis. PNAS Sptember 2002, vol 99, 19: 12031–36

Hartmann, N., 1942, The New Ways of Ontology, Chicago press, USA.

Herman Muller 1964, From *Nobel Lectures, Physiology or Medicine 1942–1962*, Elsevier Publishing Company, Amsterdam, 1964.

Huang, S., Eichler, G., Bar-Yam Y., and Ingber, DE. 2005 Cell Fates as High-Dimensional Attractor States of a Complex Gene Regulatory Network. Physical Review Letters 94 (12): 128701–4.

Jacob, F., 1976, La Logique du Vivant. Gallimard, Paris, France.

Jacob, F., 1986, Le Jeu des Possibles: Essai sur la Diversité du Vivant. LGF, Le livre de poche, Paris, France.

Körner, S., 1999, Categorial Frameworks. Cambridge University Press, Cambridge, UK.

Leaman DW, Salvekar A, Patel R, Sen GC, Stark GR, 1998, A mutant cell line defective in response to double-stranded RNA and in regulating basal expression of interferon-stimulated genes. Proc Natl Acad Sci U S A. 1998 Aug 4;95 (16): 9442–7.

Lemaire PA, Lary J, Cole JL, 2005, Mechanism of PKR Activation : Dimerization and Kinase Activation in the Abbsebce of Double-Stranded RNA. J Mol Biol (2005) 345: 81–90.

Lewin, B. 2000. Genes VII. (Oxford University Press, New York).

Neuman Y, 2004, Meaning-Making in the Immune System, Perspectives in Biology and Medicine, vol 47, 3:317–27.

Pinkstaff JK, Chapell SA, Mauro VP, Edelman GM, Krushel LA. Internal Initiation of translation of five dendritically localized neuronal mRNAs. Proc Natl Acad Sci USA 2001 Feb 27;98 (5): 27770–5.

Polli, R., 2001, The Basic Problem of the Theory of Levels of Reality. Axiomathes 12: 261–283. Kluwer Academic Publishers.

Rothenburg S, Deigendesch N, Dittmar K, Koch-Nolt F, Haag F, Lowenhaupt K, Rich A., 2005, A PKR-Like Eukaryotic Initiation Factor 2α Kinase from Zebrafish Contains Z-DNA Binding Domains. PNAS, Feb 1 2005, vol 102, 5: 1602–1607.

Salthe S., 2005, The Spontaneous Origin of New Levels in a Scalar Hierarchy. Entropy 2004, 6: 327–343.

Shannon, CE and Weaver, W. The Mathematical Theory of Communication (Univ. Illinois Press, 1948)

Sijen T, Fleenor J, Simmer F, Thijssen KL, Parrish S, Timmons L, Plasterk RH, Fire A., 2001 On the role of RNA amplification in dsRNA-triggered gene silencing. Cell. 2001 Nov 16;107(4):465–76.

Varela FG, Maturana HR., 1974 Autopoiesis: the organization of living systems, its characterization and a model. Curr Mod Biol. 1974 May; 5 (4): 187–96.

Volpe TA, Kidner C, Hall IM, Teng G, Grewal SI, Martienssen RA., 2002, Regulation of heterochromatic silencing and histone H3 lysine-9 methylation by RNAi. Science. 2002 Sep 13;297(5588):1833–7. Epub 2002 Aug 22.

Westernhoff HV and Palson BO, 2004, The evolution of meolecular biology into systems biology. Nature Biotech, 22, 10: 1249–53.

Wilson, EO. Consilience The Unity of Knowlodge (First Vintage Books Edition, 1998).

Winston FK, Weiss HB, Nance ML, Vivarelli-O'Neill C, Strotmeyer S, Lawrence BA, Miller TR., 2002, Complement factor B gene regulation: synergistic effects of TNF-alpha and IFN-gamma in macrophages. J Immunol. 2002 Sep 1;169 (5): 2627–35.

CHAPTER 15

CELLULAR SEMIOTICS AND SIGNAL TRANSDUCTION

LUIS EMILIO BRUNI

Department for Media Technology and Engineering Science, Aalborg University, Lautrupvang 15, 2750 Ballerup, Denmark., E-mail: leb@media.aau.dk

Abstract: Semiosis, the processes of production, communication and interpretation of signs – coding and de-coding – takes place within and between organisms. The term "endosemiosis" refers to the processes of interpretation and sign transmission inside an organism (as opposed to "exosemiosis", which refers to the processes of sign interpretation and transmission between organisms of the same or different species). In Biosemiotics it is customary to recognise the cell as the most elementary integration unit for semiosis. Therefore intra and intercellular communication constitute the departure point for the study of endosemiotics

In contemporary molecular and cell biology, signal transduction research has remarkably contributed to a major paradigm shift in biology in which biology is now seen as a "science of sensing". Once we recognise that sensing is one of the necessary properties of life, we cannot do without considering semiotic logic in order to construct our understanding of living phenomena. Given the central integrating role of signal transduction in physiological and ecological studies, this chapter outlines its semiotic implications.The multi-modality and modularity of signal molecules and relative "infrastructure" components poses one of the central problems for understanding metabolic codes: the occurrence of different instances of "cross-talk", "redundancy" and "categorial sensing" at different hierarchical levels. The term "categorial sensing" captures very well the essence of the "outstanding question(s)" in signal transduction; i.e., how specificity is determined, how ubiquitous signals or messengers convey specific information, how undesired cross-talk is avoided, how redundancy integrates the system. This chapter proposes a basic conceptual toolbox for interpreting empirical data that deals with such puzzling phenomena from a biosemiotic perspective

Keywords: cellular semiotics, signal transduction, cross talk, ubiquitous signals, sign-processes, categorial sensing, digital-analogical consensus, specificity, systems of correspondences, emergent interpretant, triadic logic, biological information, context, hierarchy

M. Barbieri (ed.), Introduction to Biosemiotics, 365–407.

1. INTRODUCTION

"Communication", or in semiotic terms "semiosis", is a defining property of all life manifestations (Sebeok, 1985/1976). This is one of the premises of the biosemiotic framework, which sees biological processes from a sign-theoretic perspective.

Thus semiosis, the processes of production, communication and interpretation of signs – i.e., coding and de-coding – takes place within and between organisms. The term "endosemiosis" refers to the processes of interpretation and sign transmission inside an organism. On the other hand, "exosemiosis" refers to the processes of sign interpretation and transmission between organisms of the same or different species and in general the interpretation of environmental cues. All endosemiotic sign processes are (directly or indirectly) linked to phenomena in the organism's environment. Organisms are wrapped in semiotic networks in which specific circulating signs are accessible only to complementary systems of interpretation. The exosemiotic sign processes, which transform the objective environment into subjective universes, are intrinsically related to the endosemiotic sign processes in a continuous basis (von Uexküll et al., 1993).[1]

Figure 1 summarizes some instances of endosemiosis and exosemiosis. At the exosemiotic level we have pheromones, i.e., signals released by one organism that can be picked up by the signal transduction networks of other organisms of the same species, thus informing behaviour or gene expression in the latter. This kind of communication can be encountered from bacteria (e.g., quorum sensing) to eukaryotes. Then we have inter-species and inter-kingdom signals, for example when a species of bacteria in the guts of a herbivore emits an elicitor that plants' cells are able to recognise prompting the plant to respond by emitting another signal-molecule that attracts the herbivores' predators (Baldwin et al. 2001). Would these be pheromones? In general these are being called info- or semio-chemicals. On the other hand, at the endosemiotic level, i.e., signals within organisms, we have intracellular signals (e.g., second messengers) and, in the case of multicellular organisms, intercellular signals (e.g., hormones and neurotransmitters). These latter networks are mainly the subject matter of cellular signal transduction. However it must be kept in mind that the endo- and exosemiotic codes are intrinsically linked by systems of correspondences (as will be stated later).

By recognising the cell as the most elementary integration unit, T. von Uexküll et al., (1993) differentiate four endosemiotic integration levels:

1) The microsemiotic level - sign processes occurring within the cell and between its organelles, which take place through relations between networks of genes, enzymes, signals and second-messengers.
2) Cell-to-cell communication by cytosemiotic processes in neighbouring cells (direct metabolic and electrical contact at 'gap-junctions'), including coordinated responses of group of cells that share a regulating signal.

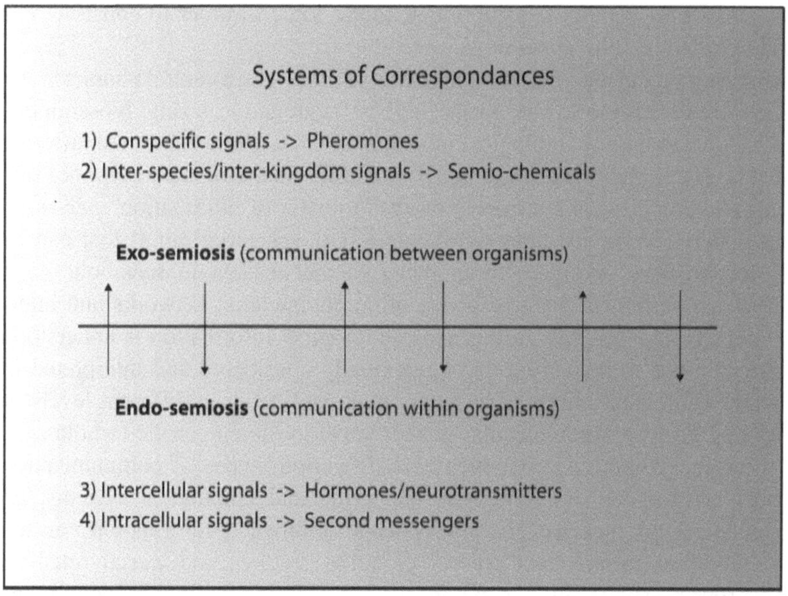

Figure 1. Instances of endosemiosis and exosemiosis. The endo- and exosemiotic codes are intrinsically linked by systems of correspondences

3) Endosemiotic networks that link the most diverse cells into functional units, including systems for short-distance sign vehicles (e.g. transmitters) and systems for long-distance sign vehicles (e.g. hormones, antibodies).
4) The combination of cells into organs and/or systems [as well as the emergence of higher order physiological codes] (von Uexküll et al., 1993).

von Uexküll et al., (1993: 9) also stated that a linear hierarchical scale cannot account for the complexity of semiotic processes. Therefore biosemiotics searches for multidimensional and ramified models as well as for circular models joining together different integration levels (von Uexküll et al., 1993: 9). So these integration levels should not to be considered as sharp frontiers, given their coextensive nature. More subtle integration levels can be identified in between these levels and these are not necessarily manifested as emerging physical structures but sometimes can also be manifested as a new complex logical product based on already existing structure. Later I will be revisiting this relation between endosemiosis and exosemiosis when considering signal transduction networks.

In Biosemiotics it is customary to recognise the cell as the most elementary integration unit for semiosis. Therefore intra and intercellular communication constitute the departure point for the study of endosemiotics. In contemporary molecular and cell biology, signal transduction research has remarkably contributed to a major paradigm shift in biology in which biology is now seen as a "science of sensing". Once we recognise that sensing is one of the necessary properties of life, we cannot do without considering semiotic logic in order to construct our understanding of living phenomena.

By considering processes of communication (semiosis) as a central characteristic of living systems from the lowest to the highest aggregation levels, biosemiotics seeks to develop a notion of "biological information" that is relevant to the different hierarchical levels of the living world and to the multiple biological disciplines that study them. The emphasis is not merely on the "transfer of information" *per se*, as if it was a material thing (i.e., the sign-vehicle is the material thing) that can be physically moved from one place to another (whether in genetic, metabolic or in ecological systems), but on the emergence of communication networks and interpretation contexts and systems. In this sense, biological information is understood as sign-action: it has to be sensed and contextually considered and interpreted in order to work. Biological information functions at and between different levels of complexity that go from the molecular-genetic level to the epigenetic (whole-cell) level up to more systemic levels which include various types of communication systems such as nervous, immunologic, endocrine and ethological systems, up to ecosystems. At all these levels and systems "biological information" as the vehicle for communication must present common features and causal relations (Emmeche, 1998). Above all, the emphasis has to be put on the "continuous chain of information" from the lower to the higher hierarchical integration levels and vice versa. In this regard, Gregory Bateson's approach to information, hierarchical contexts and analog/digital communication has been recognized as highly relevant to a more fully developed semiotic approach to biology (Hoffmeyer and Emmeche, 1991). Therefore some of the concepts that I have developed in order to characterise the logic behind cellular semiotic networks draw inspiration from Bateson's insight.

I present a "toolbox" of concepts for"mapping" semiotic networks across hierarchical levels and for relating the different emergent codes in living systems. I define "the signalome", and its necessary predecessor, the "embryonic signalome", as the substrate through which emerging codes constitute levels of integration at different physical and logical levels of the hierarchy.

What deserves to be called the "Ca^{2+} code" is presented here as an example to advance a hypothesis of how cellular systems achieve the necessary categorial sensing that allows them to avoid undesirable cross-talk by using the semiotic regularities that I have called "digital-analogical consensus" – a recurrent pattern for the creation of complex logical products that constitute specific signs.

Given the central role that the elucidation of signal transduction networks has acquired in the "integrative agenda" in biology, this chapter outlines the semiotic implications of these networks and tries to exemplify how a semiotic approach can be of help when organising the knowledge that can lead us to understand the relevance, the role and the position of signal transduction networks in relation to the larger semiotic networks in which they function, i.e., in the hierarchical formal processes of mapping, translation, transformation and transmission of information in physiological and ecological studies.

2. BIOLOGICAL INFORMATION BETWEEN LEVELS OF COMPLEXITY AND THE "INTEGRATIVE AGENDA"

When it was thought that the "information problem" was solved and put aside with the cracking of the "genetic code", biologists began talking again about information and about cracking other "codes". New problems arrived with regulation and signal transduction networks. Already in 1962, biochemist Erwin Chargaff pointed out that:

"If there is no continuous 'chain of information' from the lowest level to the highest, there is not justification in claiming that 'DNA is the repository of biological information'"
(Quoted in Sarkar, 1996: 199)

Although this argument was raised to rebut the usefulness of the notion of "biological information", in reality it only, and very strongly, rebuts the exclusivity of DNA as biological information (or more precisely as the physical support for information). Rather, the argument poses a very interesting and central challenge to contemporary biology: how can we conceive "the continuous chain of information from the lowest level to the highest" and perhaps from the highest to the lowest?

In figure 2 a hypothetical hierarchy of levels in living systems is shown. It doesn't matter here how one draws the hierarchy; the important issue is that there are processes of contextual interpretation of information at any given level and between levels in the continuum of the embedded systems.

This is the field in which biosemiotics operates. How do codes and semiotic networks emerge in this picture? How can we map the communication processes across emergent levels? Is there "downward causality"? We say we have found regulation in "signal-transduction". Have we found it? What regulates what?

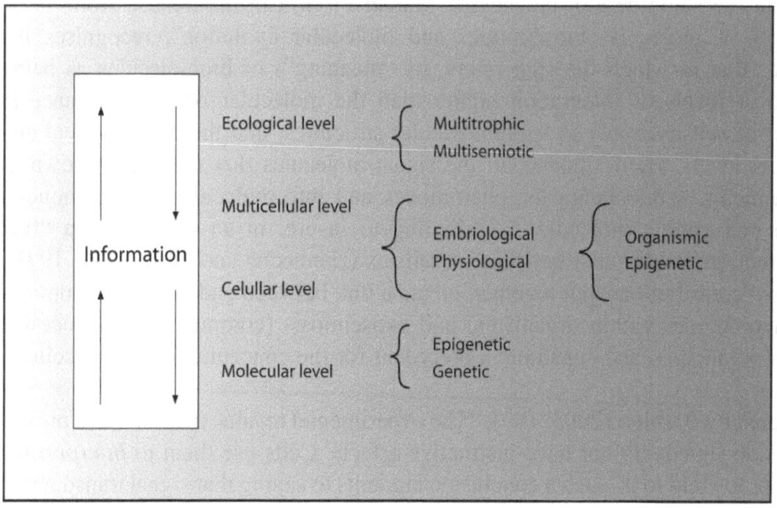

Figure 2. Biological information between levels of complexity

It is precisely the impressive results of molecular biology that have devalued the role of linear causality in biological systems. The overwhelming omnipresence of "cross-talk", "redundancy", "pleitropy", "epistasis", "polygenes", "cryptic variants" (e.g. the "jukebox" effect in development), for example, has posed serious challenges to the logical foundations of biology. Based on this experience, it is easy to foresee that a further challenge to those foundations will become evident when biologists learn to recognise (as it is happening already) the existence of semiotic processes (implicit in e.g. signal transduction, non-trophic interactions, etc.) and of emergent properties and processes in a historical and hierarchical perspective. The concatenation of emergent and hierarchical levels will require different logics to think about causality, not just renaming well-established terms and concepts. This is what biosemiotics can offer to the new epistemological developments in biology. The emergence of codes has a relation to the emergence of new causality, a different kind of logic, dependant on but different from the logic of mechanical causality. Everywhere and every time a code emerges, there is already an *emergent interpretant* (more about this below), which is logically "above" the formality of the code itself.

As early as 1975, shortly before his death, biochemist and biophysicist Gordon M. Tomkins – considered a major figure in the development of molecular biology – sketched a model for the evolution of biological regulation and the origin of hormone-mediated intercellular communication. He claimed that:

> "Since a particular environmental condition is correlated with a corresponding intracellular symbol, the relationship between the extra- and intracellular events may be considered as a 'metabolic code' in which a specific symbol represents a unique state of the environment." (Tomkins, 1975: 761).

He further argued for an apparent generality of such a code. The recent discoveries in the field of signal transduction have confirmed how right Tomkins was.

Around the same period, biochemist Marcel Florkin in his treatise from 1974, "Concepts of molecular biosemiotics and molecular evolution", recognises the signified (that to which the sign refers, its "meaning") of biomolecules as being involved at levels of integration higher than the molecular one, for instance at the level of self-assembly in supramolecular structures, and the physiological and ecological levels. He designates all the signal-molecules that today are known as info-chemicals, semiochemicals, pheromones and info-molecules as "ecomones", i.e.: the non-trophic molecules contributing to insure, in an ecosystem, a "flux of information" within and between organisms (Emmeche and Hoffmeyer, 1991). These two contributions, put together, create a link between endosemiosis (communication processes within organisms) and exosemiosis (communication processes between organisms) and constitute a precedent for the conceptualisation of cellular semiotics.

As stated by Barbieri (2003: 109), "The experimental results, in brief, have proved that outside signals do not have instructive effects. Cells use them to *interpret* the world, not to yield to it. Such a conclusion amounts to saying that signal transduction is based on organic codes, and this is in fact the only possible explanation of the data."

In this direction, the field of signal transduction networks constitutes one of the first conceptual links between hierarchical levels. It has opened the doors to the integration of molecular techniques with embryologic, developmental, physiological and ecological approaches. It has also re-dimensioned the centrality of DNA as the sole source of biological information. Whereas DNA was the dominant and central element in the conceptual and experimental framework of biology, it can easily be claimed that today its place is being taken by signal transduction.

Signal transduction research has remarkably contributed to a major paradigm shift in biology in which now the discipline is seen as a "science of sensing". Once we recognise that sensing is one of the necessary properties of life, we cannot do without considering what in semiotic terms could be called "triadic logic" (more about this below) in order to construct our understanding of living phenomena.

The "integrative agenda" has become a priority in biology. Regulation at all levels has become essential. A cross-sectoral look of current biological scientific literature reveals that at all hierarchical levels there is an increasing importance being ascribed to the "context", consideration of communication systems and information, and a general call for the integration of molecular biology with developmental, physiological and ecological approaches. The integrative agenda thus depends on the consideration of the flow of information within organisms (the genetic, epigenetic and physiological levels) and between organisms of the same or different species, i.e., the ecological level of functionally integrated multitrophic and multisemiotic systems (i.e. considering also non-trophic interactions).

Lack of proper consideration of the context is one of the main limits of reductionism and is systematically becoming a main concern in all sub-disciplines of biology. The importance of the context is a recognised challenge to all empirical endeavours, and multiple knockout strategies will have to be rethought accordingly. The context, as a meta-code, provides the key for the interpretation of codes, which are the result of habits, "crystallised" patterns of behaviour, codified actions. A code is something to be interpreted contextually. As we go up in the scale of emergent processes, empirically, the context acquires further importance and complexity, its spatial arrangement being larger and comprehending "down-stream" processes in the hierarchy (Bruni, 2003).

3. THE SIGNALOME

Let me define the "signalome" as a "frozen" picture of all the known (and yet to be known) basic physical support molecules of all the known (and possible) signal-transduction networks active or ready to be triggered in a given moment (Bruni, 2003).[2]

Besides the massive identification of genes and their functions, in order to follow the reductionist strategy, we also need an equally massive characterization and classification of regulatory elements of genes, protein regulators, signal-transduction components and the elicitors of the cascades that determine complex genetic reactions in response to variable environmental cues. With the help of global-array technologies it is possible to model gene and protein expression networks and profiles. The passage from genome sequences to higher hierarchical levels would require the generation and correlation of data about: the regulation and interactions of genes and gene products within cells, the interactions and communications between cells, and the biological responses and susceptibilities of cells and organisms to biotic and abiotic environmental cues. In sum we would need to go from a focus on one pathway at a time to the integration of multiple pathways. To additively reintegrate all the reduced parameters we need computer power and more sophisticated algorithms capable of correlating the multidimensional data pouring from expression arrays that may include more than 20,000 genes assayed in different cell or tissue types, different genotypic states, different physiological states, different developmental states (considered at different times), after different sets of cues, perturbations or stimuli. Here the central assumption is that biology "happens" from the DNA sequence, through the structure and function of proteins, through the interactions of DNA and proteins in simple pairs and as parts of complex networks involving the hundreds or thousands of genes and proteins that control complex biological responses.

In this bottom-to-top research strategy "biological information" is allegedly called to play an important role. However the conceptualizations of "biological information" offered by some of the leading approaches that are tackling biological complexity (such as Systems Biology) are not really drawing the consequences and the logic implied by biological information, i.e., the contextual interpretation of sensed differences implied in semiotic processes (a more detailed treatment about the contradictory conceptualizations of information in Systems Biology is in Bruni (2003)). A common mistake in this context is to confuse "scientific information", data for the researcher, with "biological information", differences sensed by organisms. But in reality, the Laplacean algorithms that should correlate all these massive data sets have to deal exclusively with the kinetics of molecular interactions, which are reduced and deconstructed in the lab and then reassembled in the computer. There is no place for "information" or communication within and among living systems in this model. There is only molecular kinetics, molecules poking and reacting with each other when their concentrations are statistically relevant. Even if possible, it would be of no help to know all the actors in a mega-production if we do not understand the languages in which they play.

4. TOWARDS AN INTEGRATIVE CONCEPT OF "BIOLOGICAL INFORMATION"

Let me quickly work my way towards an integrative concept of "biological information". Peirce's logical description of the way a sign functions in nature corresponds very closely to the concept of context-dependent information in biological systems developed by Gregory Bateson (1972, 1979). This concept of "biological information" departs from any paradoxical physical account of information, i.e., it holds as truth that information is information not matter nor energy, and thus that certain materials such as DNA or any of the so-called "informational" molecules are not *per se* information. It also departs from the purely probabilistic accounts of the mathematical theory of information formulated by the cyberneticists, although instead of excluding these accounts it rather overlaps with them.

In Bateson's definition, the smallest unit of information is a difference or distinction, or news of a difference. So information means a difference that makes a difference to some system capable of picking it up and reacting to it, i.e., a system with some sort of interpretative capacity. For there to be a "difference", news of a distinction, there has to be a biological system that senses it. Otherwise they would not be differences, they would be just impacts.

A sign, or in Bateson's terminology, an idea, can be a complex aggregate of differences or distinctions. It can be formed by the smallest units of informational processes, i.e., news of a single difference (Bateson, 1979: 250), as e.g., the binding of a single signal-molecule to its membrane-receptor. More elaborate signs and ideas can be formed by complex aggregates of elementary differences (which constitute more complex differences). This implies the emergence of codes: "Every effective difference denotes a demarcation, a line of classification, and all classification is hierarchic... differences are themselves to be differentiated and classified" (Bateson, 1972: 457). That is, they have to be recognised as *patterns*.

"The number of potential differences in our surroundings... is infinite. Therefore, for differences to become information they must first be *selected*..." (Hoffmeyer and Emmeche, 1991: 122); and they have to be *categorised* by an interpretative system with such capability of pattern recognition. Biological information functions like signs in the sense that it is context dependant and requires interpretation processes. There is no information without interpretation (i.e., pattern recognition), and herein the importance of the context. As it can be noticed, in this perspective biological informational molecules are not restricted only to DNA and amino acid sequences.

This way of understanding information and sign-function gives place to the following distinction between causal links:

1) On the one hand we have the world of non living billiard balls and galaxies – the material world – characterised by the kinds of regularities described in the physical sciences, where forces and impacts are the "causes" of events (Bateson, 1979, Bateson and Bateson, 1989: 211). This is what Bateson defined as the "pleroma" and corresponds to Peirce's "dyadic action".
2) On the other hand we have the world of the living – where *distinctions* are drawn and a *difference* can be a cause – all processes in which the analog of cause is information or a difference, i.e., the entire biological and social realm, the world of communication, necessarily embodied in material forms subject to physical laws of causation as well as the distinctive processes of life (Bateson, 1979, Bateson and Bateson, 1989: 207). This is what Bateson defined as the "creatura" and more or less corresponds to Peirce's "triadic action" (and hence my use of the terms "triadic causality" and "triadic logic" as the logic pertaining to

situations in which there is a response to a sensed difference in a non mechanical way, as oppose to mere impacts and energy exchanges).[3]

The Peircian definition of the sign is a logical description of the way triadic causality functions in nature. A sign is an irreducible triadic relation. It represents a relation between three factors: 1) the primary sign – the sign vehicle – i.e., the bearer or manifestation of the sign regardless of its significance (that which stands for something else) 2) the object (physical or non-physical) to which the sign vehicle refers, and 3) "the interpretant" i.e., the system, or the interpretation key, which construes the sign vehicle's relationship to its object (Hoffmeyer, 1996: 19). This relation is customarily represented as a sign triad.

The two kinds of action are irreducible, but inseparable and superimposed. "...information does not belong to the sphere of matter and energy, but to the subjective and non-dimensional sphere of structure, pattern and form... At the most fundamental level the distinction between life and non-life is dependent on this ability: the response to differences... Nothing in the world of living systems makes sense unless we include in our explanations this peculiar ability to respond to selected differences in the surroundings" (Hoffmeyer and Emmeche, 1991: 123).

This causal distinction has been very hard for biologists both to acknowledge and to avoid. For example, in Jacques Monod's view there is an acknowledgement that chemical interactions determine the behaviour of the operon, but these interactions do not explain the behaviour of the system responsible for control. Such controls "confer heightened coherence and efficiency upon the cell... the very gratuitousness of these systems... enabled it to elaborate the huge network of cybernetic interconnections which makes each organism an autonomous functional unit, whose performances appear to transcend the laws of chemistry if not to ignore them altogether" (Monod, 1971; in Sarkar, 1996: 207). Monod's definition of "gratuity" – "the independence, chemically speaking, between the function itself and the nature of the chemical signals controlling it" (Sarkar, 1996: 206) – is exactly what is implied by the emergence of a code, or one of its main characteristics, i.e., its (logical) independence from the chemical nature of the substrate. This concept is very much related to the existence of a superimposed logic to the dyadic logic of material-mechanical causality. Plainly said, "gratuity" implies triadic logic. Or, as stated by Shapiro (1999: 28): "what distinguishes cellular biochemistry from chemistry outside the living cell is that cellular events are subject to biological regulation by signal transduction networks."

5. CELLULAR SEMIOTIC NETWORKS

So let us move into cellular semiotic networks. We tend to see the process of signal transduction as beginning with an extracellular signal and ending with the transcription of a gene. But the extracellular signal is not the beginning and the mRNA molecule is not the end of the semiotic network. They are just transient signs that take the process into new developments which will produce new sets of interconnected informational pathways in an endless progression until the system ceases to be a living-semiotic system within a network, that is, when its whereabouts will be determined exclusively by physical dynamics and there will be no more room for sensing, constructing and interpreting signs out of concentrations of signal-molecules that make differences.

So where is regulation? What regulates what? Because such complex networks are not exclusively determined by mass-energy restrictions, a random event, such as the building up of a given extracellular signal's concentration in the periphery of a cell, will produce a non-random response to such an event. This non-random response is not deterministic in the physical sense because the system that reacts to the random event has a repertoire of responses of which it will select the optimal one based on a global interpretation of the context – this is why we need to think in terms of semiotic (triadic) logic instead of exclusively in terms of mechanical (dyadic) logic. Selection of responses, "choices", can be achieved at different levels. If a response can be selected at a rather higher level of integration, the alternative responses must exist as possible and "distinguishable" coded patterns in the system (Bateson, 1972: 405).

Contrary to our genetic determinism (and now our signal determinism), choices at higher hierarchical levels, determined by sensing larger aggregates of differences, will have larger restrictive or regulating effects upon the whole hierarchy by influencing a larger set of circuits and networks as opposed to a single signal-pathway mediating or contributing to the expression of a single gene, which in turn contributes to a phenotype. A hormone or a neurotransmitter does not control anything, as it is usually stated in many textbooks; it rather cooperates with something. We can say that it is a limiting or a cooperative factor, but not properly a regulator. Regulation is a continuous process and anywhere you enter the circuit you will find a sort of "local regulator" or a checkpoint, which in turn is regulated and controlled by further ramifications of the semiotic network. In other words, all the pathways involved in such "control" or "regulation" processes are themselves opportunities for further regulation and control. So every single component of the "regulating system" opens a further pathway for regulating, controlling or limiting possibilities. For example, a "mechanism" for regulating cell-to-cell signalling is modulation of the number and/or activity of functional receptors on the surface of cells. For instance, the sensitivity of a cell to a particular hormone can be down-regulated by endocytosis of its receptors (i.e., invagination of the extracellular domain of the receptors), thus decreasing their number on the cell surface, or by modifying their activity so that the receptors either cannot bind ligand or form a receptor-ligand complex that does not induce the normal cellular response (Lodish et al., 2000: 894–895). More generally, "the ability of cells to respond appropriately to extracellular signals also depends on regulation of signalling pathways themselves" (Lodish et al., 2000: 894). One can be sure that the process that leads to endocytosis of receptors in order to decrease the sensitivity of the cell is also "regulated". So every single component of the "regulating system" opens a further pathway for regulating, controlling or limiting possibilities. Are we hopelessly caught up in a process of infinite regress? Lodish et al. (2000: 886) assert that "The coordinate regulation of stimulatory and inhibitory pathways provides an efficient mechanism for operating switches and is a common phenomenon in regulatory biology". From a mechanist point of view this leads to a paradox: does the coordinate regulation provide an efficient mechanism or does an efficient mechanism provide coordinate regulation? Where is this thing? What is it?

How could we delimit a semiotic network? Once you enter the world of communication, organization, etc., you leave behind the whole world in which effects are brought about by forces and impacts and energy exchange and you enter a world in which "effects" are brought about by sensing *differences*. The whole energy relation is different (Bateson, 1972: 452). In bioenergetics it is natural and appropriate to think of units bounded at the cell membrane, or at the skin; or of units composed of sets of conspecific individuals. These boundaries are then the frontiers at which measurements can be made to determine the additive-subtractive budget of matter-energy for the given unit. In contrast, in informational physiology and ecology, the semiotic aspects deal with the budgeting of pathways, codes and of probability. The resulting budgets are fractionating (not subtractive) (Bateson, 1972: 460).

The elementary unit of information, *a difference that makes a difference*, is able to make a difference because the pathways along which it travels and is continually transformed are themselves provided with energy. The pathways are ready to be triggered (Bateson, 1972: 453). For example, let's take a mammalian cell's signal transduction pathway that transduces the signal from the cell-surface receptors to the nucleus. The first part of the journey, the arrival of the signal molecule to the vicinity of the receptor, that which will produce the first difference, is energized from "behind", by some source outside the system, and, if it comes from the environment like, e.g., an odorant, it can be said to be energized in the ordinary hard-science way (if instead the signal is generated by another living system, the network could then be extended in that direction). But once the difference is transduced inside the system, this type of travel is replaced by travel which is energized at every step by the metabolic energy latent in the protoplasm which *receives* the difference, recreates or transforms it, and passes it on (Bateson, 1972: 453). It is at this point that the mere physical-mechanical logic does not suffice for constructing our models and explanations. What is needed is a logic, which in addition to the physical-mechanical logic of dyadic causality considers additionally the semiotic logic of triadic causality. Restraints of many different kinds combine determining uniquely a given pathway or sequence of events. In biological systems these restraints, or determinants, include cues, i.e., sources of information that will guide the system in its "selection" or in its development. From the point of view of the cybernetic observer, these pieces of information are restraints in the sense that they increase the probability of a given manifestation or event to happen or a given pathway to take course (Bateson, 1972: 400). From the semiotic point of view these pieces of information are differences that make a difference forming an emergent interpretant within a hierarchical structural-functional system. Cybernetics deals with the probabilities of pathways while semiotics deals with the *choices* of pathways that the system makes, based on the global interpretation of such restraints and probabilities in relation to its internal coherence. In this sense living systems are said to be stochastic. The restraints – including sources of information – lay out the probabilities of the pathways among which the informed system, based on its global interpretation, will tend. So in biological systems, restraints do not fully determine the outcomes of events; they increase the probabilities of certain pathways over others. A specific complex configuration of cues guides the system in its development at every instant, in a continuous way. The hierarchical nature of contexts (contexts within contexts) is universal for the semiotic aspects of phenomena. Therefore we tend to seek for explanation in the ever-larger units. Without context there is no communication (Bateson, 1972: 402). It turns out that in biological systems, regulation is nearly always linked to semiotic controls and for this reason, regulation will tend to be the compound effect of many limiting factors at different levels of the hierarchy, but regulation, which is close to homeostatic balance, will always be integrated at higher levels of the system.

6. SIGNALS THAT BUILD SIGNS: ZOOMING THE SYSTEM

At the beginning of this chapter we started from the signalome at an ecological level. We mentioned signal networks between organisms of the same species, signal networks involving inter-species and inter-kingdom signals, and finally intra- and inter-cellular signals. We also stressed that the endo- and exo-semiotic codes are intrinsically linked by systems of correspondences. In all these codes we should be able to trace some semiotic regularities and continuities if we are to consider "the continuous chain of information from the lowest level to the highest" in order to pursue an "integrative agenda" in biology. Having delineated the signalome very generally, let us now narrow it down to some specific examples in order to consider some of the semiotic regularities that can be observed within the functional codes involved in signal transduction. The generalisations proposed here can be then extended to other instances of the signalome (Bruni, 2003).

Within animals, intercellular signalling involving extracellular secreted molecules have been classified into four types: 1) the endocrine signaling system 2) the paracrine signaling system, in which signaling molecules released by a cell only affect target cells in close proximity, for example neurotransmitters 3) the autocrine signaling system, by which cells respond to signals that they themselves emit, as in the case of many growth factors and 4) a fourth way of signaling that involves "fixed" signals, which are attached to the plasma membrane of a cell and which can directly enter into contact with a membrane receptor attached to the adjacent cell.

Let us take for example the well-studied endocrine systems of signal transduction, which uses hormone-signals for remote communication between cells. In these systems, hormones travel, usually through the blood vessels, and communicate a single difference by binding to a cell's membrane-receptor, or by diffusing into the cytoplasm and finding its receptor inside the cell. However, it should be kept in mind that signal transduction does not function with a single signal. The process consists in translating the *analogical* concentration of signals (sensed by the compound effect of a number of *digital* signal-receptor bindings) into an analogue concentration of single transforms that reflect the analogical information of the concentration present at the input.

Usually, what is used as classification criteria for the different types of pathways is the typology of the cell-surface receptor. Cell-surface receptors responsive to water-soluble hormones (in endocrine systems) have been classified into families that give rise to different types of pathways (Lodish et al., 2000). In my examples I will consider some of the structural differences and the formal similarities of two types of signal transduction pathways that have been characterised in detail: those that involve G protein-couple receptors (GPCRs), and those that involve receptors with intrinsic enzymatic activity, in particular receptors with tyrosine kinase activity (RTKs).

Normally these systems work as follows: a signal (a hormone), emitted by a remote cell, makes contact with the surface receptor of the "target" cell. This produces a conformational change (and dimerisation or oligomerisation of the receptor) that permits the activation of a cascade of events and components in which the "difference" created by the binding of the hormone to the surface receptor is "transduced" through different possible mechanisms. The different *intermediate* steps may include a modular arrangement of ready-to-be-activated components that give rise to identifiable codes which are implemented through different infrastructure but which share some logical principles, interfaces and cross-talking pathways. For example, it is possible to trace some analogies and relations between the following two types of codes:

- There is on the one hand a type of code implemented by the production or release of any of several second messengers, from ions to lipids, as e.g., Ca^{2+} or cAMP.
- On the other hand there is a type of code implemented through the use of post-translational modifications of cytoplasmic proteins, for example phosphorylation or proteolysis.

Both types of codes share a very general logical pattern that I refer to as *digital-analogical consensus* (see below). Such a pattern can be found to be operative when a living system needs to fine-tune specific responses to a given contextual state. In this case both codes are cofactors in a larger emergent code, i.e., these two types of codes combine – "cross-talk" – in an emergent code resulting in more complex logics and patterns because of both, larger combinatorial possibilities and larger contact with cross-talking pathways.

The cascades of patterns of second messengers and of protein interactions and modifications is what then relays the signal – sometimes amplifying or diversifying it – to the nucleus where it is finally "translated" into a cellular response. The interpretation key for each signal is embedded in the larger message that its concentration conveys. One single molecule will not be enough to transduce the necessary concentration threshold for the "last" signalling event of the cascade to happen, i.e., the transcription of particular mRNAs that will work as signs in further semiotic networks, from translation and so on. Actually, what is conveyed is news of differences in concentrations. The whole code of signal transduction is based on signs consisting in complex patterns of concentrations of different signal types and the subsequent modulations of concentrations in all the intermediary steps. In Bateson's terminology, the transform of a difference (caused for example by the binding of a single signal-molecule) travelling in a circuit is an elementary idea. The concentration of transforms is a less elementary idea. Still less elementary an idea is the difference created by cocktails of concentrations of transforms of diverse signals acting simultaneously. The distinction between components in the system has to do more with the physical modality of the mechanisms involved in the formation of a new sign – which will transduce the information further in the pathway – than with the formal logic with which such components operate.

The signalome of an endocrine system is generally constituted by:

1) signals
2) receptors
3) effectors
4) molecular switches
5) second messengers
6) adapter proteins
7) sensor molecules
8) channels, pumps and exchangers
9) buffers

Almost all the actors of signal transduction networks can be placed in one of the categories listed above. But it is worthwhile observing that some components could be placed in more than one category; or they may play a role in one moment and play a different one immediately after. Being the transductional process based on "cascades", some molecules can change their role from one step to the next, being sometimes a signal, i.e., requiring a specific concentration threshold to be recognized, and/or by being a necessary cofactor for creating some analogical consensus (as e.g., when interacting with an adapter protein), or it can be an effector by (in)forming the next signal-effector, i.e., by participating in the relay chain (as e.g., with protein kinases).

The consideration of the relativity of roles may on the surface appear as unnecessary or trivial, but it helps to decentralize our attention from any single component as the sole "regulator", as it is usually claimed when we say that such and such signal or second messenger "regulates" a given process. In a signal-cascade some of the "actors" may be in chronological order receptors, signals or effectors. This is the nature of semiotic processes: signs produce other signs in continuous and multidimensional processes. For example, specific threshold concentrations of inositol second messengers translate into specific threshold concentrations of Ca^{2+} second messengers, which in turn participate in a given "consensus" or higher logical product to establish an even higher analogical message. But it does not seem correct to claim that a given ion, molecule or protein complex is "the" regulatory element of a given process. There is no signal transduction network, which is regulated by such a single element, and there is no signal transduction pathway that stands on its own. Any "second messenger" is no less a regulator than the primary or the final signal. Primary signals come from and go to different directions and networks, all of which offer further possibilities for regulation. There is no final signal either, because the process is continuous and signal pathways do not end. Signals are only transformed within larger circuits of branching and interconnecting chains of causation. These chains of causation are "closed" or integrated in the sense that causal interconnection can be traced around the circuit and back through whatever position was arbitrarily chosen as the starting point of the description (Bateson, 1972: 404).

7. MODULARITY

When we are exploring how differences are sensed, transformed and conveyed across hierarchical levels, forming therefore higher order differences, we are mostly concerned about regularities in the formal and logical aspects of such processes rather than in the regularity of the physical structures that underpin them. The material means implicated in the formal process can be bewilderingly diverse. Since biology has focused mostly on the diversity of structures, rather than on the formal logics behind biological-semiotic processes, induction has necessarily been the norm, presenting biology as a science with very few deductive principles, generalisations or rules, and focused rather on specific, local and apparently idiosyncratic cases, putting us in front of a jungle of proteins where it is sometimes difficult to see the forest for the trees.

Let us say that the difference created by cocktails of concentrations of transforms of diverse signals acting simultaneously form an idea. The functionality of such an idea is somehow "shaped", or informed, by the context. The context poses the question and the system comes up with the idea.

A very similar "idea", the result of a complex aggregate of differences, can emerge (developmentally and evolutionarily) through different infrastructure configurations and local solutions.

Many examples can be found within signal transduction in developing systems where activation of different modular arrangements of components can give rise to the same intermediate or final responses. This is obvious at lower hierarchical levels, for example when a particular cellular response can be modulated by different kinds of signal transduction networks which exhibit different signal-receptor complexes, although in some cases the two different pathways may share a common "idea", such as a Ca^{2+} second messenger, at some step of the particular cascade, i.e., the same component can be used modularly for different purposes at different hierarchical levels, for example when some receptors that are activated by different ligands and mediate different cellular responses, nevertheless mediate a similar signalling pathway by using components and steps common to both pathways (Lodish et al., 2000: 862).

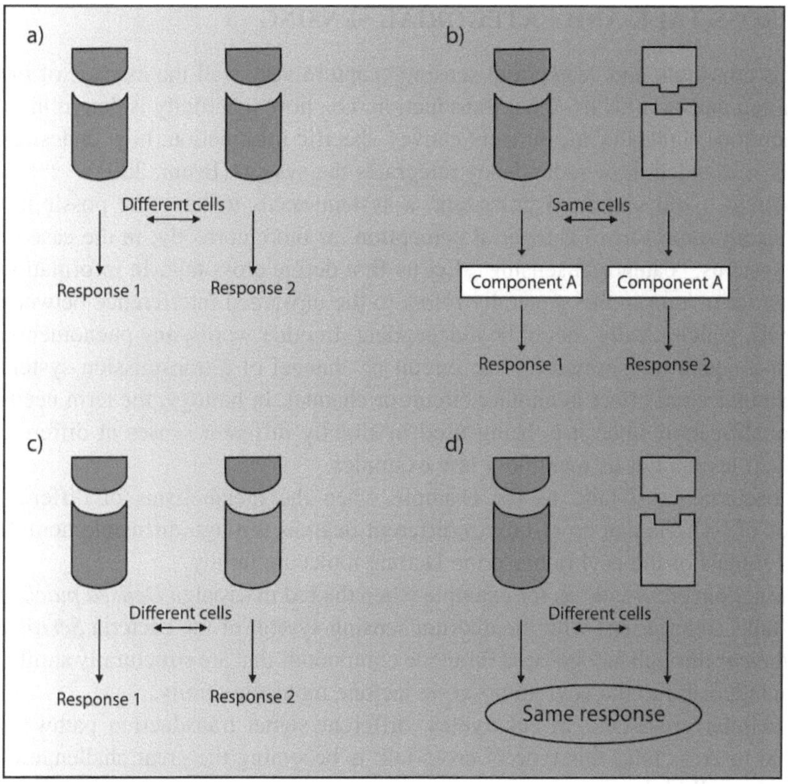

Figure 3. Modularity in signal transduction:
a) The same signal and receptor in different cells can promote very different responses (as diverse as proliferation, differentiation, and death).
b) Activation of the same signal-transduction component in the same cell through different receptors often elicits different cellular responses. The binding specificity of two receptors may differ while their effector specificity is identical.
c) Different cell types may have different sets of receptors for the same ligand, each of which induces a different response. Some signalling-molecules can function in more than one modality (e.g., epinephrine can function as both neurotransmitter and hormone).
d) Different receptor-ligand complexes can induce the same cellular response in some cell types.

We can say that such *modularity* is a central feature in signal transduction. The same result can be achieved through different "infrastructure" by combining common components and principles, or, conversely, different arrangements of the same, or very similar, components can result in very different responses (see figure 3).

However, there is always receptor-signal *binding specificity*, and the resulting receptor-signal complex exhibits *effector specificity*, i.e., it mediates a specific cellular response. How this specificity is determined is considered "an outstanding question in signal transduction" (Lodish et al., 2000: 905).

8. CROSS-TALK AND CATEGORIAL SENSING

The terms cross-talk and "categorial sensing" capture very well the essence of the "outstanding question(s)" in signal transduction, i.e., how specificity is determined, how ubiquitous signals or messengers convey specific information, how undesired cross-talk is avoided, how redundancy integrates the system (Bruni, 2003).

In order to avoid undesired cross-talk, a system needs to have the possibility for exercising some sort of categorial perception, or more correctly, in the case of cellular systems, "categorial sensing". Let us first define cross-talk. In information theory the term "cross-talk" generally refers to the unwanted interference between two signals, which ideally should be independent. In other words, any phenomenon by which a signal transmitted on one circuit or channel of a transmission system creates an undesired effect in another circuit or channel. In biology, the term needs a little qualification since it is being used in slightly different senses at different hierarchical levels. Let us mention a few examples:

- Inter bacterial cross-talk: as for example when the metabolisms of different species of bacteria can cross-talk to different degrees through diffusible homologous signals of the acyl homoserine lactone molecule family.
- Inter kingdom cross-talk: as for example when the red macroalga *Delisea pulchra* cross-talks (interfering) with the quorum sensing system of the bacteria *Serratia liquefaciens* through halogenated furanone compounds that are structurally similar (but antagonists) to the acyl homoserine lactone molecule family.
- Intra cellular cross-talk: in eukaryotes, different signal transduction pathways are said to cross-talk; this type of cross-talk is becoming the great challenge of molecular cell biology.
- Modular cross-talk: within single pathways in endocrine systems, what could be called the Ca^{2+} and the phosphorylation codes are said to cross-talk.

When you have "universal" signals that work specifically in specific pathways which communicate, sometimes it may not be completely correct to speak about cross-talk between pathways, for if the pathways are meant to be linked and function together through a second messenger, then it would only be "normal" talk (in which case we could use the term "interface"), whereas, cross-talk proper is what occurs between semiotically compatible systems, but which are not set up or prompt to communicate under "normal" circumstances, (e.g. mimicry, agonism, antagonism). But mimicry, agonism and antagonism could also enter into the category of "normal" talk at higher hierarchical levels (physiological or ecological), being enough to enlarge our functional semiotic network to include the collateral sources of cross-talk.

Cross-talk can be better understood by considering the cases of homology in signalling systems, i.e. when molecules, components or modules of very different systems, which normally are not in communication, present functional compatibilities that would allow for a component of one system to interfere, deviate or even work properly in another system. For example, some signal molecules in *C. elegans* have been found to work as neurotransmitters in humans. If by an unfortunate event a specimen of *C. elegans* could find its way to a human brain, the two systems would cross-talk.

The nature of "cross-talk" has many different implications. Sometimes the system needs, and regularly uses, cross-talk, and sometimes it needs to avoid it, and it regularly does. This means that cross-talk needs to be understood as a complex combination of signals, pathways and therefore of regulatory agents. It is very important to stress that cross-talk has to be studied and identified at different hierarchical levels.

In the case of integrated cellular signal networks, where a common signal, as e.g., Ca^{2+}, is used at different levels and in different ways, what is necessary is precisely the avoidance of cross-talk, for otherwise the semiotic system would be ruined by uniformity. A specific signal-receptor event may be transduced into a ubiquitous second messenger, which then regains the specificity of the pathway somewhere downstream, where it directly or indirectly informs a specific effector that finally canalises the specificity of the "original" signal to the specific response. At such points of the cascade, the system is vulnerable to undesirable cross-talk with other pathways (see figure 4).

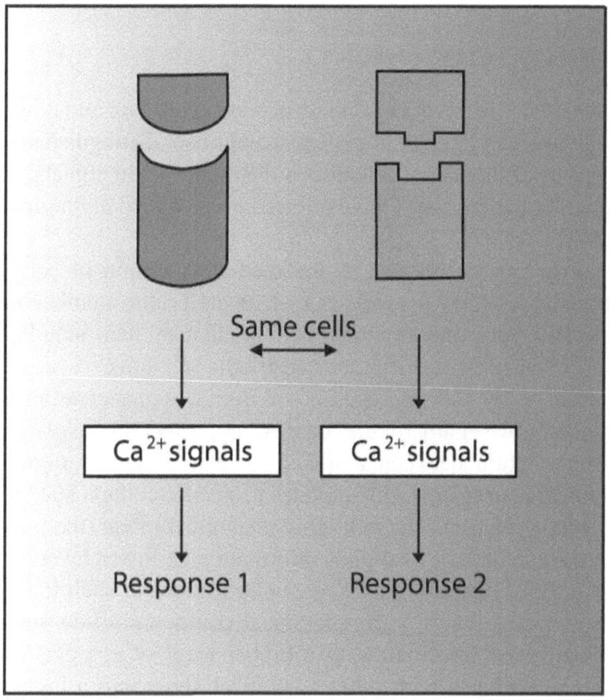

Figure 4. Different signal-receptor complexes lead to the same second-messenger, a ubiquitous signal such as Ca^{2+}. However the specificity of each pathway is maintained as to avoid undesired cross-talk of both pathways

So let us summarize the essence of the problem. The mediatory role of specific signals leads to "universal" signals, as e.g., Ca^{2+}. Specific *patterns* have to be decoded in order to maintain specificity. Such specificity is vulnerable to cross-talk. In order to avoid such undesired mixing of pathways the cellular system needs to be able to count on some sort of categorial sensing (Bruni, 2003).

Categorial perception (originally a notion from cognitive psychology) as a prerequisite for the proper functioning of semiotic networks has been discussed in Stjernfelt (1992) and Hoffmeyer (1996). By categorial perception we can understand "the ability to slot a bewildering number of impressions into categories" (Hoffmeyer, 1996: 77). Maybe, at the cellular level, it would be more appropriate to talk about categorial sensing (being perception a more complex elaboration of sensing which may lead to cognition). Let me restate this in terms of signalling systems by saying that categorial sensing can be seen as the capacity for identifying and ordering patterns out of a contextual matrix, i.e., the recognition of meaningful patterns out of ubiquitous signals. In other words, it is the capacity for pattern recognition, which is the action of extracting contextual meaning from what would otherwise be ubiquitous signals, avoiding therefore anarchic cross-talk, which would be deleterious to any "self-organised" system.

9. DIGITAL-ANALOGICAL CONSENSUS

One way to look at how elementary differences build up and are sensed up and down the biological hierarchy, and how can biological systems categorise, distinguish and obtain relevant information out of otherwise ubiquitous differences and signals, is by considering a communication pattern that I have referred to as *digital-analogical consensus*[4] (Bruni 2002, 2003).

Digital-analogical consensus can be defined as the mediatory action of codes which are formed at different hierarchical levels out of an indefinite number of dyadic causal relations, specific "lock and key" interactions, that by their simultaneous occurrence give rise to emergent specificities and triadic relations.

New analogical signs emerge by the aggregation of digital signs creating a complex lock-and-key relation between a particular context and a specific response. Such complex lock and key relation between a specific configuration of digital events (that form and analogical aggregate), which relates a specific context state to a particular response, is a triadic relation. The emerging analogical mode (the bulk of information) influences the circulation of digital information at lower levels in the hierarchy (downward causation) determining new configurations at such lower levels. The process is hierarchical because such analogical compound effect may constitute a "quasi-digital" piece of information to a higher level of aggregation ("to be or not to be"), which combines with other analogical signs into a higher order logical product. In this way the new analogical sign can be a digital contribution to a still larger or more complex analogical sign (Bruni, 2002, 2003) (see figure 5).

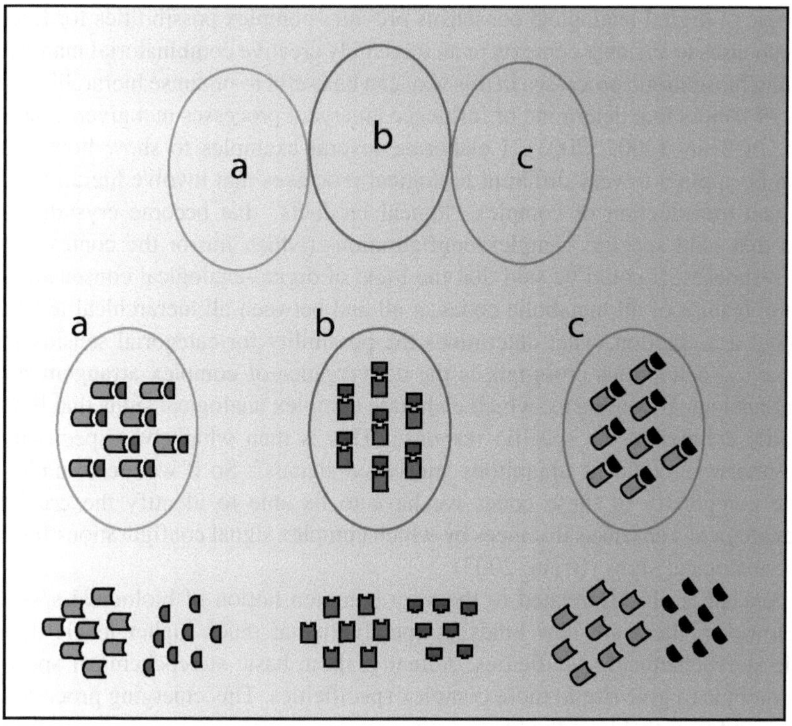

Figure 5. Digital-analogical consensus. The presence or absence of a single signal molecule or corresponding receptor is a digital event. The concentration of a signal molecule (or of a corresponding receptor) is an analogue. Equally, a given concentration of signal-receptor complexes (formed by signal-receptor digital binding events) constitutes an analogue. But a particular threshold concentration of signal-receptor complexes can constitute a digital message at a higher level as in (a) which together with (b) and (c) form a higher-level analogue. The triadic logic comes into place in the sense that (a) + (b) + (c) are sensed synchronically and an interpretant (a link) is formed which binds such specific configuration (of signal cocktails) to a specific response

An emerging "state" constitutes a difference that can be sensed by some system or part of a system with interpretative capacity, i.e., capacity for pattern recognition. Every effective difference denotes a demarcation; a line of classification, and all classification is hierarchic. In other words, differences are themselves to be differentiated and classified (Bateson, 1972: 457). But also complex aggregates of differences are to be differentiated and classified.

We have in other words the formation of an interpretant by the synchronic occurrence of a combination of discrete factors, determinants or circumstances.

A continuous process in which an indefinite set of digital messages in simultaneous occurrence form an analogical message that links the most comprehensive interpretation of the context to an appropriate response, i.e., what could be referred to as an *emergent interpretant* (Bruni, 2003).[5]

The logic of digital-analogical consensus provides complex possibilities for fine-tuning responses to variable contexts in an incredibly creative combinatorial manner. Visualising biosemiotic processes in this way can be useful to organise hierarchically the suits of factors that determine or influence emergent processes in a given causal network. In Bruni (2002, 2003), I elaborate several examples to show how this logic can be applied to very different biological processes that involve hierarchical sensing and transduction of complex "logical products" that become crystallised in codes that bind specific complex configurations (which mirror the context) to specific responses. It could be said that the logic of digital-analogical consensus is a common feature of all metabolic codes at all and between all hierarchical levels.

In signal transduction, what determines the possibility for categorial sensing in order to avoid deleterious cross-talk is the convergence of complex arrangements of digital-analogical consensus, which elaborate complex analogical signs that bind the specific context to the specific response. This is then what gives specificity to what otherwise could be ubiquitous "universal signals". So if we are to understand the complexity of these codes we have to be able to identify the crucial digital-analogical-consensus instances by which complex signal configurations form complex analogical signs (Bruni, 2003).

This concept is closely related to the very common notion of biological specificity. However, there are new kinds of specificities at much higher levels than the basic stereochemical specificities. Actually, these basic stereochemical specificities combine to give rise to more complex specificities. This emerging process is related to, and is probably at the base of, the increasing semiotic freedom exhibited by complex organisms, i.e., the extent of logical (or causal) independence that some processes can acquire with respect to the physical dynamics of the substrate that underlies such processes. The most extreme example of this would be the path through different levels of ascending complex "lock-and-key" mechanisms that goes from stereochemical specificities up to "free will" or natural language (a sophisticated emerging "system of correspondences"). (On the other hand, one extreme example of downward causation in biological systems would be the path that goes from fertilization and differential use of DNA in the fertilised egg of an embryonic scientist until he/she grows up into a full-blown biologist with the capacity for modifying the physical substrate of living systems by means of the emerging cultural product represented by biotechnology).

There are basic types of specificities which give rise to new and more complex types of specificities, for example the specificity of each DNA sequence for its complementary strand, as modulated through the specificity of DNA base pairs, or the specificity of the relation between DNA and protein, modulated by "genetic information", understood as the specification of a protein sequence, i.e., the linear amino acid residue sequence of a protein from a DNA sequence as a process of "translation", i.e., the triplet-amino acid specificity. There are more complex types of specificities such as gene-enzyme specificity, enzyme-substrate, antibody-antigen, signal molecule-receptor, activation complex-DNA, and so on. The simultaneous and complex "activation" of an indeterminate number of these "lock and key" mechanisms mediate the emergence of new informational-semiotic contexts and new and more complex "lock and key" mechanisms and specificities like for example "cocktail" of signals-cellular response, pattern of neural firing-specific cognitive response, host-symbiont specificity, organism-niche, and so on.

Specificities at different levels become an analogical message out of the complex interaction of many lower-level specificities. These complex specificities establish "systems of correspondences", "systems of ideas in circuit" (see below). The importance of considering semiotic contexts hierarchically is that sometimes at a given level what may look as an "either-or" choice of function or manifestation may be determined by the compound effect of a larger analogical message, a bulk of information, which has a causal link to the lower level. For example, whether a pathogen protein acts as a virulence or as an avirulence factor is determined by a larger gestalt at a level above the dyadic resistance-(a)virulence protein relation. Digital-analogical consensus emerges as a general pattern for sign construction, i.e., for generating complex specificities and lock and key mechanisms, creating immense combinatorial semiotic possibilities for regulating and fine-tuning complex, detailed and decentralised responses to equally complex, detailed and decentralised stimuli.

Digital-analogical-consensus in biological systems can be compared to what the engineers call "coincidence counting", which is a method of counting that employs a coincidence circuit so that an event is recorded only if events are detected in two or more sensing devices simultaneously. Such counting methods are used to reduce background noise. The ability to distinguish true coincidence events from scattered and random coincidence counts is the basis of categorial sensing in cellular systems. This true coincidence event is what I have been calling digital-analogical consensus. A consensus because it is not just any coincidence, it is an "agreement of many participants", a very specific one which, has been codified in a evolutionary habit and it links the contextual complex set of stimulus, the system and the response in a triadic logic.

10. THE CA^{2+} CODE

Ca^{2+} Ions – The Signal

As mentioned before, one of the modular components of many pathways is what deserves to be called the Ca^{2+} code.

> "Of the approximately 1,400 grams of calcium that are in the human body, less than 10 grams manage to escape being trapped in the skeleton and teeth. These few grams might be an insignificant quantity, but they are extraordinarily *significant qualitatively*. They circulate in the blood and extracellular spaces, and penetrate cells to regulate their most important activities" (Carafoli, 2003: 326, my italics).

The versatility of calcium as an intracellular "second messenger" has led some authors to talk about its "universality" as a signal. This ubiquitous intracellular signal is held to be responsible for controlling multiple cellular processes throughout the life of eukaryotic cells from fertilisation to apoptosis, including embryonic pattern formation, cell differentiation and cell proliferation (Berridge et al., 2000). "...the Ca^{2+} signal is important in cells from their origin to their death. It controls the creation of cells at fertilisation, masterfully guides them from infancy through adulthood to old age, and finally assists them at the time of their demise" (Carafoli, 2003: 331).

When the Ca^{2+} concentration rises to certain specific threshold levels many different functions can be activated. A rise in cytosolic Ca^{2+} induces a variety of cellular responses. One of the main questions researchers are asking themselves is: how can these elevations of Ca^{2+} concentration regulate so many processes? Part of the answer lies in the versatility of the Ca^{2+} signalling system in terms of speed, amplitude, and spatio-temporal *patterning* (Berridge et al., 2000: 11). But another part of the answer lies in what we have already said, that no single component of a signal-transduction network is by itself the regulator of a cellular response, it is rather one of many mediators. Actually, it is not the simple linear rise in concentration that informs the system and triggers a response. It is rather the *fluctuation* of concentrations which create differences. For this purpose cells employ a sophisticated and extensive repertoire of signalling components, which comprises a "Ca^{2+} signalling toolkit" that can be assembled in combinations to modulate signals with widely different spatial and temporal profiles (Berridge et al., 2000; Carafoli, 2003).

Sources of Ca^{2+} signals

Ca^{2+} signals are generated by using both internal and external sources of Ca^{2+}. The internal stores are held within the membrane systems of the endoplasmic reticulum (or equivalent organelle) and within the mitochondrion. The external sources come from the extracellular environment. Additional sources come from buffer molecules that may capture free Ca^{2+} ions, and from the nuclear envelope. Release from these internal stores and recruitment from the environment is achieved through various channels that respond to signals (see figure 6). There seems to be reciprocal interactions and cooperation between the different organelles and channels in modulating specific patterns of Ca^{2+} concentrations. Environmental signals indirectly induce some of the channels that let Ca^{2+} in and out of the cytosol, contributing in this way to configure specific patterns of concentrations of free ions. The digital signals represented by single Ca^{2+} ion constitute an analogical sign represented by spatio-temporal patterns of specific threshold concentrations.

De-coding Proteins

"De-coding" proteins are the ones that react to a specific concentration pattern and therefore continue the cascade towards a specific response. By binding Ca^{2+} ions, de-coding molecules undergo a pronounced conformational change that allows them to continue the cascade towards specific effectors, usually protein kinases, which

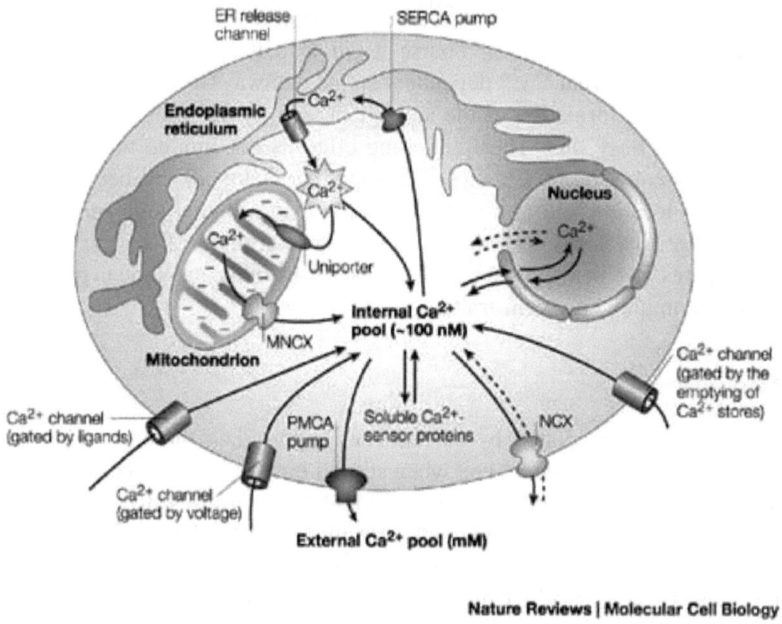

Figure 6. Internal and external Ca^{2+} sources
Reproduced with permission from Nature Reviews Molecular Cell Biology. Carafoli, Ernesto (2003) copyright (2003) Macmillan Magazines Ltd. (www.nature.com/reviews)

alter other proteins, translating in fact the calcium message into the phosphorylation code and thereby directing the cascade towards a specific response. A major family of these molecules is the family of EF-hand proteins which include hundreds of members, of which calmodulin is the most thoroughly investigated. There is a group of EF-hand proteins which are collectively called "neuronal Ca^{2+} sensors" which mediate neuronal functions such as the release of neurotransmitters (Carafoli, 2003: 330).

Buffer Molecules

Ca^{2+} buffer molecules intercept free Ca^{2+} ions in the cytosol (or in organelles) and keep them unavailable until they are required as free ions again, constituting an additional mechanism to give specificity to a given needed pattern.

For example, when buffer molecules capture a necessary given number of free Ca^{2+} ions, keeping the cytosolic concentration under a certain threshold, the probabilities of activating the different pathways that arc sensible to concentrations above such a threshold remain low. This can be compared to the example given by Bateson (1972: 403), in which a telephone exchange at a time of emergency may be "jammed" when a large fraction of its alternative pathways are busy. There is, then, a low probability of any given message getting through. Since the sign (or rather part of it) that may prescribe a certain transcriptional response is a precise concentration of Ca^{2+} ions (plus of course a whole battery of other consensus parameters), all the individual Ca^{2+} dependent signal pathways of the network may contribute or not to the formation of the analogical sign. A Ca^{2+} ion bound to the buffer molecule is a busy line. The fractioning effect is not limited to that level. The threshold concentration of Ca^{2+} (as a compound analogue), by being or not being at a certain location and at a certain time, is also a digital sign in a larger analogical message that leads for example to the transcription of a gene, whose product participates (by being or not being present) in other analogical products that give rise to complex emergent traits.

Second Messengers

Second messengers are mediated by second messengers, i.e., there are different Ca^{2+} mobilising messengers (generated when stimuli bind to cell surface receptors) that cooperate in different specific digital–analogical consensuses that activate or inhibit different mechanisms (e.g., channels) for modulating influx and outflux of Ca^{2+} in the cytosol. The different Ca^{2+} mobilising messengers can coexist in cells where they seem to be controlled by different receptors that respond to specific signals (Berridge et al., 2000: 12).

The Channels

There is a continuous fluctuation of Ca^{2+} concentrations created through many different in-and/or-out-channels that operate at the different sources of Ca^{2+}, i.e., membrane-intrinsic proteins that transport Ca^{2+} ions across membranes. Channels possess receptor domains being actually "receptor-channels". The channels are activated or deactivated (directly or indirectly) by extracellular signals (e.g., neuro-transmitters), other second messengers (e.g., inositol-1,4,5-trisphosphate), voltage differences, and by Ca^{2+} itself. Usually a *consensus* of different second messengers (and other components), plus Ca^{2+} itself, is required for such activation.

Two well studied families of channels are:

1) the Ins(1,4,5)P3 receptor channel, for short InsP3R, and
2) the ryanodine receptor channel, for short RYR. (Berridge et al., 2000; Lodish et al., 2000; Carafoli, 2003).

These channels coexist in cells as "clusters of channels", where they seem to be controlled by different receptors that respond to specific signals and where they are used cooperatively to fine-tune the formation of complex patterns of Ca^{2+} signals that inform specific responses (see figure 7).

Creating Patterns of Patterns

The different degrees of excitability and concentration of different kinds of channels, depending on the levels of the appropriate Ca^{2+} mobilising messengers, modulate

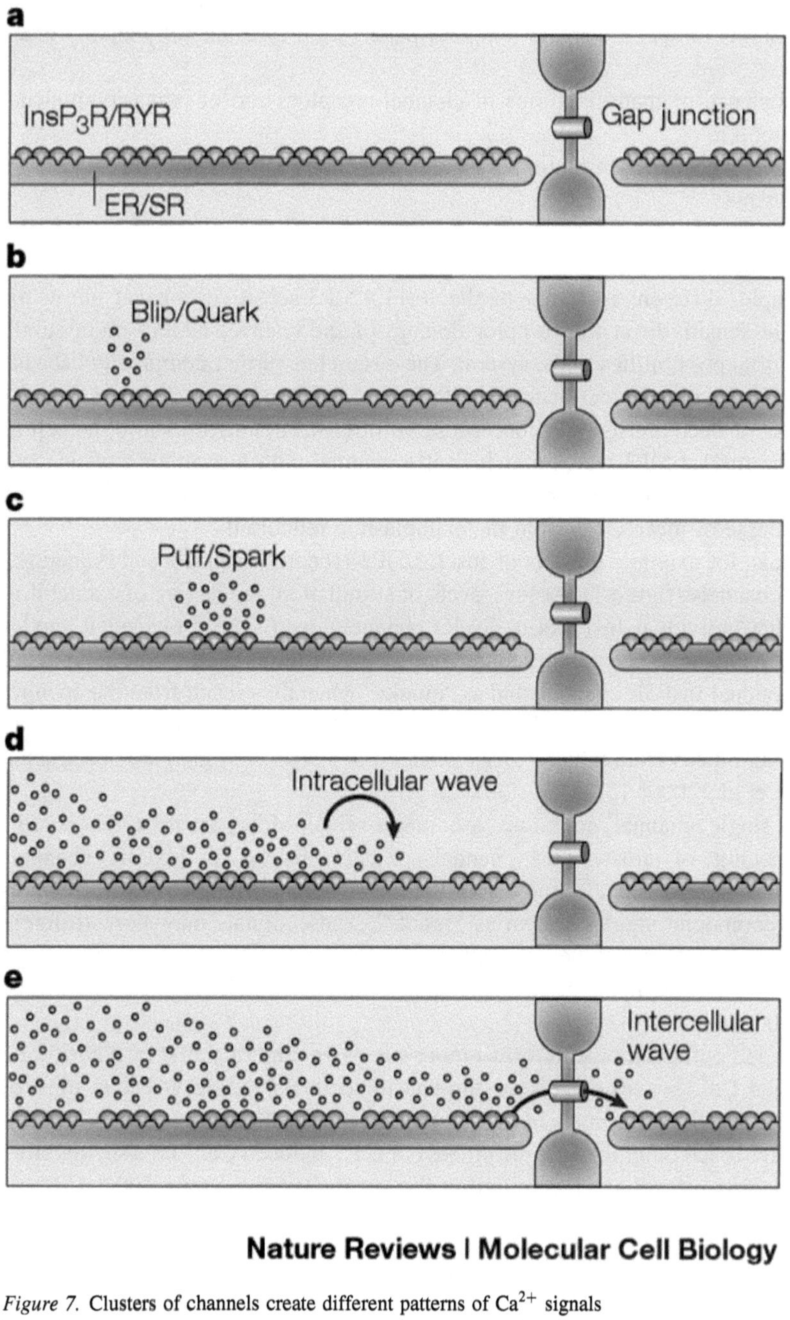

Figure 7. Clusters of channels create different patterns of Ca^{2+} signals
Reproduced with permission from Nature Reviews Molecular Cell Biology. Berridge et al. (2000) copyright (2000) Macmillan Magazines Ltd. (www.nature.com/reviews)

different kinds of spatio-temporal *patterns* of Ca^{2+} signals (Berridge et al., 2000: 15). Variations of specific spatio-temporal patterns are modulated by among other things:

- The existence of many isoforms of channel-receptors and of other components of the toolkit.
- Different levels of concentration of distinct activated channels and other necessary components.
- The different degrees of excitability or sensitivity of different kinds of channels grouped in "clusters".

For example, different isoforms of the Ins(1,4,5)P3-sensitive channel allow for differential sensitivity at the receptor domain of the channel adding more subtle combinatorial possibilities to the system. The circuit has further complexity because Ca^{2+} itself is involved synergistically with Ins(1,4,5)P3 and other *consensus* factors to increase or decrease its own concentration. Ins(1,4,5)P3 diffuses into the cell to engage the Ins(1,4,5)P3 receptor (which is a channel with a receptor domain) and together with some already present Ca^{2+} create a digital–analogical consensuses for the release of more Ca^{2+} from the endoplasmic reticulum.

Let's take for example clusters of Ins(1,4,5)P3-receptor-channels and ryanodine-receptor-channels (figure 7). At low levels of stimulation, the degree of excitability is such that individual InsP3Rs or RYRs channels open. These elemental single-channel signals have been recorded as "blips"[6] when they result from the opening of an individual InsP3R channel, and as "quarks" when they result from the opening of an individual RYR channel. These are considered the fundamental events that are the building blocks from which *more complex* Ca^{2+} *signals* (signs) are constructed (Berridge et al., 2000: 15).

These single channel emissions are rare events. More usual is the coordinated opening of clusters of channels. Such clusters of InsP3R channels generate compound signals known as "puffs" while clusters of RYR channels generate compound signals known as "sparks". These signals may show different amplitudes suggesting that there are either variable numbers of channels within each cluster or variable numbers of channels open within an individual cluster (Berridge et al., 2000: 15).

Sparks and puffs combine to form a more extensive kind of signal by constituting intracellular Ca^{2+} waves that sweep through the cell. For waves to occur, most of the InsP3R and RYR channels in the clusters must be sufficiently sensitive to Ca^{2+} to respond to each other through the process of Ca^{2+} induced Ca^{2+} release, therefore setting a positive feedback that amplifies the wave (Berridge et al., 2000: 15).

Intra-cellular waves can become inter-cellular waves by propagating through gap junctions. Such intercellular communication contributes to the coordination of many cells. However it is not yet clear how the waves traverse the gap junction or whether before being transduced to the adjacent cell, the message is translated into Ins(1,4,5)P3 second messenger or even to some other extracellular mediator such as ATP (Berridge et al., 2000: 16).

Finally, we have also frequency and temporal aspects of the signals. Ca^{2+} signals are usually presented as brief spikes. In some cases, individual spikes are sufficient to trigger a cellular response. When longer periods of signaling are necessary, spikes are repeated to give waves with different frequencies - ranging from a few seconds to 24 hours. Cells respond to changes in stimulus intensity by varying the frequency of Ca^{2+} waves. For example, spikes, which generate fluctuations, i.e., differences, can initiate gene expression more effectively than a steadily maintained level of the same average concentration (Berridge et al., 2000: 17).

According to Berridge et al. (2000: 17) "To use such a frequency-modulated signalling system, cells have evolved sophisticated 'molecular machines' for decoding frequency-encoded Ca^{2+} signals". They point out two Ca^{2+} sensitive proteins that seem to decode wave frequency (Ca^{2+}/calmodulin-dependent protein kinase II and protein kinase C).

"Now that the molecular and physiological mechanisms have been identified, the new challenge is to determine how this versatile Ca^{2+} signalling system functions in specific cellular processes. The Universality of this signalling system is evident in its *emerging function* during various developmental processes..." (Berridge et al., 2000: 20), my italics).

The point here is that even if the isoforms of the channels in a cluster are highly specific, their mediatory role still leads to "universal" signals, Ca^{2+} patterns, that have to be decoded and whose specificity as signal is partly related to a concentration threshold (which is part of the sign), but which by itself would not assure the required specificity to avoid the possibility of cross-talk or misinterpretation. There is still the need for categorial sensing.

11. WHY CATEGORIAL SENSING?

Before continuing with the calcium case let me mention some possible explanations presently being considered to explain how (what in this work has been referred to as) categorial sensing is achieved in signal transduction networks in general, i.e., how a given concentration of a versatile second messenger informs the specific response and not other components of the network sensible to the same messenger:

1) One aspect that has been considered to explain this phenomenon is that specificity may be achieved thanks to the specificity that the downstream component – which in the previous step was informed by the second messenger – has for the subsequent substrate. For example, cellular responses to hormone-GPCR-induced-rise in cAMP vary among different cell types and tissues. "In virtually all eukaryotic cells studied, the action of cAMP appears to be mediated by one or more cAPKs [cAMP-dependent protein kinases], but the nature of the metabolic response varies widely among different cells. The effects of cAMP on a given cell type depend, in part, on the specificity of the particular cAPK and on the cAPK substrates that it expresses" (Lodish et al., 2000: 887). That is,

the effects of a second-messenger on a given cell type depend, *in part*, on the specificity of the particular components that it informs and, of course, on the specificity of the components that follow after. But this is not enough to explain the crucial problem of categorial sensing. When in one cell type there is more than one signal network that leads to increases in cAMP (like e.g., in liver cells), and also leads to different responses, the hormone inducing the largest concentration of the common second messenger theoretically would also influence the activity of the networks that require a lower threshold concentration of the same second messenger.

2) Another argument to explain the process of categorial sensing of these common signals is the consideration of the role that they may play as global regulators in "regulons"[7] to coordinate the combined action of different responses by a hierarchical variation of thresholds in the concentration of the messenger that influences different components during the development of its concentration curve, in which case the "right" sequence of thresholds would be achieved by the changes of concentration derived by a combination of hormones acting simultaneously to fine tune the response (which is a sort of digital-analogical consensus). For the same reasons as in the previous explanation, this by itself is not enough to explain how categorial sensing is achieved. Rather, this fine tuning mechanisms could explain why then it would not be enough to have only the pathway with the hormone that expresses the highest concentration, which would overlap the other ones. The action of timely coordinated emission of differential concentrations makes a richer and much more differentiated concentration-development curve.
3) A third possible explanation for categorial sensing in the case of a rise in cAMP that may produce a response that is required in one part of the cell but is unwanted, perhaps deleterious, in another part is by the discovery of anchoring proteins. Recent biochemical and cell biological experiments have identified a family of anchoring proteins that localise inactive cAPKs (the kinasis effector that is informed by cAMP) to specific subcellular locations, thereby restricting cAMP-dependent responses to these locations. This family of cAMP kinase-associated proteins (AKAPs) posses one domain conferring a specific subcellular location and another that binds to the regulatory subunit of the specific cAPKs (which is to be informed by the second messenger) (Lodish et al., 2000: 888). In other words, anchoring proteins function as an efficient recruiting net for ready-to-be-activated cAPKs, but which could anyway be susceptible of activation by cAMP of different origins.

This last possibility is very interesting because it may easily be related to the role that other types of molecules may have in the process of categorial sensing like for example adapter proteins, molecular scaffolds and buffers. "Specific anchoring proteins may also function to localise other signalling proteins including other kinases and phosphatases, and thus may play an important role in integrating information from multiple signalling pathways to provide local control of specific cellular processes" (Lodish et al., 2000: 888).

These three explanations do not exclude each other but they do not suffice to explain how complex specificities are achieved.

12. THE CA^{2+} CODE AS AN EXAMPLE OF CATEGORIAL SENSING

Let us go back to the calcium code and suggest a possible explanation for categorial sensing from the sign-theoretic perspective advanced in this work.

Since Ca^{2+} patterns have to be de-coded, their specificity as signals is related to a pattern of *concentration thresholds of different elements* (in simultaneous occurrence) that form the vehicle of a more complex sign. It is because of this variability that different types of cells may exhibit very different responses to the same extracellular signal or the same second messenger (Bruni, 2003).

The inositol second messenger is generated when stimuli bind to cell surface receptors – and its concentration contributes to determining whether Ca^{2+} (already present in low quantities) can activate the Ca^{2+} channels or not. Ins(1,4,5)P3 diffuses into the cell to engage its receptor which is in a domain of the channel-protein. Together with some already present Ca^{2+}, it creates a digital–analogical consensus for the release of more Ca^{2+} from the endoplasmic reticulum. Different isoforms of the Ins(1,4,5)P3-sensitive channel allow for differential sensitivity at the receptor domain of the channel adding more subtle combinatorial possibilities to the system. Different kinds of channels have different degrees of excitability depending on the levels of the appropriate Ca^{2+} mobilising messenger (Berridge et al., 2000: 12). Increasing the level of Ca^{2+} enhances the sensitivity of the channels to the other consensus signals. This is the key to the "autocatalytic" process of Ca^{2+} induced Ca^{2+} release. But the cytosolic autocatalytic action of Ca^{2+} seems to be more complex: it can be both stimulatory and inhibitory and can vary between the different receptor-channel isoforms. This gives rise to some very interesting relationships between the level of activity of the channel (stimulation/inhibition) and Ca^{2+} cytosolic concentration (Berridge et al., 2000: 12).

At low concentrations of the Ins(1,4,5)P3 second messenger, Ca^{2+} has a stimulatory effect on the channel but it inhibits it once a certain Ca^{2+} concentration threshold is reached, giving rise to a bell-shaped function (see figure 8a). But it has been observed that sometimes the channels are not inhibited by high Ca^{2+} cytosolic concentrations – particularly when Ins(1,4,5)P3 is present, also at high concentrations. In this case, instead of a bell-shape, the relationship between channel activity and Ca^{2+} level is sigmoidal, with the peaks of the curve depending on the presence or absence of Ins(1,4,5)P3 which when present at certain threshold concentration collaborate in the digital-analogical consensus that enhances the stimulatory action of Ca^{2+} (see figure 8b).

These instances of digital-analogical consensus contribute to shaping the fluctuation curve and provide the system with the capacity for categorial sensing, which is what gives specificity to the analogical sign in order to avoid what would otherwise be anarchic cross-talk of ubiquitous signals (see figure 9). Besides a

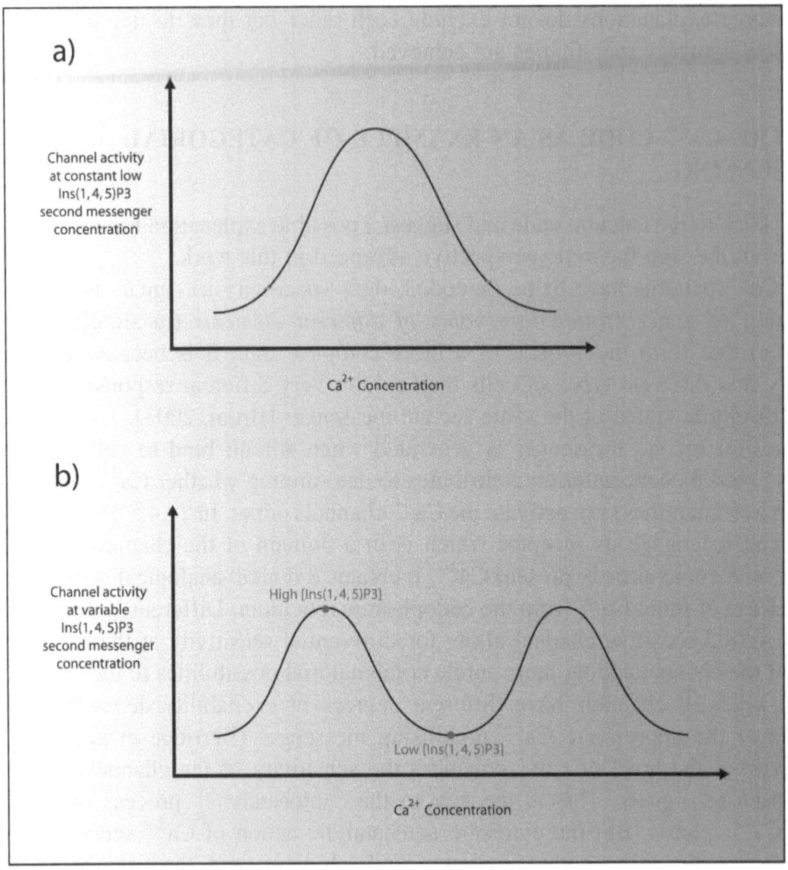

Figure 8. a) Channel activity as a function of Ca^{2+} concentration at constant Ins(1,4,5)P3 second messenger concentration. b) Channel activity as a function of Ca^{2+} concentration at variable Ins(1,4,5)P3 second messenger concentration

given concentration threshold of channels, this particular case of digital-analogical consensus requires the simultaneous occurrence of certain specific threshold concentrations of for example Ins(1,4,5)P3 second messenger, Ca^{2+} ions, other Ca^{2+} binding proteins, adapter and scaffolding proteins, etc. These parameters must all have their own fluctuation curves, and the moments at which they intercept in complex combinations of specific thresholds are the moments of digital-analogical consensus that will link a particular emission of Ca^{2+} patterns to a particular context in which the sign will be interpreted properly, i.e., categorically and hierarchically, linking a complex set of cues to a specific response (see figures 10 and 11).

Cytosolic buffers also play a major role in categorial sensing of Ca^{2+}. They are involved "... in shaping both the amplitude and duration of Ca^{2+} signals. During each spike, they act as a halfway house for Ca^{2+} by loading it up during the

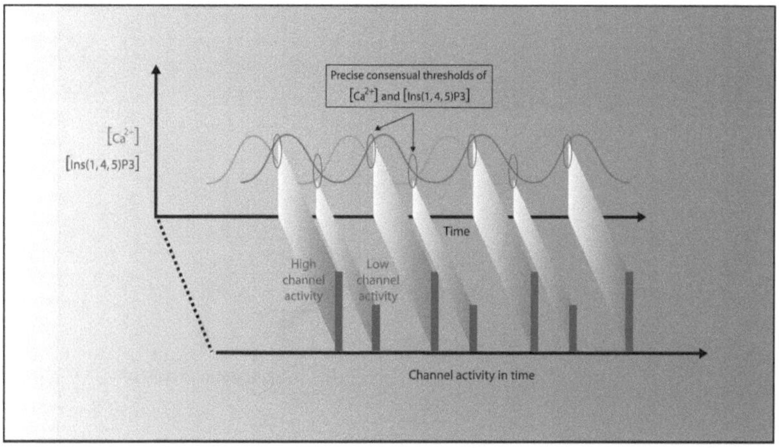

Figure 9. Specific simultaneous threshold concentrations of Ca^{2+} and Ins(1,4,5)P3 determine a specific digital analogical consensus that results in a specific channel activity. In reality the consensus involves many more variables that result in a specific complex response

ON mechanisms and then unloading it during the OFF mechanisms". They also "...limit the spatial spreading of local Ca^{2+} signals. This is particularly important in neurons that contain high concentrations of buffers, which are believed to ensure that Ca^{2+} signals are largely confined to synapses" (Berridge et al., 2000: 14).

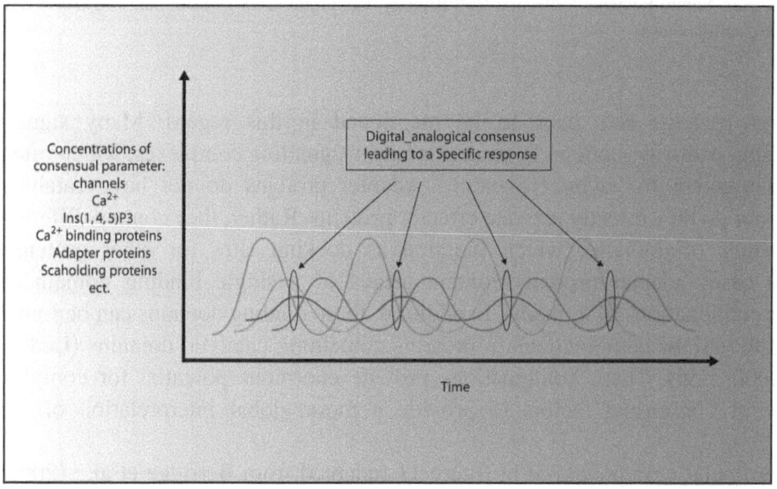

Figure 10. In the Ca^{2+} code, digital-analogical consensus requires the simultaneous occurrence of specific threshold concentrations of different elements, for example Ins(1,4,5)P3 second messenger, Ca^{2+} ions, Ca^{2+} binding proteins, adapter and scaffolding proteins among others. When the specific (digital) concentration thresholds of the various elements coincide, they form an analogical sign. In this way the Ca^{2+} signals are sensed categorically and therefore lead to a specific response

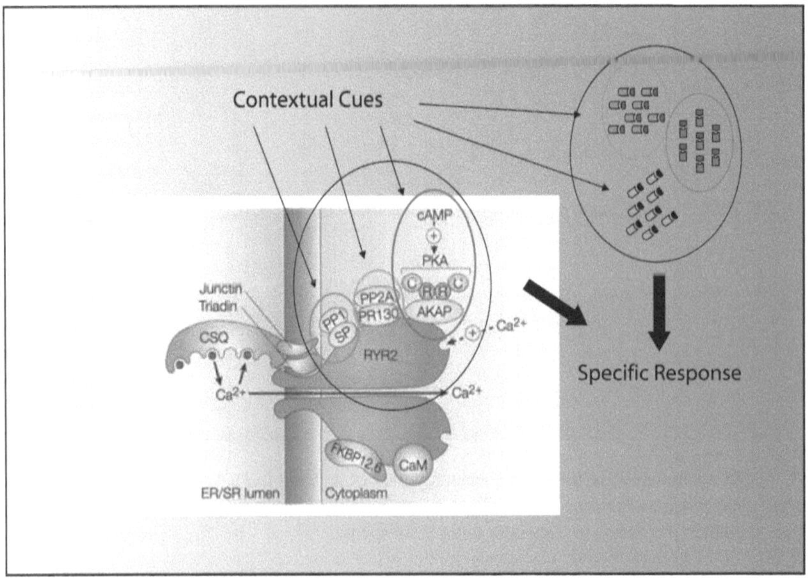

Figure 11. Digital analogical consensus for activating Ca^{2+} channels. The variety of co-factors that have to be simultaneously present imply a precise "cocktail" of signals and transforms of signals, which mirror a particular interpretation of the context, forming an analogical sign out of many coincidences of digital presences (some of which may in turn be analogues at a lower level). This particular analogical configuration for activating a cluster of channels to form a particular pattern of Ca^{2+} signals will be a digital event in a larger network (adapted from Berridge et al., (2003: 521)). Reproduced with permission from Nature Reviews Molecular Cell Biology. Copyright (2000) Macmillan Magazines Ltd. (www.nature.com/reviews)

Adapter proteins also have to be mentioned in this regard. Many signal-transduction pathways contain large multiprotein signalling complexes, which often are held together by *adapter proteins.* Adapter proteins do not have catalytic activity, nor do they directly activate effector proteins. Rather, they contain different combinations of domains, which function as docking sites for other proteins. In some cases adapter proteins contain arrays of a single binding domain or different combinations of domains. In addition, these binding domains can be found alone or in various combinations in proteins containing catalytic domains (Lodish et al., 2000: 856). These combinations provide enormous potential for complex interplay of consensual factors to provide a more global interpretation of the context.

The particular case portrayed in figure 11 (adapted from Berridge et al.,, (2003: 521)) refers to the activation of the Ryanodine receptor 2 (RYR2) for the release of Ca^{2+} in cardiac cells. The complex that constitutes an instance of digital-analogical consensus is composed of four subunits that form the channel, which is associated with various proteins that function to modulate its opening. "The endoplasmic/sarcoplasmic reticulum (ER/SR) luminal Ca2+- binding protein calsequestrin (CSQ) modulates the sensitivity of RYR2. The interaction between CSQ and RYR2 is facilitated by the transmembrane proteins triadin and junctin. The reversible phosphorylation of RYR2 by cyclic AMP (cAMP) is controlled by protein kinase A (PKA), which is composed of regulatory (R) and catalytic (C) subunits that are attached through an A kinase anchoring protein (AKAP). Dephosphorylation depends on protein phosphatase 2A (PP2A), which is attached through the isoleucine-zipper-binding scaffolding protein PR130, and on protein phosphatase 1 (PP1), which is attached through spinophilin (SP). RYR2 is also modulated by calmodulin (CaM) and by FK506-binding protein 12.6 (FKBP12.6)" (Berridge et al.,, 2003: 521).

These multiprotein signalling complexes constitute higher order specificities. A particular complex "lock-and-key" is created by which a complex configuration of concentration thresholds of signals and transforms of signals links a specific contextual demand to a specific cellular response, giving rise to the interpretant.

This principle may be behind many "auto-induction" processes in living systems, which in this way are never really "auto". They always depend on other consensus factors that modulate the "auto" effect. Otherwise auto-induction or auto-catalytic processes in living systems would be a once-in-a-life-time experience, disrupting the steady-state of the system by sending it into a positive feedback loop that could make the system collapse in the absence of any negative feedback control. This could as well add a cue to the understanding of some degenerative and pathological processes, such as abnormal cell proliferation.

The significance of certain specificity, or of a more complex "lock-and-key", lies in the triadic relation between the match of the lock, the key and the door that it opens.

13. FURTHER SEMIOTIC CONSIDERATIONS IN METABOLIC CODES

Fluctuations Versus Sustained Rise

The purpose of the fluctuations of Ca^{2+}, rather than a sustained rise in cytosolic Ca^{2+}, has not been well understood until recently. It has been suggested that one possibility is that a sustained rise in Ca^{2+} might be toxic to cells (Lodish et al., 2000: 891). On the other hand, semiotically it makes perfect sense. Fluctuations of patterns (of Ca^{2+} concentrations) are pertinent when transducing a message into a higher order code. We see here how the logical product is not necessarily quantitatively proportional to the mass that expresses it, i.e., logical products may not necessarily be formed incrementally in proportion to an increment in mass. In semiotic processes the variation of patterns is not always proportional to mass quantity. If this were the case, there would not be possibility for any kind of digitality or codification, i.e., we would be back to the world of dynamics, forces and impacts.

The fluctuations of Ca^{2+} are related to the conformation of specific signs composed by specific threshold concentrations, which cooperate in co-determining, informing and interpreting specific contexts. This fluctuation is part of a sign-network that works through a process of digital-analogical consensus. So the fluctuation is indispensable for the semiotic system. The sophistication of the code that is supported by such fluctuations and its embeddedness in a larger code gives the system remarkable semiotic plasticity for very sensible and complex fine-tuning and calibration functions.

The fluctuations are important because the different threshold concentrations at different times constitute part of a map of an equivalent territory in those precise moments. Fluctuations in thresholds are sensed as differences in the fluctuation curve. Some thresholds are important, i.e., are part of signs, and some are not, depending on the rest of the context, including other signals and other signs, i.e., consensus factors (see figure 10).

Semiotic Toxicity

If the function of Ca^{2+} is a semiotic function, it makes absolute sense that there is a fluctuation rather than a sustained rise in cytosolic Ca^{2+}. This phenomena, if understood linearly, in dyadic terms, as just a simple increase of a needed metabolite whose presence and quantity at a given moment is determined by the amount of mass that needs to be metabolized, – as opposed to Ca^{2+} that forms part of a configuration that will convey an analogical message – it would be hard to understand the sophisticated fluctuations of concentration, and why that would be the case instead of a simpler feedback curve of an average metabolite of the type: now there is too much please decrease/now there is too little please increase. This is so because we are talking about information, not about mass.

A sustained increase in Ca^{2+} probably would not be toxic in the strict chemical sense. More likely a sustained increase in Ca^{2+} would be toxic primarily because it would disrupt the communication and regulation by taking away the possibility of categorial sensing of the different thresholds that are specifically composed by cooperative components in response to a particular constellation of signals and cues. In other words, the semiotic system would be ruined by uniformity.

The Phosphorylation Code

It is not possible here to go into the details of the "phosphorylation code" but let me only enunciate its nature as a digital-analogical code for the creation of complex systems of specificities, comparable in this sense to the Ca^{2+} code with which it "cross-talks" and co-operates in a modular fashion to participate in higher emerging codes.

In this case, some effector proteins – kinases and phosphorilases – create *patterns* of phosphorylation by cyclically phosphorylating and de-phosphorylating specific residues in substrate proteins leading to sensitisation or desensitisation of cells to various stimuli. The phosphorylated form of some proteins is active, whereas the dephosphorylated form of other proteins is active. Protein kinases modulate the activity or the binding properties of one or more substrate proteins by phosphorylating serine, threonine, or tyrosine residues. On the other hand, protein phosphatases remove phosphate groups from specific substrate proteins, i.e., they de-phosphorylate them. The combined action of kinases and protein phosphatases can cycle proteins between active and inactive states.

In other words kinases and phosphorilases "sculpt" specific (digital) "differences" on their substrates, providing them with a specific (analogical) recognition pattern, i.e., phosphorylation and/or desphophorylation of specific substrates produce meaningful patterns, a compound analogical message out of different single digital phosphorilated sites. So what may change, i.e., what becomes relevant, is not the concentration of the substrate itself but the concentration of those with a specific phosphorylation pattern.

There are many possibilities for second messenger codes and the phosphorylation code to interface with each other, before, during and after the production of the second messenger and conversely before, during and after the production of phosphorylation patterns. When the concentration of the second messenger is de-coded, the message is transformed into the phosphorylation code.

Transitivity, Kinetics, Isomorphisms, Affinity, PH

The transitivity of the different concentration gradients of successive signals is of course related also to the specific biochemical rates of the different reactions that occur within the network. In this sense the biochemistry contributes to the "punctuation" necessary to convey an analogical message composed by configurations of concentrations of signals and other mediator-components. Kinetics helps to determine the time-intensity nature of the signal. The rates of certain reactions act as a timer to control the length of time of an association, e.g., molecular switches. The timing of the event, the duration of the signal, based on the kinetic rate is an analogical message, more complex than the mere digital presence or absence of the signal. To the analogical composition of the signal we have to add the specific concentration threshold necessary for the signal to be "meaningful" as well as the rest of the consensus cofactors that integrate a complex sign, which is what assures the proper "categorial sensing" of what otherwise would be ubiquitous or meaningless signals.

The system is organised in such a way as to take advantage of these given physical restraints in order to incorporate them into a functional code. The same can be said about isomorphisms, affinity (specificity), modularity, PH tolerance and other structural restraints, that rather than determinants, are structural features that can be incorporated to articulate active codes through systems of correspondences. For example, there is a relation between affinity (Kd) and concentration in signalling systems. Concentration thresholds, which are part of complex cellular signs, are influenced by the signal-receptor affinity (which can also be an analogical variable), as well as by any habitual presence of signal-analogues (agonists and/or antagonists) in the system. Furthermore, different PH domains may modulate binding specificities in some molecules. These are some of the first rules implicit in metabolic codes that have been widely recognized by biologists (Bruni, 2003).

14. FROM SYSTEMS BIOLOGY TO SYSTEMS OF CORRESPONDENCES

With the advent of Systems Biology the multidimensional nature of protein functions in time, space and context has been recognized as a major problem in biological research. Protein functions can vary with developmental stage, anatomical location, and environmental context. Systems biology has recognized that the various ways in which proteins interact with each other are often not evident in their genes. For example, when, where and the extent to which a particular protein is chemically modified in signal transduction cascades is not part of any genetic program, although the potential for modifications and the recipes for proteins needed for this activity are (partly) encoded in DNA. This is why instead of on individual genes or proteins, the emphasis should be put on *"systems of correspondences"* as the chief co-evolutionary units (Bruni, 2003).

Systems of correspondences constitute complex "lock-and-key" mechanisms which guarantee the proper match between a complex set of contextual cues and the appropriate cellular, organismic or ecological response.

What are evolving today (and since a long time ago) are not single entities but entire complexes of sophisticated networks at all levels. What are informed by the genome are integrated systems of functional domains which constitute elemental units for a great diversity of emergent codes. The different functional domains in a single protein allow its interaction in and with different directions of the network and with different actors of the system. Each functional domain represents a correspondence with other domains distributed in the products that are coded in the genome, as well as correspondences with products coded in or by the environment, including organisms of the same or different species (see figure 1).

For example, a homologous phosphorylisable sequence (i.e., a sequence which is susceptible to being phosphorilated in a particular residue) can be encountered in different proteins combined with a variety of other domains that give its particularity to the protein. In this way, the code of phosphorylation is distributed in the whole system. These correspondences at the level of functional domains are what are actually coded in the genome i.e., networks of correspondences, which are used to constitute metabolic codes; not complete complex phenotypes. What are coded in the genome are the elemental units of specificity, which are used and arranged modularly in the distributed network, as well as the "recipes" for successful structural elements.

Part of the arrangement is implicit in the complex architecture of the genome. But the model for integrating circuits must be an analogue implicit in the *embryonic signalome* (Bruni, 2003). The analogical "know-how" to ensemble and differentiate systems of correspondences must be inherited in the embryonic signalome. Once cells start dividing, the new cells get both the library and the whole system of interpretation. During differentiation the library remains the same, therefore differentiation starts by changes in the signalome. In order to start development the fertilized egg-cell must be able to decipher the DNA-code as well as many other signals that are already present in its inner and outer context. As pointed out by Hoffmeyer and Emmeche (1991: 127), this need for the participation of cellular structure shows us that a sort of 'tacit knowledge' is present in the egg cell. In other words, the "tacit knowledge" must stand in a system of correspondences with the genome architecture and many of the potential signals present in the context. The extent of this tacit knowledge is still vague and hard to evaluate, but at least it has to include positional information, embryonic "structural templates" that may set the orientation of certain epigenetic trajectories and above all it must include a sort of essential or "embryonic signalome" that is already functional in interpreting the configuration of signals that start fertilisation and which selectively operate on DNA to start differentiation and development of the (inherited) embryonic signalome itself and consequently of differential use of DNA by the organism (Bruni, 2003).

The combinatorial possibilities of domains constitute complex codes with different infrastructural organisation and mechanisms but which share common logical principles. In this view, DNA is a library of distributed architectures of integrated systems of corresponding (specific) sequences: the emergent digital units of the DNA code. The sequences or domains – be they binding sites, integrating repetitive motifs, protein domains, or regulatory sequences, etc. – are used modularly within systems of correspondences and specificities that reach beyond the organism into its niche.

The evolution of hierarchical specificities requires an evolutionary mechanism that is not so much based on single genes but on modular components of systems of correspondences. Certain forms in the context pose a question, so to speak, of which the emergent component is an answer, a "functional" idea. This can be appreciated in the evolution and development of many specificities, for example antibody and antigen in mammalians, or avirulence factors and response-determinants in plants. What evolves and develops are systems of correspondences. What survives are "systems of ideas in circuit".

15. CONCLUSIONS AND FINAL REMARKS

In line with Tomkins' intuition about the generality of a metabolic code in which a particular state of affairs or environmental condition is correlated with a corresponding set of intracellular signs (Tomkins, 1975), we could argue that the generalisations and examples proposed in this chapter could then be extended to other instances of the "signalome".

When we say that elemental signals constitute more complex signals – which function within an increasing complexity of patterning discrimination – what we really are talking about is the emergence of complex signs and patterns out of signals that have to be recognized in such patterns.

There are plenty of examples that show us how virulence and pathogenesis are context dependent. A virulence factor, and/or a pathogenic organism, may not be such if not in a specific context.That context is a semiotic niche full of signs, some of which trigger virulence out of an otherwise "neutral" factor or organism. We could say that there are no pathogens but pathogenic circumstances.

In a sense it is the context that becomes pathogenic and at the same time it becomes ill. Since the context is constantly changing so is the semiotic niche of a particular system.

Identifying the proper digital-analogical consensus instances in hierarchical systems – i.e., when particular configurations and patterns constitute specific signs that relate a particular contextual situation to a corresponding response – may have important implications for the understanding of pathogenesis. Similar generalizations can be made in the interrelation of hierarchical systems such as the immune and nervous systems in somatic, neurophysiologic and cognitive processes.

With the examples presented here it could be argued that certain pathological processes can be viewed as a communicational dysfunction at a given hierarchical level. The interesting question would be how "information" gets across levels and how a context can become pathological, how the system misinterprets a given complex pattern of signals prompting a pathological response.

NOTES

[1] It is useful to point out what in this framework could be the difference between a signal and a sign. Signals and signs imply an understanding of causality which is different from the physical causality of particles' impacts and energy exchanges. In natural systems, both signals and signs have to be sensed by a living system. Otherwise they are neither signals nor signs but just impacts and energy exchanges. The causality generated by the response to a sensed difference and not merely by the response to an impact will be referred here as "triadic causality" and this is precisly the logical understanding that the concept of sign brings to the scene. A signal is one type of sign, but not the only kind of sign active in living systems in general or in cellular systems in particular. A signal by itself can be a sign, but this is probably the most elemental level of a sign. Simplifying we could say that a signal, creating an elementary *sensed difference*, is a digital message while a sign could be considered as an analogical message - a more complex aggregate of signals. Thus, digital signals can work as signs and in turn complex signs may also have a digital effect that makes them function as elementary signals. In any case what differentiates signals and signs from *impacts* and *energy exchanges* is their triadic nature i.e.: the fact that rather than an "impulse", a signal transmits "news of a difference".

[2] A similar concept, a "signalsome", has been defined in Berridge et al., (2003: 525) as "The collection of components that constitute the different signalling pathways found in specific cell types".

[3] For practical reasons it is useful to make such a distinction between "dyadic causality" and "triadic causality", which derive from Peirce's semiotics (although it has to be said that in Peirce's writings the distinction is made between dyadic and triadic "action"; but we could without major problems substitute the word "action" by "causality" in order to be consistent with other conceptions of causality that may be integrated into a system of explanations). In this view, the Aristotelian categories of material and efficient causalities are considered to correspond to "dyadic action" while the categories of formal and final causalities are reduced to "triadic action". I will assume this practical modern re-framing of

the Aristotelian notion of causality as a useful epistemological tool for biology, and I will leave the Aristotelian formulation open to the more ontological kind of arguments (Bruni, 2003).

[4] The distinction between analog and digital codes is not a simple one. For a wider treatment of this distinction see Bateson, (1972, 1979); Wilden (1980); Heims (1991); Hoffmeyer and Emmeche (1991); Hoffmeyer (1996). The distinction depends on the hierarchical nature of contexts (and thus the existence of meta-contexts) in the multidimensionality of semiotic processes. According to Bateson (1979: 249) "A signal is *digital* if there is discontinuity between it and alternative signals from which it must be distinguished. *Yes* and *no* are examples of digital signals. In contrast, when a magnitude or quantity in the signal is used to represent a continuously variable quantity in the referent, the signal is said to be *analogic*". In digital systems response is a matter of "on-off thresholds". In analogic systems response is *graded* (i.e., varies continuously) according to some variable in the trigger event (Bateson, 1979: 122–123).

[5] It is worth to spend a few lines to make clear in what sense this notion is being understood here. "The interpretant" is the regularity by which the sign-vehicle links a particular object (e.g., a particular contextual demand, a necessity, a stress, a state of affairs) to a specific effect or response. The sign-vehicle (e.g., a signal-molecule, a blend of signals, etc.) acts as a mediator between the object to which it refers (a particular aggregate of contextual parameters) and the effect that such a sign-vehicle (and indirectly the object) produces on a system (or on a stage in the process of development of that system). The sense in which I will be using the term does not imply an autonomous entity of any kind. It can be rather viewed as a level of integration. The easiest way to grasp the meaning of the concept is to think about it as *the level* at which a complex configuration of signals or signs makes a difference to some living process or entity. But the emerging interpretant is not the entity or the process itself. In practical terms, from the observer's point of view, it is a focal level that for the purpose of our analysis we can identify as the point or the moment of convergence for different kinds of factors that acting together "select" a direction for the whole system. It will be an emerging interpretant if the resulting action is a pattern or a habit that can be observed regularly in relation to a particular state of the context. I put the emphasis on the emergent characteristic of the interpretant because it has to be considered as something that is generated at every second in a continuous basis. Actually the formation of an interpretant is the creation of meaning (and/or function) itself. "We cannot directly observe the interpretant according to which a living system codes effects of the environment on its receptors into signs. We have to infer it from the system's behavior; we have to interpret endosemiosis... by reconstructing their 'history': we take the last act or behavior of the living system as an indexical sign pointing to the interpretant which, as the coding instance, has assigned to the sign the meaning it has with regard to the system" (von Uexküll et al.,, 1993: 15). The emerging interpretant can be seen as the locus at which a goalseeking system (which can be contained in a larger system) defines its goal. It is that part of the emerging system that achieves a higher logical type of manifestation with respect to the dynamics that underpins it (Bruni, 2003).

[6] From the *Webster's Encyclopedic Unabridged Dictionary*:
"Blip": a spot of light on a radar screen indicating the position of a plane, submarine or other object.
"Quark": one of the elementary particles believed to form the basis of all matter.
"Puff": an abrupt emission of air, vapor, etc. A short, quick blast, as of wind or breath.
"Spark": ignited or fiery particles thrown of by burning wood. The light produced by a sudden discontinuous discharge off electricity.

[7] In eukaryotes, a regulon is usually defined as a genetic unit consisting of a noncontiguous group of genes under the control of a single regulator gene. In bacteria, regulons are global regulatory systems involved in the interplay of pleiotropic regulatory domains and consist of several operons.

The regulon is defined as the entirety of all genes regulated positively (enhanced transcription) and/or negatively (reduced transcription) by a specific common regulatory factor. On the other hand, the "stimulon" is defined as the entirety of all genes responding to a specific external signal (*stimulus*) by increased or decreased transcription. A second messenger is actually closer to being the stimulus of a stimulon than the regulator of a regulon, fact which emphasizes the problem of categorial sensing.

REFERENCES

Baldwin, Ian T; Halitschke, Rayko, Kessler, Andre; Schittko, Ursula (2001). Merging molecular and ecological approaches in plant-insect interactions. *Current Opinion in Plant Biology*, vol. 4, no. 4: 351–358.

Barbieri, Marcello (2003). *The Organic Codes. An Introduction to Semantic Biology.* Cambridge University Press, Cambridge, UK.

Bateson, Gregory (1972). *Steps to an Ecology of Mind.* New York: Chandler Publishing Company.

Bateson, Gregory (1979). *Mind and Nature. A Necessary Unity.* New York: Bantam Books.

Bateson, Gregory and Mary Catherine Bateson (1987). *Angels Fear. Towards an Epistemology of the Sacred.* New York. Macmillan Publishing Company.

Berridge, Michael J.; Lipp, Peter and Bootman, Martin D. (2000). The Versatility and Universality of Calcium Signalling. *Nature Reviews/Molecular Cell Biology*, Volume 1, October 2000, pp. 11–21.

Berridge, Michael J.; Bootman, Martin D. and Roderick, H. Llewelyn (2003). Calcium Signaling: Dynamics, Homeostasis And Remodelling. *Nature Reviews/Molecular Cell Biology*, Volume 4, July 2003, pp. 517–529.

Bruni, Luis E. (2002). Does "quorum sensing" imply a new type of biological information? *Sign Systems Studies*, vol. 30.1: 221–243.

Bruni, Luis E. (2003): *A Sign-theoretic Approach to Biotechnology.* Ph.D. Dissertation, Institute of Molecular Biology, University of Copenhagen.

Carafoli, Ernesto (2003). The calcium-signalling saga: tap water and protein crystals. *Nature/Molecular Cell Biology*, Volume 4, April 2003, pp. 326–332.

Emmeche, Claus (1994). The computational notion of life. *Theoria-Segunda* Epoca 9(21):1–30.

Emmeche, Claus (1998). Defining life as a semiotic phenomenon. *Cybernetics & Human Knowing*, Vol. 5, no. 1, pp. 3–17.

Emmeche, Claus (1999). The Sarkar challenge to biosemiotics: is there any information in a cell? *Semiotica* 127 - 1/4: 273–293.

Emmeche, Claus and Jesper Hoffmeyer (1991). From Language to Nature: The Semiotic Metaphor in Biology. *Semiotica* 84(1/2): 1–42.

Heims, Steve J. (1991). *The Cybernetics Group.* Cambridge, Massachusetts: The MIT Press.

Hofffmeyer, Jesper (1996). *Signs of Meaning in the Universe. The Natural History of Signification.* Blomington: Indiana University Press.

Hoffmeyer, Jesper (1997a). Biosemiotics: Towards a New Synthesis in Biology. *European Journal for Semiotic Studies* 9 (2): 355–376.

Hoffmeyer, Jesper (1997b). Molecular Biology and Heredity. Semiotic Aspects. In: Uexküll Thure von (ed.) *Psychosomatic Medicine.* Munich: Urban & Schwarzenberg, pp. 43–50.

Hoffmeyer, Jesper and Claus Emmeche (1991). Code-Duality and the Semiotics of Nature. In Anderson Myrdene; Merrell Floyd (eds.), *On Semiotic Modelling.* New York: Mouton de Gruyter, pp. 117–166.

Lodish, H., A. Berk, S.L. Zipursky, P. Matsudaira, D. Baltimore and J. Darnell (2000). *Molecular Cell Biology.* New York: W.H. Freeman and Company.

Sarkar, Sahotra (1996). Biological Information: a Skeptical Look at Some Central Dogmas of Molecular Biology. In: Sarkar Sahotra ed., *The Philosophy and History of Molecular Biology: New Perspectives.* The Netherlands: Kluwer Academic Publishers, pp. 187–231.

Sebeok, Thomas A. (1985/1976). *Contributions to the Doctrine of Signs.* Lanham: University Press of America.

Shapiro, James A. (1999). Genome System Architecture and Natural Genetic Engineering in Evolution. In: Caporale, Lynn Helena (ed.), *Molecular Strategies in Biological Evolution.* New York: Annals of the New York Academy of Sciences (Vol. 870), pp. 23–35.

Tomkins, G. M. (1975). The Metabolic Code. *Science* 189, pp. 760–763.

Stjerfelt, Frederik (1992). Categorial Perception as a General Prerequisite to the Formation of Signs. In: Thomas A. Seabeok and Jean Umiker-Sebeok, eds., *Biosemiotics: The Semiotic Web 1991.* Berlin: Mouton de Gruyter.

Uexküll, Thure von; Geigges, Werner and Herrmann, Jörg M. (1993). Endosemiosis. *Semiotica* 96–1/2, pp. 5–51.
Wilden, Anthony (1980/1972). *System and Structure: Essays in Communication and Exchange.* New York: Tavistock Publications.

CHAPTER 16

INNER REPRESENTATIONS AND SIGNS IN ANIMALS

STEPHEN PHILIP PAIN

Kongensgade, 15 1-sal, 5000 Odense C. Denmark, biorhetorics@mail.dk

Abstract: At the beginning of the twentieth century, behaviourists like John B. Watson (1878–1958) changed the focus of attention from the inside of the brain (mentalism and introspection then being the main trend in psychology at the time) to the outside (Watson, 1913). They believed that we could learn nearly everything about animals and humans by studying their performance in learning experiments, and this was both measurable and verifiable. Today in the first decade of the twenty-first century, there has been a return to the inside. The neurosciences seek physiological explanations and connections between external behaviour and the neural mechanisms within the nervous system. With the revolution in magnetic resonance imaging (MRI) technology researchers are now able to visually represent neural activity. Other researchers have developed mathematical models and programs to visualise the patterns created in the periphery prior to central integration

The author in this paper would like to distinguish these descriptive forms of representation from actual representations, i.e., those of which the animal is actually aware or conscious. Why does an animal sometimes make perceptual mistakes? (Case Study I "The Turtle and the Plastic Bag"). Is there more to dispositions? (Case Study II: "Taking Representation for a Walk. Argos and the Fake Daniel Dennett"). How is prey represented to an animal? (Case Study III "Representation of Prey in the Jellyfish/Herring Predator-Prey Dyad"). Does a simple animal feel pain or suffer? (Case Study IV: A Can of Worms. The Earthworm as Bait)

It will be argued on the basis of contemporary biosemiotic research that animals (including both vertebrates and invertebrates) represent environmental information internally, and these representations can be subdivided into i.) primary or peripheral representation and ii.) central representation which are quantitative and qualitative respectively. Sensory information is conveyed via signals, these are received as stimuli then transduced into internal signals (see *Theoretical Framework*). At this stage the animal is not aware of the quality of the information as it has not yet been integrated or processed in a ganglionic complex. One can describe the properties of this pre-integrated information as quantitative and syntactical i.e., spatial and temporal ordering of incoming signals and their relations. The sign which is the smallest unit of qualitative representation arises only after integration of information from two or more discrete sensory modalities. These findings have repercussions for current models of animal learning and behaviour,

M. Barbieri (ed.), Introduction to Biosemiotics, 409–455.

especially in lower invertebrates (the principal subject of this paper); they also challenge the development of robots based on so-called simple systems

Keywords: biosemiotics, inner representation, sign, animal behaviour & cognition, predator-prey dyad, robotics

INTRODUCTION

In a peer recognised book on communication in the *Perspectives in Ethology* series (Plenum Press 1973–) there is only one reference to semiotics and this is by the volume editor, Donald H. Owings, and Eugene S. Morton:

> Information has appealed to contemporary biologists because of its connection to two bodies of theory. Engineering Theory (Shannon and Weaver, 1949) defined information as reduction of uncertainty and measured it in terms of binomial units of choice called bits [...]. (Owings & Morton, 1997, 360)
>
> Semiotic Theory defined information as a property that permits organisms to choose among courses of action. In its appeal to messages, rather than bits of information, the semiotic treatment came closer than the engineering approach to accommodating such concepts as meaning or knowledge that are intuitively associated with information in ordinary language bits [...]. (*ibid.*)

According to the authors, biologists have "combined these two approaches" obtaining the best of both worlds, the "intuitive appeal of commonsense" and the "rigor" of mathematics, under the heading of "information perspective" (*ibid.*) This is not quite true, because the "Engineering" theory clearly dominates both the biological sciences and the neurosciences. The principal references to meaning in the same book are in an "outrigger" paper devoted to speech acts and signals. (Horn, 1997, 352–353). There are many reasons why one theory is applied more than others, but one suspects that it has much to do with the nonparsimonious nature of semiotics, the lack of testability and predictability. In this paper I shall prove this wrong by using a biosemiotic approach to "real" problems. This has required an overhaul of the "linguistic" baggage of biosemiotics, and a very necessary refinement of its parameters. I have covered this in Part Three: The Theoretical Framework. My starting point was paradoxically an engineering one. I realised that a hierarchical model used in bioengineering would be useful for establishing the working level of biosemiotics. If we look at a communication situation, say between two birds calling each other, we will see that an information perspective will describe and predict the flow of information between the two, but stop short of following the process further, into the heads of the birds. In an earlier book on animal communication, W. John Smith referred to how syntactics was like the Cheshire cat's grin without the Cheshire cat. (Smith, W. John, 1977, 18). Similarly a neuroethological approach that takes up the process at this juncture will follow the route from reception, transduction to the neural mechanisms involved, but as it is primarily cellular or neuron determined, the approach does not provide an understanding of "How a bird experiences a song." (Zupanc, 2004).

My theory of biosemiotics then intends to address this lacunae in the sciences – by placing qualitative representation into a biological context. I believe that qualitative representation requires a neurophysiological structure/process for its production. A sign in this respect can be defined as the sum of sensory afferents integrated within a ganglionic complex (or greater). I distinguish this from a signal that is an environmental phenomenon (except signals within conventional semiotic systems such as gestures). A signal is syntactical. Again this is not a linguistic-semantic conception of syntax, such as word order, it is to do with sequence and firing order within a spatial temporal frame. A signal is a peripheral quantitative representation. A sign on the other hand is an integrated unit of a central qualitative biological representation. As we shall see my research into the simplest nervous system (the cnidarian) has revealed that at the very beginning of qualitative representation there is a modularization of sensory information, and more importantly a grounding of information within a navigation module – a triangulation if you will (see Case Study III).

The principle part of this paper is taken up with the problem or possibility of "inner" representation in invertebrate cognition; a field which if we are to go by the number of internet hits is a subdivision of animal cognition still in its infancy. Part one is concerned with vertebrate cognition: with a current environmental problem – the ingestion of plastic by sea turtles, this is followed by a discussion of representation and dispositions to respond in dogs. The next part concentrates on invertebrate cognition; there are three sections including two case studies: i.) the representation of prey in the jellyfish/herring larva predator-prey dyad, and ii.) representation of pain in earthworms. In each of these cases I have focused on the integrative nature of sign production, and based my hypotheses on individual interactions. The natural sciences have emphasised the predictive element of scientific practice, but often the models and programmes are based upon extrapolations drawn from aggregate behaviour, this is especially true in lower invertebrates which are multitudinous. Charles S. Elton in his classic *Animal Ecology* (1971; orig. 1927) describes the "pyramid of numbers" based on the food-chain and sustainability:

> [...] the animals at the base of a food-chain are relatively few in numbers, and there is a decrease in between the two extremes. (ibid., 69)

For Elton the small herbivores are populous and form the "key-industries" which are "able to provide a large margin of numbers" while the larger carnivores are supported by this, their "smaller margin" results in a smaller increase. Natural relations and cycles are today often conflated with those found in economic sciences. A textbook on biological oceanography demonstrates this admirably. The first two chapters deal with the spring bloom and physiology of phytoplankton, even at this stage we come across "stocks", but it is in the third chapter we are told without reference to its economic parentage:

In the trophic-dynamic approach to ecosystems. We try to measure the production at each trophic level. "Production" is incorporation of new organic matter into living tissue, that is, biomass elaboration. (Miller, 2004, 40)

Compare this with the definition of production in an economics textbook:

Production can be defined as the creation of wealth which, in turn, adds to society's welfare. (Hardwick, P. et al., 1994, 19)

Other analogies might be the nitrogen and carbon cycles with the "*trade cycle: slump, recovery, boom and deflation*. (*ibid.*,432). In the case studies I have focused on individual behaviour and interaction at the expense of the "big" picture, and when dealing with lower invertebrates I have turned Elton's numbers pyramid on its head and treated these animals as one would deal with mammals or a single shopper. This is a deliberately nonparsimonious approach. Here one can seek support for this approach in sensory ecology as developed by David D. Dusenberg (1996) which emphasises the organism's sensory apprehension of its environment; the work of John M. Fryxell and Per Lunberg (1998) on individual behaviour ecology and its interaction with organisations. Instead of dealing with large populations, we can examine the individual relationships involved in for example the prey-predator dyad. Similarly we might find complexity in what had hitherto been reduced by economic models which worked with the dynamics of large populations, this is particularly true of navigation studies. It could be said that by inverting the pyramid we can give animals considered simple more attention. I have in short taken the ontogeny of all life forms seriously. In each of my case studies, I have taken the approach, irrespective of whether they are vertebrates or invertebrates, that they are all individual organisms capable of qualitative representation. More research and experimentation need to be carried out on animal learning in general. This is very true of lower invertebrates as some of the key learning experiments were conducted nearly a hundred years ago. I believe that while sensory ecologists and biologists may tilt at the reductionist economic models used for prediction of predation, distribution, etc., they falter at the door of cognition. In an excellent review of wolf communication, Fred H. Harrington and Cheryl S. Asa, conceded:

The integration of signaling systems has not been studied in wolves. (Harrington, Fred, H. & Asa, Cheryl S. (2003)

All animals integrate information from many sources. Yet we often choose to focus on one modality without synthesising the data from other fields. The consequences of this is unevenness in results. Several major hypotheses await experimental testing. For example the theories regarding salmon homing. If we gain a better understanding of the way an animal produces qualitative signs that go to make representations, we will also realise that their learning ability is generally greater than assumed. Cognitive ethologists have started to take invertebrates seriously, though it has mostly been the "sexier" arthropods and cephalopods, and approached these animals with nonparsimonious higher-order methodology (Burghardt, Gordon M. 2005, 359). Animals like non-colonial crickets have been found capable of "learning" from other conspecifics to avoid a predator (Coolen et al., 2005); paper wasps can recognise each other's markings, and can discriminate human faces; octopuses might be capable of "play", and so on. These reassessments have been assisted by new forms of representing information in the brain (imaging technology). However it should be emphasised that these representations (images) are not the actual neural activity, but blood-flow and there are problems with the usage of imagining. It is a question of balancing the predictive and nonpredictive. The anecdotal approach of Sir John Lubbock who kept and studied a pet wasp, a eusocial insect, seems to be far removed from the present goals of science. But the extravagant computer programmes used to predict swarming behaviour also seems to have inherent problems. A tendency to go over board with numerical modelling has been critiqued in a number of sciences. A leading figure in his field of biological oceanography, Charles B. Miller while recognising the value of models cautions about the neglect of observation. (Miller, 2004, 91).

PART ONE: VERTEBRATES AND REPRESENTATION

CASE STUDY I: The Turtle and the Plastic Bag

When I was younger I remember my mother setting the table with a brand new plastic tablecloth – I remember it vividly because the smell of vinyl chloride made me feel nauseous. Ever since that eventful day I realised that there must be something amiss about plastic. Now in the twenty-first century many decades after plastics were first used, the environment is quite literally saturated in the stuff. The US alone in the late 1980's produced 30 million tons of plastic debris annually, much of which is still around as it is not biodegradable but photogradable, and although it will eventually break up in time, it has a life-span of 1000 years! (Thompson et al., 2004; Moore, 2003, Derraik, 2002, Moore et al., 2001) Even the biodegradable plastics contain material which is harmful to the environment. While many of us are aware of the dangers to young children posed by plastic bags (they may suffocate), and to terrestrial wildlife in the countryside, it may come as a shock to know the true extent of pollution caused by plastics in the sea. A 100 billion polyethylene bags are produced annually – of these a fair number end up in the sea. On a micro-level the sea is in effect a "plastic soup" and an analysis has shown the ratio is 6:1 (6 – plastics; 1 – plankton). Tiny particles are ingested by all life forms:

> The closest interaction of an organism with its environment is the ingestion of a subset of that environment and the subsequent alteration and absorption of that subset as it passes through the digestive tract of the organism. Bjorndal, 1997, 199.

The impact of these particles on marine life is devastating, and if not immediately life threatening, are sublethal creating untold agony and discomfiture, and impacting upon reproduction. Larger fragments or plastic objects have obvious consequences (Orenstein, 2001, 270, Lutcavage et al., 1997).They can kill the largest marine mammals who ingest them because of their appearance or get entangled in them – they fool birds who think they look like squid or fish, so much that they try to feed parts to their young. However it is the transparent plastic bags that have caught the attention of wildlife conservationists because the transparent bags are ingested by endangered sea turtles who mistake them for gelatinous invertebrates (true jellyfish, comb jellies, sea squirts and salps) which make up a substantial part of their diet. (Bjorndal, 1997, 209) Picture turtle hatchlings fresh off the beach, first waddling off to the sea edge, then plunging into the water, swimming frantically and continuously for hours, and in this frenzy ingesting small prey as they go. During the pelagic stage of their lives, the majority of turtles are carnivorous, before becoming herbivore specialists in algae or sea grass. The turtle has a sensory system attuned to an aquatic lifestyle: colour vision, eyes that can discriminate objects, ears that pick up sounds in the water, and noses dedicated to certain scents, pheromones, perhaps natal sites, and prey attractants. Imagine the hydrodynamics involved, the sensations, the expectations, and a hunger driving the hatchling and juveniles to feed – and how these contribute to mistaking plastic for jellyfish. (Bartol & Musick, 2003). Tragically the largest turtle, the leatherback has evolved to subsist on an almost exclusively jellyfish diet, mixing it with tunicates and salps, all with low nutritional value (Bjorndal, 1997) This means a leatherback has to consume numerous jellyfish to maintain its body weight and temperature (it has a high metabolism). One could with this knowledge appreciate what happens when it comes across floating bags.

What has this all to do with biosemiotics? Rather than asking this question we might ask, how can biosemiotics assist researchers? Firstly a descriptive form of biosemiotics can help in structuring the causal relations between the biotic and abiotic factors. A visual and logical depiction of the relationships involved in plastic ingestion within the marine ecological systems, can perhaps be more effective in educating a public about the problems than plain statistics, it can help in power-point presentations intended for funding administrators. Biosemiotics can also help in the preparation of a phylogenetic or cladistic presentation of the relations between the animals and the plastics. For example, while leatherbacks regularly feed on jellyfish, other turtles like green turtles may consume them in the absence of their main food sources. Out of 38 dead green turtle guts examined 23 in a study contained anthropogenic debris including plastic bags (Bugoní, Krause, Petry, 2001). Today because of the depletion of fish stocks, more carnivores have had to rely on the jellyfish as part of their diet. This increases the possibility of plastic ingestion, as jellyfish unfortunately ingest minute parts of plastic in the course of feeding, as do many other macroplankton life forms. So there is the very real likelihood that a leatherback would be susceptible to two forms of danger, one immediate, such as choking, or restriction of the stomach (by ingestion of plastic bags), and the other long-term: poisoning through ingestion of particles that have accumulated high levels of DDTs and PCBs. What can biosemiotics do here? I think one possibility is to develop experiments to discover why it is that an adult turtle mistakes a plastic bag for a jellyfish. What is so attractive about a plastic bag floating in water? Let us suppose we look at it in terms of marine habitats. If it is on the surface the bag would be considered neustonic. Off the coast it will

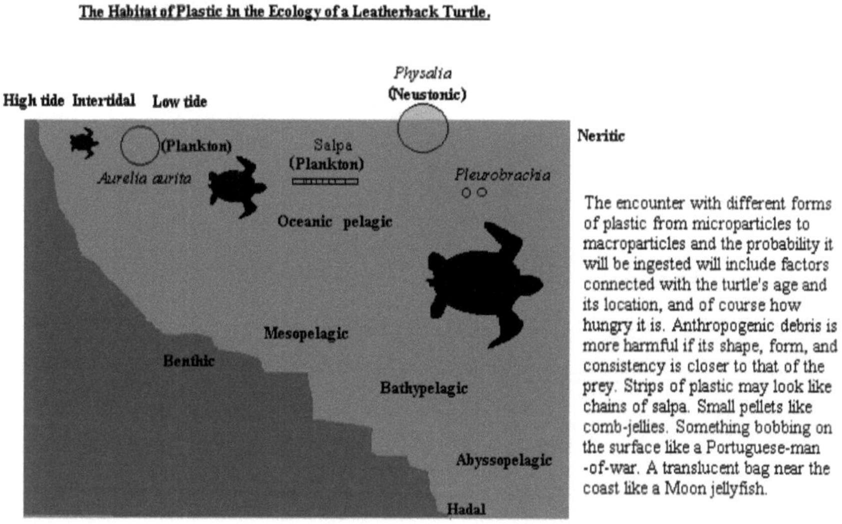

Figure 1. Habitat of Plastic. based on diagram of habitats by Levinton, 2001 and on Bugoní, Krause, Petry, 2001

be subject to local tidal harmonics, the water motion will be different according to configurations (sun, moon) and physiological features. Its movement will be greater of course at the sea's edge. While out at sea in deeper water, the natural waves (nontidal) such as gravity and capillary waves will cause the bag to move differently. We might also add factors such as geographical, climatic and so on. The motion of a bag (if full of water) drifting on the surface might look to a leatherback like the zooid float bags belonging to the Portuguese-man-of-war, a siphonophore and member of the phylum cnidaria (a phylum to which true jellyfish belong). This mistake would be more probable if the leatherback encounters the bag while swimming on the surface. Alternatively the same bag in shallower water but under the surface may be mistaken for the moon jellyfish. Strips of plastic may be taken for salps.

Among the attributes that may persuade a leatherback turtle it has encountered a real jellyfish would be:

- Visual – shape, colour, light (features)
- Acoustic – sound of water flow (propulsion)
- Olfactory – water already conditioned by jellyfish
- Gustatory – consistency (gelatinous, mucus or film of bacteria on surface)
- Tactile – gelatinous, slimy

We might also be interested in spatial and temporal data. For example at what distance would a turtle orient towards a plastic bag. What are its usual foraging habits and range? In animal behaviour and communication studies, economic and information models would be used. But what about the representation of the pseudo-prey? In other words how is a plastic bag represented internally? Suppose the turtle

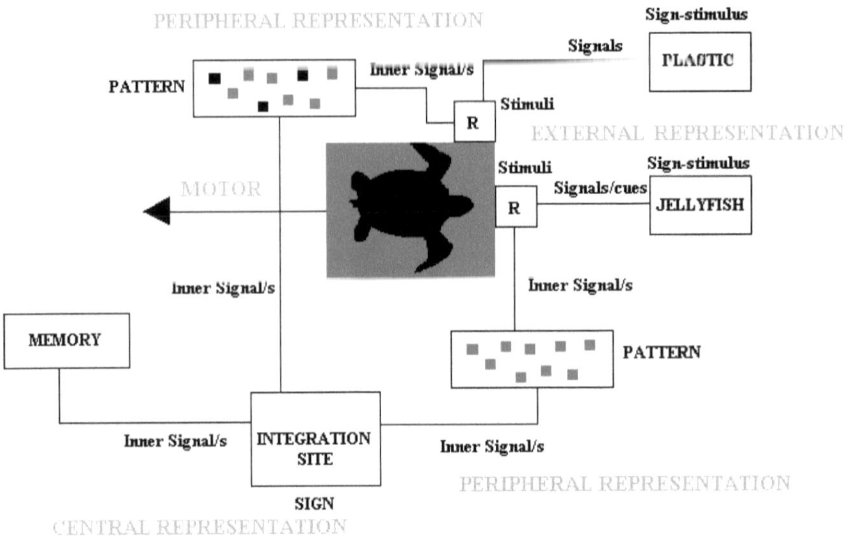

Figure 2. Representation in the Leatherback Turtle

in a laboratory tank was presented with a choice between a bag and a jellyfish. Could it tell the difference or learn to discriminate? These are questions that concern cognitive scientists. Neuroethologists on the other hand seek to tease out the quantitative relations between the motor output (orientation, hunting, feeding) and the neural mechanisms that are connected to them. How can a biosemiotician be useful here? If we opt for a mentalist explanation of a behavioural output, then the sign is a component of a representation. Two sets of sensory information (say the visual and acoustic signals) are integrated into a qualitative sign within a central processing area. Here it is important to emphasise the presence of levels in communication and interaction. We can begin by analysing the information emanating from the plastic bag. Is it a sign-stimulus? When the turtle first makes contact with the bag, does it release a feeding algorithm? Perhaps. However prior to this, one can follow the course of the sensory information (as signals) which are then received by their respective receptors, transduced, converted into internal impulses or signals, and then integrated. Which one of these pathways is prioritised? Certain features or patterns will trigger off a motor output. What are these? If we look at the visual pathway we could measure the firing patterns within in the retinal area and correlate this with the activity within the central processing areas in the brain.

We could then calibrate what are the sensory afferents required to produce a representation of a jellyfish and match these with the plastic bag. We might then discover from this comparison a means by which we can persuade turtles not to ingest plastic bags. In the above diagram it is speculated that although the patterns are not equivalent, the turtle's sensory system is fooled. Of course the other factors such as velocity, light, waves, hunger and so forth should all be combined in the experiment. Notice the types of representation, and that the properties of communication/interaction are quantitative until the site of integration. One could conduct this cognitive and biosemiotic experiment using a leatherback in a tank with non-invasive scanning technology that can measure the level of excitation within regions of the brain when shown its prey, a plastic bag, and a control. The work of Charles L. Anderson et al., (2000) on garter snakes serves as a model for this experiment. Unfortunately in private correspondence (Feb. 2006) with one of the authors, Gordon M. Burghardt I have been informed that the equipment is not currently available.

CASE STUDY II: Taking Representation for a Walk. Argos and the Fake Daniel Dennett

A dog while out for a "walk" will zigzag across a territory, sniffing here and there, and if male invariably cocking a leg to urinate. But why does he choose a particular place to urinate? Dogs will like other canids often mark or erase territorial borders. When they urinate they are stating that "they have been here" – and this includes information about their sex and status. We can see these marks clearly in the snow. Those yellowish holes with grass showing through. That is a public representation – a sign. It is a sign from our point of view, because we think in linguistic terms. But is it a sign for the dog? How is it received? The dog moves along following streams of scents. Once it has picked up a scent that has gained its attention (food, mating, territory, The dog follows an unseen olfactory plume in what could be called an olfactory landscape. From our point of view we have to be quite close to observe the "mark." A dog can pick up this conspecific "**signal**" from a distance. However the act of "cocking a leg" known as "raised-leg urination" (RLU) (Berkoff, 1979) happens when the concentration is at its highest. It is then the **mark** (a **sign** in the descriptive sense) becomes a **sign-stimulus** that releases a behavioural response. Applying my approach to this case study we would use the Shannon & Weaver 1949 model to measure the information flow in quantities and to predict outcomes, Reynolds equations for diffusion, etc., cost-benefit analysis and game theory for the value of the signal in socioeconomic terms, Tinbergen and Lorenz for the sign-stimulus, and various mathematical models for neural activity, and finally a biosemiotic model for the "qualitative" representation within the dog's brain. The billion dollar question is how do I know what is represented? For example when the dog follows a scent made by another dog with which it has had prior contact, does it have a "percept" or image of that other dog, or does it remain a featureless and fixed response? Is the referent the dog or the decoded scent? Public representations are external phenomena/objects and open to inspection. Private representations occur within the organism, they can be sub-divided into the peripheral and central. In the case of private representations it is clear that although we cannot experience what is going on – we can at least have a fairly good idea of the sensory inputs at local points. It is much more difficult to assess the inputs into the integration areas and networks. Does it need visual

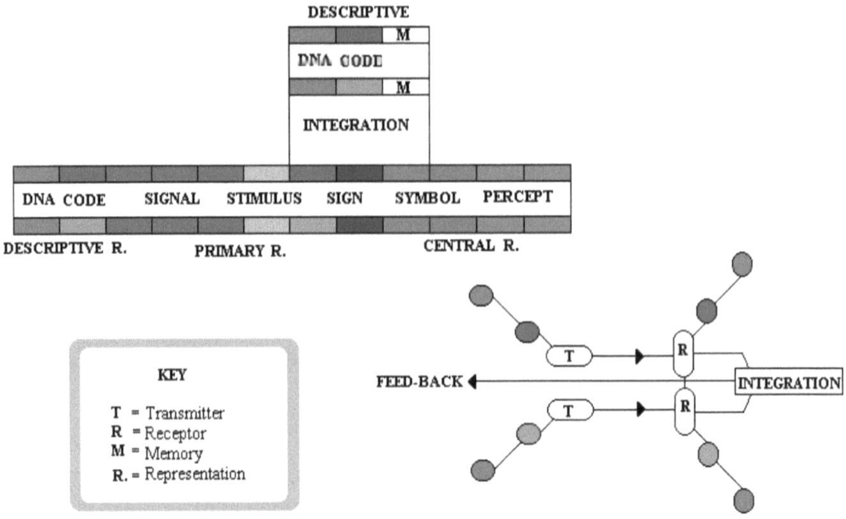

Figure 3. Semantic Relations in Canid Representation

confirmation? We can tell it is urine when the odour is strong, but after awhile when it has reached its threshold we cannot distinguish it from other spilled liquids. A dog does not need such visual evidence. This is shown by Daniel Dennett's example in a symposium about representation based on the story of Odysseus's dog in Homer's *Odyssey*.

I have a wax copy of my head, I have clothes that have my scent on it, and so forth. And I arrange to plant this out in the field where the dog will see it. And the dog sees it and comes running out wagging its tail and barking and then it freaks out. And it checks and checks and it sniffs and sniffs [...] (in Clapin (ed.), 2002, 48)

Argos has lots of, as it were, the right dispositions, vis-à-vis this dummy in the field that smells right and so forth, but there's a thing it can't get its head around, which we can get our heads around. The difference between having an unbounded disposition to believe particulars of a certain ilk, and believing the universal quantification, is that we can treat the belief in the second case as an object in its own right, and can build with it, and can even use it as a step in inference. (*ibid.*, 50)

In Dennett's example the counterfactual information is visual. From the dog's perspective the scent suggests to him that his owner has been there very recently. The dog freaks out because something is amiss. Moreover the dog has been deceived. We have to put this anecdote into a behavioural context. Olfaction is the primary mode of kin and hierarchy recognition. It is moreover a honest signal. (Harrington & Asa, 2003, 86).We do not know the status between Dennett and Argos. Is Argos the alpha or beta? Dennett argues that the dog has dispositions to respond which are inflexible, it cannot conceptualise internally the external states of affairs to draw upon an inference. What if we think of a human situation. Say for example that a father whose child has been missing for several days, he is distraught with worry, then he comes across what he assumes to be his daughter standing in a field. He runs in excitement, calling out, clearly overjoyed. All he had to go on were the physical shape and colours from afar. He believes that it is his daughter right up until he turns the dummy around to see the ghastly truth. Now this would cause a shock to anyone. He might walk in circles, sobbing, shrieking, and showing many signs of being emotionally disturbed. Can we from this example infer that he lacks the capacity to create internal representations? No. Why not? Because we assume that he is like us, and most of the medical literature supports this.

What is the difference between the two cases? Dennett probably already believed that the dog would respond in a fixed manner, and knew that the dog's brain was incapable of processing "symbolic" representations. He was arguing from two sets of generalisations: one drawn from ethology, the other from neuroscience. In fairness, the example was anecdotal, from the "top of his head" as we say, nevertheless, it is rather typical of how many of us approach these problems of representation. First impressions can be misleading. When a dog or wolf sniffs a mark and then leaves quickly we might conclude that it was the smell that led to this reaction, but

> Observing from an aircraft, we would probably conclude that some message contained in the urine of the mark and decoded by the olfactory system stimulated the wolf's prompt exit. Yet perhaps it was a visual cue, such as the height of the mark above the ground, that so unsettled the loner. (Harrington & Asa, 2003,68)

Here we might note the difference between visual, acoustic and olfactory signals. The first two are synchronic telling the receiver about the "immediate state" of the sender, whereas the olfactory signal is diachronic, providing the history of the sender. (species, individual identity, gender, breeding condition, emotional state, age, diet) (*ibid.*, 80)

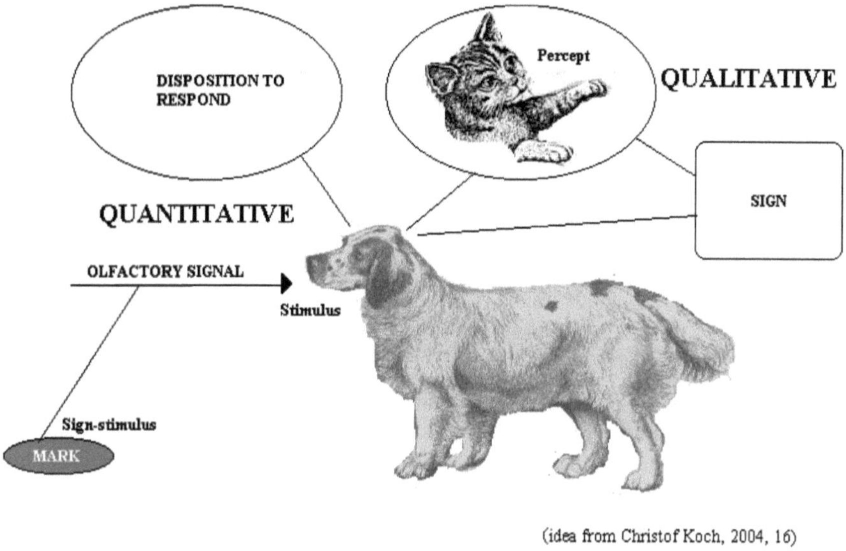

Figure 4. Canid Semantics

PART TWO: INVERTEBRATES AND REPRESENTATION

REPRESENTATION IN INVERTEBRATES

According to my theory of representation there are two main forms of representation within the nervous system, these are i.) **primary or peripheral,** ii.) **central**, the former which Lambert Schomaker has identified as:
1. Topological coding
2. Firing rate
3. Recruitment
4. Distributed representations and coherence
5. Temporal coding, vetoing, synchronisation (Schomaker, 2004, 3)
He emphasises that these representations "*located in peripheral*" *system* are "*unlikely to differ substantially from those in the intermediate (central) processing stages.*" (*ibid.*). I on the other hand believe that **central representation** differs from the peripheral form because it involves greater plasticity, different types of conduction, different combinations of neurotransmitters/modulators, and above all complex inputs from different sensory afferents. In addition there is awareness of processing – **attention** or **interest** arises in the central processing areas. Moreover these representations can be seen as components of a non-linguistic system of communication. Walter J. Freeman has another take on representation and meaning, believing that representation occurs externally, while meaning is enacted/produced internally by interlocutors. I feel this may be viable at the symbolic level with humans, but at the level of lower invertebrates it is unworkable. Still one may think of external representations as being public – and the "inner representations" as private. (Freeman, Walter J., 2000) It used be the case that the concept of "inner representation" was integral to cognitive science and even to robotics, but R.A. Brooks in the 1980's while on a visit to Thailand realised that invertebrates can achieve complex tasks without central representation, a position which Schomaker above partially refuted. Brooks believed that invertebrate behaviour could be used as the basis for designing bottom-up models that could be used in the construction of more elaborate models – the eusocial insect was also to provide the neural model for simple networks and swarm "bots" and swarm programs:

> The fundamental decomposition of the intelligent system is not into independent information processing units which must interface with each other via representations. Instead, the intelligent system is decomposed into independent and parallel activity producers which all interface directly to the world through perception and action, rather than interface to each other particularly much. The notions of central and peripheral systems evaporate – everything is both central and peripheral. Based on these principles we have built a very successful series of mobile robots which operate without supervision as Creatures in standard office environments.

I have in another paper challenged reductionist modelling that is based on the assumption invertebrates are simple living systems capable of solving complex problems (Pain, 2006). In Sidney Och's *A History of Nerve Functions* (Cambridge, 2004, 334) there is a diagram of classical conditioning of neurons of the snail *Aplysia californica.* It is modified from the work of Nobel Prize winning Eric Kandel's textbook, *Principles of Neural Science* (2000). We can learn from this diagram the "molecular nature of the learning process" (Och, 333). In picture A we see the snail's tail displayed revealing the siphon and gill. A sensitising stimulus (unconditioned) is applied to the tail which activates a sensory neuron which synapses on a facilitatory neuron that synapses on a motor neuron that controls the gill area (gill withdrawal reflex). In the same picture we can also see a tactile stimulus (conditioned) given to a sensory neuron from the mantel also resulting in synapses on the motor neuron. Some think to make a robot jellyfish, or for that matter any invertebrate robot, one need only tease out a quantitative correlation between a behavioural function and a synapse. However D.O. Hebb was himself aware of this Churchland reductionist fallacy:

What has given reductionism a bad name is the conclusion, after a theoretical analysis of a mental variable has been made and it is "reduced" to some pattern of neural activity, that the mental process in effect no longer exists. This is the nothing-but fallacy: Mental activity is a myth, what really exists is something in the brain. (Hebb, 1980,43).

The reductionist fallacy often insists on a direct correlation between an action/attitude and a neuronal site. One from this determinism can develop a neural architecture or network. From such a belief many books and institutes, even Nobel Prizes have been spawned by it. Smythies nicely summarises this:

This led to the belief that individual synapses represented more or less permanent connections between neurons that operated a binary code (i.e., producing an axonal action potential) and not firing (no action potential). Thus the computations performed by the brain could be described exclusively as the results of fixed nerve nets operating by such processes as matrix multiplication (Churchland and Sejnowski 1992). Learning was supposed to depend largely on a change in weights at individual synapses, which altered the probability that activity at that synapse would contribute to firing its postsynaptic neuron. (Smythies, John. (2002) 1)

The picture of fixed structures is repeated in the classical conception of the synapse and its elements. In a conveyor belt model:

The action of the neurotransmitter was terminated by its departure from the receptor molecule and its subsequent reuptake into the pre-synaptic terminal for reuse (e.g., glutamate, Glu) or its further metabolism to inactive products (e.g, acetylcholine, Ach, and dopamine). The receptor remained in the membrane, awaiting the arrival of another Neurotransmitter molecule, whereupon the same process would be repeated. (ibid.)

Smythies' own work and three reports i.) Hevroni et al., (1998), ii.) Husi et al., (2000), iii.) Craig and Boudin (2001) have delivered a coup de grâce to the above notions of the neuronal synapse, and in doing so undermine the theories of learning based on direct correlation. In short the synapse components are not soup-like but structured, and more importantly the cellular signalling events are in themselves extremely complex involving cross-talk and other molecular forms of communication (Cerione (2000), 556 In: Smythies, 4).

If we look at both Brooks' thesis and Kandel's approach to neuroscience we find serious contradictions and problems at all levels. These might be itemised as follows:

- Learning and memory: Unchallenged acceptance of learning theories that need to be reassessed in the light of new sciences
- Unconditional acceptance of reflex theory
- Refusal to incorporate new data from the field (plasticity in invertebrate behaviour)
- Predominance of Neurone doctrine now challenged
- Synapse theories – now challenged
- Refusal to incorporate neurochemistry – over emphasis on "circuit boards"
- Not enough work on multi-sensory perception and sensory ecology

When we analyse the hierarchical organisation involved in this work, we find that there are several inherent problems and contradictions. Each of these levels can said to be an analysable level (Hebb, 1980); often dealt by different fields which in turn employ different modes of reasoning (deductive and inductive), and involve varying degrees of certainty. At the level of behaviour: John C. Fentress in an essay "Ethology and Neuroscience." in honour of Robert Hinde the ethologist, upbraids the neuroscientists for ignoring the true complexity of ethology:

> A common problem with strict undirectional "reductionistic" references to behaviour is that the richness and diversity of behavioural phenomena are not only controlled, but subsequently ignored. Mechanisms for non-existing phenomena are not interesting, however. The ethologist has much to offer here by dissecting natural streams of expression into component properties, and then evaluating how these properties cohere in the production of higher-order patterns. (Fentress, 1991, 80)

CASE STUDY III: Representation of Prey in the Moon Jellyfish/Herring Larva Dyad

Predation is a behavioural factor or process of feeding. There are two agents or parties involved. The predator and the prey. If we look at predation in an integrative and hierarchical manner, we begin with feeding and move all the way down to the neural mechanisms that are connected with it (Tinbergen, 1951).This is the neuroethological approach which is based on the classical stimulus-response paradigm (Zupanc, 2004). Behaviour is the software, and what happens inside the blackbox is the hardware. The causal relationship involved in neuroethology is a story that is similar to "The House that Jack built." And perhaps just as surreal. Suppose we consider the act of suicide. According to this approach we can follow a component of suicide, say anxiety or the act of restless walking near a track, then establish a relationship to a sublevel mechanism, such as a motor movement of a hand, then further down to the sensory system, and connect this to a neural mechanism. However the problem is that of categories. Suicide, includes a set of social and cultural variables. There are many disagreements about motives. Yet in this case the neuroscientist accepts without reservation one theory. There are also problems with the attribution of functions at the next analysable level. We can continue all the way down to the synapse, and even at this level there is lack of consensus – some for example see structures others processes. It is possible for one scientist to see level A to be fixed, level B fixed, and level C plastic; while another might view level A as arbitrary or plastic, level B as fixed, and level C as fixed and plastic! Turning to predation, Eberhard Curio in his excellent book on the subject conceded "Despite its pervasive nature opinions differ as to what predation really is." (Curio, 1976, 3). Staying with the integrative approach, there is also a problem with cognitive agency. It has become common practice in neuroscience to attribute cognitive agency to parts of cells, and cells and structures (organs) belonging to multi-cellular life forms. While a virus can be a biological agent, it is not a cognitive agent, nor for that matter is a bacterium. A brain is certainly not a person or agent. In the first case we are considering, the herring larva and jellyfish *en toto*, are the cognitive agents. When the jellyfish makes contact, the sensory receptors receive information, but this does not mean that the jellyfish is as yet "aware or conscious" of the information, as it has not been processed. Norman B. Rushforth concluded in an article on learning in coelenterates that:

> Today, we realize that endogenous activity in neural units is widespread, ranging from simple invertebrate ganglia to complex brain structures. [. . .] The animal does not passively respond to an environmental stimulus but plays an active role in interactivity with it. (Rushforth, 1973)

In the same article although there was clear evidence that *Aurelia a.* is capable of habituation, there is little evidence to support the presence of association and higher forms of learning. No wonder that early naturalists thought cnidarians were plants. Yet, there are members of the same phylum that have eyes that are remarkably complex – the existence of which has led to discussions of which came first, the eyes or the brain? When we look at the jellyfish swimming in the sea, effortlessly, its billowing form like a marine ballerina, we see nothing that is really suggestive of intelligence, no tell-tale sensory organs (apart from Cubozoa), nothing to latch upon in an anthropomorphic turn of mind, nothing with which to identify. And looking at the beached forms, those failed soufflés that smell so much, aside from the radial shape, the gonads, and the gelatinous material (mesoglea) from which it gets its name, all we worry about are the tentacles that might sting us. Indeed it is the cnidae (intracellular barbs) on the tentacles that provide the name of the phylum, and the major apomorphic feature by which to classify its members (Nielsen, 2001, p.59). Since there are so many jellyfish during the summer, we just consider them to be a nuisance, however for fishermen and many nations that depend on fish, the jellyfish population explosion represents a serious economic and ecological disaster. It is in this context the study of the jellyfish predation is of immediate concern. From the purely zoological point of view they are of interest because their behaviour and associated neural mechanisms are indicative of early cognition.

A few years back a Japanese toy manufacturer produced an aquaroid, a robot jellyfish that was powered by a solar cell and able through primitive sensors to avoid touching the sides of a tank. According to some accounts, the *Aurelia aurita* medusae (it has a sessile form, a polyp) is not much smarter. But it is. Jellyfish have recently been shown to be capable of navigation with such precision that they dumbfound researchers. They have on a noncognitive level more impressively also been shown to predict the El Niño effect. These feats are done collectively. If we look at the individual medusa it is more difficult to ascertain what exactly is going on. One outstanding component of its ethogram is locomotion. Indeed it was the jellyfish's swimming behaviour that first attracted the attention of George Romanes, (1848–1894) a disciple of Darwin. He wanted to find out if there was a motor or control system. He managed to locate it by excising one by one the rhopalia (club like formations stationed in multiples of four around the bell's margin), and found the jellyfish could still manage to get along, like an old Dakota on one propeller. The pacemakers were located in the region of the rhopalium. Romanes also discovered a nerve net in the bell (Romanes, 1885). Research later established that there are two nerve nets (one connected with swimming), and the existence of various sensory equipment in the marginal sense organs. Suddenly the floating bag of jelly was more complicated than hitherto thought. This initial interest, much in the late nineteenth century, subsided, and experiments in this area have been much reduced. This of course has a lot to do with the interest in molluscs which have larger axons and more developed homologous organs, and with the subsequent advances in molecular neurobiology. One is more likely to read papers on neurotransmitters or synapses in cnidarians than to do with macroprocesses or functions (Grimmelikhuijzen & Westfall, 1995). Comparative psychologists tend to ignore cnidaria. Animal behaviour books might if one is lucky have a word or two on bioluminescence. But learning in cnidaria is sniggered at. Perhaps some of this is due to those early successes of Romanes which almost conclusively proved that jellyfish are not cognitive animals. It might also be a question of commercial interests. Prime research has been in toxins and in connection with fisheries.

With the above in mind I would like to propose that more research should be done on individual interactions and behaviour. We can start by looking at the predator-prey dyad, one based on research done with small medusae and very tiny herring larvae. A herring larva would be very lucky to survive into adulthood, and luckier still to be able to reproduce. The odds are naturally against it. By way of analogy, one could think of the female Atlantic cod that lays 5 million or so eggs, of those 4,999,998 will not survive (Ridley, 2004, 72). Of course over fishing has reduced those odds further, but worse still has been the increase in the jellyfish population. Jellyfish of many species feed upon the eggs, larvae and compete with the adult herring for zooplankton. Imagine one has survived to the first stage of its young life. It has already quite an array of sensory equipment, eyes (not with adult resolution), chemical, mechanical detectors as well. It needs these as it hunts in a meandering motion the copepods that are its prey. As it gets older its overall escape velocity will have increased, as would its cruising speed. The drag at the egg sac stage may be more. There are two forms of hunting strategies that have evolved in jellyfish predation. One is ambush. The other is cruising. (Arai,1997,73) The prey are caught/attached by two methods: so-called fly-paper method in which prey are caught "all over both their upper and lower umbrella surfaces in streams of mucus" and transported by means of cilia to food pouches, and from there moved by the oral arms into the gastrovascular system (Hardy, 1972,147; Southward, 1955, Orton, 1922). The second, and one more successful with herring larvae is by stinging them with the cnidae on the tentacles. This method like the use of mucus is dependent upon the prey being extremely close to the jellyfish. There are numerous problems associated with the study of jellyfish cognition. Firstly we must remember that the jellyfish is a medusa form of an animal that is primarily sessile, taking the polyp form and one that is immobile and attached to a rock etc. The level of communication differs according to what form the animal is, and what age it is. For example as a planula larva, the scyphozoa uses two sensory systems, moving toward its goal through chemotaxis that is not too different from that found in bacteria. (Müller & Leitz, 2002) Even in the medusa form there are important differences. The diet changes according to age and size (morphometrics). A jellyfish is also likely to respond differently when it is in a larger colony, and of course there are real differences between how it responds in a laboratory aquarium and in the field. The number and type of prey also must be taken into account. There are essentially two hunting strategies adopted by medusae, i) ambush, ii) cruising. The *A. aurita* hunts while on the move. The dynamics involved in hunting and capturing herring larvae of the first stage has been documented by Bailey and Batty, (1983). (cf. Arai, 74–75).

Prey capture is linked directly to locomotion. Costello (1992) (Arai,76) has shown the clear relationship between the bell pulsation, fluid motion and the subsequent prey capture. Contact is made with entrained particles and prey under the bell. Here is a description of how the *A. aurita* feeds:

> Encountered prey are paralyzed or killed by the action of the nematocysts of the medusa, embedded in mucus and transported by ciliary currents to one of eight external food pouches located at the umbrellar margin. The prey and mucus are picked up by the tips of the oral arms and are then transported towards the mouth by ciliary currents on the inner walls of the oral arms. Thus they are passed from the food pouches via the groove running inside the oral arms to one of the four gastric pouches, where the food is digested by the action of gastric enzymes. A ciliary-driven flow, from the gastric pouches through the gastrovascular canal system, distributes the digested food within the medusa. (Hansson, Lars Johan, 1998, 7–8).

At the height of behaviourism, the release of the cnidae would be termed reflexive and in combination with other patterned behaviour – seen as "a series of concatenated reflexes" that constitute an instinct. (Watson, 1913 106). Two sets of information emanating from the herring larva would be interpreted as stimuli, i.e., chemical (leaked endocrinal material) and mechanical (waves caused by swimming). Watson who opposed introspection in any form, would have never entertained the idea of biosemiotics – the work of Uexküll and the objective school was an anathema for him. (*ibid.*, p.9) He hated the concept of image, which of course is central to semiotics. Perhaps he has a point? How could the prey possibly be represented internally in an animal lacking a brain? X swims near Y. X's cues/signals are "eavesdropped" by Y. Y responds by stinging X. In this case the assumption is that there are two cognitive persons. But at this stage, is Y aware or conscious of X?

What do the cues/signals mean to the jellyfish? The notion of cognitive person is interesting, because in another member of the phylum, siphonophora, are composite animals, made of biological persons (zooids) each with different duties. Moreover, in an essay for *The Cognitive Animal* (Berkoff et al., 2002) Charles Taylor on the basis of his research and others, concluded:

> For purposes of understanding cognition, especially animal cognition, and probably subjective experience, organisms are best viewed as collections of sensors, effectors, and processors of limited abilities surrounding and located throughout the organism. These collections communicate primarily with other sensors, effectors, or processes that are mostly nearby, that mostly have a limited bandwidth, and that function as an ensemble. (Charles Taylor in Berkoff et al., 2002, p.157)

Is subjectivity scalar? In the predator-prey dyad, there are two animals, one that is a vertebrate, the other is an invertebrate. Their nervous systems and structures are very different. In the case of the herring larva, there is one major executive structure, while in the case of the jellyfish there are eight marginal centres and ganglia structures. If we were to adopt the notion of scalar subjectivity or a hierarchical and collective structure of cognition, then it would seem that in the jellyfish, the initial sensors (sensory receptors picking up the chemical and wave signals) would function at one level of subjectivity or awareness, and this information would pass onto another level of integration involving a higher form of awareness, until the information reaches the ganglia in the rhopalia regions leading to a motor response in the whole organism. Let's look at this in terms of hierarchical levels of representation and meaning. Instead of zooids as in a composite animal we could adopt a modular approach to representation. In this case there would be a tentacular/defensive

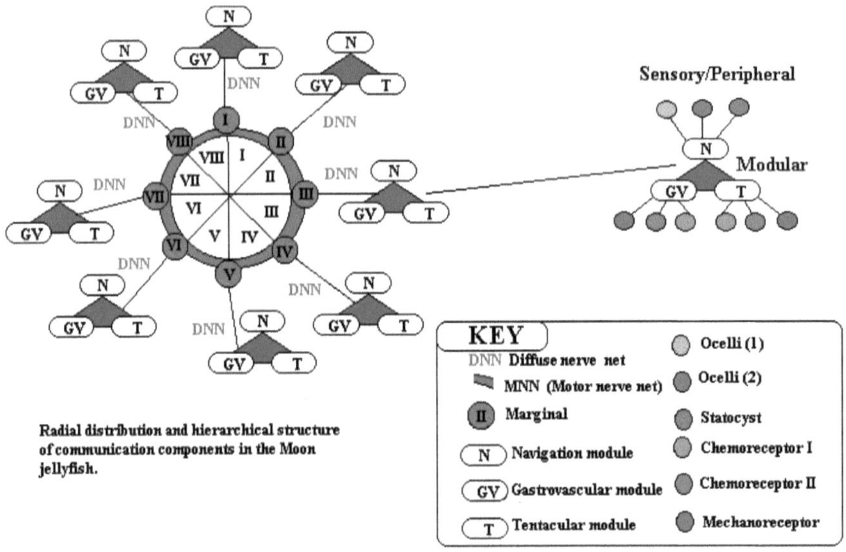

Figure 5. Communication components in the Moon jellyfish

module, a gastrovascular module, and a navigation module. Representation in these modules are peripheral, but also multimodal. Sensory afferents from the first two modules are transported along a diffuse nerve net

(DNN) to the marginal ganglia where they are integrated and later exercise a modulatory effect on the rate of the pace maker and through the motor nerve net (MNN) change the swimming rate. Since there are eight sections in the jellyfish body this means there are twenty four modules each with their own different sets of environmental information such as degree of salinity, toxins, temperature, pressure, magnetic/gravitational fields, intensity of light, prey type and concentration, conspecific proximity. Each of these modules would in addition have their own number of neurons and neurotransmitters:

In the above diagram the modular relations and general arrangement is suggestive of distributed cognition, i.e., the cognitive tasks are distributed, however it seems that there is weighting, the marginal "stations" have higher level sensory modules connected with navigation (Hollan, Hutchins, Kirich, 2002, pp.75–94). It is these rather than the tentacles or oral parts we associate with a "head" and since they are closer to the centres of integration we might see them as a stage towards cephalization. How the brain arose from these centres is of course speculative, but I think that since there was in other members of the phyla a reduction in marginals with a concurrent advancement in the optics of the eyes, it would seem there would also be a greater cross-modularity, particularly between the attack and defence module (tentacular) and the navigation module. In short the evolution of the brain and eye is linked to greater specialisation in predation and defence. The box jellyfish for example has four marginals, numerous eyes, and a faster overall swimming speed – and swims like a fish rather than around the oral arboral axis. (Nilsson et al., 2005) Yet this does not mean we have to "dumb" down the moon jellyfish. Going on the above relations and its behavioural repertoire in the field, one has to seriously question whether its upper limit of learning is habituation. It is my contention that there is already a greater amount of cross-modularity in the moon jellyfish than what is evident in the motor responses. For example in the field the moon jellyfish migrate both vertically (up and down the water column) determined by time of day and prey migration and horizontally. There has been ample evidence that in horizontal migration the jellyfish are capable of migrating to a precise location in a fjord using sun-compass navigation. (Hamner et al., 1994) This has been also observed in a number of other invertebrates, especially eusocial insects (eg: Andel, D., Wehner, R. 2004). While it is true that the jellyfish possess the necessary visual equipment for carrying out this kind of navigation, I would like to develop a hypothesis that more than one modality is involved. The basis for this hypothesis is derived from research in vertebrate homing and migration. For example it has been common knowledge for hundreds of years that salmon and eels migrate thousands of miles to their respective homes – and in the case of the salmon two hypotheses have been put forward as explanations. The first is the imprinting hypothesis in which salmon are conditioned or imprinted after birth to recognise their natal waters – the waters having specific chemicals peculiar to that

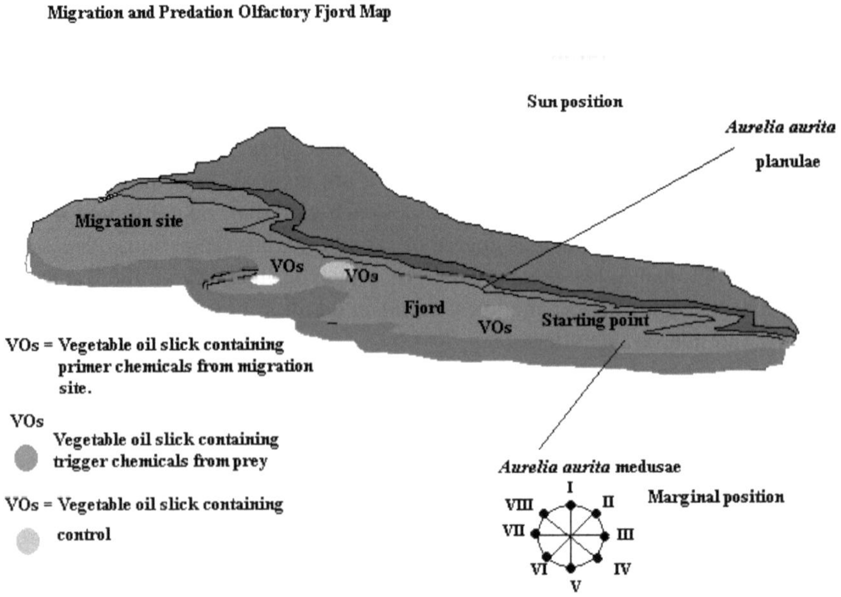

Figure 6. The Olfactory Landscape of a Moon Jellyfish

area. (Hasler & Scholz, 1983) The other hypothesis is that the salmon are following a pheromone trail (Nordeng, 1971). A researcher, Gabrielle A. Nevitt working with salmon moved onto research marine bird migration and olfaction, she discovered that petrels smell the landscape, and use a multi-sensory approach for both foraging and homing, i.e., using visual landmarks within an olfactory map. (Nevitt, 1999 (a), (b)) I have incorporated this information with the Hamner research (1994) and developed a potential experiment:

In this experiment vegetable oil slicks or buoys with slow diffusion devices would be located in areas away from the i.) the destined migration point (home) and ii.) in an area with low or no prey density. These slicks or buoys would contain chemicals connected with the prey (such as triggers asparagine,glycerine, leucine, tyrosine, oleic acid, palmitic acid, triolein, cephalin) (for sources, see Arai, 1997, 85), others connected with reproduction (pheromones?), predators and of course controls (Arai, 1997). One would then be able to see whether the jellyfish like salmon and petrels can migrate to a home, or food sites using a combination of sensory apparatus. There is evidence that medusae are attracted to short-range chemical attractants, and that in the planula stage *Aurelia aurita* will seek out areas where there are chemicals emitted by bacteria on shells and stones, and that metamorphosis is triggered once a certain threshold is reached. Given these facts, and the evolution of homing, it seems to me that there is a very strong likelihood that the sites for "home" are chosen based upon chemical information cross-referenced with visual (sun position, light intensity) and gravitational information. This experiment is based on the new field within sensory biology, namely sensory ecology:

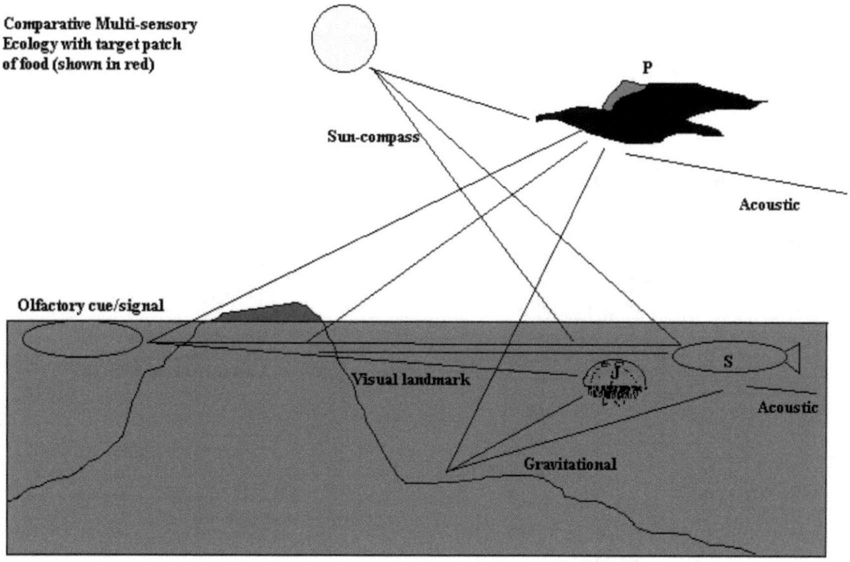

Figure 7. Multi-sensory Model

Sensory ecology occupies the interface between the inner and outer worlds of animals, but also exists at the transition between abiotic factors: perceptual systems represent a transfer function, transforming physical stimuli into ecological effects via behavioral acts of the animal. (Weissburg, Marc J. (2005) ("Introduction" by Marc J. Weissburg. In: Sensory biology: linking the internal and external ecologies of marine organisms – *Marine Ecology Press Series* Vol 287: 263–307, 2005. (eds. Marc J. Weissburg and Howard I Browman).

I have drawn a comparative diagram below to show the ecological relations between animals seeking out a patch or area of primary production. The jellyfish (J) does not use long-range acoustic information like the petrel (p) or salmon (s) but does on the other hand respond to more localised mechanical disturbances.

In the above figure the importance of the visual component is illustrated. It is my contention the sensory afferents from other modalities are either "grounded" in the visual "neural working space" or integrated with the data emanating from there. Each of the marginal navigation modules provides the jellyfish with co-ordinates of its current location with regard to the sun and magnetic fields. The information from the lower modules is then integrated or grounded within the ganglion. A single herring larva appears as the sum of the sensory afferents (from all modules), integrated and modulated within a ganglion, and this information is compared with inputs from the other marginals. The lead is established by the greatest threshold, in other words the intensity of the stimuli. In normal circumstances the mode of locomotion and feeding prevents a medusa from focusing on a single prey, unless as stated before it is particularly troublesome which is akin to the author finding a burnt cornflake in his breakfast bowl. However, in laboratory conditions, if the medusa is of a certain diameter (small) and "starved" it is conceivable the degree of

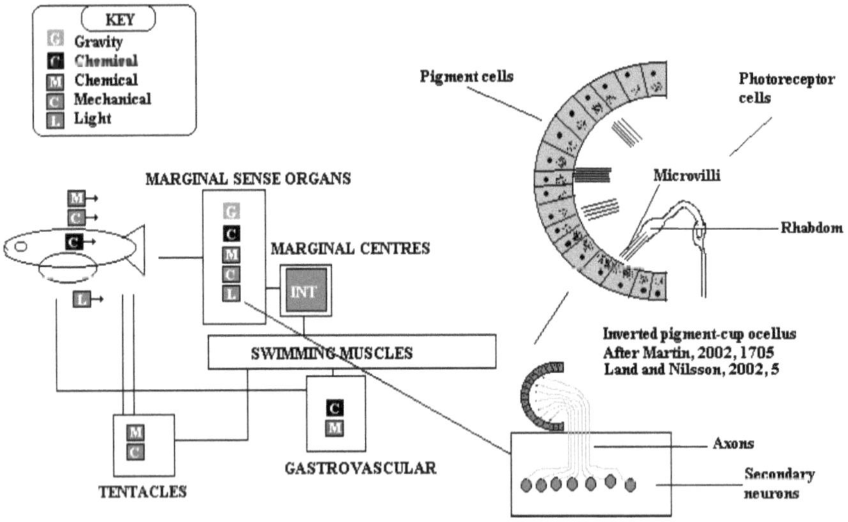

Figure 8. Visual Biology and Modules

attraction and representation is greater. In the field, a more likely scenario would be the migration of jellyfish to a patch of primary production involves a large number of prey which would be represented as the sum of afferents in an integrated ground. The qualitative sign (type M^2) arises after environmental information (signals) have arrived and been received as stimuli, processed and tranduced into intracellular impulses, modulated then integrated within a ganglion connected with a visual neural working space (the ocellis).

Case Study IV: A Can of Worms. The Earthworm *Lumbricus Terrestris* as Bait.

I was inspired to write a paper on earthworms after reading a recent report by a Norwegian scientist who concluded that in all probability earthworms do not suffer when they are used as bait (Lauritz, 2005). I must admit that I would have been the last person on earth to come to refute these conclusions, because I had once in my "Huckleberry Finn" days often gone fishing, equipped with a bamboo stick, some old fishing line, a rusty hook and a jar of freshly dug earthworms. I used the worms to catch eels off the Northumbrian coast. They were just bait, and I did not think twice about them. I also came across earthworms when digging for potatoes. I would sometimes cut through them, but heard my father tell me that it was ok, because they would grow again. Now, today I would like to question these prejudices of mine, and others in general concerning invertebrates. Why is it that we can accept the possibility of pain and suffering in vertebrates and not in invertebrates? This should be qualified as there are some invertebrates now covered by animal protection laws and laboratory guidelines – these include cephalopods, crustaceans and some eusocial insects. But why not earthworms? Don't they feel pain, and doesn't capturing, handling and impaling them on a hook constitute unnecessary suffering as in the various Acts?

After reading the report, I got out a copy of Darwin's book on earthworms to see what the great naturalist had to say on the matter. The book by the way was read in tandem with a couple of novels by the comic genius P.G. Wodehouse, and a biography of the playwright Dario Fo — a rare cocktail indeed, and as with such a literary alchemy, thoughts and images and perspectives tend to cross over. When for example I read of Darwin shouting at a flower pot of earthworms to test their sense of hearing, I thought of a surreal character in Dario Fo drama without the politics of course. In Wodehouse's novel *Love Among the Chickens*, the novelist narrator cum hero talks to a dog called Bob about the matters of the heart, but after finishing his monologue realises sadly that his affectionate companion functions on another level, an instinctive one. In this wonderful mix of science and comedy, I discovered that Charles Darwin's approach to earthworms was one of sympathy and childlike curiosity– and perhaps it is this spirit that I wish more scientists could have inherited than the cold calculating mind of reductionists. Darwin was no less a scientist for this approach. Even when he played the piano and tested if they reacted to different notes. It makes sense. Of course, his greatest accomplishment in the book is his observation of their burrowing habits over a long period of time, upon which he concluded that they have an intelligence that exceeds the purely instinctual. As to emotions, I think if the earthworms (*Lumbricus terrestris* as there are many species of earthworms) were ever excited it was through transference! Darwin did not write much about suffering or pain except to say that their wriggling movements were an exaggeration of what they were actually suffering.

> There is little to be said on this head. We have seen that worms are timid. It may be doubted whether they suffer as much pain when injured, as they seem to express by their contortions. Judging by their eagerness for certain kinds of food, they must enjoy the pleasure of eating. Their sexual passion is strong enough to overcome for a time their dread of light. They perhaps have a trace of social feeling, for they are not disturbed by crawling over each other's bodies, and they sometimes lie in contact. (Darwin, 1881, 34)

This actually sums up my own position with regard to earthworms; they do suffer pain but it is a matter of degree.

What is suffering and pain? For most of us it is a subjective matter. One can for example measure how a cilia is affected by a tactile sensation, how the relevant cells are excited and in turn the velocity of the impulse to the parts to be innervated and then the tension in the muscles or effectors, etc. These are all quantities. They do not tell us anything about the quality of what is happening. In the case of humans one can talk to them, and if for example the pain is in a thumb, we and they can discuss the probable cause, the impact, the effect, the situation, but most importantly the qualities, i.e., the kind of pain, whether it is: sharp, acute, dull, lingering, crushing, pressing, shooting, etc. While it might be true that there is a degree of commonality in the quantities involved in vertebrate pain, there is certainly no agreement in how the pain is represented in each individual, and actually, what goes on in the periphery system can be modified by the "meaning" of the pain. Let us take an example from life. Once I was staying in a house in Amami Oshima in the Ryukyu archipelago and I was sitting on the veranda with my hands dangling. Before on the trip to the island I had been told many stories about the deadly snakes (*habu*) that inhabited the island. The memory of this had a great influence on what happened next. I was bitten! I screamed and shouted for my host – who being practical was on the phone to get a doctor. I was dying. But was I? No, a younger member of the host's family pointed out that the culprit was their puppy who was shading under the veranda. I had created in my brain a virtual theatre of what was going on from the memory, experience, and of course the context. If I had seen the puppy then it would have been different. But here we see how the emotional state feeds back into the periphery and can have an impact on the intensity of the pain and local inflammation. Moreover it can also influence recovery. Wall in the introduction to a classic textbook on pain develops a model from traditional models such as the behaviourist, cybernetic and ethological. His model incorporates the experience of what actually goes on in the sensation, perception and representation of pain.

Wall like others before him had always been intrigued by the "phantom limb" phenomenon which occurs after a patient has had a limb amputated – they for awhile actually feel it, and even localised pain and cramps. This suggested to Wall that the brain creates a virtual theatre of what happens when the body is aversely stimulated. Does an earthworm have a virtual theatre – what happens when part of its body is cut off – or when an ant loses its legs or abdomen during fighting? Can invertebrates have the cognitive capacity for the "phantom limb phenomenon?" We don't usually ask such questions of invertebrates because we have a tendency to

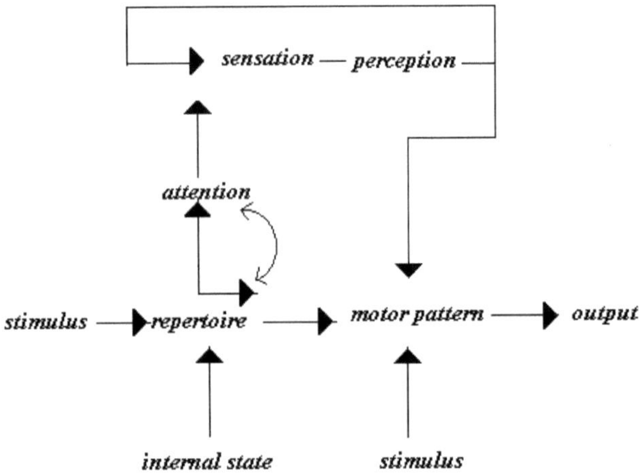

Reality/Virtual Reality Model
After Patrick D. Wall (Wall & Melzack (1994)

Figure 9. Model of Pain

dumb them down. We for example are found of bragging about the 300–500 billion neurons in the human nervous system, and then compare this with the size and quantity of other vertebrates and invertebrates. But as the shockjock Howard Stern fondly reminds us, size doesn't always count. We might remind ourselves that we have less genetic complexity than some flowers and invertebrates! In a seminal article on sentience and pain, Jane A. Smith draws our attention to the fact that the same goal can be achieved with a different design. The problem with invertebrates is not that they are simple, but that they are different ...and complex. One of the reasons that cephalopods are favoured is because of homological design – those eyes! Earthworms with very primitive eyes and photoreceptors can not compete in the beauty contest.

Let us look at the report. I can quote some of the text, please note that the earthworm *Lumbricus terrestris* belongs to the phylum Annelida:

> Annelids The segmented annelid worms are sensitive to touch. Mechanical receptors are found all over the body, and nerves from the segmental ganglia innervate the muscles of the body wall (Brusca & Brusca 2002). The reaction of annelids to noxious conditions may be reflexes. (Lauritz (2005) 26)

The ventral nerve cord of annelids sometime includes extremely long neurons, or so-called giant fibers, of large diameter. The fibers facilitate rapid conduction of impulses to the brain, bypassing the ganglia of the ventral nerve cord. The fast reaction to touch in earthworms is probably an effect of giant fibers, and result in the rapid withdrawal of the worm. The wriggling of and earthworm on a fishhook is most likely due to reflexes. (*ibid.*, 27).

> Opioid substances are also found in animals with relatively simple nerve systems. They probably play a role in sensory modulations of the earthworm Lumbricus terrestris, and injections of naloxene inhibit the wriggling and escape responses of the worms. According to Kavaliers (1988), the presence of opioids in a variety of invertebrates suggests that modulation of adverse and nociceptive responses was present at an early stage of the evolutionary history. It is unlikely that similar neuromodulary mechanisms have arisen independently in various phyla of animals. (ibid., p.33)
>
> With the relatively simple nervous system of earthworms [meitemark] and other annelids, it is very unlikely that they can feel any pain. The wriggling of earthworms on a hook can be considered as reflexes. (ibid., 36)

The exaggerated movements of the earthworm are due to reflexes. This would seem to settle the matter. We can happily impale worms on hooks because they are insentient creatures that respond to aversive stimuli by withdrawal and wriggling or moving away. Now we can look at the table of meaning in terms of the aversive stimuli (the hook):

These two types of meaning mark the divide between the current scientific view of invertebrate and vertebrate sensation of pain and suffering. It would be useful here to give a definition of vertebrate suffering.

> A negative emotional state, which may derive from various adverse physiological or psychological circumstances, and which is determined by the cognitive species and the individual being, as well as its life's experience. (D.Morton, p.1095) "Suffering" in Berkoff, Marc (Ed.) 2004. *The Encyclopedia of Animal Behavior*. New York, Greenwood Press).

Properties	*M*[1]	*M*[2]
Origin of Stimulus	Hook	Hook
Agent	Decoder	Interpreter
Location	Environment/sensory Organs	Mind/brain
Message Unit	Signal	Sign
Essence	Quantity	Quality
Structure	Fixed	Plastic
Approach	Contextualist	Mentalist

Figure 10. Meaning Types involved in hooking.

What we must prove then is a very tall order. Firstly that the *Lumbricus terrestris* can be in a negative emotional state, i.e., that earthworms have emotions, secondly (and this implied by the "cognitive species") that the earthworm has a cortex or something that functions like one, and thirdly that it has a "life's experience." I would also to make it even more controversial ask whether we can use Wall's model of pain in humans with invertebrates like earthworms?

FROM SIGNAL TO SIGN. THE EMOTIONAL WORM

We saw in the case of my experience of being bitten by a snake/puppy that memory and emotions played an important role. But what are emotions? According to Stephen Jones English Dictionary (based on Sheridan's), an emotion is "a disturbance of the mind or a vehemence of the passions, while my old standby the COD defines emotion as an agitation of the mind, feeling, excited mental state. Both these definitions mention an external force or agent that "disturbs" or "agitates" the mind. Here perhaps is the concept of the hormonal influencing the neuronal? When we feel sad, or upset, there is a neurotransmitter or neuromodulator at work. What about pain? There are several transmitters and modulators at play, the primary associated with these are the opioids which have an analgesic effect. The earthworm for future reference as with many invertebrates has to the capacity to produce opioids. Staying with the basic model of emotion i.e., that of the neuronal and hormonal here is what Jean-Dier Vincent had to say concerning invertebrates:

Perhaps because the neuronal aspect is more important than the hormonal, the mechanisms of invertebrates are easier to study in depth than are those of vertebrates. Indeed, the importance of the wiring system reveals itself in all its magnificence when we try to link a behaviour pattern to the workings of isolated groups of nerve cells.

And

The very advantages of invertebrates, however, limit their usefulness. Their simplicity and stereotyped behaviour are of no help in taking on the complex strategies observed in higher vertebrates. (Vincent, Jean-Didier, (1990) The Biology of Emotions (translated by John Hughes), Oxford: Basil Blackwell) p.87)

Why is the hormonal dimension of less importance? Is it because such a limited approach to invertebrates facilitates modelling which is the prime preoccupation of neuroscientists? Is this another example of "dumbing down." Let's now consider emotions in evolutionary terms, they after all did not come from nowhere. According to R. J. Davidson the classic Schneirlan approach/withdrawal mechanisms are actually the founding blocks of emotion:

> This research supports the view that what are basic to emotions are not universal response patterns but rather the dimensions of approach and withdrawal, "basic principally because of their phylogenetic primacy." (Raines & Greenberg, 1998, pp. 76–77)

Should we not then pay more attention to what goes on in the head of an invertebrate? Is there more plasticity in the reflexes which Lauritz goes on about? If we were to poke either the anterior or posterior ends, the earthworm will respond almost immediately by classical withdrawal. The effected segments shrink, and it is this response that further undermines the claim of consciousness or suffering in the earthworm. We can examine the neurological pathway to illustrate this point. The points of greatest sensitivity are the anterior and posterior segments where the mouth and anus are located respectively. Touch those parts and one sees a rapid withdrawal because the segmental nerves excite giant fibres (medial and lateral) that conduct an impulse along the ventral cord to segmental motor neurons which in turn excite the longitudinal muscle which results in contraction activity in the local segments, a complex of these impulses co-ordinated by the brain results in the familiar wriggling or squirming activity that constitutes part of the worm's mode of locomotion. There is nothing here to suggest sentience or consciousness, nothing more than the knee-jerk or other involuntary movements in humans. In addition to this blatant example of stimulus/response behaviour, the physiological structure of the worm, i.e., its segments works against the case of consciousness. The earthworm is divided by septa (muscular partitions) into segments, and these segments are as Wallwork states "virtually self-sufficient." (Wallwork (1983) 5) Each of them "have its own supply of blood, its own central and peripheral nervous system, its own excretory system for removing metabolic wastes, and its own gonads." (Wallwork *ibid.*). This segmentation and repetition of functions and organs is termed metamerism. This kind of reflex lends itself nicely to "boxology." However as Zupanc shows in the case of crayfish, there are other variables and factors that can as in other vertebrates and humans modify the behavioural outcome. A crayfish which is subordinate will in a territorial encounter with a dominate crayfish be initially aggressive and engage in a fight, but this is determined by its status. The duration and intensity of the fight can be increased by injecting a neuromodulator that can motivate the crayfish to engage longer. Here is a boxological illustration:
(a) Environmental Stimulus Behaviour
(b) Environmental Stimulus Behaviour

As a further twist, it has been shown in research the modulators themselves can be modulated by such factors as the "social status and social history of the animal" (Zupanc *ibid.*, 196). Can earthworms modulate modulators? Are they like crayfish territorial, and can this "emotional state" be repeated in our situations as in suffering? To answer the question whether they are territorial I would like to contend

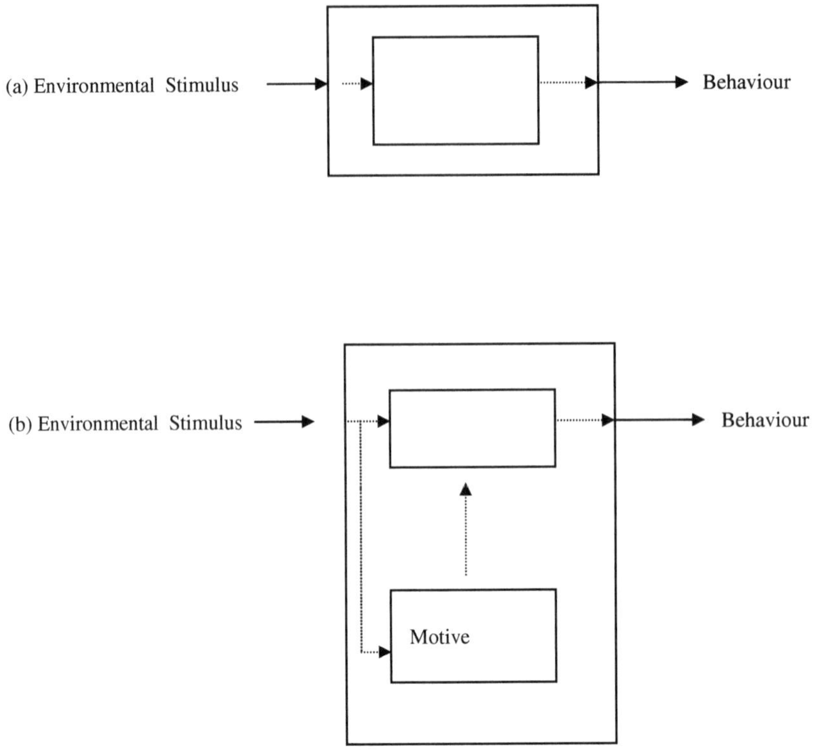

Figure 11. Modulation After Zupanc (2004) p.182

that the ideal giant reflex and other modalities of earthworms *have the capacity* to be modified by neuromodulators which in turn are affected by social factors. This occurs, although it is yet to be substantiated, when worms (*Lumbricus terrestris*) defend or assert their territorial claims. According to the Planka categorisation of animals (Planka, E. R. (1970) *American Naturalist*, 104, 592–97) *Lumbricus terrestris* is a K selection:

> Such populations are usually crowded, competition is intense and reproductive rates are kept low by this competition which increases with (density (i.e., density-dependent population). Wallwork, ibid., 47 Territorial behaviour is also a feature of individuals and species that are potential competitors may allow for the equitable allocation of resources, thereby reducing actual competition. (ibid., 50).

When we return to look at a territorial dispute we can see the interplay of mechanisms and functions suggest that there are moments when the worm can said to be in a primitive emotional state.

If, and this is of course controversial, an earthworm in a territorial dispute can modify what is an ideal giant reflex (i.e., the withdrawal movements) because of neuromodulators and social factors, then it is conceivable that an earthworm may modify fixed action patterns during other stressful occasions. According to some

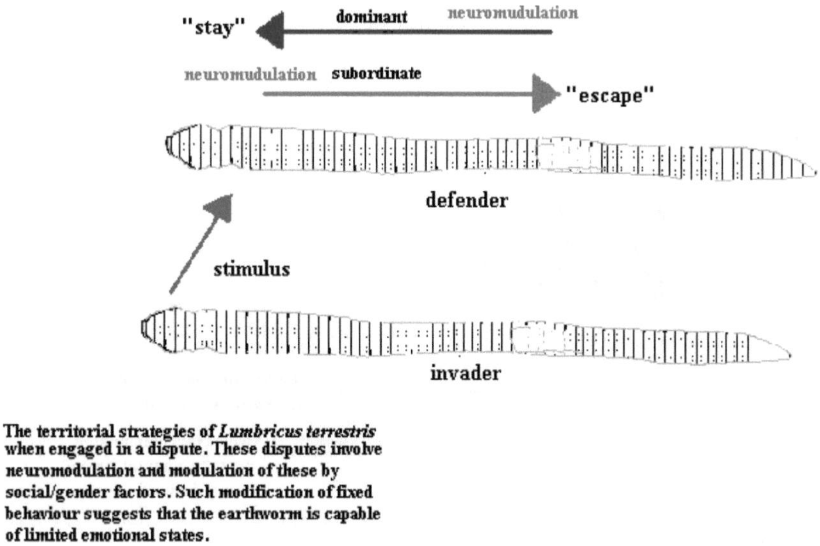

Figure 12. Territorial Strategies of the Earthworm

psychologists there is a set of six or more cardinal human emotions, these are related as they are all evolved and based upon the approach/withdrawal mechanism (the classic giant reflex). I would like to argue that the modification of aggressive behaviour (anger) in territorial disputes between invertebrates represent one of the "emotions" – and other cardinal emotions and states may also be present, including that of suffering or pain. This then seems I believe to satisfy the possibility that they can be in a negative emotional state albeit purely in the peripheral sense, as we as yet have to deal with the problem of representation of pain.

THE INTELLIGENT WORM

From this diagram we can see that the capture, handling and hooking of a worm involves numerous neural pathways, and this is very simplified! The ideal reflex connected with the approach/withdrawal mechanism has often been looked upon as a means of "robotising" invertebrate behaviour. The main reflex is not the limit of invertebrate sensitivity or intelligence. In a paper on insect communication I argued that for semiotic meaning to be present there had to be physiological evidence of an "space for inner representation" and the processing and associating of two or more signals from two or more discrete modalities. I realise now my concept of "space for inner representation" corresponds with Jean-Pierre Changeux's concept of "neuronal workspace" in *The Physiology of Truth: Neuroscience and Human Knowledge* The cerebral ganglia may not be as complex as our cortexes, but this should not be a means of dumbing down the earthworm. Darwin wrote on this very subject: *"With respect to the small size of the cerebral ganglia, we should remember*

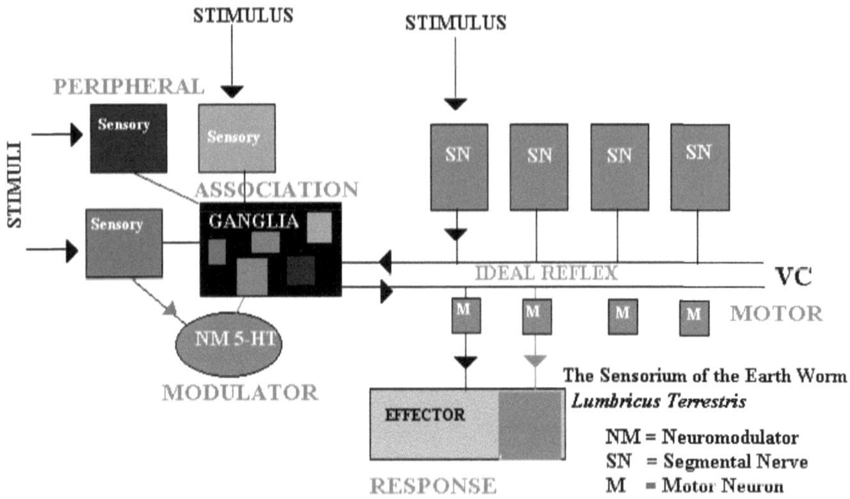

Figure 13. Sensorium of the Earthworm

what a mass of inherited knowledge, with some power of adapting means to an end, is crowded into the minute brain of a worker-ant." (Darwin, *ibid.*, 98).Comparative brain size like other sizes can be often misleading! Darwin wrote a great deal about how intelligent he thought an earthworm was in how it dealt with methods of plugging its burrow, D. O. Hebb also mentioned the worm briefly in his absorbing book, *An Essay on the Mind*:

"An earthworm can learn a a Y-maze (Y-shaped, with a single choice point) in 20 trials, about what the laboratory rat needs for the same task." (Hebb, D.O., (1980) Can an earthworm with procedural memory and a small number of neurons represent pain as we do? Does it have a virtual theatre? What happens when it loses half of its body after it is cut off by a gardener? Before looking at this more carefully, I would like to describe in semiotic and semantic terms pain or suffering:

The conception of how pain is represented is on a continuum which continues towards a symbolic representation as found in humans. We should now consider the anatomy and habits of the earthworm to understand how or if pain is represented. The earthworm belongs to the phylum Annelida and to the class Oligochaeta (Sims and Gerard, (1999), 40). It is a deep-burrowing worm which burrows vertically and leaves a cast above on the surface (i.e., the subcategory of Anécique). (Wallwork (1983), 9). Its principle food is "organic detritus, usually the decomposing leaves and stems of plants, although root material, seeds, algae, fungi and testacean Protozoa may also be ingested. (Wallwork, *ibid.*, p.10) From these organic materials, the earthworm extracts nitrogen and sugar. They ingest nearly any organic material often with soil. The ingested food contains many kinds of bacteria some of which the earthworm prefers over others – they enjoy a symbiotic relationship with some bacteria and fungi. (Sims and Gerard, *ibid.*, 20). The worm has a three-layered

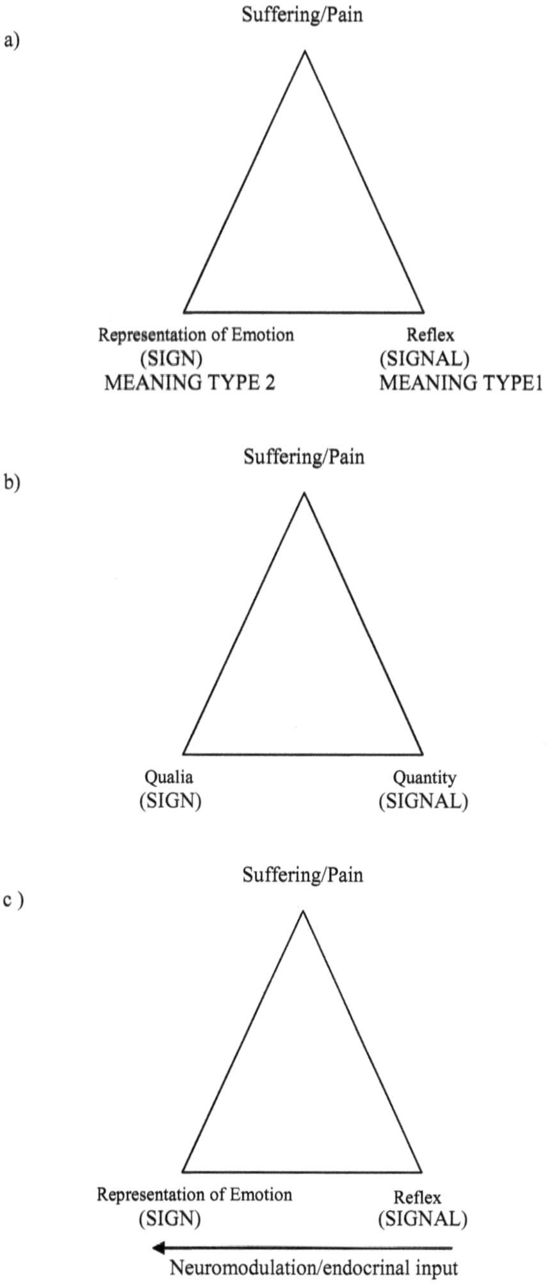

Figure 14a. Suffering Triads

body structure: an outer mesoderm, a body cavity called a coelom, and a gut. It has a hydrostatic skeleton which is ideal for burrowing. This enables the worm to squeeze into small spaces and is ideal for its peristaltic mode of locomotion. To look at a worm seems to have a symmetrical head and tail. However on closer inspection one see differences in the rings or segments and the shape. As everyone knows, an earthworm does not possess any eyes or explicit sense organs. This may help persuade us into believing it is an organism incapable of consciousness or rudimentary cognition. However, after dissection and under the microscope one will see that the earthworm does have "lens" like sense organs in the anterior segments. Indeed, it is possible to detect an array of sense organs that cover different forms of sensory reception from chemoreceptors located all over the body, mechanoreceptors or mechanosensors and photoreceptors. The ganglia complex in the third segment can said to function as a primitive brain, and as there are several sub-ganglia in the neighbourhood. Given the distribution of sensory receptions, and preponderance of major organs in the anterior and posterior regions, then it would seem plain that these areas are more "sensitive" than others. We can use this for a sensitivity map. The earthworm's lifestyle means that it is continuously faces stress from the soil (sharp, objects, etc.). We can from this conclude that its nociception system is preoccupied with desensitivitising these averse stimuli. Hence the opioids. Otherwise one would surmise, the incredible pressure felt would be so great that it would immobilise the earthworm. This is different from noxious substances and of course from stress coming from predation. It is like us pinching ourselves or accepting strain or pain as part of our occupation, and comparing this with pain inflicted by a stranger or in unexpected circumstances such as in capture, handling and hooking. My point is that these constitute unusual circumstances outside the *Umwelt* of the worm. The worm has an awareness of its location, memory of the topography, photoreceptors and other receptors provide constant information about its co-ordinates and state of well-being. This can be represented in a non-linguistic fashion. How, I am not so sure but biosemiotics can help here to describe nonvisual complexes of sensory information as it is centrally represented in the cerebral ganglia – let us say that they are "meaningful" signs which are transduced and processed signals from two or more discrete modalities. They are qualititative to the worm – we talk of qualities in the periphery but this is our own representation – not how it is represented to the worm.

PART THREE: THEORETICAL FRAMEWORK

I. The Framework

A metascience cannot and should not be expected to have applications within a natural science – and for this very reason I have sought to develop a research form of biosemiotics. Since it represents a departure from the more traditional form of biosemiotics, it can be characterised as a project, one which requires more focus and structure. To assist this project, we can enlist the aid of Peter Medawar's four levels of strata This gives us a professional basis for structuring the field of analysis according to the part, the organism and organisms:

- Molecular biologists
- Cellular biologists
- Biologists
- Biologists who work with populations of organisms (ecologists, etc) (Medawar, 1967, 101)

Furthermore I believe it should be possible to adopt the same principles as the zoologists or bioengineers with respect to the definitions. I have come up seven principles for classification of terms in biocommunication:

- Evolutionary Precedence
- Neurophysiological complexity
- Analytical properties
- Level of Representation
- Standardisation
- Decoupling
- Abstraction

Given DNA, a code, signal, sign and symbol. I believe we can use Drew Endy's three principles to assist in the structuring of our model. "Foundations for engineering biology." (Endy, *Nature* vol. 438 24th November 2005) (cf: http://parts.mit.edu) Endy's principles, the last three in the list: standardisation, decoupling and abstraction are fairly easy to understand. By **standardisation** Endy means just that, consistency in the naming and location of parts – in his project it is the parts and stages of genetic engineering. **Decoupling** is simply breaking a problem into its components or smaller problems to make it more manageable. **Abstraction** refers to the "hierarchy of abstraction" In Drew Endy's model there are four levels and three horizontal axes. From bottom to top: DNA, Parts, Devices, Systems. The "information describing biological functions might be organized across levels of complexity using abstraction hierarchies." (*ibid.*, 449–453). This model corresponds also to the Hebbian notion of analysable levels. I believe it is essential to insist on

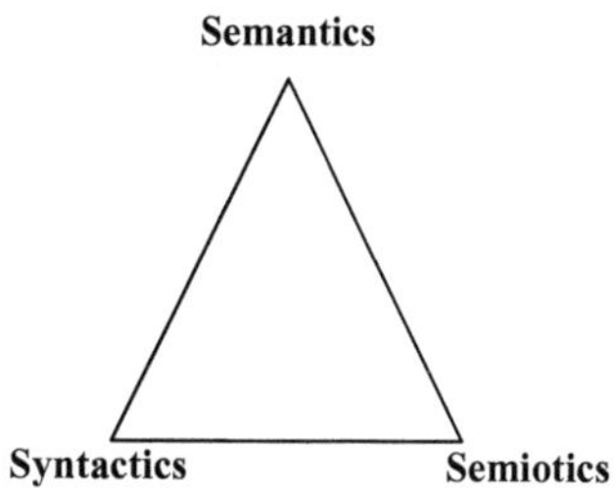

Figure 14b. The Semantic Triad

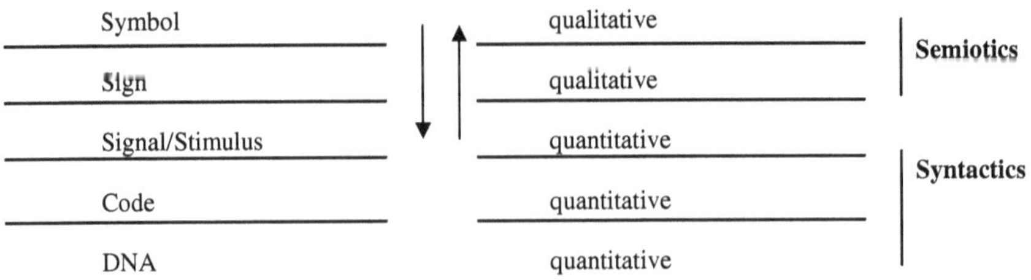

Figure 15. Hierarchy of Abstractions/Analysable Levels. (Sources, Drew Endy, 2005, D.O. Hebb, 1980)

a hierarchy in semiotic terms, because otherwise they would not be programmable and cause untold confusion for researchers. The beauty of Endy's model is that it allows the researcher to concentrate on one function or level without looking above or below. If there is a need for information from another level, then this can be asked in a discrete and specific manner without upsetting the research at all levels. It also preserves semantic integrity of the levels. The levels are based on the biology of communication and the rules of **evolutionary precedence**. If we look at the level of the signal we see that it is above a code, below a sign, it is quantitative and syntactical. This notion of syntactical is derived from the nature of a signal from point of transmission to its arrival at a cell membrane. It is essentially a sensory form of signalling. Obviously influenced by Charles Morris's "Foundations of the Theory of Signs" (1938) (in which semiotic/s is divided into three branches: syntactic, semantic and pragmatic) my notion differs in that the syntactical relations between the signals are rooted firmly in the physiological and not linguistic or Carnapian logical syntax. (Morris, 1971, 28) But at what level does representation take place? This is answered in the second figure, at the level of **primary representation** from the organism's point of view. The division of semantics into the i) syntactical (quantitative), ii) semiotic (qualitative) is based upon the nature of the **neurophysiological** structures. I developed earlier a bicameral theory of semantics, one which supports the two main threads in neuroscience and cognitive science, namely the behaviourist approach and its opposite the "mentalist" approach.

I believe they do not contradict each other if we allow that the first can be used with the signal, and the latter with communication units above such as the sign and symbol. The idea being that researchers can work in the same manner as they might in the bioengineering project at MIT. With regard to the sign itself, we can say that

Properties	*General Signal*	*True Signal*	*Cue*
Extracellular	Yes	Yes	Yes
Intracellar	Yes	No	No
Benefits	Yes/No	Yes to sender	Yes to receiver

Figure 16. Table of Signal Components

Signal Type	*Agents*	*Medium*	*Benefit*	*Example*
Intracellular	Many	Signal transduction	Cell/organism	Calcium ion
Extracellular	To cell/ cell-to-cell	Environment	Cells/organisms	"Quorum sensing"
True animal	Animal-to-animal	Environment	Animals	Dogs wagging tails
Cue	Animal-to-animal	Environment	"Eavesdropper"	Leaked pheromones
Coercion	Animal-to-animal	Environment	None	Pushing

Figure 17. Table of Signal Types

its analytic properties are semiotic and qualitative, it involves central representation in a ganglionic complex or more.

If you were now to ask whether a signal is a sign, I would say no. It helps to form a sign. And a symbol? A symbol is made up of the signs below. A signal is coded. A code is made of DNA. We might sub-divide the signal into subclasses such as a true signal, a cue, and a general signal. In behavioural sciences a signal between cells (organisms) or an intracellular signal might not be interpreted as true signals or cues both of which involve benefits (Bradbury and Vehrencamp, 1998; 2005, Barritt, Greg, J., 1992, 1–3). We might subdivide the signal into components as follows:

Do bacteria use signs? No. Why not? This is difficult to answer. But I'll have a go. Bacteria can follow a chemical trail (signal/stimulus) along a gradient because they have membrane receptors that are ligand-specific, i.e., molecules of the nutrients will bind to the receptors, this will lead to secondary intracellular activity including enzymatic activity, secondary signalling, culminating in a motor response such as tumbling etc. The input, events inside the cell, and output are measurable, even viewable. What good would it do to introduce the concept of signs here? There are no physical sites that are capable of integration necessary for central representation. I do not think for there is one reference to the "sign" in John T. Hancock's excellent book *Cell Signalling* (2004). But there is also little information theory. His book deals with the intracellular signalling in terms of amplification and cascades rather than linear flows or explicit networks. Indeed there is little maths in his book. In another book on signalling in animals by John Maynard Smith and David Harper (2004; orig. 2003) there is no mention of information theory or cybernetics, instead the emphasis is on evolution, socioeconomics and cost-benefits involved in signalling. Reading

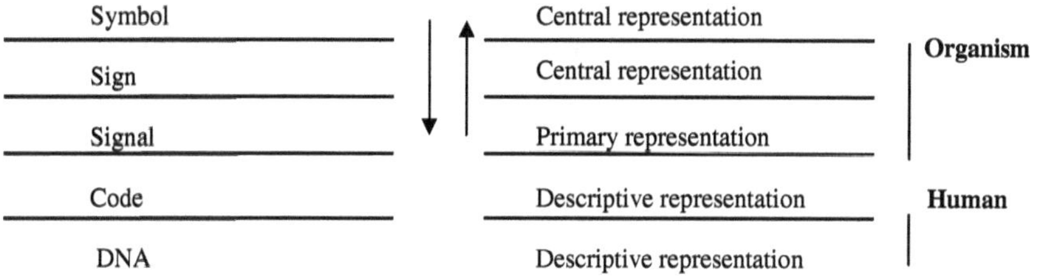

Figure 18. Hierarchy of Representations

Properties	*Meaning Type 1 (M^1)*	*Meaning Type 2 (M^2)*
Essence	Quantity	Quality
Nature	Syntactical	Semiotic
Carrier/Vehicle	Signal/Cue	Integrated Signals

Figure 19. Meaning Typology

these books I was struck by how different and contradictory the theory of signalling is. Extracellular signals have different physical features to intracellular signals (e.g., calcium ions) in their overall discreteness: coding, transmission, directionality, duration, and the agents involved. If we then look at the extracellular signals as defined by Hancock, we find that some of these may not be deemed to be true signals since according to Maynard Smith and Harper because no benefit is accrued to the sender:

> We define a "signal" as any act or structure which alters the behaviour of other organisms which evolved because of that effect, and which is effective because the receiver's response has also evolved. [...] if a signal alters the behaviour of others it must, on average, pay the receiver of the signal to behave in a way favourable to the signaller; otherwise receivers would cease to respond. [...] the signal must carry information—about the state or future actions of the signaller or about the external world—that is of interest to the receiver. (Maynard Smith and Harper, 2004, 3).

In this definition we also see that it has an implied"mentalist component.

Regarding the key problem of representation, I think this can also be discussed in purely biological terms based upon the neurophysiological structures or processes necessary for the different levels of semantic representation. Looking at Fig. 15. We can say that DNA, codes, do not entail awareness. Representations at this level are descriptive. They are from the researcher's perspective only. Representations at the level of the signal take place in the periphery and can said to be primary or peripheral representations. Representations at the level of the sign and above are of course the most controversial. Some argue that the central representations at this level are no different from those in the periphery.

Descriptive Representation	–	DNA, Code (intracellular communication e.g., calcium ion signalling)
Primary Representation	–	Extracellular Signal – sensory processing – primary awareness
Central Representation	–	CPU (ganglion like sites and more complex)

I have below provided the reader with a series of figures and tables that may further explicate my definition of meaning:

	Primary Representation		Central Representation	
Characteristic	*Meaning Type*^{0}	*Meaning Type*^{1}	*Meaning Type*^{2}	*Meaning Type*^{3}
Theory	Communication	Cybernetic	Semiotic	Linguistic
Founder	Shannon & Weaver	Norbert Wiener	Charles W. Morris	Chomsky/Pinker
Unit	Signal	Signal/Sign-Stimulus	Sign-stimulus/Sign	Sign/Symbol
Memory	None to Simple	Short-term/Long-term	Short-term/Long-term	Parallel and 0,1,2
Essence	Quantity	Quantity	Quantity/Quality	Quantity/Quality
Location	First Reception	First Reception	Second Reception Integration	0,1,2 and Higher Processing
Organs	Receptors/cells	Sensory Organs to lower brain regions	Sensory Organs to lower brain region	Brain (Higher processing regions)
Learning	Up to habituation	Habituation to classical cond.	Classical / Operant conditioning	0,1,2 and Education
Agent (Receiver)	Decoder	Decoder	Interpreter	Interlocutor

Figure 20. Meaning Type Chart

CONCLUSION

I have in this paper outlined what I consider should be the parameters of a research biosemiotics – ones which can be functional yet at the same time be compatible with the main tenets of general biosemiotics. I have situated the discussion concerning invertebrate cognition at the level of the sign which I have argued is a mentalist and therefore qualitative construct. The prerequisites for a semiotic state are that there be a communication/interaction context, cognitive agent/s capable of receiving a signal, responding to a stimulus, transducing this external or environmental information into a series of internal signals or impulses which are represented (primary and peripheral/descriptive representation) in the local area (module) and there producing M^1, before it is then carried along a pathway (nerve networks) to either a visual neural working space where it is grounded, or directly with other sensory afferents to a ganglion where it is integrated. It is in this integration site that central representation (subjective representation M^2) takes place. We might at this point distinguish this form of representation from the type current in neuroscience. For example Liqun Luo in *Nature* January 2006 succinctly describes the relationships in visual representation:

Our brain is made up of maps that organize what we sense. In the visual system, for example, on object is represented by the spatial activation pattern of retinal ganglion cells (RGCs), which form a two-dimensional sheet in the retina. RGC nerve fibres (axons) project into the brain in an orderly manner along both x and y axes, such that the two-dimensional image is recapitulised in the optic tectum region of the brain. (Luo, 2006, 23).

The fondness for maps in neurosciences and cognitive sciences is very problematic. In the above quotation we might ask who reads these "maps"? It is of course from the scientist's point of view. It is a common place to describe phenomena on the screen as a representation, but this is not **subjective representation**. Ever since E.C. Tolman came up with the concept of an internal cognitive map (1948), there have been several methodological problems connected with its usage and abuse. If we look at the original paper that is based on Tolman's experiments with rats and maze solving, we see that there is a direct correlation between the external sensory information and what is internalised. The construction of the map serves an explicit navigational function. In the same paper Tolman discusses cognitive mapping and human aggression. While we might agree that there is a direct correlation in the case of rats, we should object to the development of the concept into a metaphor. James L. Gould working with honey bees and cognitive mapping wrote:

As Tolman, who coined the term, envisioned a cognitive map, it was any mental transformation that enabled an animal to formulate a plan or make a cognitive decision. Later workers have sometimes supposed that some sort of literal map needs to be involved, but the original definition is the one used here. (Gould 2002, 41)

What Tolman did envision was even broader than this. He began his paper discussing rats and maze learning and ended as a preacher upon the mountain top!

What is the name of Heaven and Psychology can we do about it? My only answer is to preach again the virtues of reason-of, that is, broad cognitive maps. (Tolman, 1948)

Of course I understand Tolman's views in the context of the aftermath of WWII, but the broadening of his concept weakens its usefulness in natural sciences. Nevertheless, I have used below another cognitive map model developed by Treisman (1986) to illustrate how the information from a prey such as a herring larva might be represented. This model is ostensibly for higher vertebrates with brains and taken from a neuropsychology textbook. What is interesting from

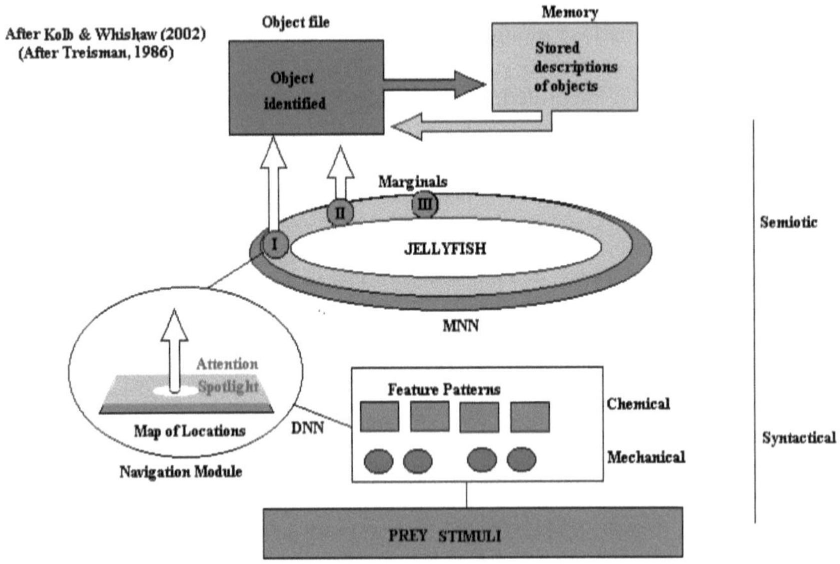

Figure 21. Attention Model of Prey Stimuli

my point of view is the notion of using models such as these to account for the relations between behaviour and neural mechanisms in lower ("brainless") invertebrates because they challenge us to consider where and how the information is processed and its evolution. They also make think about the organisation of "inner representation." My approach here is effectively the reverse of Tolman's, in that the scope of the cognitive model is narrowed down and restricted rather than broadened. It is therefore a form of ethological compatibilism. In Treisman's model the feature maps consist of the visual modality substrates of (colour, orientation, size, stereo balance) (*ibid.*, 581). What would be the equivalents and sites of these in a lower invertebrate like *Aurelia a.*?

I would like to advocate we divide the processing of information into two, the perceptual representation (this is quantitative and syntactical of which the animal is unaware) and central representation (this is qualitative and semiotic and of which the animal is aware). The basis of this division is taken from Conwy Lloyd Morgan who was influenced by the British philosopher and psychologist, G.F. Stout (1860–1944).

> We often say, for example, that interest guides behaviour in this direction or in that. But such interest must not be regarded as an impelling force; it is an attribute of the conscious situation, more or less suffused with feeling-tone. It is not easy to define; but it seems to take on its distinctive character when re-presentative elements contribute what Dr. Stout terms "meaning" to the conscious situation. The meaning in the early stages of mental development is, however, merely perceptual, and not that which comes much later—that which is implied in the phrase "rational significance. (Morgan,1900; reprint 1970, p.243)

In my theory of semiotics, qualitative meaning can only be produced by an organism that possesses ganglia able to integrate incoming signals from two more discrete modalities. The signals at the peripheral level are re-presented syntactically and expressed in quantitative terms. They become a sign complex only when the flow of information is arrested in a site of integration – they in other words capture the organism's field of attention (an assembly of neurals or neural space). This is a very limited version that corresponds to the "neuronal working space" hypothesis of Jean-Pierre Changeux in*The Physiology of Truth: Neuroscience and Human Knowledge.*

> Five principal types of relatively autonomous and specialized processors involved in perception, motricity, attention, evaluation, and long-term memory are shown as connected throughout the brain by the long-axon neurons of the workspace. [. . .] During an effortful conscious task, an association between the groups of neurons in the various processors operates in a top-down manner through the activation of a specific population of workspace neurons-principally, but not exclusively, from layers II and III of the frontal cortex. (Changeux, 2004, p.90).

I have in the course of my analysis of the two components of the respective invertebrates' ethograms, sought to explain more clearly what exactly goes on inside the marginal ganglia and cerebral ganglia – this is based on the principles outlined in the introduction. For sure I believe the internal events and types of representations are far more complex and plastic than what is suggested by the reflex theory or observed externally (motor outputs). I realise this is nonparsimonious, however

I have placed my project within the remits of ethological methodology as set out by Niko Tinbergen in his four questions regarding:

- Proximate causation and control
- Development or Ontogeny
- Function
- Evolution or Phylogeny. (Martin and Bateson, 1993, 8)

I have also chosen the "right" level of analysis – the representation of the prey in the predator-prey dyad (jellyfish/herring larva); representation of pain in the earthworm. I formulated the questions and analyses within the terms of semiotics, albeit one that is orientated towards the natural sciences. "How might a jellyfish represent internally environmental information (at the level of the sign)?" Despite the overwhelming evidence that suggests earthworms do not suffer or feel pain while being hooked for fishing bait, this can be countered by contending that both pain and suffering must have evolved. My approach is supported by new research in the field of sensory ecology that has focused more on the individual, and by traditional ethological methodology:

> A preferable approach is to master every possible type of mental aid when generating ideas and hypotheses, but to use the full rigour of analytical thought when testing them. (ibid.,1993,19)

I accept in both instances my case studies have been idealised. The jellyfish and the herring larva involved were in laboratory conditions – today the emphasis in marine biology is on the ecology and biomass, rather than the exploits of individuals. The same is true of earthworms. Scientists are interested in the relationship between the earthworm and the soil – not how an earthworm perceives its world. Yet these idealised cases have I believe illustrated weaknesses in current biology, especially in the over reliance on antiquated research. I found for example that the general theory of invertebrate learning and the types of learning and memory are woefully outdated (especially the theory of habituation and conditioning). I found books on both cnidarian and annelidan learning/intelligence referring to research conducted in the 1900's! Authors frequently refer to the paucity of research in this field, and then reference H. S. Jennings (1868–1947) who worked in 1906 (Rushforth, 1973; Jennings, 1904;1906). There has been to date still very little research in the area of chemoreception. There is also little organisation of information across disciplines. Those working with neurotransmitters do not communicate much with those working with behaviour. Biosemiotics however, while working at the level of the sign, is cross-disciplinarian.

Finally I would like to impress upon the reader about the need for research biosemioticians to maintain a distance from the classical linguistic conception of the sign and meaning. My theory of semiotics and semantics is based on the notion of a biological sign within living systems of communication and interactions that involve syntactical and semiotic relationships outside of language proper. Moreover I believe this facilitates the formalising of a qualitative component in neuroscience theory, one that can be utilised in explanations of consciousness and used in predictions of behavioural outcome, but most importantly it can help us move closer to understanding the other "citizens" of our planet.

REFERENCES

Abbot, Joan, N., Willamson, Roddy & Maddock, Linda eds. (1995) *Cephalopod Neurobiology*, Oxford: New York: Oxford University Press.

Abramson,Charles I. (1994) *A Primer of Invertebrate learning*, Washington: The American Psychological Association.

Alcock, John (2001) *Animal Behavior*, Sunderland, Mass.: Sinauer Associates, Inc.

Andel, D. & Wehner, R. (2004) "Path Integration in desert ants: how to make a homing ant run away from home". *Proc. R. Soc. London B. 271*: 1485–1489.

Andersen, Peter A.V. (1989) *Evolution of the First Nervous Systems*, New York: Plenum Press.

Anderson, Charles L., Kabalka, George W., Layne, Donna G., Dyke, Jonathan P. & Burghardt, Gordon M. "Noninvasive High Field MRI Brain Imagining of the Garter Snake (*Thamnophis sirtalis*)." *Copeia* 2000 (1) 265–269.

Arai, Mary. N. (1988) Interactions of fish and pelagic coelenterates. *Canadian Journal of Zoology*, 66: 1913–1927.

——. (1997) *A Functional Biology of Scyphozoa*, London:. Chapman and Hall.

Arbib, Michael A. & Fellous, Jean-Marc (Dec. 2004.) *TRENDS in Cognitive Sciences* 8 No.12.

Ax, Peter (1996) *Multicellular Animals,* Volume 1, Heidelberg: Springer-Verlag.

Barritt, Greg. J. (1992) *Communication within Animal Cells*, Oxford: Oxford University Press.

Barth, Friedrich G. (2002) *A Spider's World*, Heidelberg, Berlin, New York: Springer-Verlag.

Bartol, Soraya Moein. & Musick, John A. (2003) "Sensory Biology of Sea Turtles." In: Lutz. (2003) pp 79–102.

Bateson, Gregory (1979) *Mind and Nature*, New York: E. P. Dutton.

Bateson, Patrick (1991) *The Development and Integration of Behaviour*. Cambridge: Cambridge University Press.

Bentham, Jeremy (First Published 1789, New Edition 1823) (1893) *An Introduction to the Principles of Morals and Legislation,* Oxford: Clarendon Press.

Bechtel, William Mandik, Pete, Mundale, Jennifer & Stufflebeam, Robert S. eds. (2001) *Philosophy and the Neurosciences: A Reader*, Malden Mass.: Oxford: Blackwell Publishers.

Berkoff, Marc ed. (2004) *The Encyclopedia of Animal Behavior*, Westport, Conn.: Greenwood Press.

Bekoff, M., Allen C. & Burghardt, M. eds. (2002) *The Cognitive Animal,* Cambridge, Mass: MIT.

Berkoff, Marc & Jamieson. Dale eds. (1996) *Readings in Animal Cognition*, Cambridge, Mass.: MIT Press.

Berkoff, Marc (1979) "Scentmarking by free-ranging domestic dogs. Olfactory and visual components. *Bio Behav*. 4: 123–39.

Belekhova, M.G. (1979) "Neurophysiology of the Forebrain." In: Gans. (1979) pp. 287–359.

Berthold, Peter, Gwinner, Ebehard, Sonnenschein. eds. (2003) *Avian Migration*, Heidelberg, Berlin, New York: Springer-Verlag.

Bjorndal, Karen, A. (1997) "Foraging Ecology and Nutrition of Sea Turtles" In: Lutz (1997) pp.199–231.

Bonnadonna, Francisco, Benhamou, Jouventin, Pierre "Orientation in featureless environments: the extreme case of pelagic bird." In: Bethold et al., (2003).

Bouissac, Paul ed. (1998) *Encyclopedia of Semiotics*, New York: Oxford University Press.

Boyd, C.A.R. & Noble, D. eds. (1993) *The Logic of Life*, Oxford: Oxford University Press.

Bradbury, Jack W. & Verencamp, Sandra L. (1998) *Animal Communication*, Sunderland, Mass.: Sinauer Associates.

Breidbach, O. & Kutsch, W. eds. (1995) *The Nervous Systems of Invertebrates: An Evolutionary and Comparative Approach. (With a Coda written by T.H. Bullock)* Basel, Boston, Berlin: Birkäuser Verlag.

Brodeur, R.D. (1998) "*In situ* observations of the association between juvenile fishes and scyphomedusae in the Bering Sea." *Marine Ecology Progress Series*, 163: 11–20.
Brooks, R.A., (1991) "Intelligence without representation." *Artificial Intelligence* 47: 139–159.
Bugoní, Krause, Lígia, Maria & Petry, Virginia (2001) "Marine Debris and Human Impacts on Sea Turtle in Southern Brazil." *Marine Pollution Bulletin* 42 No.12: 1330–1334.
Bullock, Theodore, H, Bennett, Michael et al., (2005) "The Neuron Doctrine, Redux" *Science* 310: 791–793.
Bullock, Theodore Holmes & Horridge, Adrian (1965) *Structure and Function in the Nervous Systems of Invertebrates*. San Francisco: W.H. Freeman.
Burghardt, Gordon M. (2005) *The Genesis of Animal Play*, Cambridge, Mass.: Bradford Books (MIT Press).
Carew, Thomas J. (2000) *Behavioral Neurobiology*, Sunderland, Mass., Sinauer Associates.
Carnap, Rudolf (1946) *Introduction to Semantics*, Cambridge, Mass.: Harvard University Press.
Changeux, Jean-Pierre (2004) (translated by M.B. DeBevoise) *The Physiology of Truth: Neuroscience and Human Knowledge*, Cambridge, Mass.: Belknap Press.
Clapin, Hugh ed. (2002) *Philosophy of Mental Representation*, Oxford: Oxford University Press.
Colapietro, Vincent M. (1993) *Glossary of Semiotics*. New York: Paragon House.
Coolen, Isabelle, Dangles, Oliver, & Casas, Jérôme (2005) "Social learning in non-colonial insects." *Current Biology* 115: 1931–1935.
Corning, W.C. & Dyal, J.A. eds. (1973) *Invertebrate Learning*. (Vol. 1) New York; London: Plenum Press.
Costello, J.H. &. Colin, S.P (1994) "Morphology, fluid motion and predation by the scyphomedusae *Aurelia aurita*." *Marine Biology*, 121:327–334.
——. (1995) "Flow and feeding by swimming scyphomedusae." *Marine Biology*, 124:399–406.
Curio, Eberhard (1976)*The Ethology of Predation*. Berlin, New York: Springer-Verlag.
Dangles, Oliver, Casas, Jérôme & Coolen, Isabelle, "Textbook cricket goes to the field: the ecological scene of the neurological play." *The Journal of Experimental Biology* 209: 393–398.
Darwin, Charles (1881) *The Formation of Vegetable Mould, Through the Action of Worms with Observations on their Habits*, London: John Murray.
Davidson, R.J., Ekman, P., Saron, C.D. Senulis, J.A. & Friesen, W.V. (1990). Approach-withdrawal and cerebral asymetry: Emotional expression and physiology *I. Journal of Personality and Social Psychology*, 58(2), 330–341.
Dayan, Peter & Abbott, L.F. (2001) *Theoretical Neuroscience*, Cambridge, Mass.: MIT.
Deacon, Terrence W. (1997) *The Symbolic Species*, New York: W.W. Norton.
Dennett, Daniel C. (1993) *Consciousness Explained*, London: Penguin.
Derraik, José G.B. (2002) "The Pollution of the marine environment by plastic debris: a review." *Marine Pollution Bulletin* 44: 842–852.
Dingle, Hugh. (1996) *Migration*, New York; Oxford: Oxford University Press.
Dorsett, D.A. "Organisation of the Nerve Cord." In: Mill, P.J. (1978) pp. 115–160.
Dusenberg, David B. (1996/2) *Sensory Ecology*, San Francisco: W.H. Freeman and Co.
Dyer, Adrian G., Neumeyer, Christa & Chittka, Lars (2005) "Honeybee (*Apis mellifera*) vision can discriminate and recognise images of human faces." *Journal of Experimental Biology*. 208. 4709–4714.
Edwards, C.A., Bohlen, P.J. (1996) *Biology and Ecology of Earthworms*, (Third Edition). London: Chapman and Hall.
Elliott, Joel K and Leggett, William C. (1998) "Larval condition and vulnerability to predation: an analysis based on mixed-prey experiments." *Can. J. Fish. Aquat. Sci.* 55 (3): 626–630.
Endy, Drew (2005) "Foundations for engineering biology." *Nature* 438: 449–453.
Fautin, D. (2002)"Reproduction in Cnidaria." *Canadian Journal of Zoology*, 80: 1735–1754.
Fellous, Jean-Marc (2004) "From Human Emotions to Robot Emotions." Paper delivered at AAAI Spring Symposium on Architectures For Modeling Emotion: *Cross-Disciplinary Foundations* 2004.
Finger, Thomas E., and Wayne L. Silver, and Diego Restrepo eds. (2000) *The Neurobiology of Taste and Smell*, New York: John Wiley & Sons.
Fiske, John (1995) *Introduction to Communication Studies*,(Second Edition), London: Routledge.

Fraser, J.H. (1982) *British Pelagic Tunicates*, (Linnean Society of London, Estuarine and Brackish-water Sciences Association) Cambridge: Cambridge University Press.

Freeman, Walter, J. (2000) "A neurobiological interpretation of semiotics: meaning, representation, and information." *Information Sciences*, 124: 93–102.

Gans, Carl ed. Northcutt, R. Glenn, Ulinski, Philip S. coeds. (1979) *Biology of the Reptilia*, (Volume 10, Neurology B, Chicago: The University of Chicago Press.

Gans, Carl ed. Ulinski, Philip S. coed. (1992a) *Biology of the Reptilia*, (Volume 17, Neurology C, Hormones, Brain, and Behavior. Chicago: The University of Chicago Press.

Gans, Carl (ed.), Crews, David (coeditor), (1992b) *Biology of the Reptilia*, (Volume 18, Physiology E, Hormones, Brain, and Behavior. Chicago: The University of Chicago Press.

Godin, Jean-guy, J. ed. (1997) *Behavioural Ecology of Teleost Fishes*, Oxford: Oxford University Press.

Goldsmith, Timothy, H., (1990) "Optimization, Constraint and History in the Evolution of Eyes." *Quarterly Review of Biology*. 65. No.3.

Granda, Alan M. & Sisson, David F. (1992) "Retinal Function in Turtles." In: Gans, 1992(a), pp.136–174.

Greenberg, Gary & Haraway, Maury M. (1998) *A Comparative Psychology Handbook*, New York: Garland Publishing, Inc.

Greenfield, Michael D. (2002) *Signallers and Receivers*, Oxford: Oxford University Press.

Greenstein, Ben & Greenstein, Adam (2000) *Color Atlas of Neuroscience*, Stuttgart: Georg Thieme Verlag.

Griffin, Donald R. (1998) "From cognition to consciousness: a review" *Animal Cognition* 1: 3–6.

Griffin, D. R. ed. (1982) *Animal Mind-Human Mind*, Berlin, New York: Springer-Verlag.

Griffin, D.R. (1976) *The Question of Animal Awareness*, New York: Rockefeller University Press.

Grimmelikhuijzen, C.J.P. & Westhall, J.A. (1995) "The nervous systems of cnidarians." In: Breidbach (1995) pp. 7–24.

Halpern, Mimi (1992) "Chemical Senses in Reptiles." In: Gans (1992 b) pp. 423–323.

Hamner, W.M., Hamner, P.P. & Strand, S.W. (1994) "Sun-compass migration by *Aurelia aurita* (scyphozoa): population retention and reproduction in Saanich inlet, British Columbia. *Marine Biology*. 119: 347–356.

Hancock, John T. (2005) *Cell Signalling*, (2nd Edition) Oxford: Oxford University Press.

Handon, Lars Johan, (1998) *On the Ecology of the Jellyfish Aurelia Aurita*. Göteborg: Göteborg University.

Hardwick, Philip, Khan, Bahadar & Langmead, John (1994) *An Introduction to Modern Economics*, London: Longman.

Hardy, Sir Alister (1972, orig. 1956) *The Open Sea Its Natural History*. London: Fontana.

Harré, R. (1967) *An Introduction to the Logic of the Sciences*, London: Macmillan.

Harrington, Fred H. & Asa, Cheryl S. (2003) "Wolf Communication" In: Mech, 2003 pp. 66–103.

Harris, Roger et al. eds. (2000) *ICES Zooplankton Methodology Manual*, San Diego: Academic Press.

Hasler, A.D. & Scholz, A.T. (1983) *Olfactory Imprinting and Homing in Salmon*, New York: Springer-Verlag.

Hauser, Marc 2001.*Wild Minds*, London Penguin,

Hebb, D.O. (1980) *Essay on Mind*, Hillsdale, New Jersey: Lawrence Erlbaum Associates Publishers.

Innis, Robert E. (1982) *Karl Bühler: Semiotic Foundations of Language Theory*, New York and London: Plenum Press.

Ioalé, P., Nozzolini, M. & Papi, F. (1990) Homing Pigeons do extract directional information from olfactory stimuli. *Behavioral Ecology and Sociobiology*. 26: 301–305.

Jacob, Français (trans. Betty E. Spillman) (1974) (orig. 1970) *The Logic of Living Systems*. London: Allen Lane.

Jennings, H.S. (1906) *Behavior of Lower Organisms*, New York.

Johnson, M.C. Wuensch, K.L. (1994) "An investigation of habituation in the jellyfish *Aurelia aurita*." *Behavioral and Neural Biology*, 61: pp. 54–59.

Kandel, Eric R., Schwartz, James H. & Jessell, Thomas M. eds. (2000) *Principles of Neural Science,* (4th edition) New York: Elsevier.

Kass-Simon, G. & A.A. Scappaticci. (2002)."The behavioral and developmental physiology of nematocysts." *Canadian Journal of Zoology, 80:1772–1794.*

Keller, Evelyn Fox (2002) *Making Sense of Life*. Boston: Harvard University Press.

Kloeswal, Christian, J.W. et al., eds. (1986) *Writings of Charles S. Peirce*. (Volume 3) Bloomington: Indiana University Press.

Klopfer, Peter H. (1973). *An Introduction to Animal Behavior*, (Second Edition), Englewood Cliffs, New Jersey, Prentice-hall Inc.

Koch, Christof (2004) *The Quest for Consciousness*, Englewood, Colorado: Roberts and Company.

Kolb, Brayan & Whishaw, Ian Q. (2003) *Fundamentals of Human Neuroscience,* (5th Edition). New York: Worth Publishing.

Konishi, Masakazu (1993) "Brain and Behaviour" In Boyd, C.A.R. & Noble, D. eds. *The Logic of Life*, Oxford: Oxford University Press. (pp. 169–187)

Land, M.F. & Nilsson, Dan-E. (2002) *Animal Eyes*, Oxford: Oxford University Press.

Lauritz, S. (14.01.2005) "Sentience and Pain in Invertebrates." *Report to Norwegian Scientific Committee for Food Safety, Dept. of Animal and Aquacultural Sciences*, Sømme Ås/Oslo.

Lee, K.G. (1985) *Earthworms: Their Ecology and Relationship with Soils and Land Use*, North Ryde, N.S.W.: Academic Press (Australia).

Lehner, Philip N. (1996) *The Handbook of Ethological Method,s* (second Edition), Cambridge: University of Cambridge Press.

Levinton, Jeffrey S. (2001) *Marine Biology*, New York: Oxford: Oxford University Press.

Luo, Liqun. (2006) "Developmental neuroscience: Two gradients are better than one." *Nature*, 439 23–24.

Longstaff, A. (2000) *Neuroscience.* Oxford: Bios Scientific Publishers.

Lund, Nick (2002) *Animal Cognition*, Hove, Sussex, Routledge.

Lutcavage, Molly E., Plotkin, Pamela, Witherton, Blair & Lutz, Peter L. (1997) "Human Impacts on Sea Turtle Survival." In: Lutz (1997) pp.387–409.

Lutz, Peter, L. & Musick, John, A. eds. (1997) *The Biology of Sea Turtles*, (Vol. I) Boca Raton, Florida: CRC Press, Inc.

Lutz, Peter, L., Musick, John, A. & Wyneken, Jeanette. eds. (2003) *The Biology of Sea Turtles,* (Vol. II) Boca Raton, Florida: CRC Press, Inc.

Mackie, G.O. (2002) "What's new in Cnidarian Biology?" *Canadian Journal of Zoology*: 80 Number 10. 1649–1653.

Mackie, G.O., Pugh, P.R., & Purcell, J.E. (1987) Siphonophore biology. *Advances in Marine Biology*, 24:97–262.

Mackintosh, Nicholas (1987) "Animal Minds" In *Mindwaves* edited by Colin Blakemore and Susan Greenfield, Oxford.

Martin, Vicki, J. (Oct 2002) "Photoreceptors of Cnidarians" *Canadian Journal of Zoology*: 80 Number 10. 1703–1722.

Mason, Robert T. (1992) "Reptilian Pheromones." In: Gans, (1992) pp.112–228).

Maynard-Smith, John & Harper, David (2003) *Animal Signals*, Oxford: Oxford University Press.

Maynard-Smith, John & Eörs Zathmáry (1999) *The Origins of Life*, Oxford, Oxford University Press.

Mech, L.Davd & Boitani, Luigi. eds. (2003) *Wolves*, Chicago: University of Chicago Press.

Meglitsch, Paul A., Schram, Frederick, R. (1991) *Invertebrate Zoology*, New York: Oxford University Press.

Mill, P.J. ed. *Physiology of Annelids,* London: Academic Press.

Mill, P.J. "Sense Organs and Sensory Pathways." In: Mill, P.J. (1978) pp. 63–114.

Miller, Charles B. (2004) *Biological Oceanography*, Oxford: Blackwell.

Moore, Charles, J. Moore, S.L., Leecaster, M.K. & Weisberg, S.B. (2001) "A Comparison of Plastic and Plankton in the North Pacific Central Gyre." *Marine Pollution Bulletin* 42 No 12 1297–1300.

Moore, Charles. (2003) "Trashed: "Across the Pacific Ocean, plastics, plastics, everywhere." *Natural History* 112 n.9. November 2003.
Moore, Janet (2001) *An Introduction to the Invertebrates*, Cambridge: Cambridge University Press.
Moore, P.A. & Bergman, D.A. (2005) "The smell of success and failure: the role of intrinsic and extrinsic chemical signals in the social behavior of crayfish." *Integrative and Comparative Biology*. 45: 650–657.
Morgan, Lloyd, Conwy (1894) *An Introduction to Comparative Psychology*, London: Water Scott Ltd.
Morgan, Lloyd, Conwy, (1970) (orig. 1900) *Animal Behaviour*, New York & London: Johnson Reprint Company.
Morris, Charles (1971) *Writings on the General Theory of Signs*, the Hague: Mouton.
Morton, D. (2004) "Suffering" In: Berkoff, (2004).
Müller, Werther A., & Leitz, Thomas (2002) "Metamorphosis in the Cnidaria."*Canadian Journal of Zoology*, 80: 1755–1771.
Neumann, John von. (1974: Orig 1958) *The Computer and the Brain*, New Haven & London: Yale University Press.
Nevitt, G.A. (1999) (a) "Olfactory foraging in Antarctican seabirds: a species specific attraction to krill odors." *Marine Ecology Press Series* 177: 235–241.
Nevitt, G. A., (1999) (b), "Foraging by Seabirds on an Olfactory landscape", *American Scientist*. 87: 46–53.
Nielsen, A.S., A.W. Pedersen & Riisgaard, H.U. (1997) Implications of density driven currents for interaction between jellyfish (*Aurelia aurita*) and zooplankton in a Danish fjord. *Sarsia*, 82(4): 297–305.
Nilsson, Dan-E, Gislen, Lars, Coates, Melissa M., Skogh, Charlotta, Garm, Anders. (12 May 2005) "Advanced optics in a jellyfish eye. *Nature* 435: 129–246.
Nolfi, Stefano, Floreano, (2000) *Evolutionary Robotics*, Cambridge Mass.: MIT.
Nordeng, H. (1971) "Is the local orientation of anadromous fishes determined by pheromones." *Nature* 223: 155–159.
Ogden, C.K. & Richards, I.A. (1949 8th Edition*) The Meaning of Meaning*, London: Routledge & Kegan Paul.
Orenstein, Ronald (2001) *Turtles, Tortoises and Terrapins*. Buffalo, N.Y.: Firefly Books.
Owings, Donald H., Beecher, Michael D., Thompson, Nicholas S. eds. (1997) *Perspectives in Ethology* (volume 12: Communication), New York: Plenum Press.
Owings, Donald H & Morton, Eugene S. "The role of information in communication: An assessment/management approach." (In Owings et al. (1997) pp.359–390.
Pain, Stephen (2006) "The Ant on the Kitchen Counter." *Journal of Biosemiotics* (Forthcoming).
Paradis, A.R., Pepin, P. & Brown, J.A. (1996) "Vulnerability of fish eggs and larvae to predation: review of the influence of the relative size of prey and predator.*"Can. J. Fish. Aquat. Sci.* 53: 1226–1235.
Pechenik, Jan. A. (2000) *Biology of Invertebrates*, New York: McGraw-Hill.
Piatigorsky, Joram & Kozmik, Zbynek (2004) "Cubozoan jellyfish: An Evo/Devo model for eyes and other sensory systems." *International Journal of Development Biology*. 48: 719–729.
Purchell, J.E. and M.N. Arai. "Interactions of pelagic cnidarians and ctenophores with fish: a review. "*Hydrobiologia* 451 (*Developments in Hydrobiology*, 155): 1–9.
Pough, F. Harvey, Andrews, R.M., Cradle, J.E, Crump, M.L., Savitzky, A.H. & Wells, K.D. (2001) *Herpetology*, (Second Edition) Upper Saddle River, N.J.: Prentice Hall.
Puttnam, Hilary *Representation and Reality*, Cambridge, Mass.: MIT.
Quinn, Thomas, P. (2005) *The Behaviour and Ecology of Pacific Salmon and Trout*, Bethesda, Maryland., American Fisheries Society in association with University of Washington Press.
Quine, W.V. (1995) *From Stimulus to Science*, Cambridge, Mass.: Harvard University Press.
Raines, Susan J. & Greenberg, Gary in "Approach/Withdrawal Theory" In Greenberg & Haraway, 1998 74–78.
Revest, Patricia & Longstaff, Alan (1998) *Molecular Neuroscience*, Oxford: bios Scientific Publishers limited.

Richardson, Alan W. (1998) *Carnap's Construction of the World*, Cambridge: Cambridge University Press.
Ridley, Mark (2004) *Evolution*, (3rd Edition) Oxford: Blackwell.
Robson, E.A. (1975) "The Nervous System in Coelenterates" In: Usherwood (1975).
Romanes, George John (1885) *Mental Evolution in Animals*, London: Kegan Paul, Trench.
Romanes, G.J. (1885) *Jellyfish, Starfish, and Sea Urchins, being a Research on the Primitive Nervous System*, New York: Appleton.
Roy, Asim (2000) "Artificial Neural Networks-A Science in Trouble." *ACM SIGKDD* January, Volume I, Issue 2, p. 38.
Rushforth, Norman B. (1973) "Behavioral Modifications in Colenterates" In: W.C. Corning and J.A. Dyal (1973), pp. 123–169.
Satterlie, R.A. (2002) "Neuronal control of swimming in jellyfish: a comparative story." *Canadian Journal of Zoology*, 80:1654–1669.
Shannon, Claude E. & Weaver, Warren (1998) *The Mathematical Theory of Communication*, Urbana, University of Illinois Press.
Schomaker, Lambert, "Anticipation in Cybernetic Systems: A Case Against Mindless Antirepresentionalism." http://www.ai.rug.nl/~lambert/papers/ieee-smc-2004-schomaker.pdf.
Schrödinger, Erwin (1948) *What is Life*, Cambridge: Cambridge University Press.
Schrödinger, Erwin (1959) *Mind and Matter*, Cambridge: Cambridge University Press.
Sebeok, Thomas. & Ramsay Alexandra eds. (1969) *Approaches to Animal Communication*, The Hague: Mouton.
Sims, R.W., Gerard, B.M. *Earthworms* (Synopses of the British Fauna (New Series). Shrewsbury, Field Studies Council.
Singleton, Paul, (1997) *Bacteria*, Chichester: John Wiley & Sons.
Slater, P.J.B. (1999) *Essentials of Animal Behaviour*, Cambridge: Cambridge University Press.
Smythies, John (2002) *The Dynamic Neuron*, Cambridge, Massachusetts: MIT.
Steinbeck, John (1977) *The Log from the Sea of Cortez*, London: Pan Books.
Stewart, Amy (2004) *The Earth Moved*, London: Frances Loncoln.
Stout, G.F. (1919) *A Manual of Psychology*, (3rd Edition) London: W.B. Clive; University Tutorial Press.
Suchman, Cynthia L. & Sullivan, Barbara K. (2000) "Effect of prey size on vulnerability of copepods to predation by the scyphomedusae *Aurelia aurita* and *Cyanea sp*." *Journal of Plankton Research* 22, no. 12: 2289–2306
Thorndike, Edward L. (1911) *Animal Intelligence*, New York: Macmillan Press.
Thompson, Richard C. & Olsen et al., (2004) "Lost at Sea: Where is all the Plastic?" *Science* 304: 838.
Tibbets, Elizabeth, A., Dale, James (2004) "A socially enforced signal of quality in a paper wasp." *Nature* 432: 218–219.
Tinbergen, Niko (1951) *The Study of Instinct*, Oxford: Oxford University Press.
Tudge, Colin (2000) *The Variety of Life*, Oxford: Oxford University Press.
Uexküll, Jakob von (1982) "The Theory of Meaning." Edited and introduced by Thure von Uexkull *Semiotica*, 42–1 1–87.
Uexküll, Jakob von 1992. "A Stroll through the Worlds of Animals and Men." Edited and introduced by Thure von Uexkull, *Semiotica*, 89–4: 279–391.
Usherwood, P.N.R. & Newth, D.R. eds. (1975) *Simple Nervous Systems*, London: Edward Arnold.
Vauclair, Jacques (1996)*Animal Cognition*, Cambridge Mass., Harvard University Press.
Vincent, Jean-Didier, (translated by John Hughes) (1990) *The Biology of Emotions*, Oxford: Basil Blackwell) p-87)
Vincent, S.R. (1996) "Nitric oxide and synaptic plasticity: NO news from the cerebellum." *Behavioral and Brain Sciences* 19(3): 362–367.
Wall, Patrick D. & Melzack, Ronald eds. (1994) *Textbook of Pain*, (Fourth Edition). Edinburgh, London, New York; Church Livingstone.
Wallace, Robert L. & Taylor, Walter K. (1997) *Invertebrate Zoology: A Laboratory Manual*, New Jersey: Prentice Hall.

Wallwork, John A., (1983) *Earthworm Biology*, London: Arnold.
Watson, John, B. (1913) *Behavior*, New York: Henry Holt and Company.
Wehner, Rudiger (2003) "Desert Ant Navigation: How Miniature Brains solve complex tasks." *J. Comp. Physiol.* A. 189: 578–588.
Weissburg, Marc J. & Browman, Howard I. (2005) "Sensory biology: linking the internal and external ecologies of marine organisms" – *Marine Ecology Press Series* 287: 263–307
Weissburg, Marc J. (2005) "Introduction" In: Weissburg, Marc J. & Browman, Howard I. (2005).
Wenner, Adrian M. "The Study of Animal Communication: An Overview." In Sebeok & Ramsay, 1969, pp.232–243.
Wiener, Norbert, (1948) 1969. *Cybernetics* (Second Edition), Cambridge, Mass., MIT Press.
Wiese, Konrad (Ed.) 2002 *The Crustacean Nervous System.* New York; Berlin: Springer.
Williams, P.J. Brown, J.A., Gotceitas, V. Pepin, P. 1996 "Developmental changes in escape response performance of five species of marine larval fish. Can. J. Fish Aquat. 53 1246–1253.
Wilson, Robert, A., Keil, Frank, C. eds. 1999 *The MIT Encyclopedia of the Cognitive Sciences.* Cambridge, Mass.: MIT.
Woodger, J.H., 1929 *Biological Principles.* London: Kegan Paul Trench, Trubner & Co., Ltd.
Wyatt, Tristram, 2003. *Pheromones and Animal Behaviour*, Cambridge, University of Cambridge Press.
Wyneken, "Sea Turtle Locomotion Mechanisms, Behavior and Energetics." (In Lutz, 1997, pp. 165–198)
Young, J.Z., (1978) *Programs of the Brain.* (Based on the Gifford Lectures 1975–1977) Oxford: Oxford University Press.
Zupanc, Günther K.H., 2004. *Behavioral Neurobiology*, Oxford, Oxford University Press.

CHAPTER 17

A BIOSEMIOTIC APPROACH TO EPIGENETICS: CONSTRUCTIVIST ASPECTS OF OOCYTE-TO-EMBRYO TRANSITION*

JOHANNES HUBER[1,2,†] AND INGOLF SCHMID-TANNWALD[1,‡]

[1] *Faculty of Medicine, Ludwig-Maximilians-University, Munich, Germany Bachstr. 26, 69121 Heidelberg, Germany. johannes.huber@gmail.com*

[2] *Faculty of Philosophy, Munich School of Philosophy S.J. Hospital Campus Großhadern, Marchioninistr. 15, 81377 Munich, Germany.ingolf.schmidtannwald@med.uni-muenchen.de*

Abstract: Recent findings in molecular cell biology of mammalian reproduction demonstrate the importance of epigenetic mechanisms taking place during the oocyte-to-embryo transition. Employing a semiotic framework this epigenetic 'reprogramming' can be shown to be a subject's active achievement that enables the construction and activation of a diploid embryonic genome. Fertilization therefore is not merely the physical union of sperm and ovum as the paternal contribution is not just received passively, but has to be constructed actively by the perceiving organism: its entire structure has to be "put in form". Thus the oocyte-to-embryo transition can be regarded to be a natural semiotic process.

When applying a general model of semiosis, the biological process of mammalian reproduction displays a structural analogy with the course of perception: the oocyte – a single-celled mammalian organism capable of creating *Umwelt* – extracts information from its environment. As soon as the pronucleus of the spermatozoon is incorporated into the oocyte, active and specific transformations are performed on the paternal genome. By exchanging certain proteins the chromatin structure is altered and a significant and selective demethylation of the paternal genome takes place elucidating the creative role of the zygote organism as an interpreter. All these epigenetic modifications are realized by maternal means and do alter the content of information given by the paternal genome. Therefore, the activation of the embryonic genome seems to reveal basic patterns of constructivist epistemology.

*This paper is based on "Epigenetic mechanisms following mammalian fertilization reveal basic principles of constructivist epistemology" presented by the authors at the 5th Gatherings in Biosemiotics in Urbino, July 22, 2005.

† Hospital Campus Großhadern Marchioninistr. 15 81377 Munich, Germany Phone: +49-89-7095-4696, ingolf.schmid-tannwald@med.uni-muenchen.de

‡ St.-Martin-Str. 40 81541 Munich, Germany Phone: +49-89-69760146, johannes.huber@gmail.com

M. Barbieri (ed.), Introduction to Biosemiotics, 457–471.

As obviously the found principles are not an exclusive feature of perception performed by so-called higher organisms, but a general characteristic of cellular life, 'construction of information' seems to be a basic quality of life itself. Moreover, the proposed interpretation could be a valid argument for evolutionary epistemology since basic mechanisms of perception can even be found in a single-celled organism, as this subject is able to arrange and interpret its genome. Conclusively, it can be postulated that development of cognition starts and started within a single cell – ontogenetically as well as phylogenetically. Another benefit of our semiotic analysis may be the opportunity of overcoming genetic reductionism

Keywords: epigenetics, oocyte-to-embryo transition, constructivism

"There is perhaps no phenomenon in the field of biology that touches so many fundamental questions as the union of the germ cells in the act of fertilization; in this supreme event all the strands of the webs of two lives are gathered in one knot, from which they diverge again and are re-woven in a new individual life-history"

(Lillie 1919)

INTRODUCTION

At the very beginning of the development of every new mammalian organism lies the point where an individual genome has to be established. This process happens right after fertilization, which can be defined as the physical union of the sperm and the ovum to yield a zygote (Wassarman et al. 2001). Our analysis tries to show that forming an embryonic genome in mammals is more than just the physical union of both parental genomes within a single cell, but a complex semiotic process.

Reproduction in human beings has gained some special features due to the evolution of a highly elaborate psychosocial context surrounding this biological phenomenon. As a generation of human beings procreates the subsequent one, three different phases can be distinguished recurring over and over again (Figure 1).

These phases are defined according to the involvement of the three different system levels proposed by Thure von Uexküll and Wesiack (1998) in their bio-psycho-social model of human beings. Therefore, on the left side the cellular

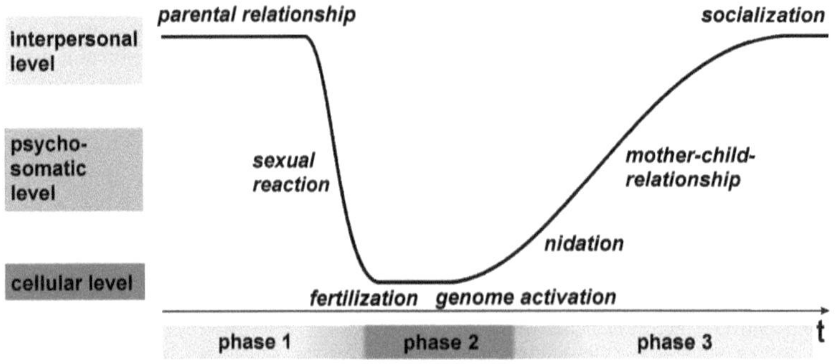

Figure 1. Context of the oocyte-to-embryo transition

level corresponds to the biological level, the psychosomatic level includes the psychological dimension, and the interpersonal level represents the social system level. The interpersonal relationship of the parents (phase 1) involving all of the three levels constitutes the previous social history of the subsequent generation (Schmid-Tannwald 2001) in which at a certain time a sexual reaction leading to offspring takes place. This sexual reaction can be regarded as a bio-psycho-social event bridging the social and the cellular level transferring the parental interpersonal relationship to an intercellular level. The process leading from complementary gametes to the early embryo can be summarized as the oocyte-to-embryo transition including fertilization and the activation of the newly established embryonic genome (phase 2). Subsequently, the biological and later the psychosocial development of the new human being increase the complexity until the social level is reached again (phase 3): nidation implements a somatic interaction between mother and child that reaches a psychological dimension during intra-uterine development establishing the mother-child-relationship. The process of socialization taking place after birth occupies many years.

In a similar way, the presented idea that human life depends on an "oscillation" between different levels of complexity was already mentioned by Barbieri (2003), Hoffmeyer and Emmeche (2005) as well as Maturana and Varela (1992). It can thus be seen from our illustration that the oocyte-to-embryo transition holds a very central position in the change of generations. This cellular process sets the precondition for reconstructing a human being from incomplete information – such as the zygote – and increasing complexity (Barbieri 2003) from a cellular level regaining a bio-psycho-social structure of organization. This important phase of the human life cycle seems to be worth a biosemiotic investigation because although "the internal structure of cells or organisms is probably describable in purely biochemical terms, this will not give us a true understanding of such structures" (Hoffmeyer 1997). Especially the epigenetic mechanisms involved during this early stage of development that are often referred to as 'reprogramming' (Dean et al. 2003; Rideout III et al. 2001) turned out to be crucial for an interpretation process that can not sufficiently be explained by a mechanistic description.

We employed a general model of semiosis put forward by Krampen (1997) in order to develop a clear methodology for analyzing the process of oocyte-to-embryo transition. The semiosic matrix offers the opportunity to differentiate the components involved, thereby assigning them their biological function and meaning in the sign process. Its importance for studying semiotics in nature has recently been stressed by Ponzio (2004).

A GENERAL MODEL OF SEMIOSIS: THE SEMIOSIC MATRIX

Basically, the semiosic matrix can be understood as a part of a general communication system (Figure 2) representing the perception of a signal, its interpretation, and behavioral consequences due to effectors of the interpreter. Shannon's (1948) model bears analogy only concerning the basic structure of the semiosic matrix, but it does not describe the components involved correctly as the features of the interpretant are far more complex than the concept of a receiver.

A slightly simplified version of the semiosic matrix is shown in Figure 3. Like the model shown above, the flow chart is to be read from left to right with a perception side (receptor) on the left and a behavior side (effector) on the right. As the crucial mediating step in between, an interpreter carries out semiosis; in higher organisms,

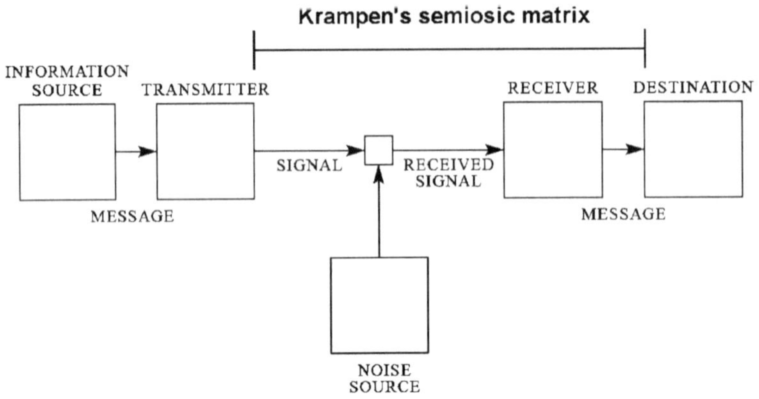

Figure 2. Schematic diagram of a general communication system (modified from Shannon 1948)

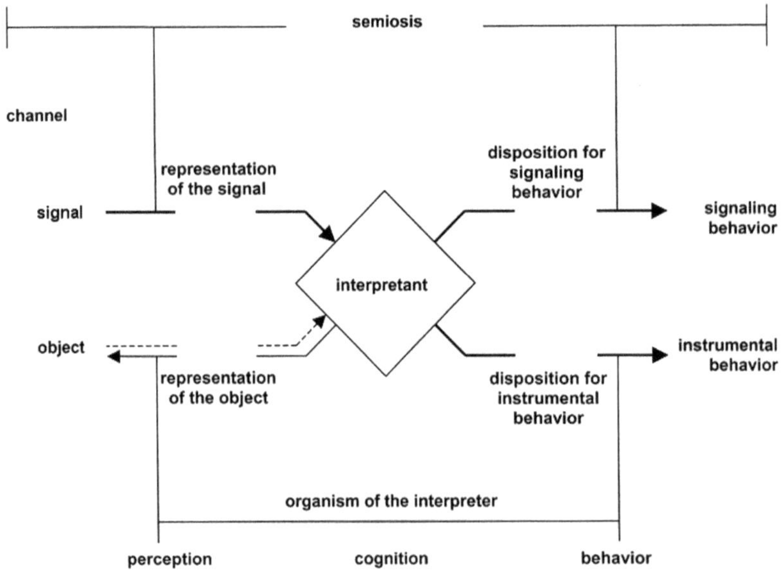

Figure 3. The semiosic matrix (after Krampen 1997)

this process is called cognition. The limits of the interpreter are indicated by the two vertical lines on either side of the rhombus representing the interpretant.

Semiosis begins with the channel-mediated perception of a signal – a material sign vehicle – that is received and represented inside the organism. As it becomes interpreted by the organism, a triadic sign relation is established according to Peirce (1931–58) between sign(al), interpretant and object. Moreover, Krampen (1997) stresses the central meaning of the interpretant in the process of semiosis as it can lead to a change in the disposition of the organism for different behavior. Following Tembrock (1971) behavior can be divided into signaling and instrumental behavior, and both can be changed by the process of semiosis due to an altered disposition for both kinds of behavior.

In contrast to the imputed relation of 'the sign standing for the object', the arrows leading from signal via interpretant to object represent a causal relationship of meaning. The dashed line between the object and the interpretant indicates that an object has to be perceived and stored within the organism (has to be "known") first, before reference to it can be made.

Semiosis as described above can, therefore, be defined (after Krampen 1997) as a channel-mediated process in which a sign is related to an object by being perceived and represented within the organism of the interpreter; due to being mediated by an interpretant the representation of the signal is connected with the representation of the object, which represents the object within the organism. Via the interpretant, this process of symbolizing and referring triggers dispositions for instrumental and signaling behavior, that are both related to the object and terminate in overt behavior of both kinds.

The pivotal role of the interpretant is indicated by its central position in the semiosic matrix. Therefore, a very important pre-requisite for semiosis is the existence of an interpretant contained inside the interpreter organism. In order to prove this feature present it is crucial to show that there is a subject capable of interpreting during the oocyte-to-embryo transition.

APPLYING THE SEMIOSIC MATRIX TO THE OOCYTE-TO-EMBRYO TRANSITION

The feature of reacting as a subject becomes obvious when realizing that gametes and zygote are mammalian cells with the ability of creating *Umwelt.* According to Jacob von Uexküll's theory (Uexküll, J. von 1996), every single-celled organism creates a primitive kind of *Umwelt* by which it is surrounded. The elements constituting this individual reality are actively chosen by the organism's receptors, proving that certain entities matter for that respective organism because it attaches a meaning to former meaningless stimuli. During the process of forming a multicellular organism this ability is lost on the cellular level because it is transferred to the organism as a whole. Therefore, single mammalian cells integrated in a complex multicellular system interfere with each other, however, are not able to create their own *Umwelt.* During gametogenesis two different kinds of short-living single-celled organisms of the same species are generated and the ability of creating *Umwelt* is regained again. This special property of the gametes is necessary for successful fertilization, therefore showing that information from outside is obtained actively. The mechanisms of recognizing the complementary gamete follow a cascade of biochemical reactions leading to the formation of the zygote. In every mammalian species this zygote represents a third kind of single-celled organisms.

As the oocyte is capable of creating *Umwelt* therefore embodying a complex interpretant, we can now apply the semiosic matrix to the oocyte-to-embryo transition and analyze the fit (Figure 4).

The oocyte-to-embryo transition can be regarded as a perception process mediated by the fertilization cascade: the sperm cell (signal) is received by the oocyte and transformed into the male pronucleus (representation of the signal). Due to protein biosynthesis representing a part of the cellular organization of the oocyte, the paternal genome is linked to its gene products by employing the base sequence of certain encoding domains. By this means, a biosemiotically well described sign relation is established (Hoffmeyer 2002). Another crucial feature of the interpretant is the ability to regulate the activity of certain gene loci. This interplay of activation and silencing (Felsenfeld and Groudine 2003) is achieved by epigenetic mechanisms (see below). Thereby, the disposition of the interpreter for behavior is changed: by activating gene loci, which encode short-chained proteins functioning as hormones (such as β-HCG or the 'Early Pregnancy Factor'), the disposition for sending embryonic signals is

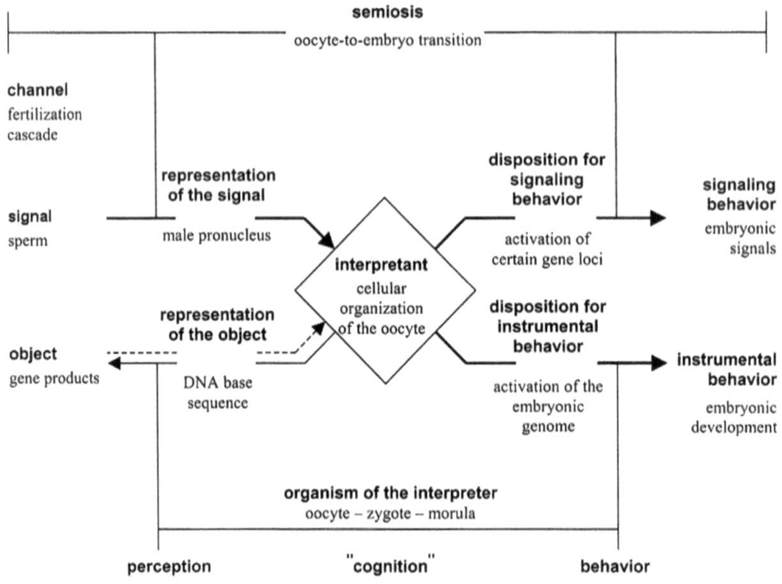

Figure 4. The semiosic matrix applied to the oocyte-to-embryo transition (after Krampen 1997)

created and an embryo-maternal-dialogue becomes initiated (Herrler et al. 2003). Moreover, the proper activation of the embryonic genome is essential for further development.

One of the key features of fertilization appears to be the perception of information from outside the organism. There is no other event in biological processes during mammalian development in which such a significant amount of genetic information is acquired from outside the perceiving organism. Nevertheless, the use of the term "cognition" is surely wrong in a cellular context, but its equivalent on this level of organization seems nevertheless to be a complex process of semiosis. Some of the elements decisive for the oocyte-to-embryo transition are in need of closer investigation within the given scope. For a detailed discussion of all the elements involved please refer to Huber (2006a).

The signal perceived by the oocyte is the sperm consistent with the conception of an organism functioning as a message (Sharov 1998). The sperm cell is received by a quite complex channel made up by the fertilization cascade on the cellular level allowing to perceive the male pronucleus as the main representation of the signal inside the organism of the interpreter. This representation is more than just a shifting between different compartments as will be shown in the next section. Moreover, recent data indicates that in addition to the well-known paternal contributions such as the haploid genome and the second centriole even small amounts of paternal RNA could have important effects on early embryogenesis (Krawetz 2005).

The interpreter in this context is the oocyte. As it develops towards the early embryo its maternal interpretant stays the same until about the 8-cell stage although the name given to the cellular organism embodying it changes from oocyte via zygote to morula (cf. Figure 6). The reason therefore lies in the maternal origin of the interpretant, which is not significantly altered before the 8-cell stage. Translation of embryonic mRNA begins not before the cleavage of the 4-cell embryo, and consequently no expression of the new-established embryonic genome takes place until then (Schultz et al. 1999). Only after this developmental stage there is an embryo-specific protein pattern distinguishable from the one of the oocyte (Schultz 2002). As actual embryonic proteins do not exist, and as the entering sperm did not carry a significant amount of proteins, the oocyte-to-embryo transition is sustained by maternal means. This interpretation is strongly supported by recent findings that parthenogenesis is possible also in mammals – at least in principle (Kono et al. 2004).

The maternal interpretant is extremely complex and includes the whole cellular organization of the oocyte. This broad conception of the interpretant finally relies on the well-known axiom *omne vivum e vivo* (all living from living) as only a complete cellular organization can initiate the development of a new living being. The two main fields of interpretation performed during early development are protein biosynthesis and the activation of the embryonic genome (Pesce and Schöler 2001). Some of their necessary components comprise biomembranes, enzymes, structural proteins, educts of cellular metabolism (Harvey et al. 2002), mitochondria, energy in form of ATP, nucleic acids and ribosomes. All of these components are structures and molecules that have been produced and stored inside the oocyte during oogenesis allowing the possibility of early embryonic development (Gosden 2002).

Another feature of the interpretant is its enormous potential and flexibility that becomes obvious in somatic cell nuclear transfer experiments (Wilmut et al. 2002). These show that even differentiated somatic nuclei can be reprogrammed into embryonic nuclei by the oocyte. Great technical difficulties and quite limited success rates demonstrate the complexity of the process on the other hand.

Besides the activation of the embryonic genome, protein biosynthesis is an important part played by the interpretant that realizes DNA-coded information by transcription and translation. This is also a biosemiotic process (Emmeche 1999; Lumsden 1986) essential for further development. The necessary system therefore consisting of RNAs, ribosomes and energy constitutes the 'ribotype' (Barbieri 1981) and represents part of the interpretant. But how do the construction of the embryonic genome and its activation take place?

EPIGENETIC MECHANISMS FOLLOWING MAMMALIAN FERTILIZATION REVEAL BASIC PRINCIPLES OF CONSTRUCTIVIST EPISTEMOLOGY

The mechanisms enabling establishment and activation of the embryonic genome do not change the base sequence of the parental genomes; however, the gene expression patterns are altered massively. Therefore, we deal with epigenetic modifications reprogramming the pronuclei. 'Epigenetic' in the present context can be defined as "any heritable influence (in the progeny of cells or of individuals) on gene function that is not accompanied by a change in DNA sequence" (Li 2002).

The existence of these epigenetic mechanisms becomes easily evident: the fact that all of the hundreds of different cellular phenotypes in multicellular organisms arise from the same genome highlights the importance of some kind of regulation above the genetic level in addition to mere environmental factors. Only a powerful cellular epi-genetic (in ancient Greek 'επι' means 'on top of') control can lead to the expression of a cell-specific subset of genes in a defined quantity generating such different cells like hepatocytes, fibroblasts and neurons. It was one of Waddington's (1957) great merits to realize that the connection between genes and development must be multi-causal and highly cross-linked. Consequently, he coined the term epigenetic in its present meaning.

During the last decade, epigenetics became an important discipline in molecular biology and medicine, as correct epigenetic regulation is essential for successful development. This is shown by a great number of epigenetic errors in cancer cells, epigenetic diseases or the low efficiency rate in cloning. Consistent epigenetic mechanisms have great impact on cellular key functions such as the regulation of gene expression (Jaenisch and Bird 2003), imprinting (Reik and Walter 2001), and silencing of tumor suppressor genes or foreign (e.g. viral) DNA (Wolffe and Matzke 1999). Therefore, a deeper understanding of epigenetics might enable progress in medical treatment of several diseases (Miyamoto and Ushijima 2005) and reasonable measures concerning novel techniques in reproductive medicine and gene therapy (Huber 2006b; Johnson 2005).

Epigenetic regulation takes place on two principle levels with the cellular context affecting both strongly, so that it could be considered a third one. The first level consists in covalent chemical modifications of the DNA molecule mainly of methylation of cytosine bases. This mechanism commonly referred to as DNA methylation is described below in more detail to exemplify our interpretation. On a second level, the packaging of the DNA double helix is changed, thereby altering the accessibility of certain parts of the genome. DNA and histones are the basic components of a chromosome, in which the DNA helix is wrapped around core histones and then folded into higher-order chromatin. Chemical modification of DNA-associated proteins like histones (Turner 2000) and chromatin remodelling (Li 2002) lead to changes in the spatial structure of the genome and thus influence the quantitative degree of transcription. According to the effect on gene expression, the underlying chromatin structure is characterized to be either permissive or repressive (Eberharter and Becker 2002). Moreover, even the spatial organization of these different chromatin stages inside the cell nucleus seems to reflect a topological architecture with functional relevance for gene expression (Cremer and Cremer 2001). Finally, the cellular context has great impact on these regulations as they are influenced by cytoplasmic phenomena such as cellular metabolism, signaling pathways and enzymes. Therefore, the cytoplasm of a mature oocyte is characterized as a reprogramming milieu (Solter et al. 2005).

But what happens on these levels during the oocyte-to-embryo transition? The cellular context is changed dramatically especially for the male pronucleus: the paternal genome enters the reprogramming milieu of the oocyte, and the formerly quiescent egg metabolism is activated due to fertilization. The former sperm nucleus is unwrapped from protamine – the histone equivalent in sperm that enables an even tighter packing of the DNA – and remodelled by using maternal histone proteins. This process hits very much the mechanistic thinking of putting "in form" (Uexküll, Th. von and Wesiack 1998) as the packaging of the paternal genome is massively altered and therefore the spatial appearance is changed. Moreover, both pronuclei are then transformed from a genetically quiescent structure into a functional embryonic genome. It has been shown that the regulation of higher-order chromatin structures by DNA methylation and histone modification is crucial for genome reprogramming during early embryogenesis (Dean et al. 2003). But even on the level of DNA modifications massive changes are performed.

The active modifications in DNA methylation patterns demonstrate impressively how the cellular organism of the oocyte constructs the perceived information by reprogramming the male genome (Reik and Dean 2001). In order to comprehend the far-reaching consequences of changes in DNA methylation, we present a strongly simplified model of how this epigenetic modification works (Figure 5).

Fundamental for gene regulation in mammals (Jones and Takai 2001), DNA methylation is basically just the addition of a methyl group at the hydrogen position of the nuclear base cytosine, which results in the modification of cytosine into

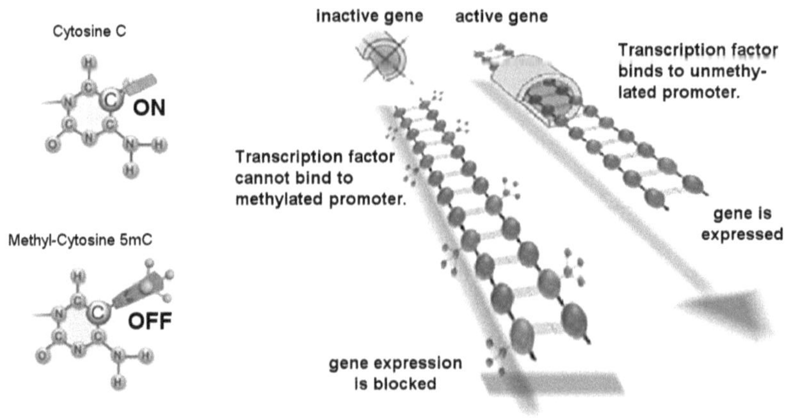

Figure 5. Cytosine methylation can be understood as a signal influencing gene expression (reproduced with the kind permission of Epigenomics AG, http://www.epigenomics.com)

5-methyl-cytosine. Methylation is performed more often in certain base sequences that are quite frequent in promoter regions of genes (CpG islands). If many of the cytosine bases are methylated, the transcription factors initiating transcription cannot bind to the DNA strand, and the gene is not expressed. In contrast, a low methylation level in promoter regions leads to a significantly higher expression rate. Therefore, a low degree of methylation means 'on' concerning gene expression and a high degree stands for 'off'.

These marks regulating gene expression and contributing to the information contained within the genome are actively changed in the paternal genome during the oocyte-to-embryo transition. Experimental data from the mouse model provides evidence for the active demethylation of the paternal genome, and in most mammalian species including humans these findings have been confirmed (Beaujean et al. 2004). Only in sheep and rabbit there seem to be divergent dynamics in the change of overall methylation during reproduction (Young and Beaujean 2004), which might be related to different timing of embryonic genome activation.

After fertilization in mice, demethylation of the parental genomes takes place (Figure 6). However, this does not affect them to an equal extent. The paternal genome is significantly and actively demethylated within 6-8 hours after fertilization, whereas the maternal genome looses its methylation slowly and passively as existing methylations are not re-established after DNA replication during several cleavage divisions. A repressive chromatin structure is thought to protect the maternal genome against the extensive epigenetic modifications imposed on the paternal genome. Active demethylation of the paternal genome involves a genome-wide erasure of DNA methylation and then a period of locus-specific *de novo* methylation that might establish parent-specific developmental programs during early embryogenesis (Dean et al. 2003).

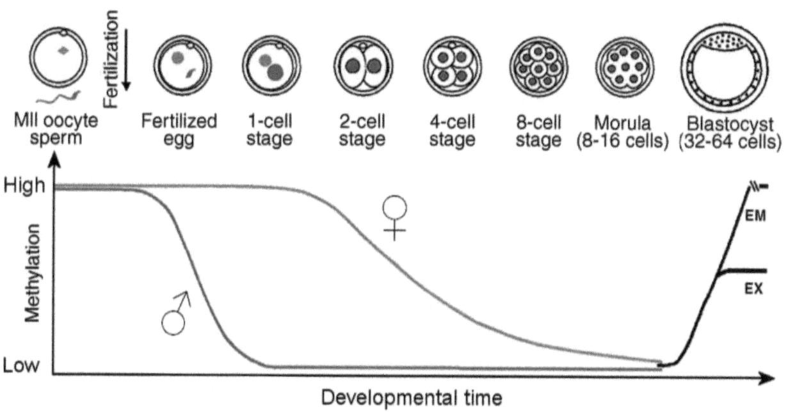

Figure 6. Dynamic reprogramming of global methylation in mammalian pre-implantation embryos (reprinted from Dean et al. 2003 with permission from Elsevier).
The paternal mouse genome (♂) undergoes active demethylation while the maternal genome (♀) is passively demethylated. *De novo* methylation (black line) is observed after the morula stage differing in embryonic (EM) and extra-embryonic (EX) lineages

Thus, immediately after fertilization the oocyte's cellular mechanisms create the informational content of the received paternal genome actively within the process of perception. This is exactly what the key statement of constructivist epistemology is all about: perception is not considered to be a reflection of an independently existing reality, but an active transformation of received signals into a perceived and thereby constructed subjective reality; elements of the world surrounding us are actively "put in form" (Uexküll, Th. von and Wesiack 1998), which correlates with the hermeneutic explanation of 'information' deriving from the Latin word *in-formare* (put in form).

The position of constructivist epistemology argues that our complete knowledge about the world is a construction performed by our brain that relies on sensually perceived information. As this input is not an image of the world, but mere data subjected to the restrictions of our sense organs, the resulting construction allows only limited access to the world. Therefore, objective knowledge is impossible leaving us with inter-subjectivity as the highest achievable degree of certainty. Although criticism of a positivist world view dates back until ancient Greek philosophy, some of the most important founders of this epistemological position were Kant (1787), Berkeley (1710) and Vico (1710), whose dictum "*verum ipsum factum*" (the true is the same as the made) already pointed in the direction of knowledge construction.

As we have seen, even single-celled organisms like the oocyte do not perceive a signal (like the sperm) just as it is, but modify its content actively and extensively in order to imbed this constructed information in some kind of "cellular subjective reality". Therefore, epigenetic mechanisms following fertilization reveal that elements of the world outside are actively put "in form" by even a single-celled organism (Uexküll, Th. von and Wesiack 1998).

DISCUSSION

By applying a general model of semiosis (Krampen 1997) we tried to 'operationalize' our scientific proceeding providing a comprehensible methodology. At the same time, we kept ontological premises at an absolute minimum. According to our semiotic analysis the oocyte-to-embryo transition embodies a crucial act of interpretation performed by the oocyte organism. As soon as the pronucleus of the spermatozoon is incorporated into the oocyte, active and specific transformations are performed on the paternal genome by maternal means, thus altering the content of information given by the paternal genome. We agree that fertilization is not an exclusive example for single cells constructing information, but it seems to be a quite evident one.

Talking of 'perception' on a cellular level certainly should not implicate the existence of a mind or cognition as it is found in so-called higher organisms, but emphasize the fact that every living organism follows semiotic causality. This means that emergent phenomena in living organisms cannot sufficiently be comprehended by a pure mechanistic description. A strong argument for this claim is that a perceiving subject is imperative in order to attach different meanings to the same material entity. This circumstance can be regarded a process of signification as it is a main characteristic of signs to contain the potential of meaning different things in different contexts. The paternal genome can have different meanings depending on the interpretation performed by a cellular subject: a male pro-nucleus transferred into a somatic cell (e.g. an epithelial cell) will never gain the same biological meaning as it does during the oocyte-to-embryo transition, because different signal processing structures influence the content of the signal perceived.

This capacity of giving different meanings to the same material entity is founded within the cellular organization of living organisms (Huber and Kummer in press a, b), because even on a cellular level – representing the basic level of life – "biological reality" is constructed depending on a subject's perception. Living organisms do not passively receive signals from their surrounding, but construct their *Umwelt* actively. Therefore, construction and processing of information seems to be a basic quality of life itself and should be defined as one of its characteristics.

Moreover, a deeper understanding of evolutionary epistemology is rendered possible since basic mechanisms of perception – suggested to be the central one by constructivists – can even be found in a single-celled organism. Construction of information did not start in the heads of persons (Glasersfeld 1995), but it began evolving already with the origin of life. Thus, development of cognition can more coherently be postulated to have started in a single cell – ontogenetically as well as phylogenetically.

Keeping this evolutionary perspective helps understanding how "conventional codes" (Barbieri 1985) might have been established. The feasibility of fertilization depends essentially on the complementary nature of the gametes. As sexual reproduction is considered to have evolved from an asexual mitotical reproduction in single-celled organisms, these conventions must have been established synchronously with the developmental switch towards sexual reproduction and are therefore relying on the common origin. This thought is consistent with our proposed interpretation because in "communication the interpretant mostly functions according to a conventional code" (Krampen 1997). Also Bateson and Bateson (1987) mentioned that there must be some kind of fine adjustment happening between the gametes: "At the moment of fertilization (...) each gamete is a validating template for the other. What is surely tested is the chromosomal constitution of each, but no doubt the similarity of the whole cellular structure is also verified" – and verifying a perceived signal undoubtedly is also an active process of interpretation.

The realization of "subjectivity" in biology might also help overcoming the still quite popular fallacy of genetic reductionism. Sadly, this misconception of the genome is still very common although it lacks scientific foundation. As can be seen from our analysis of oocyte-to-embryo transition, the importance of a subject that is able to arrange and interpret its genome must be highly esteemed. An individual genome is not a "blueprint" of the adult organism and it is not even something *like* a "blueprint"! It can rather be compared to a mail-order catalogue that supplies the cell with enzymes and structural proteins according to the orders given by the cellular subject (Seidel 2001). This image provides a better understanding of the existing hierarchy: the cellular organism (subject), which of course is strongly influenced by environmental factors, regulates gene expression (what and how much is ordered) by epigenetic means (ordering form).

Epigenesis is all about activating or silencing parts of the genome and thereby constructing different phenotypes from the same genome. Therefore, wrong metaphors like 'book' or 'blueprint' should be avoided and substituted by functional descriptions that pay respect to the importance of the interpreting subject. Batesons' intuition suggests that, too: "It seemed to me that we might think of the state of the egg immediately before fertilization as a state of a question, a state of *readiness to receive a certain piece of information*, information that is then provided by the entry of the spermatozoon." (Bateson and Bateson 1987)

ACKNOWLEDGMENTS

Part of this study was developed within the scope of Johannes Huber's dissertation performed at the Faculty of Medicine, Ludwig-Maximilians-University, Munich (Huber 2006a).

REFERENCES

Barbieri, M. (1981) The ribotype theory on the origin of life. *Journal of Theoretical Biology* **91**: 545–601

Barbieri, M. (1985) The semantic theory of evolution. London: Harwood Academic Publishers

Barbieri, M. (2003) The organic codes: an introduction to semantic biology. Cambridge, New York: Cambridge University Press

Bateson, G., Bateson, M.C. (1987). Angels fear: towards an epistemology of the sacred. New York: Macmillan

Beaujean, N., Hartshorne, G., Cavilla, J., Taylor, J., Gardner, J., Wilmut, I., Meehan, R., Young, L. (2004) Non-conservation of mammalian preimplantation methylation dynamics. *Current Biology* **14**: R266-R267

Berkeley, G. (1710/1982) A treatise concerning the principles of human knowledge. Winkler, K. (ed.) Indianapolis: Hackett Publishing Company

Cremer, T., Cremer, C. (2001) Chromosome territories, nuclear architecture and gene regulation in mammalian cells. *Nature Reviews Genetics* **2**: 292–301

Dean, W., Santos, F., Reik, W. (2003) Epigenetic reprogramming in early mammalian development and following somatic nuclear transfer. *Seminars in Cell and Developmental Biology* **14**: 93–100

Eberharter, A., Becker, P.B. (2002) Histone acetylation: a switch between repressive and permissive chromatin. *EMBO reports* **3**: 224–229

Emmeche, C. (1999) The Sarkar challenge to biosemiotics: is there any information in a cell? *Semiotica* **127**: 273–293

Felsenfeld, G., Groudine, M. (2003) Controlling the double helix. *Nature* **421**: 448–453

Glasersfeld, E. von (1995) Radical constructivism: a way of knowing and learning. London, Washington D.C.: The Falmer Press

Gosden, R.G. (2002) Oogenesis as a foundation for embryogenesis. *Molecular and Cellular Endocrinology* **186**: 149–153

Harvey, A.J., Kind, K.L., Thompson, J.G. (2002) REDOX regulation of early embryo development. *Reproduction* **123**: 479–486

Herrler, A., von Rango, U., Beier, H.M. (2003) Embryo-maternal signalling: how the embryo starts talking to its mother to accomplish implantation. *Reproductive BioMedicine online* **6**: 244–256

Hoffmeyer, J. (1997) Biosemiotics: towards a new synthesis in biology. *European Journal for Semiotic Studies* **9**: 355–376

Hoffmeyer, J. (2002) The central dogma: A joke that became real. *Semiotica* **138**: 1–13

Hoffmeyer, J., Emmeche, C. (2005) Code-Duality and the Semiotics of Nature. *Journal of Biosemiotics* **1**: 37–91

Huber, J. (2006a) Fertilisation und embryonale Genomaktivierung: eine semiotische Interpretation. Dissertation: Ludwig-Maximilians-University, Munich

Huber, J. (2006b) Do germline interventions justify the restriction of reproductive autonomy? In: Niewöhner, J., Tannert, c. (eds.): Gene therapy – prospective technology assessment in its societal context. Amsterdam: Elsevier

Huber, J., Kummer, C. (in press a) Leben. In: Brugger, W. (ed.): Philosophisches Wörterbuch. Freiburg, Basel, Wien: Herder

Huber, J., Kummer, C. (in press b) Organismus. In: Brugger, W. (ed.): Philosophisches Wörterbuch. Freiburg, Basel, Wien: Herder

Jaenisch, R., Bird, A. (2003) Epigenetic regulation of gene expression: how the genome integrates intrinsic and environmental signals. *Nature Genetics* **33**: 245–254

Johnson, M.H. (2005) The problematic in-vitro embryo in the age of epigenetics. *Reproductive BioMedicine online* **10**: 88–96

Jones, P.A., Takai, D. (2001) The role of DNA methylation in mammalian epigenetics. *Science* **293**: 1068–1070

Kant, I. (1787/1998) Critique of Pure Reason. Guyer, P., Wood, A. (eds.) Cambridge, New York: Cambridge University Press

Krampen, M. (1997) Models of semiosis. In: Posner, R., Robering, K., Sebeok, T.A. (eds.): Handbuch Semiotik: Ein Handbuch zu den zeichentheoretischen Grundlagen von Natur und Kultur. Berlin, NewYork: Walter de Gruyter, 247–287

Kono, T., Obata, Y., Wu, Q., Niwa, K., Ono, Y., Yamamoto, Y., Park, E.S., Seo, J.S., Ogawa, H. (2004) Birth of parthenogenetic mice that can develop to adulthood. *Nature* **428**: 860–864

Krawetz, S.A. (2005) Paternal contribution: new insights and future challenges. *Nature Reviews Genetics* **6**: 633–642

Li, E. (2002) Chromatin modification and epigenetic reprogramming in mammalian development. *Nature Reviews Genetics* **3**: 662–673

Lillie, F.R. (1919) Problems of fertilization. Chicago: University of Chicago Press

Lumsden, C.J. (1986) The gene and the sign: giving structure to postmodernity. *Semiotica* **62**: 191–206

Maturana, H.R., Varela, F.J. (1992) The tree of knowledge: the biological roots of human understanding. Boston: Shambhala Publications

Miyamoto, K., Ushijima, T. (2005) Diagnostic and therapeutic applications of epigenetics. *Japanese Journal of Clinical Oncology* **35**: 293–301

Peirce, C.S. (1931–58) Collected Papers of Charles Sanders Peirce. Hartshorne, C., Weiss, P., Burks, A.W. (eds.) Cambridge, Mass.: Harvard University Press

Pesce, M., Schöler, H.R. (2001) Oct-4: gatekeeper in the beginnings of mammalian development. *Stem Cells* **19**: 271–278

Ponzio, A. (2004) Dialogism and biosemiotics. *Semiotica* **150**: 39–60

Reik, W., Dean, W. (2001) DNA methylation and mammalian epigenetics. *Electrophoresis* **22**: 2838–2843

Reik, W., Walter, J. (2001) Evolution of imprinting mechanisms: the battle of the sexes begins in the zygote. *Nature Genetics* **27**: 255–256

Rideout III, W.M., Eggan, K., Jaenisch, R. (2001) Nuclear cloning and epigenetic reprogramming of the genome. *Science* **293**: 1093–1098

Schmid-Tannwald, I. (2001) Schwangerschaft als Ausdruckszeichen des Subjekts "Frau". In: Schmid-Tannwald, I., Overdick-Gulden, M. (eds.): Vorgeburtliche Medizin – zwischen Heilungsauftrag und Selektion: S.233–270. München: Zuckschwerdt

Schultz, R.M. (2002) The molecular foundations of the maternal to zygotic transition in the preimplantation embryo. *Human Reproduction Update* **8**: 323–331

Schultz, R.M., Davis, W. Jr., Stein, P., Svoboda, P. (1999) Reprogramming of gene expression during preimplantation development. *Journal of Experimental Zoology* **285**: 276–282

Seidel, J. (2001) Das Genom als Quelle-Katalog. *Stimmen der Zeit* **9**: 591–600

Shannon, C.E. (1948) A mathematical theory of communication. *The Bell System Technical Journal* **27**: 379–423, 623–656

Sharov, A.A. (1998) From cybernetics to semiotics in biology. *Semiotica* **120**: 403–419

Solter, D., Evsikov, A., Peaston, A., de Vries, W., Holbrook, A., Knowles, B. (2005) The oocyte as a reprogramming milieu. Abstract. *The Society for the Study of Reproduction:* **38***th* Annual Meeting. Quebec City, Canada

Tembrock, G. (1971) Grundriß der Verhaltenswissenschaften. Berlin: Akademie-Verlag

Turner, B.M. (2000) Histone acetylation and an epigenetic code. *Bioessays* **22**: 836–845

Uexküll, J. von (1996) A stroll through the worlds of animals and men. In: Wolfgang Schirmacher (ed.): German Essays on Science in the 20th Century. New York: Continuum, 171–178

Uexküll, Th. von, Wesiack, W. (1998) Theorie der Humanmedizin. Grundlagen ärztlichen Denkens und Handelns. 3. Aufl. München, Wien, Baltimore: Urban & Schwarzenberg

Vico, G. (1710/1988) On the most ancient wisdom of the italians. Ithaca, New York: Cornell University Press

Waddington, C.H. (1957) The strategy of the genes. London: Allen & Unwin

Wassarman, P.M., Jovine, L., Litscher, E.S. (2001) A profile of fertilization in mammals. *Nature Cell Biology* **3**: E59-E64

Wilmut, I., Beaujean, N., de Sousa, P.A., Dinnyes, A., King, T.J., Paterson, L.A., Wells, D.N., Young, L.E. (2002) Somatic cell nuclear transfer. *Nature* **419**: 583–586

Wolffe, A.P., Matzke, M.A. (1999) Epigenetics: regulation through repression. *Science* **286**: 481–486

Young, L.E., Beaujean, N. (2004) DNA methylation in the preimplantation embryo: the differing stories of the mouse and sheep. *Animal Reproduction Science* **82-83**: 61–78

CHAPTER 18

LANGUAGE AND INTERSPECIFIC COMMUNICATION EXPERIMENTS: A CASE TO RE-OPEN?

DARIO MARTINELLI

Docent of Semiotics and Musicology – University of Helsinki, Ilomäentie 11 A 5 – 00840 Helsinki – Finland, dario.martinelli@helsinki.fi

Abstract: Goal of the present article is to re-open the question of language and Interspecific Communication Experiments (ICE, from now on) from a biosemiotic point of view, starting from the realisation that crucial aspects of the issue have been so far underrated or even missed. In particular:

- The specifically semiotic achievements of the ICE;
- The connection between these achievements and the notion of language;
- The consideration of ICE, as research carried out on individual non-human subjects, with specific background, attitudes, etc., in relation to the results achieved;
- The several ethical implications of ICE, in terms of both methodology and outcomes.

Although I am aware that most of the reflections proposed in this article are to say the least controversial, I am by all means convinced that the discussion on these matters should be restarted, as it deserves a more thorough scrutiny, and – perhaps – a less anthropocentric approach

Keywords: – Language, Anthropomorphism, Sebeok, Interspecific Communication, Umwelt

PROBLEMS AND THESES

The semiotic discussion on language, and its supposedly human species-specificity, relies – for most of its points – on Sebeok's reflections about language itself, and about the ICE carried out by several psychologists and primatologists in the second half of the 20th century. Sebeok's arguments, later reinforced by many of his followers, can be summarised in the following points:

1. Language is a species-specific human device. The ability of language acquisition is a result of the mental capacities of the human being. "The word 'language' is sometimes used in common parlance in an inappropriate way to designate a

M. Barbieri (ed.), Introduction to Biosemiotics, 473–518.

certain nonverbal communicative device. Such may be confusing in this context where, if at all, 'language' should be used only in a technical sense, in application to humans. Metaphorical uses such as 'body language', 'the language of flowers', 'the language of bees', 'ape language', or the like, are to be avoided" (Sebeok 2001 p. 14);

2. First than being a communication tool, language must be considered a modelling system, specifically, the, very human, secondary modelling system. "Language is, by definition, a secondary cohesive modelling system providing humans with the resources for extending primary forms ad infinitum. [...] From a biosemiotic perspective, the language code can be defined as the cohesive system providing the modelling resources for converting what von Uexküll (1909) called 'concrete living existence' into 'active plans'. (Sebeok, Danesi 2000: 108)". Identifying language with communication was, according to Sebeok's literary words, "a vulgar error" (1991, p. 71);
3. The difference between language and non-human forms of communication (and modelling systems) is of qualitative type. Quite simply, humans have language, other animals do not, rather than saying that language is a more refined form of communication (or modelling system) as compared to other animals' devices, as a gradualistic/darwinian approach might suggest. Exactly its nature as secondary modelling system makes language a uniquely human feature: "All the animals paleontologists classify generically as Homo, and only such, embody, in addition to a primary modelling system ..., a secondary modelling system, equivalent to a natural language. The difference amounts to this: while the Umwelten of other animals model solely a (for each) 'existent world', man can, by means of the secondary system, also model a potentially limitless variety of 'possible worlds' ". (Sebeok, 1996, p. 106);
4. As I understand it from personal conversations with scholars of the likes of Kalevi Kull and John Deely, such a capacity to model possible worlds is expressed via three major (and again, exclusive) characteristics of language: what I call distant time semiosis, narrativeness, and linking signs. By distant time semiosis, I mean the ability of keeping track, transmission and reconstruction of both recent and remote past events, and the ability to articulate projects and expectations regarding both immediate and remote future events. Such an ability is both direct and indirect, the former being related to the personal experience of the subject, and the latter referring to experiences that the subject has not lived or will hardly live personally. In other words, language allows human beings to talk not only about their childhood and about their intention to live in a country-house when they will finally retire from working, but also to discuss the defeat of Napoleon in Waterloo and to wonder about the day when UFOs will finally land on Planet Earth. By narrativeness, it is not only meant story-telling (which is still an important feature itself), but mostly the general capacity of accessing and describing alien umwelten, either imaginary or not. Regardless whether these descriptions are trustworthy or not, what matters here is the fact that in principle, any dialectic account

of a given context, environment or reality is in the human semiosis possible. The descriptions of both the Umwelten of the species Pekin Duck *Anas platyrhynchos*, and of the fictional character Donald Duck (inspired by the very same species) are possible only thanks to the existence of such a device like language.Finally, with linking signs (or linking words) I intend to point out a specific aspect within the broader concept of syntax, namely those signs that in verbal speech are known as conjunctions, transitions, and prepositions. Linking words are a form of para-signs that do not refer to any other existing entity than themselves, and whose function is to create meaningful relations among signs that, on the contrary, stand for something else than themselves only. In other words, linking words are yet another confirmation of the capacity of language to create possible worlds, even when those 'worlds', as in this case, are simply signs that are untied to tangible entities. Any other claims concerning the uniqueness of language shall be dismissed, once and for all, as anthropocentrically-biased semiological, rather than semiotic, blunders. Those include the use of symbolic signs, the ability to consciously deceive the receiver of a message, the ability (and the tendency) to put into action semiosis of aesthetic type, and so forth (more details, from the zoosemiotic point of view, can be found in Sebeok 1990: 77–98, Sebeok 1981: 210–259, Martinelli 2004, and Cimatti 1998: 59–106, 179–189 and 205–210).

5. The ICE, carried out throughout the last decades, were either a failure or, when apparently successful, fatally biased by misinterpretations of the Clever Hans Effect (CHE, from now on) type. However controversial such a position may be (the experiments conducted were of the most diverse types, and one might suggest that to draw one single conclusion out of them is at least not extremely accurate), there seem to be an almost absolute consensus around this point, especially among Sebeok's followers who do not hesitate to label those experiments as anti- or pseudo-scientific. Felice Cimatti (1998: 107–165) seems to be the only exception to this rule.

During the present paper, I will try to argue that these five pillars built upon our (i.e., human) concept of language, are either inaccurate or incomplete, needing additions in some cases and revisions in others. The following will be the thesis I plan to defend:

1. *Language is a species-specific human need.* That is to say, the capacities (cognitive and practical) to acquire language are not inaccessible to other species, however human being has been and is the species who appears to be the most interested in pursuing it, and – by consequence – the best disposed to. Moreover, even though not specifically denied by semioticians, *the role of the constitution of the human vocal apparatus in the acquisition of language seems to be highly underrated*, in favour of the idea that the entity that is most responsible for such a process is exclusively the human brain;
2. Language must be certainly considered a modelling system more than a communication device. However, *the phylogenetic and ontogenetic bases for language to be acquired remain those of the communication system.* In other

words, language proves to be more efficient as a modelling system, and other sign systems prove to be more efficient than language as communicative forms. *But*, it should not be forgotten that a) language was originated as a communication system, and b) communication is the primary reason why language is handed parents to offspring.
3. Still, if language is – also or mostly – a modelling system, then, in evaluating interspecific communication experiments, thus, *what shall be analysed more thoroughly is if and how non-human animals use language to map their Umwelt, and – possibly – to access the human Umwelt.*
4. *Linking signs seem to be the only specific characteristic of language that is mastered by human beings only. Distant time semiosis and narrativeness, on the other hand, are not exclusive of humans, although undeniably reinforced and improved by their employment of language.* More specifically, linking signs are the very upgrade, added to a natural communication system, that allows an impressive improvement of the narrative and distant-time semiosic potential of an individual.
5. *The discussion conducted by semioticians on ICE is surprisingly generic and approximate, and requires an almost radical revision.* Topics in urgent need to be re-discussed are at least the following: evaluation of the methodologies and results of the ICE in themselves; distinction within different ICE; the CHE and the entire discussion on anthropomorphism; and the consideration of otherwise ignored aspects and outcomes of ICE.

I shall concentrate most of my discussion on the last point of my list, for my impression is that several conclusions about language, and its supposed human species-specificity, did actually depart from this very argument.

Before starting my argument, however, a little statement is needed. To me, Thomas Sebeok is to zoosemiotics what Paul McCartney is to pop music. I like every single album McCartney issued in his long carreer, with the Beatles, with the Wings, and as a solo artist. All of them, except *Driving Rain*, issued in 2001: that album for me is weak, it is badly produced and the quality of the songs is quite below his standards. Similarly, I truly admire (and am inspired from) every single theoretical formulation or reflection that Sebeok produced in zoosemiotics during his long carreer. All of them, except the ones on language and ICE, which I shall very humbly take the liberty to criticise in this article. As my admiration for the ex-Beatle has always been out of discussion, I hope that this article will not mean to anybody that my often explicitly-stated admiration (and personal gratitude, as a zoosemiotician) for Sebeok has vanished or anything.

A BRIEF HISTORY OF INTERSPECIFIC COMMUNICATION EXPERIMENTS

For sake of clarity, I shall take the liberty to summarise the most important ICE carried out throughout the last century, focusing on what I consider the most relevant achievements (or failures) in any of them.

In most of the instances, the scholars working on ICE advanced the basic claim that there is no unbridgeable gap between human language and other communication systems. Such a claim relied on the Darwinian idea of evolutionary continuity between *homo sapiens* and other species, so that language should be considered only a more and better refined system for communication. However, this was not the sole point of departure. Together with the Darwinian one, another stand, of behaviouristic type, would define language as a conative tool, whose function is basically that of transmitting thoughts from a sender to a receiver. It is no coincidence that the boom of these experiments was in the 1960's. Within such a framework, language was seen as an absolutely 'normal' behaviour, although a little more complex than usually, that could be learnt as any other behaviour.

The very first experiments were affected by the at least arguable assumption that the vocal-acoustic channel was the only reliable source for a communication system to be considered true language. The idea probably originated from writings by Samuel Pepys (*The diary of Samuel Pepys*, 1661), Johann Conrad Amman (*Surdus loquens*, 1692 and *Dissertatio de loquela*, 1700) and Julien Offray de La Mattrie (*L'homme machine*, 1748), who were all very keen on the opinion that monkeys could be easily taught to speak. The plan was thus to teach non-human animals to vocally *pronounce* human words. The first victims of such attempts were the chimpanzees **Peter**, raised and observed by Lightner Witmer around 1909, **Joni**, raised and observed by Nadia Kohts between 1913 and 1916, **Wiki**, raised and observed by Keith and Catherine Hayes between 1952 and 1958, and **Gua**, raised and observed by Winthrop and Luella Kellog during the late 1960's. A separate treatment of these four cases is not necessary, as the outcomes were equally poor. All of them – after years of long intensive training – learned maximum the not-exactly-encouraging amount of 3–4 words, that incidentally would happen to be the same: *mama,papa,cup* and *up*. All of them, at the same time, proved to understand dozens of words. Similar attempts were made also with other apes, like the orangutan trained by William Furness in 1916 (and whose name, curiously, was never reported), and the gorilla Toto (Maria Hoyt, in 1942). Of course, the real problem – as I will point out further on, and as studies like Liebermann, Crelin & Klatt 1972 demonstrated once for all – lies in the neurological and anatomical incapacity of non-human bodies to articulate sounds in the way that humans do. The difficulties of these apes were *expressive*, not *cognitive*.

Much later in time, Irene Pepperberg thought that the idea could be resat and applied to a species that, in fact, does not have these expressive limitations (if anything, it might have cognitive handicaps, as it is not a great ape, and not even a mammal). That was the case with the parrot **Alex**. Started in 1977, this experiment was somehow meant also to remove the prejudice that parrots are just able to repeat things like, well, parrots. Alex soon proved to be able to understand and to pronounce about 100 English words. He answered correctly about the 80% of the questions he was asked (including pretty specific ones), while his mistakes were basically due to two factors: 1) *misunderstanding*: sometime Alex would react to words that are similar in sound to the ones actually pronounced, (cable/table, fork/cork etc.); 2) *Umwelt*: for instance, Alex would classify a piece of paper and a piece of leather in the same way. Possibly his criteria for categorisation had more to do with the bi-dimensional and squared shape of the two objects. I shall later return to the issue of Umwelt-based categorisations.

An example of Alex's skills are reported in Cimatti 1998 (page 132, my translation). Note that the experiments were structured in the so-called double-blind way. One trainer would ask the questions to Alex, and another one, who was unaware of the questions and thus impossible to be biased in his/her interpretation, would detect Alex's answers::

Trainer1 – Ok Alex, let's begin. What's this? (*holds a wooden button*)
Alex – button wood
Trainer2 – He said button wood
T1 – Right, here is a wooden button. Alex, you're smart (*lends the button*)
A – (*takes the button, then let it fall*) Want cork
T1 – Right, here is the cork
A – (*plays with it for a while*)
T1 – Ok, Alex, it's enough, give me the cork (lends his hand, and Alex gives him the cork)...what's this? What's its shape? (*holds a wooden red triangle*)
A – Three...(*hesitates*) corners wood
T2 – He said three corners wood, but I'm not sure
T1 – Alex, speak clearly, please
A – three corners wood
T2 – "three corners wood"
T1 – Right, bravo Alex
A – Want nut (*T1 lends a nut, he eats*)
T1 – Look, what's this? (*holds a grey piece of leather*)
A – Want nut!
T1 – First you tell me what is this, then I'll give you another nut
A – Leather grey...want nut!!!

What we understand from this amusing conversation is that:

1 – Alex seems to have a full comprehension of the questions he is asked. The hesitation, followed by a more resolute statement, appear to be the result of a careful evaluation of the object he is asked to describe. Plus, he does not hesitate to refuse an object, when it is different from his requests.

2 – He does not necessarily answers the questions to please trainers, and rather tries to get what he wants (which, together with the presence of another trainer, according to the double-blind test, reveals that the conversation between trainers and trainee is not affected by the CHE). In fact, the idea that we get from the last few lines is that Alex at some point gets bored of the training and wants to eat, and answers the last question only because he understands that this is the only way he can achieve his beloved nut.

3 – He shows understanding of quite articulated sequences of words. Especially the last sentence by the trainer requires second-order type of reflections.

4 – He can count at least up to three.

In any case, as it became clear that not only verbal communication had to be considered language, scholars became more and more interested in interspecific communication, also because of the growing fame of behaviourism. In 1966, Allen and Beatrix Gardner started an IC project with a female chimpanzee named **Washoe,** who is nowadays the 'speaking' animal *par excellence* (up to became a sort of synecdoche for all other experiments). It is maybe the case to say something more about this project, not only because, among all others, this is the one I had the opportunity to study more in detail, thanks also to a personal corrispondence with the current leader of the project, Dr. Roger Fouts. The Washoe-project was in fact a real waetrshed between the past and the present of ICE. It was the first one to be a complete success, it set several methodological indications that were followed in the next ICE programs, and it inaugurated a series of ICE based on linguistic sign systems alternative to speech. Washoe was indeed trained with an *ad hoc* version of the American Sign Language (ASL). The main characteristic of ASL is that its semantic and syntactic aspects are totally comparable to the *normal* verbal language.

The Gardners had set two main targets for their project: to prove that chimpanzees are able to learn a human language, and – perhaps mostly – that a real interspecific communication between humans and other animals is possible. The training was organised so that Washoe could imitate her trainers' gestures, and – at the same time – through direct manipulation of her arms. The training environment was set in a cosy, relaxing, non-laboratory-like atmosphere, an aspect which the Gardners (plus Roger Fouts, who joined the project one year later, and finally carried it out as leader) considered very important in order for Washoe to be trained properly (Gardner-Gardner 1969: 666)

The first signs she was taught were the *passepartout* "more" and "come here", the latter being not only a request for a trainer to approach her, but also for objects and – associated with other signs – for certain actions (she would say "come here tickle" when she would want to be tickled). Very soon she was taught more signs, and showed increasing curiosity in learning them. Some of these signs were in fact very close to the natural intraspecific gestures by which a chimpanzee interacts with his/her fellows. The sign for "come here" for instance was homologous to the "grooming" sign, by which chimps ask to be cleaned from bugs. Far from considering that a problem, the Gardners in fact encouraged such mixtures, in order to facilitate Washoe's work, and claimed that the very same process occurs in human language, when abstract signifiers quite often carry some characteristics of the signifieds (as in the case of onomatopoeic words). And, after all, their resemblance to nature sounds do not make these words *less words* than – say – "wardrobe", or "acknowledgment".

The results of the Washoe project were rather remarkable: Washoe used spontaneously the signs, even only to name things, rather than just making requests,[1] then started to associate the ASL signifiers to more complex meanings (for instance, the words *flower* or *car* were used not only for real flowers and real cars, but also for pictures portraying those items – see Gardner-Gardner 1969: 667). Signs were not anymore strictly related with the context in which they were produced for first. The sign "open" was used not only for doors, but also for refrigerators, windows, taps: in other words, Washoe had turned those signs into *cognitive entities* (Gardner-Gardner 1969: 670). Further, Washoe immediately learned that signs had purely abstract meanings, and were not necessarily connected with perceptive resemblances: the sign for "flower" represented in the beginning all the smelling object, and started to designate a flower only when Washoe was taught the sign "perfume".

As her ASL competence grew, Washoe showed great interest in adjectives, up to compiling object-attribute relations of metaphoric type (e.g., the term 'dirty' soon acquired a moral connotation: Washoe would use the expression "Roger dirty" every time she wanted to insult her trainer for not having granted one of her requests). In formulating sentences of the object-attribute or object-action type, it is intriguing to note that Washoe had her personal rule on how establishing syntactic orders: with no exception, the perceptively most relevant or most dynamic entities were signed for first. So, it was "Roger good" and not "Good Roger"; "Bottle red" and not "Red bottle"; "Look there" and not "There look"; "Come here tickle" and not "Here come tickle"; etc. This is a rather important point, as it appears in most ICE with other animals. It really seems that the elementary, perceptive syntax of animals is of gestaltic type.

In the following years, Washoe was also given ASL-trained company (equally-talented chimpanzees of Washoe's gang were Moja, Tatu, Ally and others, who used ASL quite regularly in intraspecific communication, as well[2]) and was also allowed to form a family, with both natural and adopted offspring. Her adopted infant, Loulis, by Gardner's deliberate decision, was not taught ASL in the first 5 years of his life, but nonetheless acquired more than 50 signs by watching the other chimps, thus reflecting the manner in which human children acquire language (Kosseff 2000, e-text). Later in time, to her natural infant Sequoyah, Washoe taught ASL by her personal initiative (ibid.).

Another very succesful ASL-based ICE program was started in 1972 by Francine Patterson with a gorilla named **Koko**. The main point, in this case, was a direct challenge to CHE: the trainer wanted to establish a solid emotionally-close relationship with the ape. Patterson's theory was that children gain motivation in learning language most of all because of the interaction with parents. To deprive them of such a fundamental element would mean to deprive them of language itself. Indeed, results with Koko proved very satisfying, just like Washoe's case. Koko could comprehend spoken words and use the correspondent symbols (which means that the signified was a semiotic entity in her mind). Moreover, she could create new signs to describe objects she did not know the name of (like "white tiger" for "zebra"). Very important was also the referential use of the signs. She would not make just simple requests, but she would talk to Patterson about what she could see around her. Basically, she chatted with her trainer, once more dismissing the (semioticians') widespread belief that language-trained apes could not go beyond a basic, conative use of the signs. Well-documented, as with Washoe, is also Koko's large use of lies.[3]

An ASL-based program was also successfully attempted with an orangutan named **Chantek**, starting from 1978, by Lyn Miles. The distinctive feature of this ICE was that the trainee was put in a context in which all trainers communicated with ASL, even among themselves, and nobody was making specific attempts to train him directly. When Chantek was told something, he was told by ASL, and when he wanted to communicate, he had to use ASL: simple as that. And, indeed, very soon, the orangutan learned to use the sign system spontaneously and efficiently.

Finally, the American Sign Language was also the sign system used in the very ICE about which semioticians really show extensive knowledge: Herbert Terrace's **Nim** program. After few years of apparent success, Terrace had one of the most famous second thoughts in the history of science: he realised that the whole training was heavily affected by the CHE, and – very honestly – he admitted it. I shall later add further comments about this instance.

In any case, ASL was not the sole alternative to speech. For the sake of CHE-free research, in fact, sign-languages have the limitations of being highly analogical, i.e., gestures are not totally neat signs, and a wrong sign may appear as the right one just because it is represented by a similar gesture: a strictly digital sign system would avoid such an inconvenience by designing very precise boundaries between one sign and another. David Premack, whose project started in the same year as the Gardners', wanted to teach human language to a chimpanzee named **Sarah** by using coloured plastic symbols that had no whatsoever iconic (or indexical, for the record) relation with the object represented (e.g. an apple was represented by a blu triangle). Depending on which side one looks at it, this project was a partial success or a partial failure.

Apart from concrete objects, whose acquisition was fast and efficient, Sarah showed ability to learn abstract concepts like "name of" (used as "what's the name of x?" when she would not know how to call a certain object), colours, big-small, equal-different, squared-rounded, etc.. She also could answer to questions like "what is the colour of...". The problem with Sarah, however, was again in the CHE, that seemed to be much more evident here than in other occasions: Sarah was probably manipulating trainers more than how they were manipulating her.

It was exactly the risk of CHE that led scholars to elaborate ICE that excluded, the more as possible, any interaction between trainer and trainee. Duane Rumbaugh, in the mid 1970's, attempted to teach language by employing symbols. The trainee was again a chimpanzee, whose name was **Lana**. As it was considered important to avoid any contacts with humans, Rumbaugh felt that computer automation could prevent anyone from cuing the animal (besides, the training would have required less people). The machine was programmed to perform certain tasks, like dispensing food or displaying an image in response to pressing of the proper symbol on the keyboard. Lana was then given a special keyboard which had symbols (called lexigrams) in place of regular keys. The trainer was then connected to Lana with another, equally equipped, computer. The interaction would thus occur only through the respective monitors. Again, the results were controversial: if Lana clearly learned to ask for things, and to engage into conversations of relative complexity, it is also true that she would do so only in response to a specific need (usually, hunger or thirst).

Rumbaugh was followed by his wife, Sue Savage-Rumbaugh, who carried out what is perhaps, together with Washoe's and Koko's, the most interesting and successful ICE program. Savage-Rumbaugh felt that the core-question was to establish whether apes understand the signs they produce, rather than finding out about their grammatical capabilities. Therefore, the entire methodology of the program deviated from the trends of that time (for instance by moving away from some of the design features of language - productivity and displacement, most of all - which on the contrary were so popular in the other studies). Another interesting aspect is that Savage-Rumbaugh's work departed from a quite critical attitude towards other ICE. In all previous programs, she would maintain, the primates were not using signs symbolically, but rather indexically: they learned to associate certain behavioral patterns with certain consequences. In her view, indeed, the main point of human language is that it uses symbols, which implies three factors: 1) the physical external substance of a word, e.g. "door", as either written down or spoken with a given linguistic pattern; 2) the relationship between that and a real door. When the word "door" is pronounced one conjures up a mental representation of a door; 3) the capability of symbols to make one think about things that are not present, or even not existent, like unicorns.

The experiment, performed on a small community of bonobos, among which **Kanzi** is by far the most famous, achieved excellent results. Kanzi, who was born in 1980, when the project had already begun, took everybody by surprise from the very beginning of his training, when he showed he already knew some lexigrams and could already understand a remarkable amount of spoken words (evidently, this competence was achieevd by observing his fellows apes, especially his adoptive mother **Matata**). Kanzi's competences grew rapidly:

> Once it was understood that Kanzi needed no specific training to grasp the meaning of spoken words, or their written symbols, reward based training was completely abandoned. As additional symbols were added to his keyboard, they were used in conversation with Kanzi but he was not trained to associate them with specific objects. Instead, all symbols were employed in conversations with Kanzi, and every attempt was made to aide Kanzi in understanding the conversations. Sometimes this entailed accompanying words and lexigrams with gestures, with pictures, with video tape and, of course, with behavioral activities that made manifest the intentions which underlay the communications. Most such conversations centered around travel, finding food and playing. Kanzi's vocabulary steadily increased. Today it is well over 200 words productively and 500 words receptively (Washburn 2001, e-text)

Kanzi showed an unquestionable comprehension of spoken words, what was always a weak point of the other ICE. In various tests he was presented (including recorded words played to him through headphones), he was nearly 100% accurate on all words that were part of his vocabulary at any given age. He was also able to respond to speakers with different accents as well as to artificially-produced words. Far from succeeding only with single words, Kanzi showed clear comprehension of sentence structures, word order and grammar rules. Similar results were achieved with two other bonobos of Kanzi's community, **Panbanisha** and **Panzee**.

One last ICE I shall mention here, that is a bit sui generis, in terms of both methodology and species involved, is the program conducted by Lou Herman on two dolphins, **Phoenix** and **Akeakamai** (in Hawaian "he who loves wisdom"), in the 1980's. The training consisted in two simple sign systems: Phoenix learned an acoustic language generated by an underwater speaker, and Akeakamai an adapted version of ASL. In both cases, the words produced consisted mostly in concrete entities, actions, modifiers (like "right", "left", "bottom", "surface") plus metalinguistic signs, such as "yes", "no" and "erase" (used as a signal to interrupt an activity). In order to avoid the CHE, the trainers wore a mask that prevented them to show facial expressions. In addition, the experiment was set in such a way that one trainer would give the message, and another one, without knowing the message, would record the dolphins' answers. The syntax of the messages were organised in sequences, like "surface frisbee bring-to basket" (i.e., on the surface there is a frisbee: bring it into the basket). When a task was not possible to be performed, the dolphins would touch a panel correspondent to "no", if it was, then, after performig it, they would touch a panel correspondent to "yes". This project, too, achieved interesting results. In addition to what already mentioned, Phoenix and Akeakamai proved to understand messages referred to a different time. They would memorise the instructions, and – as the conditions would be favourable for performing the task – they would do it.

SEMIOTIC SCEPTICISM

It appears quite clearly that the history of ICE is pretty long and articulate: it has involved various animal species, several different tools and methodologies; it has departed from the most diverse working hypotheses and has ended up with the most diverse results; and, scientifically speaking, it has quite often been the result of the time they were performed, going hand in hand with contemporary theoretical trends in psychology, ethology and other disciplines. Given such a complex scenario, it is at least surprising that the account given by semioticians of these experiments has most of the time been generic, reductive, and – from time to time – merely inexact. I will never forget a lecture on the subject, given by one of the greatest living semioticians, in which subjects and methodologies of different ICE were quoted at absolute random: Washoe, who was trained with American Sign Language, was reported as the one who was communicating with lexigrams (which, on the contrary, was the case with another chimpanzee, Lana); Koko, a gorilla, was quoted as a chimpanzee; and still Washoe, who was trained by Allen and Beatrix Gardner and Roger Fouts, was reported as the one trained by Herbert Terrace[4] (who, on the contrary, was training another chimpanzee, Nim Chimpsky). Nobody is asking for undisputable precision, of course, but how seriously should one take – say – a musicologist who, during his lectures, claimed that *Hey Jude* was written by the Rolling Stones; Jimi Hendrix was the drummer for Led Zeppelin; and Abba was a German duo?

This was surely an extreme case, but the general picture is anyway pretty disappointing. Most of all, what surprises is how generically ICE are treated. Very hardly are they taken case by case (and subject by subject, a point which I consider of extreme importance): what one normally is offered is one single pot with all the experiments in, and a set of conclusions applied to all of them indistinctively. Sebeok almost seems the only one to have acquired his knowledge from direct first-hand sources. For quite a few of his followers, it appears that all they learned they learned it from him.[5] What are these conclusions? R.L.Trask (1995: 21–22) was able to summarise them in four basic points:

1) Much of the evidence in ICE is purely anecdotal: "... it consisted of reports that some particular animal on some particular occasion had been observed to do something-or-other pretty damned impressive" (*ibid.*: 21). Anecdotes, Trask points out, are not of scientific relevance;
2) The criteria for evaluating IC tests were too broad to be reliable. "For example, if a signing chimp was shown an apple and asked (in ASL) 'What is this?', the experimenters frequently counted as a correct response any sequence of signs including the sign for 'apple', which is a far cry from the sort of response usually heard from a human child learning a first language" (*ibid.*: 21);
3) In many cases, the claimed ability of certain non-human animal to actually perform the linguistic task assigned to him/her was reported by the experimenters only, so we have "nothing more than the experimenters' own word for it that the apes were making any signs at all" (*ibid.*: 21);
4) The most important reason why ICE should all be considered misleading and anti-scientific is their incapacity to avoid the CHE and – more generally – anthropomorphic interpretations of the results of the experiments: "the critics discovered that the experimental procedures typically used to test the animals were so slipshod that an animal under test could often see its human handler unconsciously forming the required response with her or his own hands, so that it could see what to do" (*ibid.*: 21–22)

To Sebeok, the "pervasive, insidious penetration of Clever Hans" (1990: 68) in ICE is the main problem, and he did not fail to remind it in every single essay he wrote on the subject. Perhaps, thus, the issue of anthropomorphism deserves a deeper discussion, which is what I will provide later on in this essay. First, however, the first three points of Trask's list, however naïve they may appear, should be commented.

The first point reveals a certain lack of attention in reading the actual texts reporting ICE. A more careful scrutiny would have suggested that most of the anecdotes are not episodic at all, both within the same experiments and, more importantly, across different experiments. For instance, the use of signs in a creative manner to name unknown objects starting from similar known ones is an 'anecdote' reported in several cases: Koko is a specialist in the field, creating such names as "white tiger" for a zebra, "finger bracelet" for a ring, "elephant baby" for a Pinocchio doll, or "eye hat" for a mask (Patterson-Linden 1981). In homologous fashion, Lana asked for an "oranged-coloured apple", not knowing the lexigram for "orange" (Von Glasersfeld 1978: 732), while Washoe is reported as performing this activity on a regular basis, one example being "drink fruit" for a watermelon, or "water bird" for a swan (Hill 1980: 336). Then again, it is true that only once did Koko name "white tiger" a zebra, or Lana name "orange-coloured apple" an orange. But that is because they were soon taught the signs for "zebra", and "orange", so why bothering to still call those objects in their own temporary way? What the scholars represented by Trask seem not to grasp here is that what counts is not the redundancy of a specific example, but rather the redundancy of a rule (or, more precisely, a modelling strategy): when Koko sees a zebra, but does not know the ASL sign for it, she starts reasoning upon it, a sort of: "Well, it's clearly an animal, it runs... it has stripes like a tiger, but it's white...". Quite exactly, Koko is modelling her experience through the language. It must be also mentioned that other ICE subjects, like Sarah, were trained to use signs (plastic symbols, in her case) for "Name of", through which they could ask how to call a certain object. This way, Sarah did not need to create white tigers or water birds, but simply she would directly ask for the right plastic symbol representing the unknown object.

As for the second point of Trask's list, we have a typical example of the situation I will explain more thoroughly in the next sections: here is a case where the fear to make anthropomorphic mistakes becomes a greater mistake itself. Everyone who is learning a new language (or any sign system) goes through a transitional period when confusion is more regular than precision. It took me months to correctly pronounce the name of the street in Helsinki where I used to live (Tähkäkuja: I challenge any non-Finnish to pronounce it correctly at the very first attempt): several pizzas were delivered at wrong addresses, just because I was not able to spell my address in the correct way to the pizzaiolo at the other end of the phone. Moreover, when I have not properly and definitely learned a word, I may easily confuse it with other semantically-related signs: still nowadays it occurs to me that I ask for an appelsiini (the Finnish word for "orange") when I want an apple ("omena" in Finnish), or that I confuse and exchange the words "maito" and "Valjo" (the former standing for "milk", and the latter being a brand that produces it). I am obviously (and famously) not gifted for Finnish, but I suppose most people have experienced similar occurrences. Not to mention the infamous "false friends".

The situation described by Trask of the ICE subjects randomly performing signs until they catch the right one (an instance which is anyway pretty rare, but Trask omits to point it out) is absolutely normal, and absolutely normal is the criterion applied by the researcher in evaluating these attempts: it is exactly what is done among humans, with children or with foreign language students. We encourage the right answer, as it occurs, with a "yes" or a smile, and we are keen to accept a few wrong attempts before the right word is pronounced when the trainee is just at the beginning of his/her learning process.

The third point is not very serious, I guess. So, we cannot really trust Premack, Gardner, Patterson and company, since they were the only ones observing their trainees performing the experiments and they may be lying. Never mind that a lot of endage documenting the experiments is available (a lot is on-line, too, so one does not even need to pay for it. Check Koko's website, at www.koko.org); and never mind that the projects were never carried out by one single researcher, but by entire teams, whose members would partly or totally change throughout the years (to mention one, Washoe's project has been going on since 1966), and young assistants, whose name would otherwise remain obscure, could easily gain newspaper headlines with scoop interviews of the type "My boss is bluffing about the apes". What is suggested here is that, since we cannot check, there is a reason to think that ICE are not reliable. If that is the principle, one cannot even start listing the many unreliable researches that nowadays infest the scientific world.

THE CASE FOR ANTHROPOMORPHISM

The ancient Greek *morphé* stands for 'form', 'shape', 'appearance'. According to Webster's Dictionary, anthropomorphism can be defined in two ways: 1) the representation of the Deity, or of a polytheistic deity, under a human form, or with human attributes and affections; and 2) the ascription of human characteristics to things not human.

In ethological research, anthropomorphism, together with his apparent opposite (in fact, close relative, both having their origin in anthropocentrism) zoomorphism, characterises a well-known, little-appreciated and often-debated methodological approach to the study of non-human animals. The dictionary of ethology, edited by one of the most authoritative Italian ethologists, Prof. Danilo Mainardi, defines anthropomorphism in the following way:

> [Anthropomorphism] is the tendency to interpret animal behaviours by ascribing to them typically human motivations and goals. Occasionally, this led to the evaluation of animals' behaviour in moral terms, up to persecuting them.[6] The wrong habit of humanising animals has long affected the interpretation of their behaviour, and the most evident examples are medieval bestiaries. Research in child behaviour has demonstrated that the tendency to anthropomorphise animals arises spontaneously, and that it disappears only in adult age, thanks to appropriate education. The same tendency is detectable in primitive civilisations and plays an often relevant role in the relation between humans and pets. Dangerous anthropomorphic interpretations have been, and still are, present in many cases of zoological divulgation. (Mainardi 1992: 48, my translation)

At least three points in this definition should be pointed out and eventually provided with deeper explanation:

1. Anthropomorphism is considered, in all respects and without exception, a scientific mistake, which, historically, seems to have damaged or delayed a fair interpretation of animal behaviour,
2. According to the terminology I used elsewhere (Martinelli 2002: 61–62), it seems rather clear that the attitude emerging from this definition is of a binary and qualitatively anthropocentric type.[7] Such a framework should be further taken into account.
3. Although Mainardi considers anthropomorphism a mistake to correct with "appropriate education", he recognises the spontaneous emergence of anthropomorphic attitudes, both psychologically (during human childhood) and anthropologically (in primitive populations).

On a very general level, it is evident that anthropomorphism is perceived by the scientific community as a serious mistake, unable to provide ethological knowledge with any useful contribution. As emphasised by Tom Regan (1983/1990: 29–30), if other animals are described as having characteristics that are also human, but not only human, such as "being alive", such attribution does not constitute an anthropomorphic mistake. The problem concerns more critical characteristics (Regan mentions the example of awareness). In such cases, to be anthropomorphic apparently means to commit one of the worst scientific crimes. Not without a touch of irony, Jeffrey Masson entitles a paragraph of his bestseller *When Elephants Weep* (1995/1996: 71–73), "Contagious anthropomorphism", and BQs several authoritative scientific essays claiming that anthropomorphism is a "pathology", a "trap", an "illusion", a "womanly attitude" (*sic!*), or simply a "lie".

The real milestone of the enemies of anthropomorphism is exactly the Clever Hans Phenomenon. Hans was a horse, owned by a certain Mr. von Osten during the first half of the 20th century. Hans came to be quite well-known for his purported mathematical computing skills. When asked to solve any kind of arithmetic task, from simple addition up to calculating the square root of numbers like 103.684,[8] Hans would amaze everyone by striking his hoof against the ground as many times as the exact result was. His fame grew greater as ever-more difficult tasks were asked of him and different verification tests were set (e.g., the tasks were written on a blackboard instead of being vocally pronounced). This went on until a scholar named Oskar Pfungst realised that when Hans was interrogated by someone who did not know the right answers, his hoof strikes were wrong. Departing from this observation (repeated and confirmed several times), Pfungst realized that Hans's arithmetic competence was in fact quite low, but that his psychological skills were superior to those of most humans. Clever Hans, indeed, did not have the slightest idea of how to solve arithmetic problems, but was remarkably skilful in detecting the exact moment when the interrogator was content of his response (i.e., of the right number of hoof strikes). Unintentionally, the interrogator would send some kind of sign (facial or postural, mainly) which displayed his/her surprise in seeing that Hans had actually reached the right amount of hoof strikes.[9] At that very moment Hans would stop hitting the ground.

Unfortunately, instead of pointing out how exceptional such psychological skills were, scholars only use this episode to recall how dangerous and misleading anthropomorphism can be (in this case, to attribute mathematical skills to Hans).[10] From then on, in experimental contexts, one speaks of the Clever Hans Effect every time the animal ends up being the manipulator, rather than the subject, of the experiment.

John Andrew Fisher's "The myth of anthropomorphism" (1990: 96–116) represents a very efficient attempt to classify the several nuances of the concept of anthropomorphism. Fisher is aware of two crucial aspects: 1) there is extreme confusion in the use and definition of the term; and 2) however one may define it, every scholar takes great care to avoid being affected by it, as one would normally do when dealing with diseases:

> Anthropomorphism is usually regarded as an embarrassment to be avoided. Philosophers and scientists often approach anthropomorphism as an obstacle to be overcome by those who wish to attribute cognitive or emotional states to non-human animals. Thus Donald Davidson suggests that 'Attributions of intentions and beliefs to animals smack of anthropomorphism'. Even those who favour animal rights try to avoid being accused of it. Annelle Baler, for example, feels obliged to say, 'I see nothing at all anthropomorphic or in any other way absurd in saying that one may break faith with an animal, exploit its trust, disappoint expectations one has encouraged it to have' [...]. And Mary Midgley asserts: 'There is nothing anthropomorphic in speaking of the motivation of animals'. Contrary to this loose consensus, I will argue that there is a considerable amount of confusion about anthropomorphism. I will argue that the mistake or fallacy of anthropomorphism is neither well-defined nor clearly fallacious. There are many different conceptions of anthropomorphism and the common ones do not support their common rhetorical use (Fisher 1990: 96).

A first separation is made between *interpretive anthropomorphism* and *imaginative anthropomorphism*. Interpretive anthropomorphism refers to "all of the usual cases of ascribing mentalistic predicates to animals on the basis of their behaviour" (Fisher 1990: 100). Typical in this case is the description of certain behavioural patterns as intentional actions. Imaginative anthropomorphism is in turn defined as "the productive activity of representing imaginary or fictional animals as similar to us" (ibid.). Such is the case with the whole tradition of fables, myths, cartoons and fiction in general.

Interpretive anthropomorphism is in turn divided into *categorical* or *situational*. The first case consists of "ascribing mentalistic predicates to creatures to which the predicates don't ever in fact apply" (ibid.: 101). The problem is precisely categorical, for given behavioural patterns cannot be attributed to given species without one falling into the anthropomorphic mistake. In the case of situational anthropomorphism, an animal's behaviour is interpreted "in ways that could possibly apply to that animal in other circumstances, but which do not in the situation in question" (ibid.). It is thus anthropomorphic to ascribe a given pattern to species X in situation X_1, but would not be in the situation X_2.

Lastly, Fisher divides categorical anthropomorphism into *anthropomorphism by species* and *anthropomorphism by predicate*. The former consists of situations when "application of mentalistic predicates could be counted as anthropomorphism depending on the species. What woul.0dn't be anthropomorphism concerning a chimp might be concerning a worm" (ibid.). The latter describes situations when "application of mentalistic predicates could be counted as anthropomorphism depending on the predicate. I have in mind applying the wrong types of predicate" (ibid.).

How dangerous is then anthropomorphism for scholars? It is useful, I believe, to take a look at the methodological precautions taken by scientists in order to avoid the CHE or other anthropomorphic mistakes, and to consider the results of these precautions. The so-called *Morgan's canon* (named after the English psychologist Conwy Lloyd Morgan, whose life bridged the 19th and 20th centuries) is one of the most classical strategies. It says that "in no case should actions or behaviours be interpreted as the result of a superior psychic faculty, when it is possible to interpret them as a result of an inferior faculty". I shall comment upon this statement further on in the present article. Morgan's canon, anyway, is not the only proposed antidote to anthropomorphism: another method, quite typical of ICE programs, is the creation of the so-called *Emotion-Free* experimental context. Every possible unwanted input from the experimenters that might result in a CHE (facial expressions that might unconsciously suggest the subject of the ICE the solution for a specific task, affective – therefore, not anymore scientific – relation established with the subject of the ICE, etc.) is made sure to be avoided. Researchers then wear masks, observe the experiment unseen from another room, leave the subject of the ICE alone in the laboratory, avoiding any kind of interaction, and so forth. As a result, the completely Emotion-Free ICE built up throughout the last decades ended up in a failure, which, according to many semioticians, is the ultimate prove that: a) non-human animals are not able to learn language; and b) The CHE is the real core of the problem. When avoided, the ICE reveal their true nature, i.e., misleading non-scientific practices.

FURTHER CRITICAL REMARKS ON THE ICE

Sebeok's criticism on ICE was not limited to CHE, and – by all means – was articulated in a much deeper way than Frisk's summary (and I do not blame Frisk, exactly because he was just summarising a number of positions on the topic). In 1980, a collection of essays entitled *Speaking of Apes*, edited by Thomas and Jean Sebeok, was published and soon became the major point of reference for semioticians on the topic of ICE.

It is in the extensive sixty-page long introduction (Sebeok 1980: 1–60) to this book that we find the most detailed illustration of Sebeok's views on ICE. In the text, the Sebeoks go through nearly the entire panorama of ICE, presenting their doubts in the softest cases, and their resolute rejection in the hardest ones. Their arguments are in all cases well-formulated and convincing, and it is probably this characteristic (together with the undeniable sacrality of Sebeok's work in general[11]) that made the approach of Sebeok's followers on the topic as acritical as previously described. Semioticians' next observations on ICE proved to be fully restful upon this very text, entitled "Questioning Apes", showing little, if any, advancement from there, in a time (the 1980's and then the 1990's, up to nowadays) when ICE programs were growing rapidly and – in most cases – successfully.

More than twenty-six years after, it seems to me reasonable to take a slightly more active attitude towards that text, in the light of what has happened in the meanwhile (and I will never get bored to repeat how important was the birth – in 1976 – and the definitive establishment – in the 1990's – of the cognitive brench of ethology, within animal studies), but also as a consequence of a more critical reading. Maybe something, although still well-formulated, is not so convincing as it seems.

To start with general remarks, "Questioning Apes" is not fully respectful of the non-written etiquette of scientific compilations. *Speaking of Apes* is a collection of essays that intends to gather the main positions in favour of and against ICE. The editors of such a type of work, in writing the introduction to the book, should be careful to summarise and contextualise all these positions, but, at the same time, not to take too strong a stand in one direction or another. Or, if they do, that should be made by referring to external sources, in order not to discredit the book itself, as a collection made of good essays for one half, and rubbish for another. In other words, the editors should not exploit the chance of writing their introduction *after* all the other texts have been delivered to them.

In spite of this, "Questioning Apes" takes all possible advantages to make it clear how ICE programs are here regarded as anti-scientific and fallacious, and how naive the pro-ICE essays within the same book are. The reader is thus already instructed at – say – page 20 that going through the essay at page 200 would be a mere waste of time.

Not only the editors want the last word: they want it for first.

The only IC scholar spared from Sebeoks' harsh criticisms, and in fact highly praised, is Herbert Terrace, who conducted an ASL-based ICE program on the chimpanzee Nim Chimpsky (a wit pun with Noam Chomsky). The program – as I already mentioned in the section on ICE history – ended up as a complete CHE-affected failure, and Terrace – with rare professional honesty – admitted it with no excuses, becoming the hero of anti-ICE supporters:

> The honesty of Terrace and his colleagues in reporting this and other procedural steps which others, we suspect, omit from their accounts, is to be applauded. It makes it possible for the reader to assess more accurately the reliability and import of the data presented. (Sebeok & Umiker-Sebeok 1980: 13)

The procedures that the Sebeoks are here referring to is an indeed serious bias of the ASL trainers of Terrace's program: they had the tendency to revise their reports after discussing among themselves, thus they ended up to agree on many more points than in their previous observations of Nim's behaviour. Now, even if one might say that – sometimes – to compare each other's observations may be useful to correct (rather than provoke) individual mistakes, we shall agree that the chance of messing up the data collected is higher than the chance of improving them. However, what puzzles here is the confidence with which the Sebeoks take Terrace's case as a synecdoche for all the other ICE programs. As I was hopefully able to illustrate, the history of ICE reports of both unsuccessful and successful programs, the latter being more recent and numerous. Terrace was simply the only one to admit his own mistakes. That makes him a highly-respectable and honest scholar, not the proof that ICE programs are all biased and deceiving.

A last general remark, before getting to more specific considerations, is the fact that the Sebeoks seem not to have been extremely interested in visiting the sites of those research programs that they so much criticise. The only instance they mention is a 1969 visit to the Gardners' house in Reno, Nevada, where the Washoe-project was taking place at the time. The description of that visit does not seem too far from how Soviet authorities would selectively guide delegations of western communist parties through the (few) decorous areas of USSR:

> During our stay in Reno, in 1969 [...] we were, as guests, necessarily at the mercy of our gracious hosts, the Gardners, in terms of where and when we could view Washoe, and what additional experiences of the project – in our case, heavily edited films of the chimpanzee, casual testing of her signs in our presence using a box of well-worn objects, and discussions with some of the animal's trainers – we were allowed to have. (Sebeok & Umiker-Sebeok 1980: 29)

Now. Let alone that in 1969 the Washoe-project was at the very beginning (it started in 1966 and it is still going on, Washoe being still alive, and many other chimps – including Washoe's offspring – being introduced in the project), and it was only during that year that the first report of the project was published (Gardner & Gardner 1969). Let alone that a research laboratory, even if located in a private house (as in Gardners' case), has its rules and limitations, and people cannot simply do and check what they want just because they do not trust their hosts. Let alone that the "heavy editing" of the films is simply meant for sparing the spectators from hours and hours of totally uninteresting material. Let alone that a chimp is after all still a chimp, i.e., a wild animal, and it is not always predictable how s/he would react to the presence of a stranger.[12] Let finally alone that we are talking about a scientific experiment, not a circus show, and the chimpanzees are not trained to *perform* before any audience, but they rather need to trust their interlocutors before actually interacting with them. What is really not convincing is how energically the Sebeoks express their (aprioristic, in most cases) reservations about not only the experiments themselves, but also about the honesty of the IC researchers.

More importantly, we are not informed about any other visit to any other IC project. Not that they were not invited, in fact. In a 1998 interview for the New York Times, Sue Savage-Rumbaugh, responsible for the Kanzi project, to the question "Many in the scientific community accuse you of over-interpreting what your apes do" answered as follows:

> There are SOME who say that. But none of them have been willing to come spend some time here. I've tried to invite critics down here. None have taken me up on it. I've invited Tom Sebeok (of Indiana University) personally and he never responded. I think his attitude was something to the effect that, 'It's so clear that what is happening is either cued, or in some way over-interpreted, that a visit is not necessary.' I would assume that many of the people associated with the Chomskyian perspective including Noam Chomsky himself have the same approach: that there's no point in observing something that certain doesn't exist. (Dreifus 1998: newspaper article)

I must say that the number of inferences and assumptions contained in Sebeok's text are a bit too many for someone who has not really checked.

Anyway, let us go through some of these (unfortunately several) assumptions that are, in my opinion, not fully accurate. At page 10, a quite severe position is taken against the ICE program carried out by Allen and Beatrice Gardner (whose essay appears at page 287 of the book), and best-known as Washoe-project (even if it included several other chimpanzees, most of them not less linguistically-talented than Washoe herself). The methodological foundations of the project are put into question, as very likely to producc "distortions", altering the results of the experiment, or – even more often – focusing on a wrong pattern or an irrelevant priority. One example occurs as

> One observer may prime another to inadvertently create situations in which a newly reported sign might be likely to recur by, for example, using a certain tool when the sign for that object has been reported by another observer. This outcome is especially likely owing to the fact that trainers were instructed by the Gardners to record the context in which a new sign was observed as well as the sign itself. Should the ape actually produce a sign in this sort of context, it would not be considered the result of outright prompting, molding, or the CHE, all of which the Gardners deny played a role in the reporting of new signs, and yet such innocent provision of opportunities for corroboration of other observers' records would certainly influence the overall course of the study (Sebeok 1980: 10).

So, seems to be the point here, two things are bad: encouraging – in whatever form – the production of a sign instead of another; and the presence, during the ICE, of contexts or situations that *facilitate* the emission of signs. Now, if such procedures are anti-scientific, then somebody should tell parents from all over the world that the way they teach language to their kids is anti-scientific. Because, that is exaclty what parents do: a) they encourage the emission of certain words instead of others, and much more insistingly than any of the Gardners (typical examples are the words "Mum" and "Dad"); and b) they continuously create situations and contexts that facilitate the emission of signs and words (a recurrent instance being those language-related toys, such as picture books whose images are named in big fonts).

Besides, what is so wrong in facilitating the emission of signs? If – say – a dolphin is attempting to teach me (i.e., a human being) the language of dolphins, I would really not mind if I was somehow facilitated in emitting 'dolphinesque' signs. In establishing how scientific is an ICE program, one should always take into account two main factors: 1) all the difficulties related to learning a communication system that is not only different: it is alien, it belongs to another species. It is much more difficult than learning a foreign, but still intraspecific, language; and 2) simply enough, the trainee subject might just happen not to be a genius. Not only Washoe may be facilitated in emitting human signs: she *must* be facilitated, as any person who is learning an alien language. After all, are we after teaching language to Washoe, or are we after teasing her?

What I am trying to argue is that the scientific cleanliness of any experiment should not be exercised for pure cleanliness' sake. The target of the experiment is much more important than its aesthetics. If what we want to check is a) whether non-human animals are able to communicate with human language; b) whether they can ask what they need/want through the use of human language; and c) whether they are able to model their perception (also) with human language; then our ICE program shall achieve a degree of methodological cleanliness that does *not* interfere with or negatively affect these targets. On the contrary, we shall create the most facilitating and inspiring environment for these target to be reached. If a chimpanzee, or a specimen of another species, learns the human language despite all the obstacles that Sebeok claims as absolutely necessary for the scientific validity of the test, than what we prove is simply that that very chimp is an abnormal genius, not that chimps are able to learn human language.

At page 14, again commenting upon the Washoe project, the Sebeoks agree with Eric Lenneberg (1980: 80) in negatively evaluating Washoe's spontaneous ASL-signs productions. If the Gardners look favourably at the creation of new signs in Washoe, as symptoms of enthusiasm towards and cognitive appropriation of the new communication tool, Lenneberg does not hesitate to apply the Ockham's Razor and the Morgan's Canon to such instances, and finds that "we are simply testing our own ingenuity to assign interpretations to productions that might, for all we know, have been emitted randomly".

I shall not comment here upon the nature of these razors, mainly because I am discussing it at length in one of the next paragraphs of this essay, therefore I will concentrate my efforts in wondering why such a paradigmatically creative entity as language is here contested in its very nature. Language is by definition something flexible and imaginative: its codes – so useful for associating a given sign to a given meaning – are constantly challenged by neologisms, metaphors, loans from other codes, and other factors. When I speak a foreign language with some native-speaker and I have problems in finding the right word, it is exactly my creativity to rescue me from misunderstanding: I will use the word I have in mind in my own language hoping that it is somewhat similar to the other idiom, I will make gestures, I will point my finger towards something similar, I will describe that one word with an entire sentence, I will make up a new word if necessary... certainly I will come up with something, and in the end – rest assured – my interlocutor, although maybe a bit amused, will understand what I have in mind.

It is this very quality that makes language an always-in-progress entity. In Italy, there was not such a thing like a straight transition from Latin to Italian. Dozens of hybrid idioms were created in the meanwhile, and even today what we call "Italian" is not entirely the same bunch of words that only few years ago we would call the same way.

In sum, my opinion is that the spontaneous creation of new ASL-signs is to be welcome as a very positive reaction of Washoe to the human language. She might not have learned the right sign for the right word, but there she goes attempting to make her point anyway. And this is especially remarkable when the so-called "innovations" take place. I already mentioned these cases in the paragraph entitled "Semiotic scepticism": an innovation takes place when the trainee does not know the sign for a given object, therefore simply makes up a new sign, based on a cognitive association (*Orange-coloured apple* for an orange, *White tiger* for a zebra, etc.). This, as I already stated, is a convincing proof that language is used precisely as a modelling system, other than a mere communicative device. Of course, for someone who is at the same time opponent of ICE and promoter of the conception of language as modelling system, this is unacceptable:

> Unfortunately, these instances are generally reported in such a way that not enough is learned of the context of occurrence to enable us to rule out the possibility of either trainer suggestion or overinterpretation. [...] there is available a more parsimonious explanation for Fouts' report that Washoe was creating a new compound lexical item when she produced "water" plus "bird" in the presence of a swan and was asked, "what that?". Since Fouts provides no evidence that Washoe characterised the swan as a bird that inhabits water, we may just as well assume that Washoe, who was very familiar with the question form what that?, was merely responding by first identifying a body of water, then a bird. (Sebeok & Umiker-Sebeok 1980: 15)

Of all Sebeok's arguments, this is possibly the weakest one, at least for the following reasons:

1) It is simply untrue that the instances of linguistic inventions are badly or insufficiently reported. Check the respective texts for confirmation;
2) It was quite cunning of the Sebeoks to discuss one of the very few instances of linguistic inventions that is liable to ambiguity: "water" and "bird" are indeed rather general concepts, so it is in principle possible that Washoe could have referred to the body of water first, and to the bird eventually. But then, what about "drink fruit" for a water-melon? "Drink" is an action, not a tangible object. What could Washoe have possibly meant, if not that kind of fruit that is actually so juicy that its consumption is more similar to drinking than to eating (at least in comparison with other fruits)? What about Koko's white tiger, then? Did she see a "body of white"??? And where did she see the tiger?;
3) Even considering the case of "water bird" only, can we really accept such a simplistic (and thus, again, Morgan's canon-friendly) explanation like that of Washoe that, once asked "What that?", refers to two different objects, one of which is at least as big as a pool (if not as a lake, or even as the sea)? How did it happen that, all of a sudden, Washoe became so generic and scarcely-selective? Imagine: one takes Washoe out in the garden and points at a ball, asking "What that?". Should we really expect that she comes up with something like "Grass ball", or "Garden ball"? And why the birds that fly are not called "Sky birds"?; plus,
4) Was there nothing else than just a body of water and a bird? Not a tree, not the sky itself, not some clouds, not another animal, not a single element that might have caused Washoe to call the poor swan something like "water tree bird"?; finally
5) I hope that the incidental comment "Washoe, who was very familiar with the question form *What that?*" did not go unnoticed. That is quite a statement, if we consider that Sebeok does not believe that Washoe (or any other language-trainee) does actually understand the questions that she is asked. His idea, as

firm supporter of the CHE, is that Washoe reacts to that question by randomly making gestures, until she notices that her trainer is satisfied. How come she is now “very familiar” with the question? Maybe she understands it (and please do not tell Morgan)?

At pages 15–16, Sebeok’s review focuses on the question of language manipulation for amusement purposes, what has been nicknamed “Monkey humour”. In the dialougue reported in Plate 1, the ASL-trained gorilla Koko teases her trainer Dr. Francine Patterson, by pretending she wants her apple juice in the most unexpected places (nose, eye, ear, and finally mouth).

The manipulation of a code for personal amusement, as applied to non-human animals, is something that does not convince the Sebeoks, who once again seem to be more at ease by calling into question the CHE:

> If a sign or other response produced by an ape appears to be inappropriate [. . .], human trainers appear all too willing to stretch their imagination in order to make the animal’s performance “fit” conversationally. [. . .] Thus, anomalous chimpanzee or gorilla signs may be read as jokes, insults, metaphors, or the like, much as the not infrequent offenses against the very elements of counting and the fundamental arithmetical proecsses made by Clever Hans were regarded in part as intentioonal jokes and by an authority in pedagogy as a sign of independence and stubborness which might also be called humour. (Sebeok & Umiker-Sebeok 1980: 15–16)

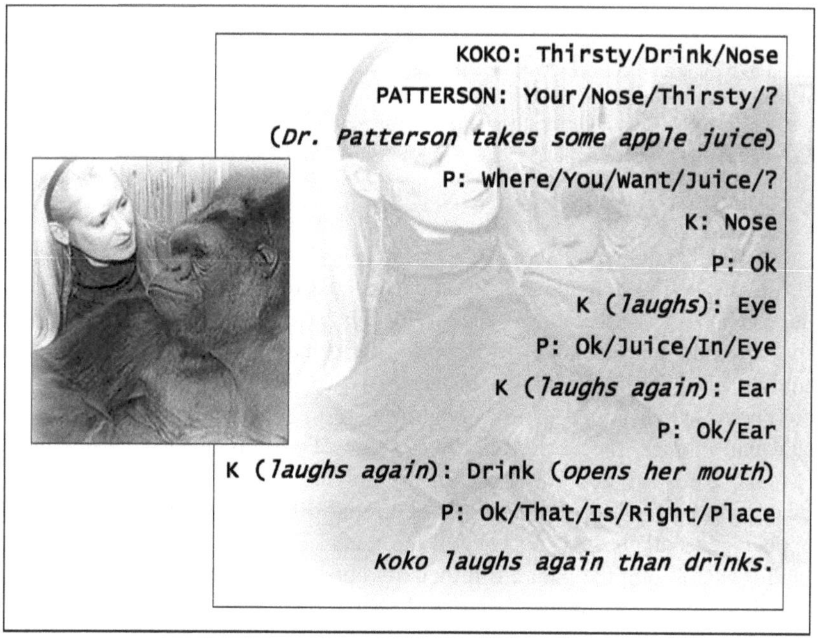

Plate 1. – ASL conversation between gorilla Koko and her trainer Dr. Francine Patterson

Further on, the very example of Plate 1 is reported as a typical CHE-affected misinterpretation of the in fact clear inability of Koko to perform the correct signs. Again, the counter-arguments are many:

1) as the request for food or drink is by far the most recurrent ASL inetraction between Koko and Patterson, it is extremely unlikely that Koko does not master it properly, and that this episode is something else than just an amusing diversion to a routine action (Patterson reports anyway other cases of her gorilla's humour, like when she was asked "to place a toy animal under a bag, and she responded by taking the toy and stretching to hold it up to the ceiling" – Patterson 1978: 456);
2) These examples show a typical and basic feature of humourism: turning things upside down. You ask me something, and I do the exact opposite. I have fun by teasing you, and I tease you by having fun. Nothing is more elementary than this, in humourism;
3) If Koko was just making mistakes, why on earth would she laugh?

Still about Koko, and about signs of "independence and stubborness", there is another critical remark at page 37. In commenting one of the ASL tasks proposed to Koko (it is not necessary to specify what kind of test, as it is not relevant in the point I am making now), Francine Patterson (1977: 10) noted that the gorilla was not awfully attracted by the exercise, and after a while (five trials per day, maximum, and two sessions per week) she would stop being collaborative and start performing the same sign over and over, or not performing any sign at all, or generally displaying boredom. Excluded these instances – said Patterson – Koko was able to perform the right signs in the 60% of the cases. The Sebeoks claim that the counting was incorrect:

> Koko is said to have scored correctly 60% of the time, or above chance, on the series of double-blind tests which were administered to her. It is reasonable to ask, however, how Koko's instances of avoiding the test, e.g., by responding to all objects with the same sign, were scored. We are not told whether or not such inappropriate responses were discounted as avoidance measures or counted as errors. [...] We can only guess whether or not Koko's performance would have been below chance were a less biased accounting to have been made (Sebeok 1980: 37)

From my understanding of Patterson's research, I can confirm that the Sebeoks are right in suspecting that she did not count Koko's reluctance as a mistake. But my question is, why should have she? Why the lack of enthusiasm to perform a given task should be equalled to a wrong performance of it? Let us compare the situation to an ordinary human instance of two people playing a game, for example chess, which is one of my favourite games. As it happens, I make both good and bad moves: once I am able to get one of my opponent's castles, next move I make a mistake and I lose one of my knights. And so on. At some point, 60% of my moves have proved to be correct, and 40% definitely have favoured my opponent. So, all in all I am leading the game. Except that, well, I get a bit bored and I want to quit the game. But then my opponent insists and say we should continue: there is no way for me to persuade her. I decide not to make any move, so the point is clear. Nothing: she insists. We go on like this for a while, I get a bit irritated so I decide to move my queen back and forth all the time, so I can quickly lose the game and finally put an end to it. At last, my opponent understands and allows me to leave.

Now, my attitude can certainly be criticised from several points of view – most of all I was not very polite, was I? – but I am positive that no one could dare saying that my reluctance to play corresponded to a series of wrong moves on the chessboard. Not even when I started moving the queen redundantly: it was very evident that I was not in the game anymore, and I was just trying to persuade my opponent to let me be. An ideal observer recording the successfulness of my playing will definitely agree that my percentage of good moves is still 60%, and that my final sabotage does not really count.

Page 16. Now the target is mistakes and imprecisions:

> The determination of trainers [the Premack's team, this time] to show their animal [the chimpanzee Sarah] in the best light has even led to the denial that errors are mistakes, again bringing to mind one of the assumptions which sets research on psychic phenomena apart from normal scientific procedures. The so-called psychic's mistakes on tests of her powers are frequently used to prove that those powers are real, based on the assumption that if the performer was using mere tricks she would be correct every time. (Sebeok & Umiker-Sebeok 1980: 16)

It is not to be excluded that I am deeply biased, as well, because I totally agree with Premack: if my trainee would perform all his/her tasks without a single error, I would really think that s/he found a way to cheat. On the contrary, if the performance displays a minimal percentage of mistakes, I would be more reassured that my ape, who – I assume again – is more likely normally-gifted intellectually, rather than a genius, has done his/her job in the best possible way.

Besides, what we are talking about here is tasks that were performed correctly in a significant percentage, i.e., more than the 60-70%: if I answer correctly 6–7 questions out of 10 in any school exam, I normally get a good mark.

Since we are now into school recollections, I obviously cannot agree on a remark made few pages further:

> A related issue concerns the personal preferenecs exhibited by the apes for certain project members. All the animals used so far in the ape "language" projects have shown some preference for certain trainers, performing better for these "favorites" than for the rest, but there has been no systematic attempt to account for these differential responses on the part of the animals. (Sebeok & Umiker-Sebeok 1980: 22–23)

Once again, the Sebeoks seem to forget that the trainees are first and foremost subjects, with their characters and preferences. They cannot be the infallible machines that the Sebeoks are demanding in order to take their learning-efforts seriously. I would have nowadays much better than a mediocre knowledge in Chemistry and French, two subjects I do like, if my two high-school teachers were not the two witches that they were.[13] The equal and contrary applies to subjects like Economy and Law, which I never thought I would like, and which I managed to learn rather decently thanks to nice and competent teachers. Finally, and most importantly, the very reason I became interested in semiotics during my University days (I was enrolled in Musicology) is because I met Gino Stefani, Professor in Musical Semiotics. He is the point of departure that eventually led a musicologist to write about Anthropological Zoosemiotics (not to mention all the other fundamental mentors I met on the way).

Therefore, I am by all means positive that having preferences for one trainer instead of another is not only normal and scientifically-acceptable. It is also healthy.

Incidentally, it is also mentioned (Sebeok & Umiker-Sebeok 1980: 23) that also the trainers have their own preferences, among the trainees. Do I need to point out that the same principle applies the opposite way?

There is still one more attack to ICE to be mentioned. It is at page 30, and consists in the remark that, every now and then, the apes use signs from their own non-human repertoire in order to support their ASL, what makes an observer "unable to determine which of the actions he sees performed by such an animal are part of its natural repertoire and which are the results of special training" (Sebeok & Umiker-Sebeok 1980: 30). I apologise for being repetitive, but, again, I fail to understand why this should be a problem. In fact, to say it all, I see it as a plus. It shows active (possibly enthusiastic) participation of the trainees to the communication process, and – most of all – it shows very clearly that they have fully understood that it *is* a communication process. That is why, when unable to make themselves clear by means of the human code, they do not hesitate to use their own sign system, hoping, perhaps in vain, that their interlocutor will finally get their point.

In conclusion to this section, I have to say that I am left quite perplex by Sebeok's attitude. It does not surprise me that the man who so convincingly argued in favour of the fully-semiotic qualities of non-human communication is here so much against the extension of the notion of language to other animals. However contradictory that may appear. I am not surprised because his definition of language has been always very clear, as something whose communicative characteristics are not at all important, in comparison to its features as modelling system. Therefore, talking about communication is an entirely different issue than talking about language.

What surprises me is something else: it is the inaccuracy and the apriorism of his arguments against ICE, like those of someone who has already made up his mind about a topic, and is not going to change it, whatever it takes. If we had to sum up the arguments I have described in few words, it would turn out that, for Sebeok, a language must be learned the following way:

1) Avoiding any encouragement or facilitation during the learning period;
2) Prohibiting creativity and imagination in producing signs;
3) Taking code manipulations as linguistic mistakes;
4) Taking boredom and insubordination as linguistic mistakes;
5) Accepting only instances of communication where not a single error is committed;
6) Prohibiting any form of personal preference towards any trainer;
7) Prohibiting the use of one's own original sign repertoire to support the learning of the new idiom.

In such conditions, I am afraid to say, not only Washoe would not learn human language. But neither Umberto Eco would.

One last general note, about the beloved CHE. Even, for the sake of argument, accepting that ICE are CHE-biased (and I am not), the fact is, there is a very thin line between imitating and understanding. Even imagining a fully-CHE-biased interaction, I find it hard to find a significant difference between performing a sign because "this is how I get the apple", and performing it because "I know this stands for 'give me an apple'". As a matter of fact both approaches prove that the trainee has learned how to achieve what s/he wants through the code he is expected to use. As soon as s/he has more than one sign (let us say, just to be banal, A for apple, B for banana, and C for candy), and process the information in such a way that s/he performs B when s/he wants the banana, and not when s/he wants a candy, then the task is successful, whatever the exact configuration of his/her reasoning is, either it goes like "I will perform B, so I get a banana", or like "I will perform B, which stands for banana". The only difference is that the former relation is indexical, the latter is symbolic, but the connection sign-object has happened anyway, which is exactly what ICE detractors object. I think we may accept that a conative type of communication proceeds by indexes, especially when the trainee proves s/he can use this foreign code in a symbolic way, as well (I hope by now I have provided enough evidences in that respect, the example of Kanzi being probably the most remarkable one).

THE ROLE OF THE SUBJECT IN ICE AND OTHER UNDERRATED ISSUES

Another aspect that somehow surprised me about Sebeok's review is that he seemed to care a lot about aspects that I do not see as so important. The aim of this section is therefore to point out what ICE-related issues have been, in my opinion, either underrated and/or not-properly dealt with. In that respect I will not spare some criticism to the ICE scholars themselves. The claim of this article is that several aspects involved in these projects were underrated or misunderstood. Now, that does not apply only to Sebeok and other semioticians, but to the performers of these projects too.

To start with, even though many of these scholars seem to have genuine love and compassion towards the non-human animals they are dealing with (what is certainly the case with Jane Goodall, Roger Fouts, Sue Savage-Rumbaugh and Francine Patterson, to mention the cases I know best[14]), it must be still emphasised how ethically questionable is the idea that a non-human animal may be stolen from his/her own environment and put in a cage for most of the time, forced to establish relations and perform tasks that s/he probably does not want or need, and exposed to a great degree of boredom and depression, when not – unfortunately – violence and deprivation.[15] It is no chance that the most successful ICE (Washoe, Koko, Kanzi, Alex, etc.) were those where the trainees were given the best possible quality of life.

Apart from strictly ethical questions, there are other issues that probably deserve a more careful scrutiny. One of those is certainly the application of the Umwelt theory. It has been said several times (also, of course, by Sebeok himself) that belonging to different Umwelten is enough a reason for a non-human animal not to be able to access a sign system that relates exclusively to the human Umwelt. Not only. The human Umwelt is, on the contrary, able of accessing any other Umwelt, at least in principle, thanks to the inherent characteristics of language itself, in particular what I have called here "Narrativeness". And if we consider, somehow, past and future (especially future) as 'possible worlds' (after all, possible-to-happen future or possibly-happened past are *things* that are not yet or not anymore objects), then also the quality that I called "distant-time semiosis" belongs to this discussion.

Moreover, even if this has been emphasised much less often, there is anyway an intrinsic difficulty in configuring an Umwelt that is different from one's own (and that applies both in sensu stricto, i.e., accordingly to Uexküll's formulation of the concept, and in sensu lato, for instance when we talk about cultural or personal Umwelten).

Now, much is to be agreed about all these statements, even though I shall take the liberty to reformulate them in what I consider to be a more accurate way. Firstly, it is certainly true that the decoding of alien Umwelten is almost a contraddiction in principle. Taking Uexküll's theories very faithfully, one shall deduce that if we, members of the Umwelt A, manage to decipher the Umwelt B, then we cannot really talk about different Umwelten anymore, because the occurrence of this very process would prove that the Umwelt B is simply part of the Umwelt A, therefore not 'alien', and in fact not Umwelt, either. Talking about *things*, in a way, makes them not *things* anymore, but *objects* already, or at least philosophical *objects*.

Secondly, the Narrative and Distant-time semiosic qualities of language (the former, in particular) seem to create a certain unicity in the human Umwelt, in that they allow a strong dialectical, philosophical and even rhetorical configuration of alien Umwelten. *But*: dialectics, philosophy and rhetorics do not give *full* access to another Umwelt, even though they certainly allow a good deal of reflections and some conclusions about it. This is important to remind, especially on an ethical level: sometimes, indeed, we take too much for granted that we understand from other animals *all* that is to be understood. To make a specific zoosemiotic example, we may understand, speculate and talk about echolocation, but a complete, senso-motorial, psycho-physiological access to echolocation (at least so far) is not possible, and that makes a huge difference: this is when Thomas Nagel is right in wondering *How does it feel to be a bat?*

More importantly, for the purposes of the IC topic, the question about Narrativeness and Distant-time semiosis is: are we sure that these features of human language are *qualitatively* different from other animals? If ICE turned out as successful attempts to impart human language on other animals, as they did, in mostly everybody's opinion except semioticians', and if language belongs to the human Umwelt, then it goes automatically that Kanzi & c. *did* access the human Umwelt, in fact they accessed what is commonly regarded as the quintessence of it. The way I see it, an extensive mastering of such a powerful device as language, as humans do have, allows a drastic increase of narrativeness and distant-time semiosis, therefore the more refined and articulated our linguistic knowledge, the deeper our narrative abilities. However, the ICE outcomes seriously puts into question our exclusivity at this regard, and – at the same time – reinforces the hypothesis that narrativeness and distant-time semiosis are language-specific features. Which makes a subtle, yet dramatic, difference. Because these features are now marking a difference between language and other communication and modelling systems, rather than between humans and other animals. If other animals were able to learn language, at least to some extent, that categorically means that it is in their cognitive potential to learn it.

Then, why it was humans, as a matter of fact, to invent language? And why language does not arise spontaneously in other animals (since they were able to acquire it only after human's guidance, in contexts of captivity)? Obviously, those questions deserve a separate and thorough treatment, but I shall not renounce providing one or two reflections. In this case, I can only attempt a purely speculative hypothesis (but it comforts me to be aware that neither opposite hypotheses have any empirical support), not before apologising for the genericity of the next statements. Language – it is my claim – arose in the species that more than any other *needed* it, not in the only one that *could* invent it. The existence of some 3000 different idioms led to the common place that language is purely a cultural phenomenon. In actual fact, no less than forty universal fonems exist, and a selective attention towards them among infants from all over the world have been empirically observed, proving the spontaneous, therefore biological, basis of symbolic language (see Malacarne 1992: 452). To accept this point makes it easier to accept that language must have been born as a result of an evolutionary need.

My opinion (which is anyway not so isolated, see also Malacarne 2005: 243–247) is that the need for a language came to be as a solution for the many physical limitations of the human being. In most of the problems that any other animal can solve alone (feeding being the most obvious example), human beings need cooperation. In that respect, language could have been subsequent to strictly adaptive functions of communication in general, such as the referential or the connative one, that required more complex specifications, due to the inability of humans to perform a given task on their own. The more efficient and articulated the communication, the more successful the cooperation. A signal arised with a specific function, e.g., to obtain from the receiver the handing of a given tool that the latter had to go and collect from another place. Later in time, such a function must have developed in more articulated ways (passing through stages of generalisation, abstraction, etc.), going hand in hand with natural selection and phylogenetic evolution of the species. The most successful human subjects were those who could communicate in the most precise and articulated way. Operatively speaking, the very *selection* consisted in the development of two brain areas, the so-called Broca (that controls the motor expression of sounds) and Wernike (that is involved in the perception of sounds), and – above all – the vocal apparatus. When scientists say that we are similar to

Chimpanzees for the 99% circa of our genetic heritage, they normally omit to add that most of that 1% of difference is exactly located in the vocal apparatus and in these two brain areas. The causal relation between these elements and the creation of language is reciprocal: the former were responsible for the development of the latter, and the latter was responsible for the development of the former.

And it was of extreme importance to have that kind of resources in order to develop a kind of device such as language: the vocal apparatus, in particular, made it possible to exploit the richness, the flexibility and the practicality of the vocal-acoustic channel. A cognitive use of the language was thus made possible, and the process of extension and enrichment started, up to the virtually limitless articulation of nowadays: it was during this process of expansion that – it is my guess – language became more important as a secondary modelling system. The very same process, as applied to hands, is described by Friedrich Engels to tool-fabrication and work (1896: 545–554): hands became at the same time the cause and the product of work.

Related to this issue is also my next consideration, which I consider quite crucial: so far, ICE have been evaluated only at an ontogenetic level, and not at a phylogenetic one. When we think about the transition from the birth of the genus *Homo* (4–5 millions of years ago), when the first – very rudimental – traces of a language presumably appeared, to the *Cro-magnon* (125,000 years ago), with whom we probably have the appearance of a less-embryonic form of language, we understand that it took thousands and thousands of years for language to take a recognisable shape. The above-mentioned anatomycal and neurological features developed gradually and slowly (and randomly, if we take a strict Darwinian perspective): it is a bit too much to demand perfection from the first generation of non-human animals who have been introduced to human language. Alex, Koko, Kanzi and the others have shown that it is in the potential of many species to learn a sign system of that kind: if we are asking for *Planets of the Apes*-type of scenarios, then we must be patient.

Leaving aside these speculations on the origins of language, remains the claim that narrativeness and distant-time semiosis are not features that characterise exclusively the human language. *Sic stantis rebus*, the very qualitative difference between human language and other communication systems is probably the existence of linking words, that is, as I already mentioned, those solely-interoceptive signs whose exclusive function is that of connecting signs that have also, and mainly, exteroceptive meanings. Curiously enough, the big advantage of linking words, and language by consequence, is not the ability of making meaningful signs, but exactly the opposite, i.e., creating meaningless signs. Signs that do not refer to anything when taken alone (unless a specific metaphoric use is made of them, but that is an entirely different issue), but can make a great deal of difference when combined together with meaningful signs, in certain numbers and sequences. Steven Hockett, in defining the famous (sometimes infamous) sixteen design features for human language (1958), had a very similar concept in mind when describing the 12th feature, duality. Non-human semiosis seems to target most of all economy and ergonomicity: signs, when used to communicate, must all make sense. Put roughly, a non-human animal emits five signs if s/he has five things to say. Humans may employ a great deal of signs just for expressing one thing. But it is exaclty this 'meaninglessness' factor that increases exponentially the semiotic potential of human language: it creates specifications, differences, appartenances, separations, distinctions, and so forth. In one word, it creates a virtually unlimited number of relations.

Coming back to the issue of Umwelt, in the strict sense, I find a bit weird that both IC researchers and semioticians gave so little importance to the way the perceptual field of a given specimen (therefore, probably, of an entire species) affects the actual reception of a linguistic sign. The fact that Washoe, once learned the ASL sign for *flower*, took to name that way everything with strong smell is quite interesting in this respect. As humans, we tend to establish an immediate connection between object and representamen on the basis of the visual channel. Therefore, if we are presented with a flower, and we are told "this is called 'flower' ", we tend to think of flowers in iconical sense, and – if we are given an image of Van Gogh's sunflowers – we can spot the objects represented as flowers again, even if their sole resemblance with the real thing is exclusively of iconical type. Washoe had probably reflected in a totally different manner. The first thing that caught her attention was the perfume of the object, i.e., she paid more attention to the chemical channel, in this case. As a consequence, she had a reason to associate the sign "flower" to the semantic category of 'smelling' objects, rather than look-alike flower-shaped ones. Cigars and aftershaves were thus as 'flowers' as roses and orchids, while the non-smelling Van Gogh picture was obviously something else. If we forget for a moment the human way of categorising reality, we must admit that this way of reasoning is by all means legitimate, and – naturally – intimately semiotic.

For the benefit of the sceptics, this instance is not isolated. A similar episode happened when Washoe, who is also one of the famous chimpanzees-painters, was asked to draw a ball. Again, as humans, our first idea for representing a ball is iconic: we trace a circle on the paper, and perhaps – if we are football fans – we add black pentagons and white hexagons inside it. But to Washoe, it seems, the most relevant characteristic of the entity 'ball' is kinetic, not iconic, as she drew a series of progressively smaller arcs, i.e., the trajectory of the ball (Fouts 2000, personal communication). Once more, such a reasoning process is by all means acceptable: in fact, it is telling us precious information on the chimpanzee's Umwelt (or perhaps of Washoe's Umwelt only, but that would be still very interesting).

In this respect, the fact that many of the ICE trainee subjects have proved in several and different cases, a high degree of flexibility in accessing the human Umwelt, conforming their own accordingly (for instance, accepting to represent symbolically a series of objects that they are used to represent indexically), shall be saluted as a very (if not the most) remarkable achievements of ICE.

Finally, however banal it may read, it is important here to clearly underline that the identity of the ICE trainee as individual, and not as a member of a given animal species, is by all means crucial, and yet very underrated by both IC scholars (at least some of them: I already listed some exceptions), and semioticians. Before than being taught to chimps, the human language have been taught to Sarah, Washoe, Moja, Lana, etc. Koko and Mike are Koko and Mike, before than being gorillas. And Kanzi is first of all Kanzi, before than being a bonobo. This list of tautologies is here quite important, since it reminds us that:

1) The trainee is first of all someone subject to emotions and feelings, and exposed to experiences and events, *then* s/he is someone on whom an ICE program is performed;
2) The individual responses of the trainees may be related to factors that are external to the experiment environment, and may simply be the result of a personal experience, or be part of the cognitive-emotional background of the subject in question; in particular
3) Observations and results (either negative or positive) of ICE have been often inappropriately generalised to entire species, at least, when not to entire orders, or to the whole Animal Kingdom *tout court*. In simple words, when Moja fails to perform the right ASL gesture, it does not mean that all the chimpanzees have failed. Likewise, if Moja will someday compose an ASL sonnet in rhyme, that does not mean that we should expect the same from Lana, Lucy, Sarah and the others;
4) An extensive account of the personal history of the subject should always be integrated with the rest of the ICE, as fundamental and unavoidable element for the correct interpretation of the trainee's response. The success of Fouts and the Gardners with Washoe and other chimps, and of Patterson with Koko and Mike, is in my opinion mostly due to the fact that this aspect was fully taken into account.

Still, many seem to forget these apparently basilar notions. In my opinion, the most evident instance where the individuality of the trainee is totally ignored concerns our expectations towards his/her behaviour. It is what I half-jokingly call "thirdness-only expectations", with obvious reference to Peirce's notions of Firstness, Secondness and Thirdness. We expose a non-human animal to a problem-solving task, by – for instance – introducing a variable in his/her usual environment. We not only expect that the trainee, who *does not* know s/he is supposed to perform a task, focuses quickly and only to that very variable, but we also expect that, while doing that, s/he privileges a line of reasoning which is by no means the most obvious and spontaneous one. I will exemplify this so far ambiguous concept through the words of someone who was able to depict it with extraordinary efficacy.

Sometimes, in fact quite often, phiosophers and artists are able to see much further than scientists, and this is mostly because they do not see empathy and sensibility as biases. They rather use them as devices for a better comprehension. It should not surprise too much, therefore, that no one better than the Nobel-awarded writer John M.Coetzee has succeded in picturing the paradoxical situation I have now described. I shall apologise for the long quotation here, but it really makes the point clear. Plus, it is a piece of great literature:

Let me recount to you some of what the apes on Tenerife learned from their master Wolfgang Köhler, in particular Sultan, the best of his pupils, in a certain sense the prototype of Red Peter. Sultan is alone in his pen. He is hungry: the food that used to arrive regularly has unaccountably ceased coming.

The man who used to feed him and has now stopped feeding him stretches a wire over the pen three metres above ground level, and hangs a bunch of bananas from it. Into the pen he drags three wooden crates. Then he disappears, closing the gate behind him, though he is still somewhere in the vicinity, since one can smell him.

Sultan knows: Now one is supposed to think. That is what the bananas up there are about. The bananas are there to make one think, to spur one to the limits of one's thinking. But what must one think? One thinks: Why is he starving me? One thinks: What have I done? Why has he stopped liking me? One thinks: Why does he not want these crates anymore? But none of these is the right thought. Even a more complicated thought — for instance: What is wrong with him, what misconception does he have of me, that leads him to believe it is easier to reach a banana hanging from a wire than to pick up a banana from the floor? — is wrong. The right thought to think is: How does one use the crates to reach the bananas?

Sultan drags the crates under the bananas, piles them one on top of the other, climbs the tower he has built, and pulls down the bananas. He thinks: Now will he stop punishing me? The answer is: No. The next day the man hangs a fresh bunch of bananas from the wire, but also fills the crates with stones so that they are too heavy to be dragged. One is not supposed to think: Why has he filled the crates with stones? One is supposed to think: How does one use the crates to get the bananas despite the fact that they are filled with stones?

One is beginning to see how the man's mind works.

Sultan empties the stones from the crates, builds a tower with the crates, climbs the tower, pulls down the bananas. As long as Sultan continues to think wrong thoughts, he is starved. He is starved until the pangs of hunger are so intense, so overriding, that he is forced to think the right thought, namely, how to go about getting the bananas. Thus are the mental capabilities of the chimpanzee tested to their uttermost.

The man drops a bunch of bananas a metre outside the wire pen. Into the pen he tosses a stick. The wrong thought is: Why has he stopped hanging the bananas on the wire? The wrong thought (the right wrong thought, however) is: How does one use the three crates to reach the bananas? The right thought is: How does one use the stick to reach the bananas?

At every turn Sultan is driven to think the less interesting thought. From the purity of speculation (Why do men behave like this?) he is relentlessly propelled toward lower, practical, instrumental reason (How does one use this to get that?) and thus toward acceptance of himself as primarily an organism with an appetite that needs to be satisfied. Although his entire history, from the time his mother was shot and he was captured, through his voyage in a cage to imprisonment on this island prison camp and the sadistic games that are played around food here, leads him to ask questions about the justice of the universe and the place of this penal colony in it, a carefully plotted psychological regimen conducts him away from ethics and metaphysics toward the humbler reaches of practical reason. And somehow, as he inches through this labyrinth of constraint, manipulation, and duplicity, he must realize that on no account dare he give up, for on his shoulders rests the responsibility of representing apedom. The fate of his brothers and sisters may be determined by how well he performs. (Coetzee 1999: 126–127)

Let alone the specific case of Köhler and Sultan. What matters here is the line of reasoning, which, in Coetzee's case, could not be more brilliant. Hardly, if ever, the emotional condition of the trainees, as subjects, is considered important. We do not expect Firstness from them: we have a straight expectation for Thirdness. We expect them to react to our stimulus in the way we have planned (and received funding for), while the rest is either uninteresting or wrong. Why? How can we be so blind, or – if you still prefer scientific jargon – methodologically biased? The same principle applies to many of my critical remarks on Sebeok: why should a trainee not have his/her own preferences? Why should s/he not be supported in the learning process? Why should we expect him/her to be a language-machine when s/he has also a life and a personality, other than a good or bad predisposition for learning ASL or lexigrams, or whatever they are?

CONCLUSION: A MARTINELLI'S CANON?

During August 2004 I took part to a summer seminar in Puhtu, Estonia, at the former country house of Jakob von Uexküll, now a biological station, owned by University of Tartu. A piece of heaven, except for the mosquitoes and a rather approximate concept of toilet. In my paper, on anthropomorphism, I discussed the Morgan's Canon. Half jokingly, I asserted that, if anything, Morgan's Canon should be turned upside down, i.e., "in no case should actions or behaviours be interpreted as the result of an inferior psychic faculty, when it is possible to interpret them as a result of a superior faculty". During the discussion, Kalevi Kull gave my provocation a more serious touch and labelled it *Martinelli's Canon*. Before I could go "Come on, don't kid me", he made a really interesting point, and my narcissism finally prevailed. Mmm... a Martinelli's Canon... It was the very same year of my friend Kristian's *Bankov's Razor*, so why not?

The problem, not a secondary one, was that, at that point, I was morally obliged to give shape and contents to 'my' canon, and not only a simple formulation. First of all, what is the main contraddiction in Morgan's Canon, such that an alternative Canon is needed? Personally, I see at least three:

1. When Morgan talks about the 'possibility' of interpreting animal behaviour in a given way, he is referring to those cases where a *speculative* choice is necessary, that is, where a non-human behavioural pattern that is analougous to a human one, does not give empirical and unquestionable evidence that it is also an homologous pattern. In other words, he refers to those cases where the animal *seems* to behave or think in a certain way (similar to the way a human would behave or think in the same situation), so the point is: does the animal *seem*, or is s/he *really* behaving like that?

In these instances, evidently, all we have is this similarity between the human and the non-human behaviour. Thus, logically speaking and when counterproofs are missing, the starting hypothesis should be: the two patterns *are* homologous, exactly because they are similar, and that is the only thing that we know at this point. Then, we make all kinds of research needed, and try to understand if the hypothesis is confirmed or not. Still, at the very beginning, and the very beginning is the research phase that Morgan's Canon is referring to, we shall think of an homology.

Let us make an example: on one side we have a car, on the opposite side we have an object that has a slightly different shape, that *seems* to move, *seems* to have four wheels, *seems* to have seats, steering wheel, windscreen wipers and a car-stereo. At the moment, that is all we know. What are we going to do? In my humble opinion, we should start from the idea that this mysterious object is actually a car, maybe from a different brand than the one we have on the other side, but still a car, and our research should be oriented in that direction (are those really wheels? Are those really seats? Is there an engine, a brake pedal, etc.?). That seems to me a more convenient procedure than thinking that the object we are observing is in fact a skateboard, so those things that looked like seats are in fact a simple flat board, and the wheels are much smaller than they seem.

Still, this is what Morgan's Canon does. Not only it suggests us not to trust appearences, it also tell us that reality is simpler (or reduced) than it appears.[16] This way, inevitably, something, of what we observe, remain unexplained. For instance, what is that thing that look like a car-stereo? What about those look-alike windscreen wipers? How do they fit in the idea of 'skateboard'? Scholars in animal studies invented an *ad hoc* solution for these disturbing problems: a big, fat black box where they put everything that cannot be explained. The most common name for this black box is "istinct".

Apparently the most easily definable concept, at least in non human-related contexts, instinct is in fact the most complex and tricky one. Decades of lively discussions among scholars have yet to produce even an approximate definition. Instinct has been defined as a "voluntary but never-previously-put-into-action behaviour", a "behavioural impulse for the accomplishment of biological functions", a "stereotyped behavioural pattern", and a "pattern put into action without a precise idea of its results". It is already apparent that these definitions are more hiding than showing: they just describe an action, but they do not give account of causes and articulation. Moreover, some of them are in contradiction to each other. What seems to be certain, is that instinct is often a functional and easy-to-use theoretical tool for the explanation of behaviours difficult to account for. As Gregory Bateson (1969) properly puts it in one of his metalogues, instinct hence ends up being a real explanatory principle:

Daughter:	Daddy, what is an instinct?
Father:	An instinct, my dear, is an explanatory principle.
Daughter:	But what does it explain?
Father:	Anything – almost anything at all. Anything you want to explain.
Daughter:	Don't be silly. It doesn't explain gravity.
Father:	No, but this is because nobody wants instinct to explain gravity. If they did, it would explain it. We could simply say that the moon has an instinct whose strength varies inversely as the square of the distance...
Daughter:	But that's nonsense, Daddy.
Father:	Yes, surely. But it was you who mentioned 'instinct', not I.
Daughter:	All right – but then what does explain gravity?
Father:	Nothing, my dear, because gravity is an explanatory principle.
Daughter:	*Oh.*

(Bateson 1969: 11)

Now, to me, no better explanation of the concept can be given than Bateson's. It is clever, witty, and ironic enough to emphasise how exaggerated a meaning we provide "instinct" with. Considering the number of actions, from the simplest to the most extremely complex ones, that are described as instinctive, it shall be much easier to define the term in one of the following ways: "Supernatural power", "Invincible weapon", or – more humbly – "Ultimate solution to every problem". To which, it should also be added a) where one can buy some instinct, b) how much does it cost, and c) if it is legal.

The problem with instinct, I fear, is its absolute conceptual flexibility. Once established that it is an impulse that drives animals (including humans) to perform a (limited) number of actions, and I definitely agree that there are actions that we perform quite *before* having a mental representation of them, it becomes extremely easy to apply it to every case where a mental representation is not detectable (i.e., it is not verbalised by those who experience it).

Here lies the difference: if we produced our hypotheses in accordance with our observation, we would look at a beaver carefully cutting its pieces of wood in such a way that they all weigh the same (which is what actually happens: they *do* weigh the same), and we would say that, indeed, the beaver is carefully cutting its pieces of wood in such a way that they all weigh the same. On the contrary, with Morgan's Canon, we are faced with a problem: our hypothesis is now implying a superior psychic faculty. And this is very bad, if it is *possible* to interpret the same pattern as a result of an inferior faculty. We need an inferior one, that is, a faculty that does not require such an articulated mental activity (provided that estimating the weight of a piece of wood is "an articulated mental activity"). So, we buy a dose of instinct: the beaver does not have the slightest clue of what is going on, but she has her supernatural power, that drives her to chose this piece of wood instead of that one, which is too heavy and too long.

Now, apart from the general theoretical remarks on instinct itself[17] (which I hope I have sufficiently illustrated, by now), the question is: was it *possible* to apply an inferior psychic faculty, in this case? Because that is what Morgan says: we do not have to apply a superior faculty, when it is *possible* to apply an inferior one. Otherwise, one might say, we shall just leave the superior one, right? On the basis of which principle are we allowed to apply the notion of instinct on the beaver's weight estimations? Was it *possible*? Well, of course, everything is possible if we want it: but that is exactly how Bateson's explanatory principle works. It explains something just because we deliberately designate it to explain it. It does not explain it in a very scientific sense. So, what did Morgan mean by *possible*? Was it a *scientific* possibility or simply a *rhetorical* one?

2. What do we mean by 'superior' and 'inferior' psychic faculty? Do we mean 'more' or 'less' similar to the human psychic faculties? If yes, as I am ready to bet, then we are typically facing that anthropocentric bias that anyone who has fully accepted Umwelt theories, and it is my case, should refuse a priori. If there is a transpecific paradigm to interpret psychic faculties, that, to me, should be adaptative skills. The more adaptive the subject, the more 'superior' his/her

psychic faculties. Still Gregory Bateson has an excellent point in this respect. In experimental contexts, he says (Bateson 1972: 368–370), the animal's intelligence is tested through four main points: 1) the animal "may or may not perceive a difference between the stimulus objects X and Y" (ibid.: 368); 2) the animal "may or may not perceive that this difference is a cue to behaviour" (ibid.); 3) the animal "may or may not perceive that the behaviour in question as a good or bad effect upon reinforcement" (ibid.), that is, that doing the right action will be granted with (usually) food; and 4) the animal "may or may not choose to do 'right', even after he knows which is right" (ibid.). As it can easily be deduced, the four points are set in order of priority, that is, 4 is possible if 3 is accomplished, 3 is possible if 2 is accomplished, etc. Therefore, step 4 reveals a superior psychic faculty than 3, and so forth. Now, are all these steps really tested in experimental contexts? Do trainers really go through the whole procedure? Bateson is pretty doubtful.

Let me now consider for a moment the art of the animal trainer. From conversations with these highly skilled people – trainers of both dolphins and guide dogs – my impression is that the first requirement of a trainer is that he must be able to prevent the animal from exerting choice at the level of step 4. It must continually be made clear to the animal that, when he knows what is the right thing to do in a given context, that is the only thing he can do, and no nonsense about it. In other words, it is a primary condition of circus success that the animal shall abrogate the use of certain higher levels of his intelligence. The art of the hypnotist is similar. (Bateson 1972: 369)

3. Any action, even the simplest one, never has a unique impact, and by consequence can never be interpret as monolith. A Canon that promotes simplicity is in practice a canon that prefers to 'chain' observation in such a way that it gives the least possible disturbance.

Clever Hans was undoubtedly unable to perform mathematical calculations, so, to Morgan, all we need to know is that this horse was not a good mathematician (what would have been a superior psychic faculty), but simply manipulated his observers (what is an inferior psychic faculty). But I wonder, why are we so superficial towards that episode? And why to evaluate the events only in the negative sense, as something (mathematic skills) that *did not* happen? What we have is also a horse that, in front of dozens of observers, was always – or almost always – able to a) detect an even microscopic facial or postural expression, b) process these expressions mentally, and c) understand which one, among the many, was to be interpreted as a message of approval.

And maybe there was even more than this. It was not only the ability to detect approval, it was also detecting the right degree of approval. Let us picture the scene: somebody asks Hans to indicate the square-root of 441 (which makes 21, as I again ignored and just checked with a calculator). The spectators gathered around the horse are very sceptic and determined not to applaude or congratulating unless the horse will really stop his counting at 21. Thus, Hans starts to hit his hoof on the ground. After two or three hits there is already a quite interesting range of expressions in the observers: somebody is chatting, somebody is laughing, somebody is betting that the horse will not even reach 10, and so on. Everything is accompanied by some non-verbal language that varies according to the related emotional state, but also to the subject involved (just think of how many ways to laugh exist). Hans understands that all these expressions are not relevant, and keeps on hitting his hoof. After some ten hints, somebody already goes like "Well... not bad!", and in general there is an increasing attention towards the scene. In this case, too, Hans could not care less, and proceeds with his counting. Here we are at 18–19 hits: people grow excited and surprised: "hey, he's doing it! He's really making it!". These are all expressions that one might easily mistake for a 'definitive' approval, the one that Hans is looking for. But no, the horse does not stop yet. He reaches 21, and then, only then, he understands that the type of expression he was looking for is now printed in almost everybody's face: it is the final approval, the one that suggests him to stop, because people are happy of his performance.

To me, such a faculty looks exceptional, definitely 'superior'. An hypnotist or a psychologist could not do any better: this is for sure. Still, to Morgan (and to Sebeok), all this seems not to have any importance: what matters is that Hans was not an Einstein. To be a Freud is not enough. Why?

The contradictions of Morgan's Canon certainly provides solid ground for a concept like Martinelli's Canon to grow. However, the development of the concept should not be exclusively in opposition to something, but also in favour of something else. The issue of anthropomorphism, the way I was hopefully able to illustrate it, presents several obscure points. What is really anthropomorphic, given that many characteristics are human but not only human? Is anthropomorphism really to be condemned in toto? And most of all, Could the fear of anthropomorphism be more dangerous for scientific research than anthropomorphism itself? In the last 20–30 years, several scholars in animal-related studies decided that they were not afraid of the big bad (anthropomorphic) wolf anymore, and basically answered "no" to the second question, and "actually, yes" to the third one. They realised that promoting a certain empathy between humans and other animals does not necessarily constitute an anthropomorphic mistake, and that neglecting it might be consistently more misleading. An example that is really worth mentioning, called "critical anthropomorphism", is analysed by Luisella Battaglia, in her essay *"Etica e diritti degli animali"*:

> Critical anthropomorphism aims to use human experience in a critical manner, in order to recognise emotional manifestations, putting into relation our most immediate subjective intuitions with comparable notions and data provided by neurophysiology, ethology, zoology etc. This way, empathy, typical of classical anthropomorphism, is integrated with the most recent scientific research on animal life and behaviour. In support of such an approach, we could say that if human-animal similarities are accepted for scientific experimentation purposes, they should be accepted also in the field of emotional sensibility. In any case, any doubt about emotional sensibility should be gauged so as to benefit the weakest subject. In particular, the presupposition of similarity, when there is no clear counterproof, should be interpreted in favour of the animals. (Battaglia 1997: 123–124, my translation)

In other words, if, as often happens, scientific research involving non-human animals – either aiming to study animals themselves, or aiming to study humans *through* other animals, as in medical research – is based on presuppositions of similarity between humans and other animals, then it shall be more fair to apply the same principle on a more consistent basis.

It is not fair [. . .] to maintain that there is a resemblance between humans and animals when we use [the latter] in medical laboratories in order to establish our rights over them, and – at the same time – to maintain [that animals are different] as moral subjects, in order to avoid our duties (ibid.: 124, my translation)

Battaglia's main point, here, is that philosophical speculation on animal's behaviour (as on other issues, as well) should concert scientific data and ethical reflections, in such a way that the latter support the former, when these are not able to provide an exhaustive response. When science is not able to give a 100% reliable answer on a given behavioural phenomenon, than ethics should be allowed to say the final word. And ethics, as conceptual basis for modern jurisprudence, is based on the principle that the weaker subjects, those who are not able to verbalise what they are doing and what they mean by that, should be favoured and protected. Such is the case with categories like infants or mentally-disadvantaged persons.

Going back to the specifically scientific aspect, Battaglia's conclusion is that all the generally emotional manifestations previously considered dangerous for scientific research (empathy, most of all, as the generator of CHE par excellence), if handled with care, become useful and plausible clues for the whole theoretical apparatus. Jeffrey Masson and Susan McCarthy (1995/1996: 75–78) summarise the issue, by bringing to attention the very old issue, whether or not should researchers give names to the subjects of their research. Or, should they promote empathy? The claim is that not only it should be allowed, but that it would in fact be necessary, for at least three reasons:

1. Because, by definition, empathy helps comprehension; hence it is easier to interpret a given phenomenon if one identifies him/her self with it to some extent;
2. Because it has been scientifically proven that many animal species give names to each other (the most impressive case being dolphins);
3. Because animals are not only subjects of human scientific research: they are also the category humans belong to. Humans are not *more or less* similar to other animals: they *are* animals. To ignore this fact means to ignore a crucial part of the story.

To add a specific semiotic note on it, what the concept of critical anthropomorphism, and empathy in particular, are promoting is nothing else than abduction, as a valuable method for scientific inquiry. Abduction is a very ancient method of investigation, which gained scientific legitimation thanks to Peirce. Consider the following dialogue between two very well-known characters:

"You amaze me, Holmes," said I. "Surely you are not as sure as you pretend to be of all those particulars which you gave."

" There's no room for a mistake," he answered. "The very first thing which I observed on arriving there was that a cab had made two ruts with its wheels close to the curb. Now, up to last night, we have had no rain for a week, so that those wheels which left such a deep impression must have been there

> during the night. There were the marks of the horse's hoofs, too, the outline of one of which was far more clearly cut than that of the other three, showing that that was a new shoe. Since the cab was there after the rain began, and was not there at any time during the morning – I have Gregson's word for that – it follows that it must have been there during the night, and, therefore, that it brought those two individuals to the house. (Doyle 1887/1999: 50–51)

However bold Sherlock Holmes appears to be in formulating his theories, they are in fact no more than simple hypotheses – probable hypotheses, though, which give the impression that events really developed that way. In theory, things could have gone differently, and nothing guarantees that they went precisely in the described way. But in practice, all the elements at Holmes' disposal, when considered as a whole, can be coherently explained only the way Holmes puts it. Thus, we are not dealing here with absolute certainty, but with an extremely significant possibility that things went a certain way, and – most of all – a relevant step forward in the investigation and a point of reference for further research, which, little by little, will prove that Holmes is right.

This is abduction. A surprising phenomenon, X, is observed. Among hypotheses A, B, and C, A is capable of explaining X. Hence, there is a reason to pursue A. Peirce ascribes the origin of the term to Aristotle, precisely to the word ἀπάγωγη, *apagoge*, translated as *abduction* (erroneously, according to Peirce, who thought the right translation should be "retroduction"; in his writings, he uses both terms[18]). The usage and contents of "abduction" are in unmistakable and somewhat ideological conflict with "deduction" and "induction". If deduction consists of proceeding from a rule and a case to a result, and if induction proceeds from a result and a case to a rule, then abduction proceeds from a rule and a result to a case. To illustrate the situation another way, I shall give a banal example (inspired, in case you wondered, by the first album of the British rock-band Franz Ferdinand):

Deduction
Rule – All songs from this CD are danceable
Case – These songs are from this CD.
Result – Therefore, these songs are danceable.
Induction
Case – These songs are from this CD
Result – These songs are danceable.
Rule – Therefore, all songs from this CD are danceable.
Abduction
Rule – All songs from this CD are danceable
Result – These songs are danceable.
Case – Therefore, these songs are from this CD.

Logically speaking, the abductive method does not seem totally trustworthy – obviously, the songs may easily belong in another album. Nevertheless, Peirce maintains, abduction is precisely the reasoning procedure that scientists use most. Further, and more significantly, it allows real progress to be made, for it is the only one of the three methods to *explain* phenomena rather than simply classify them. Abduction is thus a "logic of discovery", a kind of critical thinking which opens doors of opportunity for scientific research and which, in the end, confirms the classifications provided by inductive and deductive procedures. In practice, according to Peirce, scientific investigation should go through the following phases:

1. Observation of an anomaly;
2. Abduction of hypotheses that explain the anomaly;
3. Inductive testing of the hypotheses in experiments;
4. Deductive confirmation that the selected hypothesis predicts the original anomaly.

In the study of non-human animals, a specimen of a given species quite often displays a new, previously unconsidered behavioural pattern (phase 1); for example, an elephant using her proboscis in a non strictly utilitarian way, such as drawing on sand (several elephants have been observed doing this, the most famous case being the Indian elephant Siri). The scholar may decide to interpret this pattern according to already established theories, for instance using the Morgan's Canon. Alternatively, the scholar may try to consider the elements at disposal and to *explain* them, formulating coherent hypotheses which might possibly illustrate all the elements in an acceptable way. For instance, the scholar may formulate the hypothesis of the existence of an aesthetic sense in the elephant. Next the scholar conducts experimental research, uses direct observation and other methods in order to verify whether the hypotheses are actually true (phase 3), including, as it was actually done with Siri and company, giving the elephant paper and pencil. If the data collected during this stage are convincing, there is nothing left but to deduce the results (phase 4), i.e., that the aesthetic sense is detectable in elephants. Of course this is an extreme simplification of the whole process, but it might at least give an idea of its functioning.

Abductive strategy tends towards an empathic, emic perspective, because it is a form of investigation that truly allows for the explanation of phenomena "internal" to the subject observed, in accordance with manifest, recognisable data.

In the light of all these points, I think that a fair formulation of Martinelli's Canon (MC) is now possible. So, the rule: *In no case should actions or behaviours be interpreted as the result of an inferior psychic faculty, when it is possible to interpret them as a result of a superior faculty*. Comments:

1) Unlike Morgan's, *MC approaches the problem of inferior and superior psychic faculties in terms of abductive, not rhetorical, possibility*. A behavioural phenomenon X (e.g., a beaver carries pieces of wood that weigh all the same) is observed in non-human animals; X can be explained with hypothesis A (the beaver is mentally estimating the weight of the pieces) or B (the beaver is driven by instinct); X resembles A on the basis of perceived similarity with A_H (i.e., A as performed by human beings); hence there is a reason to pursue A. Pursuing A does not mean automatically *stating* A as the final word on the matter. It means that our *research hypothesis* is now A, and the next step of our inquiry should be *oriented* towards A. If that proves to be a failure, that is if pursuing A is not *possible*, then further hypotheses shall be considered. In this sense

2) *MC aims to economic and ergonomic research.* Pursuing A is more economic and ergonomic than pursuing B, because A already provides clues and research tools, therefore a path to follow is already indicated.
3) *MC refuses the criteria for establishing differences between psychic faculties in the way Morgan's Canon does.* Such criteria, rather than anthropocentric, should focus on the notions of Umwelt, adaptation, free will and semiosis in general. Interpretive hypotheses that rely on concepts like instinct are by principle anti-semiotic, as they deny any interaction between the animal and her Umwelt.
4) As a semiotic-centred formulation, MC treats any behavioural phenomenon as complex and multi-layered and refuses by principle reductive one-sided interpretations. An instance like the one involving the horse Hans represents a paradigmatic example of a multi-faced semiotically-rich subject for scientific inquiry, that scientists reduced to a quasi-joke by refusing to approach the episode in all its aspects. Once established that the horse was not capable of mathematic computing, they ignored his still-astounding ability of interpreting body-language, therefore missing a good chance to deepen the knowledge on horses' communication in relation to their Umwelt.
5) MC fully supports the notion of critical anthropomorphism (Battaglia 1997: 123–124), and therefore rejects reductive hypotheses on an ethical basis, too. If the interpretation of the behavioural phenomenon X is made difficult by the impossibility to take a totally reliable scientific choice between the hypotheses A and B, then – in addition to the motivation provided in point 1 of the present list – another reason to pursue A is the ethical one: "the presupposition of similarity, when there is no clear counterproof, should be interpreted in favour of the animals" (Battaglia 1997: 124, my translation).
6) On a more strictly scientific level, *MC promotes methodological empathy and*, more generally, *supports a more balanced and less radical use of anthropomorphism.* Anthropomorphism should be avoided when it applies human species-specific predicates on non human species and when produces distorted perceptions of the reality analysed. It should be not avoided, in fact encouraged, when, through empathy, it helps the comprehension of the given phenomenon, when it favours biocentric approaches to non human species, and when it is employed for didactic purposes, in order to facilitate attitudes of interspecific acceptance and compassion.

NOTES

[1] This is a rather interesting difference with what later reported by semioticians, who still nowadays insist that all these animals were capable of doing with ASL was asking for bananas. However, talking about bananas, definitely true is the fact that laboratories are distressing, boring and depressing contexts, and there is hardly anything really stimulating for the apes: we can easily guess that food becomes soon a major interest.

[2] It is to be inferred, anyway, that the intraspecific use of ASL was for the chimps meant more like a game, or like a rehearsal of the idiom that is necessary in order to communicate with the humans.

[3] As I argued elsewhere (Martinelli 2004: 77–78), "the presence of a lie in a communication system implies the following:

a) There is a code, which is socially shared, that associates a sign to a meaning. Since lies break this association, the latter had to exist previously.
b) The sender of the sign must be able to break this link.
c) In order to lie, the sender must be able to mentally represent not only the receiver's behaviour, but also her mind. That is why we can only cheat entities provided with minds (not sunglasses, for instance).
d) The sender must be able to take advantage of the broken link, making the receiver believe something that is not happening in the way she thinks.

This means that lying, when structured as a semiotic process, is a cognitive act, and cannot be seen as the result of instinct or of a simple behaviourist stimulus-reaction process".

[4] And, I am afraid, it is no coincidence that the only ICE scholar mentioned in that occasion was Herbert Terrace, since he is the one most often mentioned by Sebeok in his writings. It is unfortunately to be inferred that that lecture was exclusively based on second hand material.

[5] The most remarkable exception to this rule, as I mentioned, is represented by Felice Cimatti, who indeed should not be considered a Sebeokian or a Peircean (his theoretical bases deriving more from the likes of Jakobson, Gozzano and others), and whose book *Mente e linguaggio negli animali* (1998) I still consider one of the most brilliant pieces of work in the entire semiotic panorama. A short look at his references list easily proves that Cimatti did his homeworks on Savage-Rumbaugh, Premack, Pepperberg & Co. very well and accurately.

[6] See the extremely interesting Evans, 1906.

[7] A brief explanation of these concepts is demanded. According to usual definitions, anthropocentrism interprets Nature as (a) an entity existing *apart from* and *for the benefit of* humans, so that (b) nothing in Nature can be considered in itself, autonomously from humans; and (c) it is ethically acceptable for humans and non-humans to be treated in different ways. In other words, Nature is not of interest (e.g., to conservationists and preservationists) because of its hypothetically *intrinsic value*, but just because of its *instrumental value*, i.e., the values it has for and to humans.

Most criticism against animal-related studies tends to emphasise that a totally impartial interpretation of animal behaviour is not possible, for observations are external to the subject of study and cannot avoid frames of reference that are typical of human interpretation of reality. In this sense, the approach is anthropocentric, i.e., concentrated on and mediated by the fact of being human. Such a statement deserves however specific reflections.

First, such criticisms are a little simplistic, and merely constitute a comfortable and socially shared (thus, stereotypical) way out of facing a problem that is in fact quite complex. It may be easy to speak of anthropocentrism as an apparently unavoidable form of interpretation of reality that affects scientific research; however, to mix all its nuances in the same big pot reveals quite a lack of knowledge on the topic. It is more proper to dissect the question into all its components in order to re-interpret anthropocentrism more accurately.

Secondly, I have the feeling that those who doubt the scientific validity of animal-related studies, because of the difficulty of avoiding anthropocentrism, often seem to be sceptical about only part of the story, while in a few other cases, animal-related studies seem to enjoy everyone's confidence. Very well known is the scepticism that surrounded and partially still surrounds Darwin's theories, but where are the sceptics when it comes to evaluate the very probable anthropocentrism of pharmacological research? Should they not be at least suspicious about transferring given data from non-human species to the human one so easily?

Lastly, these kinds of criticisms are a little too defeatist. It is true that there is no way to avoid some elements of anthropocentricity, but is this an absolutely unbridgeable gap between scientific research and a correct interpretation of reality? Things are never all black or white: the impossibility of being totally objective and impartial towards a topic is not really a good reason to give up scientific research in general. Different degrees of impartiality, according to specific cases, can be achieved. The challenge is to *tend* towards absolute impartiality. Otherwise, not only animal-related studies but also 99% of scientific fields would not be scientifically believable.

The above considerations appear rather simple, if not banal. Yet when animal studies are involved, scholars tend quite often to forget them.

Hence, the very first question, Is there just one type of anthropocentrism, or are there more? In other words, How many ways exist to observe reality according to the criteria of interpretation and classification proper to the human being? My research suggests me that such criteria should be distributed on at least two layers: *default anthropocentrism* and *binary anthropocentrism*. The latter, in its turn, can be divided into *quantitative* and *qualitative* types.

The first elementary level, *default anthropocentrism*, consists in the banal consideration that the subject who observes a given animal species is evidently a human being, with all its resources, limits and modes of categorisation. What we understand about a dog, for instance, is what we are able to understand, given the means that allow us to do. Technology does not (yet?) allow us to understand a dog the way, say, a pigeon would understand it. Such a consideration is not very different from statements like "Alvar Aalto is a great architect". Quite evidently, in pronouncing such statements, we are reporting one of our forms of interpretation of reality, founded on personal experience, education, culture, perceptive sources and so on. Now, this looks to me obvious, inevitable, and not dangerous. The other way round, however, could be dangerous, for it could mean the expressing of opinion without any point of reference or any code, resulting in a sort of perceptive anarchy. As long as an anthropocentric attitude is reduced to this very basic expression, no kind of scientific research risks being taken little seriously.

The second type is *binary anthropocentrism*. Here, the fact of being a different entity from the object observed (human, rather than another animal) produces a dualistic interpretation of reality, based on criteria of *difference* (*qualitative anthropomorphism*) and/or a strongly *hierarchical identity* (*quantitative anthropomorphism*), which puts the observer, and the group s/he belongs to, in a superior position in relation to the group observed. In the case of qualitative anthropocentrism, the observer-human being tends to distinguish him/herself from the non-human animal by means of either/or qualities, which is almost a causal relation (i.e., "humans do, *ergo* animals don't"). In the case of quantitative anthropocentrism (which is a post-Darwinian anthropocentrism in a way), the difference between human beings and other animals is expressed by means of quantities (more/less). Within this framework, a statement like "Unlike Gropius, Alvar Aalto *is* a great architect" is of qualitative type, while the statement "Walter Gropius is a good architect, but Aalto is definitely better" is quantitative.

[8] Which makes 322, as of course I did not have the slightest idea before checking with a calculator.

[9] Possibly, a typical and evident display would be the opening of the eyes and mouth.

[10] One day, I would not mind being explained why being a Freud is not as remarkable as being an Einstein.

[11] I say it with no negative connotation whatsoever. Only the greatest scholars become academic saints.

[12] As a general rule, not only in Gardners' case, no visitor is allowed inside the enclosures where chimpanzees are kept. Chimpanzees are some seven times stronger than humans, and they also have denser bones and thicker skin. In most instances of interactions, chimpanzees must restrain themselves to avoid hurting humans. When playing, chimpanzees normally throw, slap and playbite each other. However, if these actions elicit laughter among them, a human would be seriously hurt. The fact that they may master a bit of human language, does not unfortunately mean that they will use rethorics and diplomacy when the situation becomes critical.

[13] I still happen to dream them in the night, and the next morning I wake up with Pink Floyd's *Another Brick In The Wall (Part II)* in mind: "We don't need no education... Teachers, leave us kids alone!"

[14] The mentioned scholars are also the ones who have the greater media exposure.

[15] An entire chapter of Roger Fouts' beautiful best-seller *Next of Kin* is devoted to the extrmely cruel conditions in which Washoe had to live during the period she moved from the Gardners' house in Nevada, to William Lemmon's laboratory in Oklahoma.

[16] Felice Cimatti (1998: 147–151) proposes an interesting and somewhat ironic example to illustrate the applications of this canon to the case of Washoe. According to the supporters of Morgan's canon, at the very moment Washoe is asked to perform the ASL gesture corresponding to a given object, she does not have the slightest idea of how to perform such a task, since she is not able to understand the relation between signifier (the ASL gesture) and signified (the object). Thus, she proceeds by attempts,

moving her arms at random, until – unconsciously or not – her trainers display apparent satisfaction, which lets her know what is the right gesture. Here is a typical example of the CHE, which supporters of Morgan's canon find the most suitable interpretation. Unless one is able to invalidate any risk of the CHE, they say, then ICE are a waste of time. Most of all, the more or less emotional relation established between researcher and animal should be avoided by creating an "emotion-free" context. Of course, to consider the use of a communication system as unrelated with social and emotional interaction is a self-evident contradiction and a more serious mistake than anthropomorphism. What is the point in learning a communication system if one has no one to use it with, does not receive adequate stimuli, and, in the end, does not have any reason to learn it? In such case, says Cimatti, the mistake is not only methodological, but also theoretical: "if language is a system based on emotive interaction [...], then it is simply impossible to study it without considering this aspect" (Cimatti 1998: 149, my translation).

[17] The discussion on instinct should not be ended with this note, as I consider the over-use of the notion as a real threat to scientific inquiry. Scholars should take a very critical approach towards those colleagues who so easily refer to this black box. In general, every application of the notion of instinct to complex multi-phased actions should be forbidden or accepted ONLY IF very convincingly and empirically motivated: it is all too easy to open the black box and put processes and actions that we are not able to explain otherwise. Except that this is not science: it is much closer to religion. What is the difference between explaining a complex phenomenon with the existence of instinct and explaining it with the existence of God? And why not aliens? I am very firm on this also because I want to protect the notion of instinct itself from massive trivialisation. THERE IS a number of human and non-human phenomena that originate at a pre-cognitive level and produce an actual physical action. But if we want to make sense out of these instances, we shall first of all circumscribe them: extemporaneous escape from danger, movements supporting physiological needs, innate competencies on immediate parental care... This is what we call instinct. Hunting strategies, elaborate parental care, establishment of social relations... if we call those instances "instinct", we do not really need any ethologist or zoosemiotician. A priest is enough.

[18] Together with the terms "Hypothesis" and "Presumption".

REFERENCES

Bateson, Gregory (1969). "Metalogue: What Is an Instinct?". In *Approaches to Animal Communication*, ed. T. Sebeok and A. Ramsay. 11–30. The Hague: Mouton.

Bateson, Gregory (1972). *Steps to an ecology of mind.* Chicago, Illinois: University of Chicago Press.

Chauvin, Remy and Bernardette, M. (1977) *Le monde animal et ses comportements complexes.* Paris: Plon.

Coetzee, John M. (1999). *The lives of animals.* Princeton: Princeton University Press.

Dreifus, Claudia (1998) "A CONVERSATION: With Emily Sue Savage-Rumbaugh; She Talks to Apes and, According to Her, They Talk Back". *The New York Times*, April 14.

Engels, Friecrich (1896). "The part played by labour in the transition from ape to man". In *Die Neue Zeit.* XIV/2.

Gardner, R. Allen and Beatrice T. (1969). "Teaching sign language to a chimpanzee". In *Science*, 165: 664–672.

Hill, Jane H. (1980) "Apes and language". In T.A. Sebeok, J. Umiker-Sebeok (eds.), *Speaking of apes.* New York: Plenum Press.

Kosseff, Lauren (2000) Primate use of language. E-text, retrieved 12 September 2005 from http://www.pigeon.psy.tufts.edu/psych26/language.htm

Lenneberg, Eric H. (1980) "A word between us". In T.A. Sebeok, J. Umiker-Sebeok (eds.), *Speaking of apes.* New York: Plenum Press.

Malacarne, Giorgio (1992). Entry "Linguaggio". In D. Mainardi (ed.), *Dizionario di etologia.* Torino: Einaudi.

Malacarne, Giorgio (2005). "Il linguaggio simbolico negli uomini e nelle scimmie". In E.Bellone (ed.), *Il comportamento degli animali.* Novara: De Agostini.

Martinelli, Dario (2004). "Liars, players and artists: a triadic zoosemiotic relation", in *Semiotica* 150 – 1/4: 77–118.

Patterson, Francine – Linden, Eugine (1981). *The education of Koko.* New York: Holt, Rineheart & Winston.

Patterson, Francine (1978). "Conversations with a gorilla". *National Geographic.* 154 (4).

Patterson, Francine (1977). "Linguistic capabilities of a young Lowland Gorilla". Paper presented at a Symposium of the American Association for the Advancement of Science: "An account of the visual mode: Man vs. Ape". Denver, 1977.

Sebeok, Thomas A. (1981). *The Play of Musement.* Bloomington, IN: Indiana University Press.

Sebeok, Thomas A. (1990). *Essays in zoosemiotics.* Toronto: Toronto Semiotic Circle.

Sebeok, Thomas A. (1991). *A Sign is Just a Sign.* Bloomington: Indiana University Press.

Sebeok, Thomas A. (1996). "Signs, bridges, origins". In J. Trabant (ed.), *Origins of Language.* Budapest: Collegium Budapest.

Sebeok, Thomas A. (2001). "Nonverbal communication" in P. Cobley (ed.), *The Routledge Companion to Semiotics and Linguistics* London: Routledge.

Sebeok, Thomas A.; Danesi, Marcel (2000). *The Forms of Meaning: Modeling Systems Theory and Semiotic Analysis.* Berlin: Mouton de Gruyter.

Sebeok, Thomas A.; Umiker-Sebeok, Jean (eds.) (1980). *Speaking of apes.* New York: Plenum Press.

Trask, R.L. 1995. *Language: the basics.* London: Routledge.

von Glasersfeld, Ernst (1978). "Les chimpanzés et le langage". In *La Recherche*, 9: 725–732.

Fouts, Roger (2004). "Frequently Asked Questions" in *Chimpanzee and Human Communication Institute* Retrieved 15th July 2005 from http://www.cwu.edu/~cwuchci/faq.html

Washburn, David (2001). *Kanzi.* Retrieved 4th November 2005 from http://www2.gsu.edu/~wwwlrc/biographies/kanzi.html

AUTHOR INDEX

SUBJECT INDEX

MIX
Papier aus verantwortungsvollen Quellen
Paper from responsible sources
FSC® C105338

If you have any concerns about our products,
you can contact us on
ProductSafety@springernature.com

In case Publisher is established outside the EU,
the EU authorized representative is:
Springer Nature Customer Service Center GmbH
Europaplatz 3, 69115 Heidelberg, Germany

Printed by Libri Plureos GmbH
in Hamburg, Germany